D1288316

Elementary Differential Equations

Elementary Differential Equations

Lyman M. Kells

Professor Emeritus
United States Naval Academy
Annapolis, Maryland

Sixth Edition

McGraw-Hill Book Company

New York St. Louis San Francisco
Toronto London Sydney

Elementary Differential Equations

Library of Congress Catalog Card Number 64-8278

ISBN 07-033530-3

67890 HDBP 75432

Cover photograph: **An artist's rendering of an advanced orbiting solar observatory satellite. Courtesy of the National Aeronautics and Space Administration.**

Preface

Differential equations comprise, in a very real sense, one of the most powerful of mathematical tools, with application to all fields of scientific endeavor. In this text, the theory and its applications are given equal emphasis. Thus the student learns to express physical laws in the language of differential equations, to solve these equations by modern techniques, and to interpret his results in terms of the original problem. He adds to his mathematical repertoire a transcendentally versatile and powerful tool for understanding and coping with the modern world, a tool which will not become obsolete.

Outstanding features of this book are its simplicity, clarity, conciseness, and teachability. The subject is divided into logically arranged topics of convenient learning length, each of which is carefully explained, illustrated profusely by examples and figures, and supplemented by lists of exercises. Great care has been taken to arrange the exercises so that the student becomes familiar with fundamental concepts while solving simple problems (with suggestions for attack) at the beginning of each list; afterward he meets more difficult problems, many of which will challenge the ingenuity of the best student. Many of the problems deal with such modern phenomena as rockets, isotopes of atomic materials, nuclear phenomena, and satellites; others relate to the more traditional areas in physics, chemistry, electricity, heat transfer, and wave motion. In this edition each list of exercises is divided into two groups: The first, distinguished by a light-blue background, furnishes sufficient material to provide understanding and assimilation; the second group contains additional problems, some rather difficult, for drill, deeper comprehension, challenge, and special assignments. This arrangement will help the teacher determine quickly how he may best use the large number of exercises, and it will aid the student by setting him two goals: one of practical assimilation, and the other of more profound understanding and greater manipulative power.

The color blue has been used throughout the book to draw attention to features of outstanding importance, such as basic equations, definitions, theorems, and important parts of figures. The student will find the emphasis achieved by the use of the color an invaluable aid to visualizing quickly, especially in review, the important ideas to be assimilated.

Most of the material of the previous edition has been reworked to obtain greater clarity and precision; wording has been clarified, and explanatory material, illustrative examples, and new problems have been added. Some sections have been completely rewritten, and significant improvements have been made in Chaps. 1, 2, 3, 5, 6, 7, 11, 12, and 15. Major improvements include: more critical discussion of existence theorems and results; a more pedagogical approach; a concentrated and clear treatment of the c- and p-discriminants and Clairaut's equation in Chap. 5; a greatly improved treatment of linear independence, operators, and variation of parameters in Chap. 6; a practically new treatment of Laplace transforms in Chap. 7; a new section in Chap. 10 on orbits of satellites; in Chap. 12 a change in the method of solving equations in series, as well as two new sections, one on basic theorems and another on expansion in decreasing powers of x; in Chap. 14 an expanded discussion of wave motion and applications; in Chap. 15 a more complete discussion of Fourier series, and a new section on Laplace transforms applied to the solution of boundary-value problems of partial differential equations.

The new treatment of Laplace transforms is simple, concise, and easy to understand. It deals not only with the ordinary topics of differential equations but also with the gamma function, the error function, and numerous functions compounded by translation, periodic repetition, and rectification of simple functions. The logical tone of the text is improved by consideration of discontinuities of solutions and critical examination based on general theorems, such as existence theorems. However, in all cases, care has been taken to preserve those qualities which were considered the strong points of earlier editions.

It is a pleasure to express my deep appreciation and sincere thanks to users of the earlier editions for their very helpful and penetrating discussions and suggestions.

Lyman M. Kells

Contents

3 Differential Equations of the First Order and the First Degree 50

4 Applications Involving Differential Equations of the First Order 81

5 First-order Equations of Degree Higher than the First 105

6 Linear Differential Equations with Constant Coefficients 126

7 Laplace Transforms 161

8 Applications of Linear Equations with Constant Coefficients 196

12 Solution by Series 289

13 Numerical Solutions of Differential Equations 322

14 Partial Differential Equations 345

Elementary Differential Equations

1

Definitions and Elementary Problems

1 General remarks

Differential equations furnish a very powerful tool for solving many practical problems of engineering and science generally, as well as a wide range of purely mathematical problems. While this book treats of the most important types of differential equations and gives strong emphasis to outstanding applications to problems of a physical nature, it also takes up many interesting applications to geometry.

The applications to engineering, physics, and science generally are of the greatest importance. A law is conceived and set forth as a system of differential equations; the solution of these equations tells a rather complete story of the states and motions to be expected of the materials obeying that law. For example, we assume the law, suggested by experiment, that radium disintegrates at a rate proportional to the amount present and express this in mathematical symbols by the equation

$$\frac{dQ}{dt} = kQ$$

By solving this equation for a 100-g lump and using facts found from experiment, the equation

$$Q = 100e^{-0.041t}$$

is easily derived. This tells us that 20 centuries hence there will be $100e^{-0.041(20)} = 44.0$ g left and that the deposit laid down 20 centuries ago was $100e^{0.82} = 227$ g; and, in general, it describes quantity and many other relations at any specified time.

Newton conceived the law of gravitation and then solved the corresponding system of differential equations to show that the earth moves about the sun approximately in an ellipse with the sun at one focus. Today we use the same theory to learn about satellites, their orbits, and methods of guiding them. Around 1865, Maxwell conceived a relation between an electric current and the corresponding magnetic field, expressed the relation as a system of partial differential equations, solved them, and from the result predicted the waves of radio. Differential equations have played a prominent role in the development of the theories of radio, radar, television, and electricity generally. Similar remarks apply to nearly every important branch of science. The many applications in this book will show the great power of differential equations and give methods of using it.

2 Differential equation. Order. Degree

The student has already met differential equations of an elementary type in his study of the calculus. Thus,

$$\frac{dy}{dx} = x^2 + 3 \tag{1}$$

is a differential equation. In general, a differential equation *is an equation containing derivatives.* If the equation contains total derivatives but does not contain partial derivatives, it is called an ordinary differential equation; if it contains partial derivatives, it is called a partial differential equation. Thus,

$$x^2\frac{d^2y}{dx^2} + 2x\frac{dy}{dx} + y = x^2 + 2 \tag{2}$$

$$\left(\frac{d^3y}{dx^3}\right)^2 + 2\frac{d^2y}{dx^2}\frac{dy}{dx} + x^2\left(\frac{dy}{dx}\right)^3 = 0 \tag{3}$$

$$\left[1 + \left(\frac{dy}{dx}\right)^2\right]^{3/2} = k\frac{d^2y}{dx^2} \tag{4}$$

$$(x + y^2 - 3y) + (x^2 + 3x + y)\frac{dy}{dx} = 0 \tag{5}$$

are ordinary differential equations, whereas

$$\frac{\partial z}{\partial x} = y \tag{6}$$

$$\frac{\partial^2 u}{\partial x^2} + \frac{\partial^2 u}{\partial y^2} + \frac{\partial^2 u}{\partial z^2} = 0 \tag{7}$$

are partial differential equations.

The order *of a differential equation is the order of the highest-ordered derivative involved in its expression.* Of the differential equations numbered (1) to (5), equations (1) and (5) are of the first order, (2) and (4) are of the second order, and (3) is of the third order.

The degree *of an ordinary differential equation algebraic in its derivatives is the algebraic degree of its highest-ordered derivative.**

Consider, for example,

$$\sqrt[3]{\left(\frac{d^2y}{dx^2}\right)^2} = \sqrt{1 + \left(\frac{dy}{dx}\right)^2} \tag{8}$$

Here the highest-ordered derivative is d^2y/dx^2, and the order of the equation is 2. Equating the sixth powers of the members of (8), obtain

$$\left(\frac{d^2y}{dx^2}\right)^4 = \left[1 + \left(\frac{dy}{dx}\right)^2\right]^3 \tag{9}$$

Here 4, the degree in d^2y/dx^2, is the degree of equation (8). Equations (1), (2), (5), and (6) are of the first degree; (3) and (4) are of the second. Equation (4) is of the second degree; for d^2y/dx^2 appears to the second degree in the equation resulting from clearing (4) of the radical represented by the 2 in the exponent $\frac{3}{2}$.

Exercises

State the order and the degree of each of the following differential equations:

1. $\dfrac{dy}{dx} = 5y$

2. $\left(\dfrac{dy}{dx}\right)^2 = \dfrac{3x}{4y}$

3. $\left(\dfrac{dy}{dx}\right)^3 = \sqrt{1 + \left(\dfrac{dy}{dx}\right)^2}$

4. $\dfrac{d^2y}{dx^2} = 3\dfrac{dy}{dx} + xy$

5. $\sqrt{\dfrac{d^2y}{dx^2}} = 3\dfrac{dy}{dx} + x$

6. $\dfrac{d^2y}{dx^2} = \sqrt{1 + \left(\dfrac{dy}{dx}\right)^4}$

7. $\dfrac{d^3y}{dx^3} = \sqrt{\dfrac{dy}{dx}}$

8. $\left(\dfrac{d^2y}{dx^2}\right)^{1/3} = k\left[1 + \left(\dfrac{dy}{dx}\right)^2\right]^{5/2}$

* A differential equation may not have a degree. Thus $\sin(dy/dx) = dy/dx + x + 3$ has no degree.

3 Elementary concept of the solution of a differential equation

Derivatives with respect to x will be denoted by primes in this chapter. Thus

$$y' = \frac{dy}{dx} \qquad y'' = \frac{d^2y}{dx^2} \qquad \text{etc.} \tag{1}$$

Any nonderivative relation between the variables of a differential equation which satisfies the equation is a solution *of it.* The graphs of the solutions of a differential equation are called its integral curves. Thus $y = cx^2$ is a solution of $xy' = 2y$, for, substituting cx^2 for y and $2cx$ for y' in $xy' = 2y$, we get the identity

$$x \cdot 2cx = 2cx^2$$

Observe that the integral curves of $xy' = 2y$, represented by $y = cx^2$, embrace all parabolas through (0,0) that have the Y axis as their axis (see Fig. 1). Through (a,b), $a \neq 0$, passes the integral curve for which $c = b/a^2$, and we observe that just one integral curve passes through (a,b), $a \neq 0$. Note that no integral curve passes through $(0,c)$, $c \neq 0$, and that every integral curve passes through (0,0), but that y' is not defined by $xy' = 2y$ at (0,0).

Example 1. Prove that $y = Ae^x + Be^{-2x} + x^2 + x$, A and B constants, is a solution of $(d^2y/dx^2) + (dy/dx) - 2y = 3 - 2x^2$.

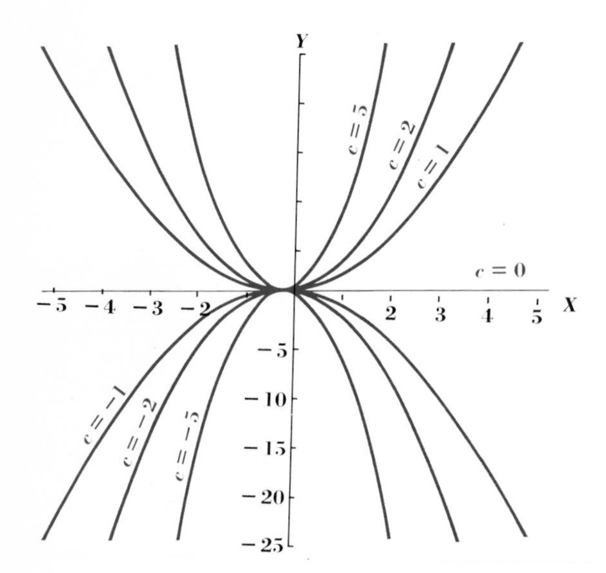

Figure 1

Proof. From $y = Ae^x + Be^{-2x} + x^2 + x$, we obtain

$$\frac{dy}{dx} = Ae^x - 2Be^{-2x} + 2x + 1 \qquad \frac{d^2y}{dx^2} = Ae^x + 4Be^{-2x} + 2$$

Substituting these values in the differential equation, we get the identity

$$Ae^x + 4Be^{-2x} + 2 + Ae^x - 2Be^{-2x} + 2x + 1 - 2Ae^x - 2Be^{-2x} - 2x^2 - 2x = 3 - 2x^2$$

Example 2.* Prove that $\ln y + (x/y) = c$ is a solution of

$$(y - x)\frac{dy}{dx} + y = 0 \tag{a}$$

Proof. Using the regular process of differentiating an implicit function, we obtain from $\ln y + (x/y) = c$

$$\frac{1}{y}\frac{dy}{dx} - \frac{x}{y^2}\frac{dy}{dx} + \frac{1}{y} = 0$$

or, solving for dy/dx,

$$\frac{dy}{dx} = \frac{-y}{y - x} \tag{b}$$

Substituting in (a) the value of dy/dx from (b), we obtain

$$(y - x)\left(\frac{-y}{y - x}\right) + y = -y + y = 0$$

Exercises†

(1–13) Prove that each equation is a solution of the differential equation written opposite it:

1. $y = x^2 + x + c$ $\qquad y' = 2x + 1$

2. $y = x^2 + cx$ $\qquad xy' = x^2 + y$

* The symbol *ln x*, with no base specified, indicates throughout this text a natural logarithm, that is, a logarithm to the base $e = 2.7183$ approximately. Also, the letter e will often be used, without explanation, to represent this base of natural logarithms.

† In each set of exercises the first group, overcast in blue, furnishes sufficient material to provide fair assimilation, and the second group contains additional problems, some rather difficult, for drill, deeper comprehension, and special assignments.

3. $y = c$	$y' = 0$
4. $y = A \sin 5x + B \cos 5x$	$y'' + 25y = 0$
5. $y = c_1 \sin 3x + c_2 \cos 3x + 9x^2 - 2$	$y'' + 9y = 81x^2$
6. $y = (x + c)e^{-x}$	$y' + y = e^{-x}$
7. $\ln y = c_1 e^x + c_2 e^{-x}$	$yy'' - y'^2 = y^2 \ln y$
8. $x = 2t + c,\ y = ct + 3$	$2y'^2 - xy' = 3 - y$
9. $y = x^4 + ax^2 + bx + c$	$y''' = 24x$
10. $y = c_1 e^{2x} + c_2 e^{-4x} + 2xe^{2x}$	$y'' + 2y' - 8y = 12e^{2x}$
11. $y^{-3} = x^3(3e^x + c)$	$xy' + y + x^4y^4e^x = 0$
12. $x = -\ln t,\ y = c_1 t + c_2 t^2$	$y'' + 3y' + 2y = 0$
13. $x = at + b,\ y = b^2t^2$	$2yy'' = y'^2$

4 Variables separable

In the differential equation

$$M(x,y) + N(x,y)\frac{dy}{dx} = 0 \tag{1}$$

each part has a definite numerical interpretation. However, for convenience and suggestive quality, (1) is often written in the differential form

$$M(x,y)\,dx + N(x,y)\,dy = 0 \tag{2}$$

The suggestiveness of (2) is exemplified below in (4).

As a first illustration of a method of solving a differential equation, we consider a type, called variables separable, which is easily changed to the form

$$f_1(x)\,dx + f_2(y)\,dy = 0 \tag{3}$$

where $f_1(x)$ is a function of x alone and $f_2(y)$ is a function of y alone. Direct integration of equation (3) gives the solution

$$\int f_1(x)\,dx + \int f_2(y)\,dy = c \tag{4}$$

where c is an arbitrary constant. Note that (4) satisfies (3) and is therefore a solution of (3) no matter what value constant c may have. Hence we can replace c by $\ln c$, $\tan^{-1} c$ or any other function of c in attempting to simplify a solution. A solution of (1) containing an arbitrary constant c of integration will be called the general solution, whereas a solution containing no arbitrary constant is called a particular solution

Example 1. Solve $y' = -y/(x - 3)$. Also find the equation of the integral curve through (4,1).

Solution. The given equation in form (2) is

$$(x - 3)\,dy + y\,dx = 0 \qquad (a)$$

Dividing this through by $y(x - 3)$ to separate the variables, and setting the integral of the left member equal to $\ln c$, we get

$$\frac{dy}{y} + \frac{dx}{x - 3} = 0 \qquad (b)$$

$$\ln y + \ln (x - 3) = \ln c \qquad (c)$$

Since $\ln y + \ln (x - 3) = \ln [y(x - 3)]$, we get from (c)

$$\ln [y(x - 3)] = \ln c \qquad \text{or} \qquad y(x - 3) = c \qquad x \neq 3^* \qquad (d)$$

since, if the logarithms of two numbers are equal the numbers are equal.

Equation (d) represents the hyperbolas shown in Fig. 1. To each value of c is associated a hyperbola.

To find the equation of the integral curve through (4,1), substitute 4 for x and 1 for y in (d) and obtain

$$1(4 - 3) = c \qquad \text{or} \qquad c = 1$$

Hence $y(x - 3) = 1$ is the required solution. The $c = 1$ hyperbola of Fig. 1 represents the corresponding integral curve.

* Answers to illustrative examples appear in color throughout the book.

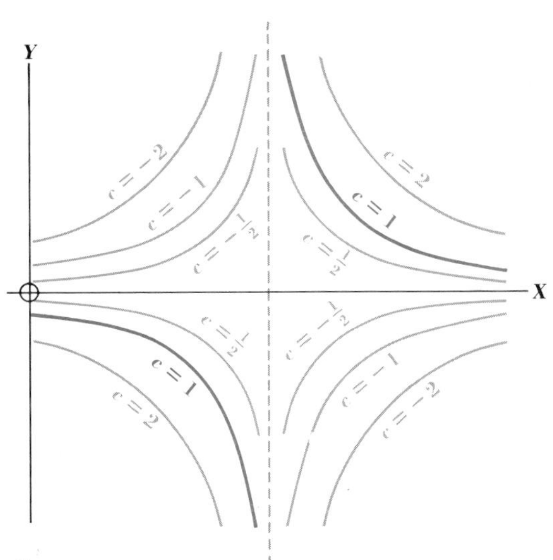

Figure 1

Consider the three equations

$$\text{I: } \frac{dy}{dx} = -\frac{y}{x-3} \qquad \text{II: } \frac{dx}{dy} = \frac{x-3}{-y}$$
$$\text{III: } (x-3)\,dy + y\,dx = 0 \qquad (e)$$

Also consider the solution (d) with $c = 0$; then

$$y(x-3) = 0 \qquad \text{or} \qquad y = 0 \qquad x - 3 = 0 \qquad (f)$$

Evidently $y = 0$ satisfies (e)I if $x \neq 3$; it does not satisfy (e)II because dx/dy is undefined if $y = 0$; it satisfies (e)III since $(x-3)\,d(0) + (0)\,dx = 0$. Similarly $x = 3$ does not satisfy (e)I, but satisfies (e)II if $y \neq 0$, and (e)III. Hence, in dealing with degenerate cases of solutions, we must consider the form of the given equation. Note that (e)III is satisfied by (d) without the reservation $x \neq 3$, or $y \neq 0$.

Example 2. Find the general solution of $a[x(dy/dx) + 2y] = xy(dy/dx)$, and then find a particular solution in which $y = a$ when $x = 2a$.

Solution. Clearing of fractions and grouping the terms containing dx and those containing dy, we get

$$2ay\,dx + (ax - xy)\,dy = 0 \qquad (a)$$

or

$$2ay\,dx + x(a-y)\,dy = 0 \qquad (b)$$

Dividing through by xy and integrating, we obtain

$$2a\int\frac{dx}{x} + a\int\frac{dy}{y} - \int dy = \text{constant}$$

or

$$2a\ln x + a\ln y - y = a\ln c \qquad (c)$$

Dividing by a, replacing $2\ln x$ by $\ln x^2$, and combining the logarithmic terms, we get

$$\ln\frac{x^2y}{c} = \frac{y}{a}$$

Remembering that $e^{\ln N} = N$, we obtain

$$e^{\ln(x^2y/c)} = e^{y/a} \qquad \text{or} \qquad x^2y = ce^{y/a} \qquad (d)$$

To find c so that $y = a$ when $x = 2a$, substitute $2a$ for x and a for y in (d), and solve for c to obtain

$$4a^3 = ce^{a/a} \qquad \text{or} \qquad c = 4a^3e^{-1}$$

Substitute this value of c in (d) to obtain the required particular solution

$$x^2y = 4a^3e^{-1}e^{y/a}$$

Exercises

(**1–15**) Find the general solution of each differential equation, and, when initial conditions are given, the corresponding particular solution:

1. $x\,dx + y\,dy = 0$
2. $dx + dy = 0$; $y = 1$ when $x = 0$.
3. $x\,dy + y\,dx = 0$; $y = 1$ when $x = 2$.
4. $(x - 1)\,dy + (y - 2)\,dx = 0$
5. $2x(1 + y^2)\,dx - y(1 + 2x^2)\,dy = 0$
6. $\dfrac{dS}{dt} = 15 - 16S$
7. $y' - 2y = y^2$; $y = 3$ when $x = 0$. Does any integral curve contain point $(a,-2)$?
8. $y' = y/[x(x - x^3)]$; $y = -2$ when $x = 2$. Show that y' is not defined when $x = 0$, 1, or -1.
9. $\theta\,dr/d\theta = -2r$; $r = 9$ when $\theta = -\frac{1}{3}$.
10. $2y\,dx + x^2\,dy = -dx$; $y = \frac{7}{2}$ when $x = 1/\ln 2$.
11. $L\,di/dt + Ri = 0$, where L and R are constants.
12. $e^x e^y\,dx - e^{-2y}\,dy = 0$; $y = 0$ when $x = 0$.
13. $\sqrt{1 - y^2}\,dx = \sqrt{1 - x^2}\,dy$*; $y = \frac{1}{2}$ when $x = 1$.
14. $4\,dy + y\,dx = x^2\,dy$; $x = 4$ when $y = -1$. Are lines $x = 2$, $x = -2$, and $y = 0$ integral curves of: (a) $dy/dx = y/(x^2 - 4)$? (b) $dx/dy = (x^2 - 4)/y$?
15. $x^3\,dy + xy\,dx = x^2\,dy + 2y\,dx$; $y = e$ when $x = 2$. Are lines $x = 0$, $x = 1$, and $y = 0$ solutions of $dy/dx = y(2 - x)/(x^3 - x^2)$?

5 *Geometric considerations*

Any differential equation of the first order and first degree may be written in the form

$$\frac{dy}{dx} = f(x,y) \tag{1}$$

* $\sqrt{u} \geqq 0$ if $u \geqq 0$, $\sqrt{u}$ is imaginary if $u < 0$.

It associates to each point (x_0,y_0) a line having as slope $(dy/dx)_0 = f(x_0,y_0)$; in other words, (1) associates to each point a direction. Observe that every point on the curve

$$f(x,y) = m \tag{2}$$

is associated with the slope m. The curves (2) are called the isoclines of (1).

Figure 1 shows some points with a line through each to indicate the direction associated with the point by

$$\frac{dy}{dx} = -\frac{x}{2y} \tag{3}$$

The isoclines of (4) are defined by

$$\frac{-x}{2y} = m \qquad \text{or} \qquad x = -2my \tag{4}$$

and this represents all straight lines through (0,0). Thus, for $m = 0$, every point on $x = -2(0)y = 0$ has zero as associated slope and every point on $x = -2(1)y$ has 1 as associated slope Note in Fig. 1 that any set of points associated with the same direction lie on a line through the origin, and every line through (0,0) is an isocline. Lines $A'A$, $B'B$, and $Y'Y$ in Fig. 1 are illustrations. Note that when $y = 0$ in (3), dy/dx has no value.

Since (3) is satisfied by any solution of it, any corresponding curve must have at each of its points the slope dy/dx defined by (3); that is, the tangent line to the curve at any point on it has the direction associated with this point by (3). Testing shows that

$$x^2 + 2y^2 = c \tag{5}$$

satisfies (3) and therefore represents solutions of it. Figure 1 shows

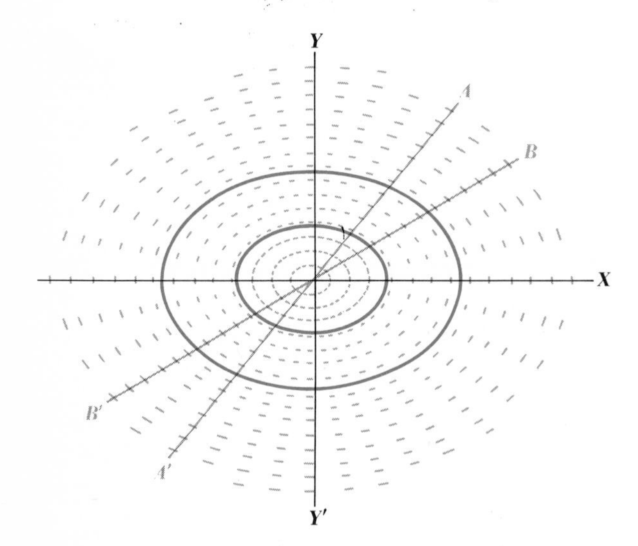

Figure 1

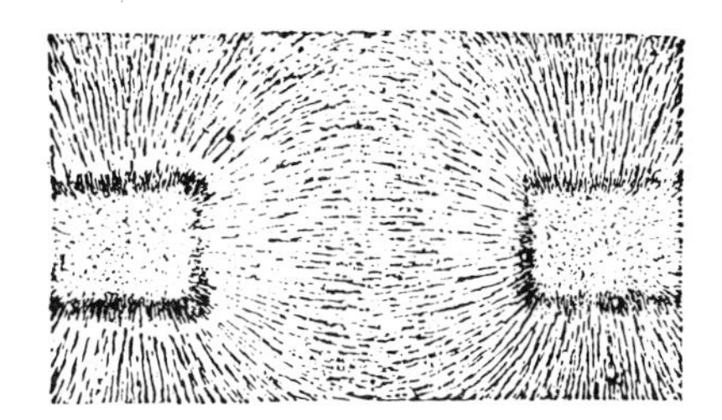

Figure 2

some ellipses representing solutions of (3). Since c in (5) may be any constant, (5) represents a family of ellipses. Any point $P(x_0,y_0)$ in the plane, except (0,0), will lie on the ellipse represented by equation (5) with $c = x_0^2 + 2y_0^2$, and through each point, except (0,0), will pass one and only one of these ellipses. The two arbitrary constants in the solution $x^2 + 2y^2 = x_0^2 + 2y_0^2$ are really equivalent to only one since all solutions could be obtained by taking $x_0 = 0$ and assigning values to y_0. The small ellipse is associated with the value $c = 1$ and the large one with $c = 4$.

Figure 2 showing the distribution of iron filings under the influence of a magnet, exhibits the same type of relationship. The iron filings serve as direction lines, and their distribution is such that the curves to which they belong are suggested. A map indicating the directions of ocean currents and winds by means of barbed lines suggests the same situation.

Exercises

1. By testing show that $y = \frac{1}{4}x^2 + c$ is a solution of $dy/dx = \frac{1}{2}x$. Find c for the curve through: (*a*) (0,0); (*b*) (0,−2); (*c*) $(0,y_0)$.

2. Write equations of the isoclines of $dy/dx = \frac{1}{2}x$. Draw short lines to indicate directions associated with points on the lines $x = 0$, $x = \pm 1$, $x = \pm 2$, and $x = \pm 3$. Draw on the same graph solution curves of $dy/dx = \frac{1}{2}x$ through points (0,0), (0,1), (0,2) and (0,−1).

3. Verify that $y = x + c$ is a solution of $dy/dx = 1$. Why is any straight line inclined 45° to the X axis an isocline? Are the lines represented by the solution the isoclines? Discuss the isoclines and solutions associated with $dy/dx = 2$.

4. Verify that $y = 2\sqrt{x} + c$ is a solution of $dy/dx = 1/\sqrt{x}$. Show that the isoclines are represented by $x = m^{-2}$. Draw the isoclines for $m = 0$, $m = 1$, $m = \sqrt{2}$, $m = \sqrt{3}$, and $m = 2$, and indicate on each line by short strokes the associated directions. Sketch part of the solutions $y = 2\sqrt{x}$, $y = 2\sqrt{x} + 1$, and $y = 2\sqrt{x} + 2$. Why does no integral curve extend to the left of the X axis?

5. Check that $y = x^2 + cx$ is a solution of $x(dy/dx) - y = x^2$. Write the solution having a graph through: (*a*) (1,2); (*b*) (−1,3), (*c*) Is there an integral curve through $(0,g)$? Is there one and only one integral curve through (a,b), $a \neq 0$?

6. (*a*) Show that $y = \ln |x| + c$ is a solution of $dy/dx = x^{-1}$. (*b*) Show that there is an integral curve through any point (a,b), $a \neq 0$, in the XY plane but that none passes through $(0,g)$. (*c*) Write the equations of the isoclines of $y' = x^{-1}$. (*d*) Sketch the integral curves of $y' = x^{-1}$ for $c = 0$, $c = 3$, and $c = -3$. (*e*) What translation is required to move the integral curve for $c = -3$ into that for $c = 7$?

7. Write the equations of the isoclines for:

(*a*) $y' = f(x)$ (*b*) $y' = f(y)$ (*c*) $y' = 1$

6 Existence theorem for $dy/dx = f(x,y)$*

An existence theorem of differential equations states the conditions under which they have solutions.

THEOREM. *If f denotes a function of x and y, then*

$$\frac{dy}{dx} = f(x,y) \tag{1}$$

has a unique continuous solution $y = \varphi(x)$ such that $y_0 = \varphi(x_0)$ provided that all values of x and y considered lie in a region S defined by

$$|x - x_0| \leqq a \qquad |y - y_0| \leqq b \qquad a > 0,\ b > 0 \tag{2}$$

and that

$$f \text{ and } \frac{\partial f}{\partial y} \text{ are real and continuous in } S \tag{3}$$

The proof of the theorem is complicated and is omitted.† The theorem specifies that one and only one integral curve of (1) passes through the point (x_0,y_0) provided conditions (3) hold in a region (2). To get evidence for this consider that $f(x,y)$, being a function, is single-valued and, by hypothesis, it is real; therefore an integral curve of (1) through (x_0,y_0) of S has a real, unique slope $f(x_0,y_0)$ at (x_0,y_0), and, because of the continuity of f, it must pass smoothly through (x_0,y_0). Hence two integral curves of (1) could not meet in (x_0,y_0) at an angle $\alpha \neq 0$; this suggests that only one curve passes through (x_0,y_0). This

* Existence theorems will be considered more completely in §98.

† For existence proofs consult: R. Courant, "Differential and Integral Calculus," vol. II, pp. 459–462, Interscience Publishers, Inc., New York, 1936; and E. L. Ince, "Ordinary Differential Equations," pp. 62–68, Longmans, Green & Co., Ltd., New York, 1927.

is generally, but not necessarily, true unless $\partial f/\partial y$ is continuous at (x_0,y_0). Consider, for example,

$$y' = \sqrt{y} \tag{4}$$

Here $f = \sqrt{y}$ is real and continuous if $y \geqq 0$, but $\partial f/\partial y = 1/(2\sqrt{y})$ is continuous if and only if $y > 0$. Hence condition (3) holds for b sufficiently small, and we conclude that a unique integral curve of (4) passes through (x_0,y_0) provided $y_0 > 0$. The solution of (4) consists of two parts, $y = 0$ and $2\sqrt{y} = x + c$. Its graph is represented in Fig. 1. Observe that through each point $(x_0,0)$ on $y = 0$ passes two integral curves $y = 0$ and $y^{1/2} = x - x_0$, but one and only one integral curve passes through (x_0,y_0), $y_0 > 0$.

Again consider $y' = x/\sqrt{25 - x^2}$. Here $f(x,y) = x/\sqrt{25 - x^2}$ and $\partial f/\partial y = 0$ are real, continuous, and single-valued provided $-5 < x < 5$. Hence, through any given point (x_0,y_0) passes a unique solution curve of $y' = x/\sqrt{25 - x^2}$ provided x lies in the region $-5 < x < 5$, $-M < y < M$, where M is any positive number however large it may be. Figure 2 represents the solution.

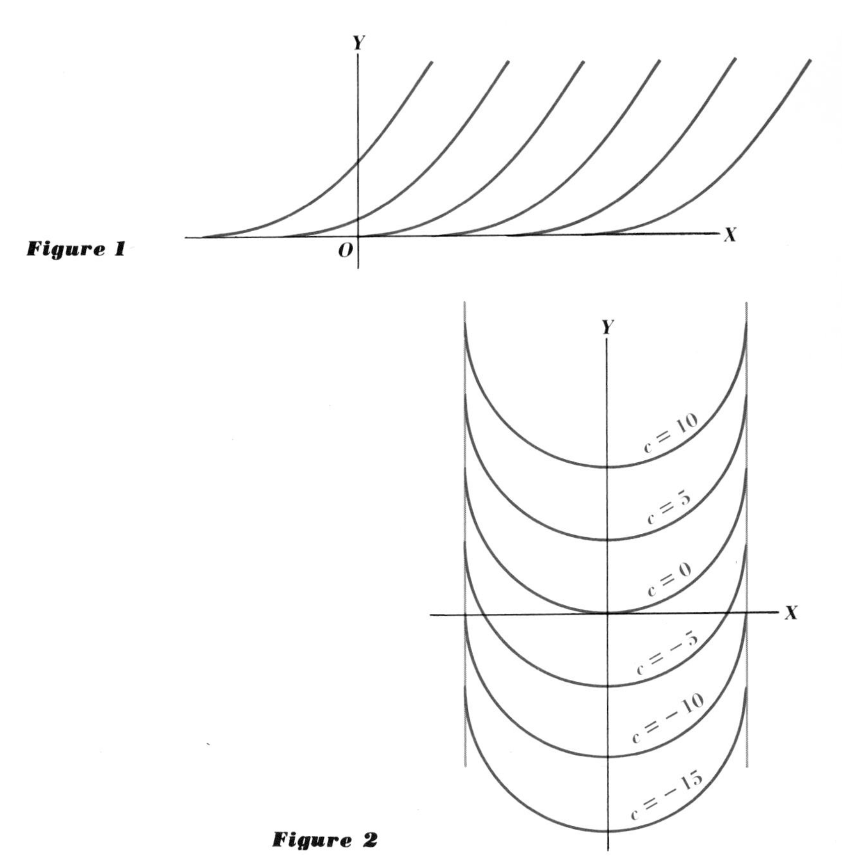

Figure 1

Figure 2

Exercises

1. Show that $y = cx$ is the general solution of $y' = y/x$. Explain why the fact that all lines $y = cx$ pass through (0,0) does not contradict the theorem. Does $y = 0$ satisfy $y' = y/x$?

(2–4) For each equation define the region in which the corresponding conditions of the theorem are satisfied and also the regions in which solutions exist:

2. $y' = \dfrac{x}{(x-1)y}$ **3.** $y' = \dfrac{\sqrt{4-y^2}}{y}$ **4.** $y' = \sqrt[3]{y \sin x}$

5. (*a*) Solve the equation $y' = x\sqrt{25-y^2}/(y\sqrt{25-x^2})$. Figure 3 represents the integral curves, and $\partial f/\partial y = -25x/(y^2\sqrt{25-x^2}\sqrt{25-y^2})$. (*b*) State the region in which the theorem is satisfied, and (*c*) the region in which $y' = x\sqrt{25-y^2}/(y\sqrt{25-x^2})$ is satisfied. (*d*) What points lie on two integral curves?

6. Define the region in which solutions of $y' = \sqrt{x}/\sqrt{y}$ exist.

★7.* The solution of $y' = \sqrt{y-x}$ is

$$x - 2\sqrt{y-x} + \ln|\sqrt{y-x} - 1| = c$$

and $y - x = 1$. In what region are the conditions of the theorem satisfied? What integral curve passes through $(a, a+1)$?

* A solid star ★ indicates a difficult problem or a complicated solution.

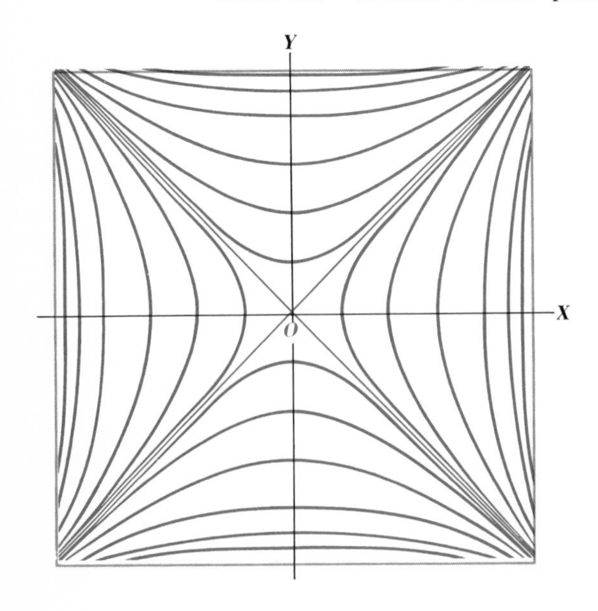

Figure 3

7 Finding the differential equation from the general solution

A solution of an ordinary differential equation of the nth order involves n independent* arbitrary constants and such a solution is called the general solution. A solution obtained by replacing the arbitrary constants by specific numbers is called a particular solution. For example $y = c_1 \sin 2x + c_2 \cos 2x$ is the general solution of $d^2y/dx^2 + 4y = 0$, and $y = 7 \sin 2x$ is a particular solution. Often an equation has a so-called singular solution not obtainable by assigning definite numbers to constants in the general solution.† Thus $y' = \sqrt{y}$ has the general solution $2y^{1/2} = x + c$, and the singular solution $y = 0$.

This section deals with the problem of finding the differential equation when its general solution is given. The procedure involved will show clearly the relation between the number of constants in the general solution of a differential equation and its order. Consider $y = ax + bx^3$. We have

$$y = ax + bx^3 \qquad y' = a + 3bx^2 \qquad y'' = 6bx \tag{1}$$

From the second and third equations of (1), it appears that we must have

$$b = \frac{1}{6}\frac{y''}{x} \qquad a = y' - 3x^2\left(\frac{1}{6}\frac{y''}{x}\right) = y' - \tfrac{1}{2}xy'' \tag{2}$$

Hence substitution of these values of a and b in $y = ax + bx^3$ gives the desired equation

$$y = xy' - \tfrac{1}{2}x^2y'' + \tfrac{1}{6}x^2y'' = xy' - \tfrac{1}{3}x^2y'' \tag{3}$$

The student may check that the first equation of (1) is the general solution of (3).

Generalizing the procedure above we obtain the following rule.

RULE. *To find the differential equation when the general solution is given: Differentiate the general solution, differentiate the derived equation, differentiate the second derived equation, etc., until the number of the derived equations is equal to the number of independent arbitrary constants in the general solution; finally eliminate the constants from the general solution and the derived equations.*

A few examples will illustrate the process.

* Linear independence is considered in §45.

† Singular solutions are considered in Chap. 5.

Example 1. Find the differential equation whose general solution is $y = c_1e^{2x} + c_2e^{-x} + x$.

Solution. The general solution and the first two derived equations are

$$y = c_1e^{2x} + c_2e^{-x} + x \qquad (a)$$

$$\frac{dy}{dx} = y' = 2c_1e^{2x} - c_2e^{-x} + 1 \qquad (b)$$

$$\frac{d^2y}{dx^2} = y'' = 4c_1e^{2x} + c_2e^{-x} \qquad (c)$$

Eliminating c_2 from (a) and (b) and then from (b) and (c), we get

$$y' + y = 3c_1e^{2x} + x + 1 \qquad y'' + y' = 6c_1e^{2x} + 1 \qquad (d)$$

Multiplying the first equation of (d) by 2, subtracting the result from the second, and simplifying slightly, we get

$$y'' - y' - 2y = -2x - 1^*$$

Also by determinants we obtain from (a), (b), and (c)

$$\begin{vmatrix} y - x & e^{2x} & e^{-x} \\ y' - 1 & 2e^{2x} & -e^{-x} \\ y'' & 4e^{2x} & e^{-x} \end{vmatrix} = e^{2x}e^{-x} \begin{vmatrix} y - x & 1 & 1 \\ y' - 1 & 2 & -1 \\ y'' & 4 & 1 \end{vmatrix} = 0$$

Example 2. Find the differential equation of the system of ellipses having their axes along the X axis and the Y axis.

Solution. The equation of the system may be written

$$\frac{x^2}{a^2} = 1 - \frac{y^2}{b^2} \qquad (a)$$

where a and b are arbitrary constants. By two differentiations we obtain from (a)

$$\frac{2x}{a^2} = -\frac{2y}{b^2}y' \qquad (b)$$

$$\frac{2}{a^2} = -\frac{2yy'' + 2y'^2}{b^2} \qquad (c)$$

* Observe that three equations, one given and two derived, were used in eliminating two arbitrary constants and that this led to a second-order differential equation having as solution the given equation in two arbitrary constants. Similarly, for an equation in n arbitrary constants, we would expect to differentiate n times, eliminate the n constants, and thus find an nth-order differential equation having the given equation in n constants as solutions.

Dividing (*b*) by (*c*) member by member and simplifying, we get

$$x[yy'' + (y')^2] = yy'$$ *

Exercises

(1–19) Find the differential equations having as solutions:

1. $y = x^3 + c$ **2.** $y = cx^2$ **3.** $y = cx + 2c$

4. $y = cx^2 + cx$ **5.** $y = ce^x$ **6.** $y = c_1e^x + c_2e^{-x}$

7. $y = c_1x^2 + c_2x + c_3$

8. $y = c_1 \sin 2x + c_2 \cos 2x$

9. All lines through the origin.

10. All circles with center (0,0).

11. $y = ae^{bx}$ **12.** $y = \sin x + c_1 \sin 2x + c_2 \cos 2x$

13. $y = a \sin (x + b)$ ★**14.**† $y = x \sin (x + c)$

15. All straight lines.

★**16.** All circles through (0,0) and (2,0) (see Fig. 1).

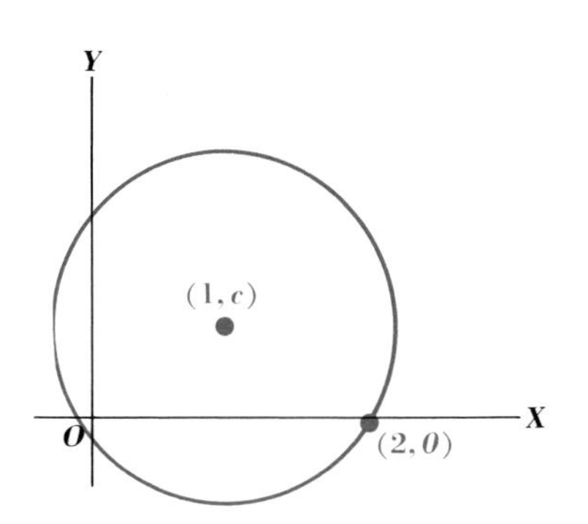

Figure 1

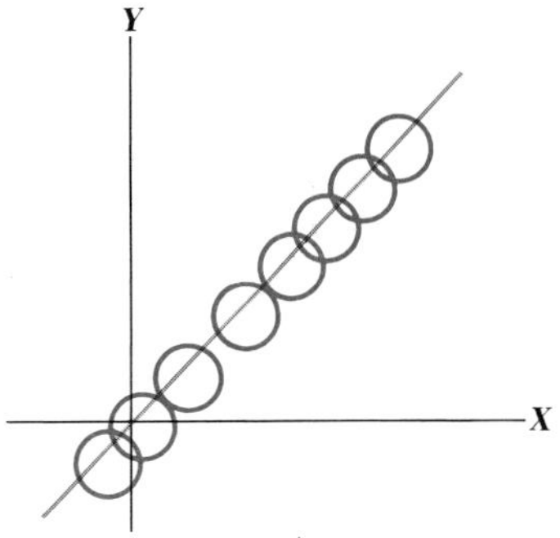

Figure 2

★**17.** All circles of radius 1 and centers on line $x = y$ (see Fig. 2).

★**18.** All tangent lines of parabola $y^2 = 4x$.

★**19.** Show that a differential equation of $ax^2 + bxy + cy^2 = 1$ is

$$\begin{vmatrix} xy''' + 3y'' & yy''' + 3y'y'' \\ xy' + x^2y'' - y & xyy'' + x(y')^2 - yy' \end{vmatrix} = 0$$

* Throughout the text answers to examples appear in color.

† A solid star ★ indicates a difficult problem or a complicated solution.

8 Differential equations of the first order and second degree

The type* of equation considered here is represented by

$$A(y')^2 + By' + C = 0 \qquad A \neq 0 \tag{1}$$

where $y' = dy/dx$ and A, B, and C are functions of x and y.

Solving (1) for y' by algebra we obtain the two roots:

$$y_1' = \frac{-B + \sqrt{B^2 - 4AC}}{2A} \qquad y_2' = \frac{-B - \sqrt{B^2 - 4AC}}{2A} \tag{2}$$

Hence we may write equation (1) in the form

$$\left(y' - \frac{-B + \sqrt{B^2 - 4AC}}{2A}\right)\left(y' - \frac{-B - \sqrt{B^2 - 4AC}}{2A}\right) = 0 \tag{3}$$

A solution of (3) will be obtained by equating either factor in parenthesis to zero and solving the resulting differential equation for y in terms of x; for any solution thus obtained when substituted in (3) will satisfy it because one factor of the result will be zero. Of course a solution of (3) is also a solution of (1).

To solve an equation of type (1): *Factor it by using* (2) *or by direct factorization, equate each factor to zero, and solve each of the two resulting differential equations.*

Example 1. Find two solutions of $y'^2 + 2xy' - 3x^2 = 0$.

Solution. $y'^2 + 2xy' - 3x^2 = (y' + 3x)(y' - x) = 0$.

Solving the two equations $y' + 3x = 0$ and $y' - x = 0$, we get

$$y + \tfrac{3}{2}x^2 = c_1 \qquad y = \tfrac{1}{2}x^2 + c_2$$

Example 2. Solve $9x^2\,dy^2 = 16y^2\,dx^2$. (*a*)

Solution. Equating the square roots of the two members of the given equation and solving the resulting differential equations, we get

$$3x\,dy = 4y\,dx \qquad 3x\,dy = -4y\,dx$$
$$y^3 = c_1x^4 \qquad y^3x^4 = c_2$$

* Chapter 5 treats more generally equations of first order and degree higher than the first.

Exercises

1. Solve $(y')^2 + 7y' - 18 = 0$ by equating to zero each factor of the left member and solving the results for y in terms of x.

(2–11) Solve the following differential equations:

2. $(y' - 3x^2)(y' - 4x^3) = 0$
3. $16y^2\,dy^2 - 9x^2\,dx^2 = 0$
4. $xy(y')^2 + (x^2 + y^2)y' + xy = 0$
5. $x^2(y')^2 + xyy' - 2y^2 = 0$

6. $x(y')^2 - (3x^2 - 1)y' - 3x = 0$
7. $xy'(y' - 2x) = y'(2xy' - x^2 - 2) - 2x$
8. $(y' - 1)(y' - 2x) = 0$
9. $x\dfrac{dy}{dx} - y\dfrac{dx}{dy} = 0$

10. $x(y')^2 + (x^2 - 2\sqrt{x})y' + 1 - x^{3/2} = 0$. In what part of the plane do integral curves exist?

11. $x(y')^2 + (\sqrt{x} - 2\sqrt{x}\sqrt{y})y' + y - \sqrt{y} = 0$; $y = 4$ when $x = 1$ (two solutions). In what part of the plane do integral curves exist?

9 Review exercises

In the following exercises derivatives with respect to x are denoted by primes.

Give the order and degree of each differential equation:

1. $x(y')^3 + y' = 7xy$ **2.** $x^2 + yy' - (y')^{3/2} = 0$

3. $\sqrt{y'} = 3(y'')^{2/3}$

Prove that each equation is a solution of the differential equation written opposite it:

4. $y = c^2 + cx^{-1}$ $\quad y + xy' = x^4(y')^2$
5. $x^2 + y^2 = cx$ $\quad 2xyy' = y^2 - x^2$
6. $e^{\cos x}(1 - \cos y) = c$ $\quad \sin y y' + \sin x \cos y = \sin x$

Find the differential equations of the following systems of plane curves:

7. $y = cx + c^3$ **8.** $y = ce^{5x}$ **9.** $(x - c_1)^2 + c_2 y = c_3$

10. All circles with centers on the line $x + 2y = 0$. *Hint:* $(2c, -c)$ represents all centers.

Define for each equation: (a) the region in which the conditions of the theorem of §6 are satisfied, and (b) the region in which solutions exist:

11. $y' = \dfrac{\sqrt{y}}{\sqrt{x - 1}}$ **12.** $y' = \dfrac{\sqrt[3]{y}}{\sqrt{x + 2}}$ **13.** $y' = \sqrt{y^2 - x^2}$

Solve the following differential equations, and determine the constants of integration when initial conditions are indicated:

14. $ay' + ay = y - xy'$ **15.** $x\,dy - 2y\,dx = y^2\,dx$

16. $(y^2 - y + 1)y' - y = y^3$; $y = 1$ when $x = \frac{1}{4}\pi$.
17. $xy' - 3y = x^2y' - 2y$; $x = 2$ when $y = 1$.

Find the differential equation of each system of plane curves:

18. $ax^2 + y^2 = 25$ **19.** $y = ax + b$

★20. $\dfrac{x}{c - 1} + \dfrac{y}{c + 1} = 1$

21. All circles with centers on $x + 2y = 0$ and passing through $(0,0)$.
22. The tangents to $x^2 = 4y$.

Solve the following differential equations and determine the constants of integration when initial conditions are given:

23. $xyy' + (1 - y^2) = 0$; $y = 4$ when $x = 1$.
24. $3e^x \tan y + (1 + e^x) \sec^2 y\, y' = 0$; $y = \frac{1}{4}\pi$ when $x = \ln 2$.
25. $x \ln y \ln x\, dy + dx = 0$; $x = e$ when $y = 1$.

26. $(y')^2 + 2y' - 3 = 0$ **27.** $x^2 \dfrac{dy}{dx} - y^2 \dfrac{dx}{dy} = 0$

28. $(y')^2 - 4x^2y' + 4x^4 - x^2 = 0$

2

Applications

10 Geometric applications using rectangular coordinates

A great number of geometric problems can be solved by using the process of expressing a geometric relation in the form of a differential equation and solving it. Take, for example, the problem of finding the equation of the curve through $(3, -4)$ having at each point (x,y) on it a slope of $2y/x$. Since dy/dx represents the slope of the curve, we have

$$\frac{dy}{dx} = \frac{2y}{x} \tag{1}$$

Integrating equation (1), obtain

$$\ln y = 2 \ln x + \ln c = \ln cx^2 \tag{2}$$

hence,

$$y = cx^2 \tag{3}$$

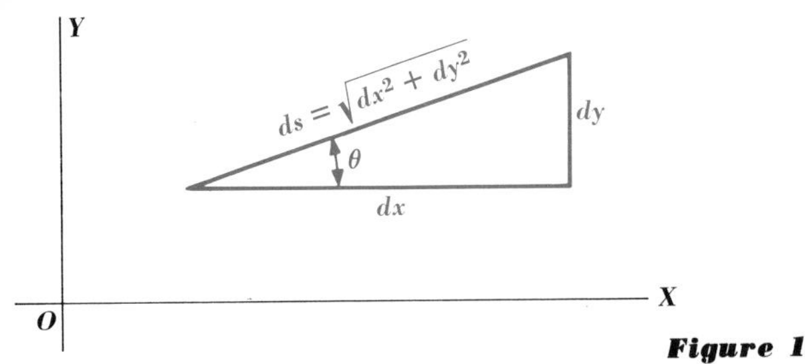

Figure 1

Since $(3, -4)$ lies on the curve, substitute 3 for x and -4 for y in (3) to obtain

$$-4 = 9c \qquad \text{or} \qquad c = \frac{-4}{9} \tag{4}$$

Therefore, the required equation is

$$y = \frac{-4}{9}x^2$$

or

$$4x^2 + 9y = 0 \tag{5}$$

Occasionally the student will need basic formulas from calculus. Therefore it is well to keep Fig. 1 in mind. It is based on the fact that dy/dx at any point (x,y) on a curve equals the slope of the tangent to the curve at (x,y). A glance at Fig. 1 recalls the following formulas:

$$ds = \sqrt{dx^2 + dy^2} = \sqrt{1 + \left(\frac{dy}{dx}\right)^2}\,dx = \sqrt{1 + \left(\frac{dx}{dy}\right)^2}\,dy \tag{6}$$

$$\tan\theta = \text{slope} = \frac{dy}{dx} \qquad \sin\theta = \frac{dy}{ds} \qquad \cos\theta = \frac{dx}{ds} \qquad \text{etc.} \tag{7}$$

Example. Find the equation of a curve if the part of each of its tangent lines from the point of contact to the intersection with the X axis is bisected by the Y axis.

Solution. To solve a problem of this kind, the student should *first draw a figure representing the curve with any point (x,y) on it and showing the essential relations involved in the problem; then try to find the value of the slope of the required curve or of some expression containing the slope, form an equation, and integrate it.*

Figure 2 relates to the problem under consideration. From it we see that $AO = OD = x$ and that the slope of the tangent at P is $y/2x$. Hence,

$$\frac{dy}{dx} = \frac{y}{2x}$$

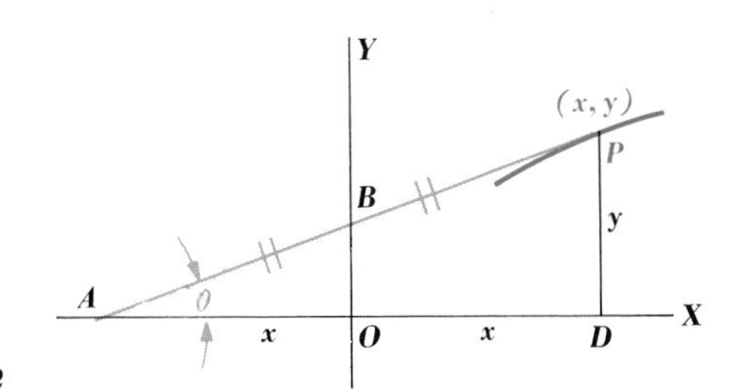

Figure 2

The solution of this equation is

$$y^2 = cx$$

Exercises

1. Use the law of sines on Fig. 1 to obtain $dy/\sin\theta = dx/\sin(90° - \theta) = ds$. Also, obtain from the law of cosines

$$dy^2 = dx^2 + ds^2 - 2\,dx\,ds\cos\theta$$
$$dx^2 = dy^2 + ds^2 - 2\,dy\,ds\sin\theta$$

2. Find the equation of the family of curves and the equation of the particular curve that passes through point (3,4) if the slope of the tangent at any point (x,y) is:

(a) $2x - 2$ (b) $\dfrac{1-x}{1+y}$ (c) $\dfrac{y-1}{1-x}$

3. Prove that a curve having a constant slope is a straight line.

4. Find an expression $u(x)$ which has a derivative with respect to x that is: (a) equal to $u(x)$; (b) 6 times $u(x)$; (c) 4 less than $u(x)$.

5. Find the most general kind of curve such that the normal at any point of it coincides in direction with the line connecting this point to the origin. Use Fig. 3.

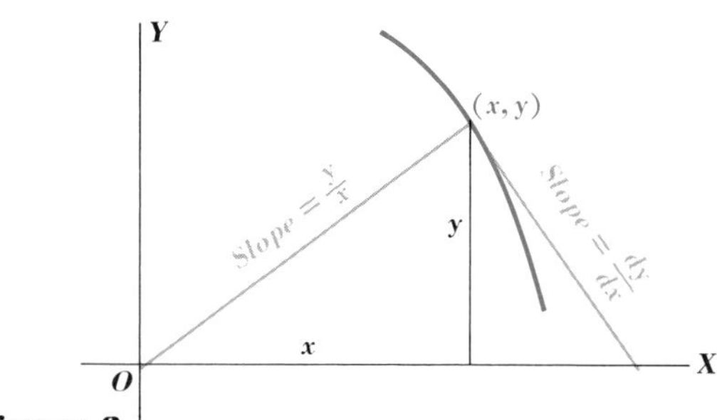

Figure 3

6. The part of the normal to a curve, at any point (x,y) on the curve, between (x,y) and the point where the normal meets the Y axis is bisected by the X axis. Find the equation of the curve.

7. For a certain curve the point of contact of each tangent to it bisects the part of the tangent terminating on the coordinate axes. Find the equation of the curve.

8. Find the equation of the curve so drawn that every point on it is equidistant from the origin and the intersection of the X axis with the normal to the curve at the point.

9. The area bounded by a curve, the X axis, a fixed positive ordinate, and a variable ordinate is proportional to the difference between the ordinates. Find the equation of the curve.

10. Figure 4 represents a curve C with $P(x,y)$ any point on it. The tangent and the normal of curve C at P cut the X axis in A and α and the Y axis in B and β, respectively. Point $(x,0)$ is E, and θ is the angle that AP makes with the X axis. Consider all line segments as directed so that $OE = -EO$, $A\alpha = -\alpha A$, etc. Verify that: (*a*) $\tan\theta = dy/dx$, $OE = x$, and $E\alpha/y = dy/dx$; (*b*) $AE = y\,dx/dy$; (*c*) $OA = x - y\,dx/dy$; (*d*) $O\beta = y + x\,dx/dy$.

(**11–18**) Read exercise 10, and then find the equation of curve C in Fig. 4 if:

11. AE is a constant k.
12. The area of triangle $PE\alpha$ is a constant k.
13. P bisects $\alpha\beta$. Note that $OE = E\alpha$; therefore, $x = y\,dy/dx$.
14. A bisects BP. Note that $OA = AE$.

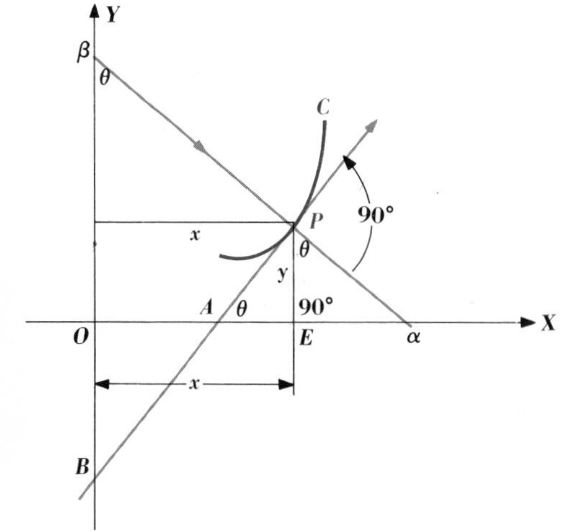

Figure 4

15. B bisects AP. Note that $AB = BP$; therefore, $-OA = OE$.

16. β is between α and P and is twice as far from α as from P. Note that $\alpha\beta = 2\beta p$, or $2OE = \alpha O = -O\alpha$.

17. B is between A and P and twice as far from P as from A.

18. $AP/PB = 2a$, a constant.

19. Find two sets of curves along which

$$3x\, ds/dx = \sqrt{9x^2 + 25y^2}$$

where s represents the arc length of the curve.

11 Isogonal trajectories. Orthogonal trajectories

If G represents a given one-parameter family of curves in a plane region R, and T a required one-parameter system such that at every point (x,y) in R a curve of T intersects a curve of G at constant angle α, then curves T are called isogonal trajectories of G in R. If $\alpha = 90°$ curves T are called orthogonal trajectories of G in R. For example, all straight lines $y = x + c$ are isogonal trajectories of lines $y = c$ parallel to the X axis, and all lines $y = cx$ through $(0,0)$ are orthogonal trajectories of all circles $x^2 + y^2 = c^2$ having $(0,0)$ as center. The region R in the first case is the whole xy plane, and the same with $(0,0)$ excluded in the second case. The parameter is c in each case.

For isogonal families G and T we have at (x,y) a curve of T with slope m_t intersecting a curve of G with slope m_g at angle α; hence, from analytic geometry

$$\tan \alpha = \frac{m_t - m_g}{1 + m_t m_g} \tag{1}$$

Now the slope of a curve is given by dy/dx. Accordingly we write

$$m_t = \left(\frac{dy}{dx}\right)_t \qquad m_g = \left(\frac{dy}{dx}\right)_g \tag{2}$$

Thus we have from (1) and (2) at (x,y):

$$\text{Isog. Traj.} \qquad \tan \alpha = \frac{(dy/dx)_t - (dy/dx)_g}{1 + (dy/dx)_g (dy/dx)_t} \tag{3}$$

If $\alpha = 90°$, m_t is the negative reciprocal of m_g, and

$$m_t = -\frac{1}{m_g} \tag{4}$$

Hence, we have from (4), for orthogonal trajectories

$$\text{Orthog. Traj.} \qquad \left(\frac{dy}{dx}\right)_t = \frac{-1}{(dy/dx)_g} \tag{5}$$

To get the system T, find $(dy/dx)_g$ at (x,y) from given data, substitute it in (3) or (5) according to requirements, and solve the resulting differential equation. Note that $(dy/dx)_g$ at (x,y) is a function of x and y. For example, if G is defined by $y^2 = cx^3$, then for the curve through (x,y) $c = y^2/x^3$ and

$$\frac{dy}{dx} = \frac{3cx^2}{2y} = \frac{3x^2}{2y} \cdot \frac{y^2}{x^3} = \frac{3y}{2x}$$

Example 1. Find the orthogonal trajectories of

$$y^2 = cx^3 \tag{a}$$

Also find the equation of the particular trajectory through (2,4).

Solution. From (a) we get by differentiation and replacement of c by y^2/x^3,

$$\left(\frac{dy}{dx}\right)_g = \frac{3x^2c}{2y} = \frac{3x^2}{2y} \cdot \frac{y^2}{x^3} = \frac{3y}{2x} \tag{b}$$

Therefore, from (5)

$$\left(\frac{dy}{dx}\right)_{\text{Orthog. Traj.}} = \frac{-2x}{3y} \tag{c}$$

The solution of (*c*) is

$$2x^2 + 3y^2 = c \tag{d}$$

This represents a family of ellipses. To find c for the particular curve through (2,4), substitute 2 for x and 4 for y in (*d*), and obtain $c = 56$. Hence,

$$2x^2 + 3y^2 = 56$$

A brief consideration of Fig. 1 will serve to clarify essential relations.

Example 2. Find the isogonal trajectories intersecting at 45° the hyperbolas

$$y(x + c) = 1 \tag{a}$$

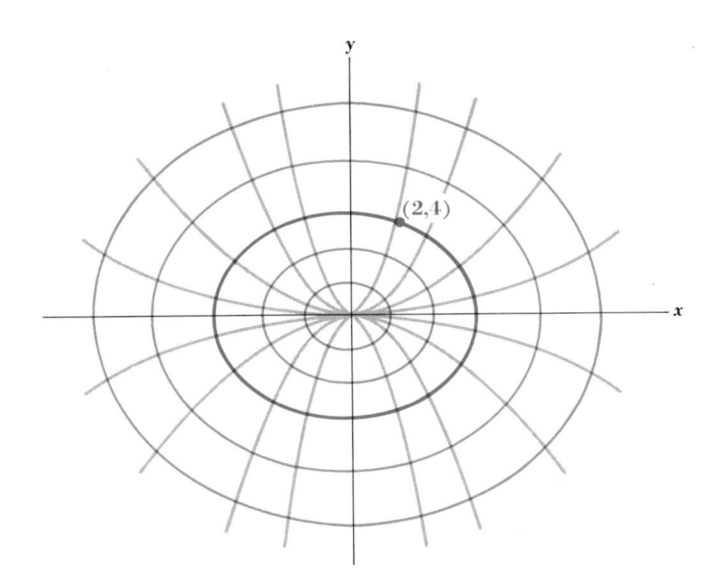

Figure 1

Solution. To find $[(dy/dx)_g \text{ at } (x,y)]$, find dy/dx from (a) in terms of x, y, and c; in the result replace $x + c$ by y^{-1}, and get

$$\left[\left(\frac{dy}{dx}\right)_g \text{ at } (x,y)\right] = \frac{-1}{(x+c)^2} = -y^2 \tag{b}$$

Now substitute 45° for α and $-y^2$ for $[(dy/dx)_g \text{ at } (x,y)]$ in (3) and obtain

$$\frac{(dy/dx) + y^2}{1 - y^2\,(dy/dx)} = 1 \tag{c}$$

The solution of (c) is

$$\ln\left|\frac{y+1}{y-1}\right| = x + y + \ln c \qquad \text{or} \qquad y + 1 = (y-1)ce^{x+y} \tag{d}$$

Exercises

Find the equations of the orthogonal trajectories of:

1. The hyperbolas $y^2 = x^2 + c$. 2. The parabolas $x^2 = 2cy$.
3. The cubics $x^2 = 4cy^3$.
4.* $y^2 + y = ce^x$; $y \neq 0$, $y \neq -1$, $y \neq -\frac{1}{2}$.

* In exercise 4 the restrictions are designed to remove all considerations of infinite slopes.

Find the equations of the isogonal trajectories cutting at $\tan^{-1} 4$ the family of curves defined by:

5. $4y = 3x + c$ **6.** $y^2 = x + c;\ y \neq 0,\ y \neq -\frac{1}{4}$.

7. $(x + c)y = 1;\ y \neq 2,\ y \neq -2$.

8. Prove that the differential equation of the orthogonal trajectories of a system $f(x,y) = c$ is $(\partial f/\partial x) - (\partial f/\partial y)/y' = 0$.

Find the isogonal trajectories at the indicated angle α for each family of curves:

9. $x^2 = cy^m;\ \alpha = 90°$. **10.** $y = ce^{3x};\ \alpha = 90°$.

11. $(x + c)y^2 = 1;\ \alpha = \tan^{-1} 4$.

12. Show that the orthogonal trajectories of the family $F(y) = f(x) + c$ are represented by $\int [1/f'(x)]\, dx + \int [1/F'(y)]\, dy = c$.

13. Let φ and ψ represent two functions of x and y.

From §8 we know that the equation

$$(y')^2 - (\varphi + \psi)y' + \varphi\psi = (y' - \varphi)(y' - \psi) = 0 \qquad (a)$$

represents two families of curves which we call the φ system and the ψ system. If in (a) we replace y' by $-1/y'$ and solve the result, we get two systems, say φ_1 and ψ_1, respectively. What relations hold among the φ, ψ, φ_1, and ψ_1 systems?

12 Geometric applications using polar coordinates

A basic formula relating to curves in polar coordinates ρ and θ is

$$\tan \psi = \frac{\rho\, d\theta}{d\rho} \qquad (1)$$

where, as indicated in Fig. 1, ψ represents the angle between the tangent

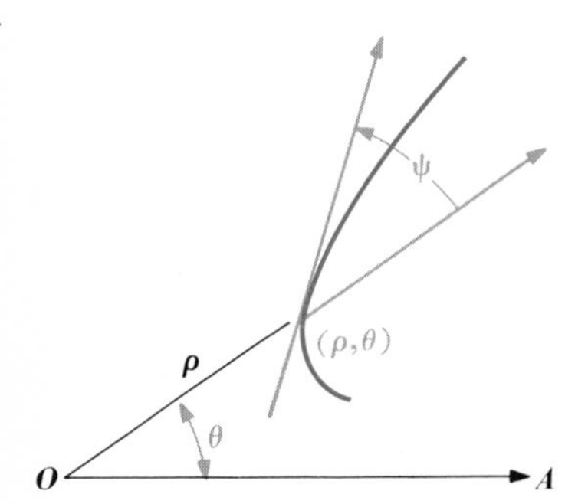

Figure 1

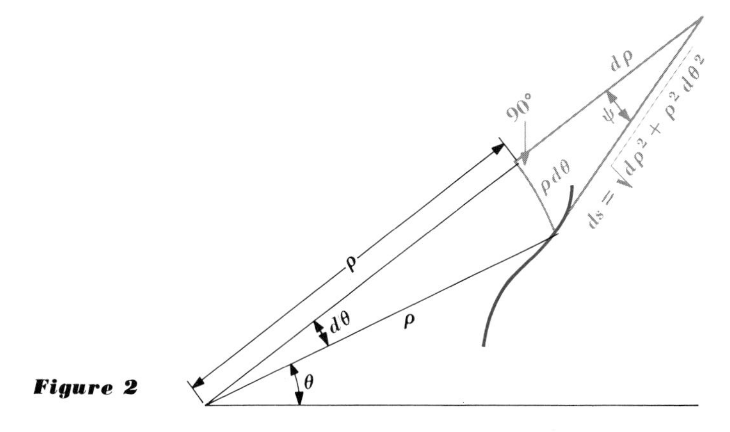

Figure 2

to the curve at (ρ,θ) and the radius vector to (ρ,θ). This formula enables us to obtain the equations of many curves having interesting geometric properties. Figure 2 suggests equation (1) and also the following equations:

$$ds = \sqrt{d\rho^2 + \rho^2\, d\theta^2} = \sqrt{\left(\frac{d\rho}{d\theta}\right)^2 + \rho^2}\, d\theta \tag{2}$$

$$\tan\psi = \frac{\rho\, d\theta}{d\rho} \qquad \cos\psi = \frac{d\rho}{ds} \qquad \sin\psi = \frac{\rho\, d\theta}{ds} \tag{3}$$

If G represents a one-parameter family of curves, if curve T cuts at each of its points (ρ,θ) a curve of G at a constant angle α, and if ψ_t and ψ_g are the respective angles ψ of equation (1) (see Fig. 3), then

$$\psi_t - \psi_g = \alpha \qquad \text{and} \qquad \frac{\tan\psi_t - \tan\psi_g}{1 + \tan\psi_t \tan\psi_g} = \tan\alpha \tag{4}$$

The family of all curves T satisfying (4) are the isogonal trajectories of G. From (1) and (4), we get

$$\frac{(\rho\, d\theta/d\rho)_t - (\rho\, d\theta/d\rho)_g}{1 + (\rho\, d\theta/d\rho)_g(\rho\, d\theta/d\rho)_t} = \tan\alpha \tag{5}$$

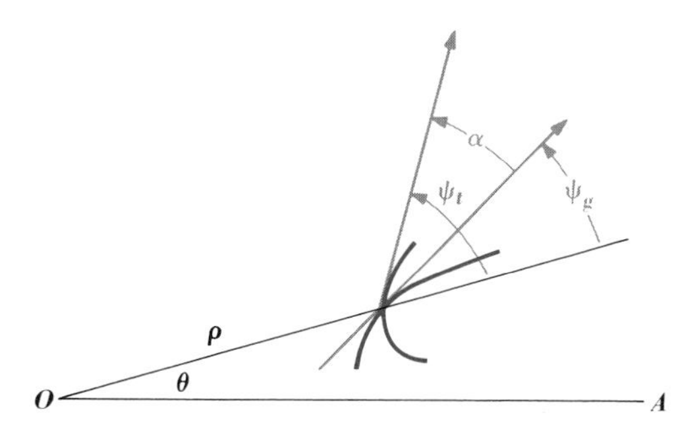

Figure 3

If $\alpha = 90°$, we get from the first equation of (4)

$$\tan \psi_t = \tan\left(\tfrac{1}{2}\pi + \psi_g\right) = \frac{-1}{\tan \psi_g} \tag{6}$$

or, from (1), for orthogonal trajectories,

$$\left(\frac{\rho\, d\theta}{d\rho}\right)_t = \frac{-1}{(\rho\, d\theta/d\rho)_g} \tag{7}$$

Example. Find the equation of the orthogonal trajectories and that of the isogonal trajectories of the family of curves $\rho = a \sin \theta$.

Solution. For the family $\rho = a \sin \theta$,

$$\frac{\rho\, d\theta}{d\rho} = \frac{\rho}{(d\rho/d\theta)} = \frac{a \sin \theta}{a \cos \theta} = \tan \theta \tag{a}$$

Substituting $\tan \theta$ for $(\rho\, d\theta/d\rho)_g$ in (7) and dropping the subscript t, we get

$$\frac{\rho\, d\theta}{d\rho} = \frac{-1}{\tan \theta} \tag{b}$$

The orthogonal trajectories of $\rho = a \sin \theta$ are defined by the solution of (b); that is, by

$$\rho = c \cos \theta \tag{c}$$

To get the equation of the isogonal trajectories substitute $\tan \theta$ from (a) for $(\rho\, d\theta/d\rho)_g$ in (5), drop the subscript t, and solve the result. The process follows:

$$\frac{\rho\, d\theta/d\rho - \tan \theta}{1 + (\rho\, d\theta/d\rho) \tan \theta} = \tan \alpha \tag{d}$$

$$\left(\frac{\rho\, d\theta}{d\rho}\right)(1 - \tan \alpha \tan \theta) = \tan \alpha + \tan \theta \tag{e}$$

$$d\theta \left(\frac{1 - \tan \alpha \tan \theta}{\tan \alpha + \tan \theta}\right) = d\theta \cot(\theta + \alpha) = \frac{d\rho}{\rho}$$

$$\rho = c \sin(\theta + \alpha) \tag{f}$$

Exercises

1. Find the equation of the curve through the point ($\rho = a$, $\theta = 0$) and cutting all lines through the pole at a constant angle α. *Hint:* $\tan \psi = \tan \alpha$.

2. In Fig. 4, O is the pole, OA the polar axis, $P(\rho,\theta)$ any point on curve C, TP the tangent to curve C at (ρ,θ), and OT the perpendicu-

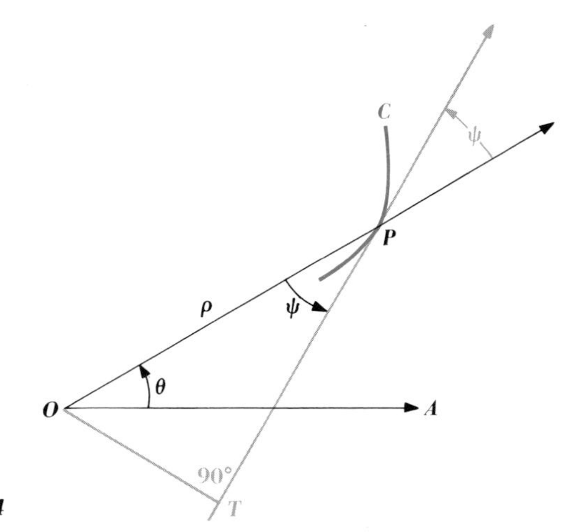

Figure 4

lar to TP. (*a*) Express OT and TP in terms of ρ and ψ. Find the equation of C if: (*b*) $OT = h$, a constant; (*c*) $OT = \rho \sin \theta$; (*d*) $TP = 2\,OT$.

(**3–9**) Find the equations of: (*a*) the orthogonal trajectories, and (*b*) the isogonal trajectories of:

3. $\rho = a \sin (\theta + \frac{1}{6}\pi)$ **4.** $\rho = a(1 - \cos \theta)$ **5.** $\rho = \dfrac{a}{1 + \sin \theta}$

6. $\rho = c\theta,\ 0 < \theta \leqq 2\pi$. **7.** $\rho = c \sec (\theta + \frac{1}{3}\pi)$

8.* $\rho = c \sin (n\theta)$ **9.*** $\rho = \dfrac{c}{2 - \cos \theta}$

10. A perpendicular at the pole P to the radius vector of any point Q on a certain curve meets the tangent at Q in the point T and the normal in the point N (see Fig. 5). Express each of the lengths NQ, TQ, NP, and TP in terms of ρ and ψ.

* Express the answers to 8(*b*) and 9(*b*) in the form $\ln \rho = \int f(\theta)\, d\theta + c$ but omit the actual integration.

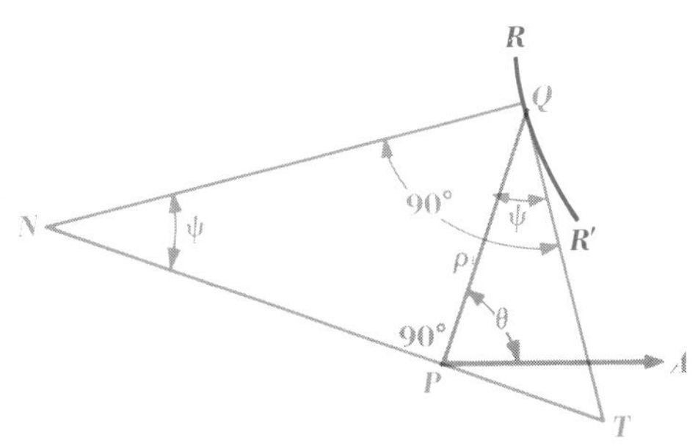

Figure 5

Using (1) and the results of exercise 10 find the equation of the curve RR' of Fig. 5 if:

11. $NP = h$, a constant. **12.** $NQ = h$
13. $NP = PT$ **★14.** $NT = 3$

13 Use of limits

In many cases it is convenient to use limits instead of determining the constant of integration and other constants. Integrating the differential equation

$$f_1(x)\,dx + f_2(y)\,dy = 0 \tag{1}$$

we obtain

$$F_1(x) + F_2(y) = c \tag{2}$$

where we get $F_1(x)$ and $F_2(y)$ by integrating $f_1(x)dx$ and $f_2(y)\,dy$, respectively. If we know that $x = a$, $y = b$ and $x = l$, $y = m$ are pairs of numbers satisfying (2) for some value of c, then

$$F_1(a) + F_2(b) = c \qquad F_1(l) + F_2(m) = c \tag{3}$$

From (2) and (3) we easily obtain

$$F_1(x) - F_1(a) + F_2(y) - F_2(b) = 0 \tag{4}$$
$$F_1(l) - F_1(a) + F_2(m) - F_2(b) = 0 \tag{5}$$

Equations (4) and (5) may be written in the forms

$$\begin{aligned} \int_a^x f_1(t)\,dt + \int_b^y f_2(t)\,dt &= 0 \\ \int_a^l f_1(t)\,dt + \int_b^m f_2(t)\,dt &= 0 \end{aligned} \tag{6}$$

For example, we find the solution of $5x^4\,dx + 2ky\,dy = 0$ satisfied by $x = 0$, $y = 5$, and $x = 2$, $y = 10$ by writing from (6)

$$\int_0^x 5t^4\,dt + \int_5^y 2kt\,dt = 0 \qquad \int_0^2 5t^4\,dt + \int_5^{10} 2kt\,dt = 0 \tag{7}$$

evaluating the indicated integrals and deducing from the results

$$x^5 - \tfrac{32}{75}(y^2 - 25) = 0 \tag{8}$$

Note that from a given differential equation and conditions we form equations of type (6), solve them, and deduce required results from them.

Exercises

1. If $2x\,dx + 2y\,dy = 0$, use the first equation of (6) to find a nonintegral relation between x and y such that $y = 3$ when $x = 2$.

2. Given $3x^2\,dx + k2y\,dy = 0$, $y = 5$ when $x = 0$, and $y = 3$ when $x = 2$, use the second equation of (6) to find k and then the first to find a nondifferential relation between x and y.

3. Given $y\,dx + k\,dy = 0$, $y = e^2$ when $x = 2$, and $y = e^3$ when $x = 5$, use (6) to find k and a nondifferential equation in x and y.

14 Physical applications

Most situations in nature are so complicated that they cannot be dealt with exactly by mathematics. The regular procedure is to apply mathematics to an ideal situation approximating the actual one and having important features of it.

The results are approximations having a practical importance which depends upon the closeness of approximation as verified by reasoning and experiment. Consider, for example, the procedure for the flight of a projectile. The forces of gravity and air resistance acting upon a large projectile rotating while moving forward are very complicated. If we assume that gravity is a constant vertical force and neglect both air resistance and rotary motion, a simple solution is easily obtained; but it is practically worthless. If, as a better approximation, we assume that air resistance is proportional to velocity and acts opposite to the direction of motion and if we get a good factor of proportionality based on experiment, the solution will give a better approximation to the actual motion and may be useful for some purposes. Finally, if a group of mathematicians, physicists, and technicians are supplied with powerful computing machines and a proving ground permitting extensive experimentation, they can get results accurate enough for any practical purpose. They would investigate all forces involved, devise a theory, and then apply methods in the development of which differential equations would play a prominent role.

In this treatment, the laws obtained by observation, experimentation, and reasoning are given. The student is required to express them in mathematical symbols, solve the resulting differential equations, and interpret the solutions.

15 Quantities varying exponentially

We easily deduce that if

$$\frac{dq}{dt} = kq \qquad \text{then } q = ce^{kt} \tag{1}$$

The second equation of (1) indicates that q varies exponentially and the first one indicates that q varies at a rate proportional to its size. Among the many quantities which vary exponentially, under certain conditions, may be mentioned the following: the temperature of water cooling in a container, electric currents, substances involved in chemical reactions, radium, and some isotopes produced by atomic explosions.

Example. Radium decomposes at a rate proportional to the amount present.* If of 100 mg set aside now there will be left 96 mg 100 years hence, find how much will be left t centuries from the time when the radium was set aside, how much will be left after 2.58 centuries, and the half-life of radium.†

Solution. Let q be the number of milligrams of radium left after t centuries. Then, since dq/dt is the rate of increase,

$$\frac{dq}{dt} = kq \qquad \text{or} \qquad \frac{dq}{q} = k\,dt \tag{a}$$

We have as pairs of corresponding values

q	100	96	Q	50
t	0	1	2.58	T

(b)

From (a) and (b), obtain

$$\int_{100}^{q} \frac{dq}{q} = k \int_{0}^{t} dt \qquad \int_{100}^{96} \frac{dq}{q} = k \int_{0}^{1} dt$$
$$\int_{100}^{q} \frac{dq}{q} = k \int_{0}^{2.58} dt \qquad \int_{100}^{50} \frac{dq}{q} = k \int_{0}^{T} dt \tag{c}$$

From the first equation of (c), obtain

$$\ln q - \ln 100 = kt \qquad \text{or} \qquad q = 100e^{kt} \tag{d}$$

* Radium does not disintegrate continuously as here indicated; very small particles radiate so that decrease of quantity takes place atom by atom, that is, discontinuously. However, the results obtained by the method of the example are reliable when fairly large amounts of radium are considered. If the method were applied to a single atom of radium, the result would be meaningless.

† The half-life of a decaying quantity q is the time required for one-half the contents of a sample to decay.

Since $q = 96$ when $t = 1$, obtain, from (d), (e)

$$96 = 100e^k$$

Replacing e^k in (d) by its value from (e), obtain

$$q = 100(0.96)^t \qquad (f)$$

Now using the Keuffel and Esser log-log duplex slide rule (any log-log slide rule may be used):

Set index of scale C opposite 0.96 on scale LL01.
Opposite 2.58 on C read 0.90 ($= Q/100$) on LL02.
Opposite 0.5 on LL02 read 16.9 ($= T$) on C.

Hence $Q = 90$ mg, and $T = 17.0$ centuries.

Instead of using the slide rule, we could solve the last three equations of (c), after supplying the logarithms, to obtain

$$k = -0.041 \qquad Q = 90 \qquad T = 17.0$$

Exercises

1. Assume that a body cools according to Newton's law $d\theta/dt = -k\theta$, where t is the time and θ is the difference between the temperature T of the body and that of the surrounding air. Find the temperature T at time t of a boiler of water cooling in air at 0°C if the water was initially boiling at 100°C and the temperature dropped 10° during the first 20 min. Also, find the time for the temperature of the water to drop from 90°C to 80°C and the temperature of the water after 90 min.

2. Replace 0°C for the temperature of the air in exercise 1 by 20°C, and solve the resulting problem.

3. For the illustrative example assume as initial conditions 100 mg now and 50 mg when $t = 17.0$ centuries, and show that $q = 100(\frac{1}{2})^{t/17.0}$. From this show that one-half the radium left at any time will decay in 17.0 centuries. Then show that if $q = q_0$ now and $q = \frac{1}{2}q_0$ at time T, then

$$Q = Q_0(\tfrac{1}{2})^{t/T} \qquad (a)$$

4. Assume that the radioactive elements mentioned below decay exponentially. Use equation (a) of exercise 3 to write the equations giving the amount Q at time t from an initial gram of: U^{238} (half-life 4.5×10^9 years), U^{235} (half-life 8.8×10^8 years), polonium (half-life 10^{-5} sec).

5. In a chemical transformation, substance A changes into another substance at a rate proportional to the amount of A unchanged. If initially there was 40 g of A and 1 hr later 12 g, when will 90 per cent of A be transformed?

6. When the electromotive force (emf) is removed from a circuit containing inductance and resistance but no capacitors, the rate of decrease of current is proportional to the current. If the initial current is 30 amp and it dies down to 11 amp in 0.01 sec, find the current in terms of the time.

7. If I represents the intensity of light which has penetrated water to a depth of x ft, then the rate of change of I is proportional to I. If I at depth 30 ft is $\frac{4}{9}$ the intensity at the surface, find the intensities at depths 60 ft and 120 ft.

8. Assume that the rate of change of air pressure with altitude (distance above the earth) is proportional to the air pressure.* If the air pressure on the ground is 14.7 lb/in.^{2} and if at an altitude of 10,000 ft it is 10.1 lb/in.^{2}, find air pressure in terms of altitude, and find the air pressure at an altitude of 15,000 ft.

9. Water leaks from a cylinder with axis vertical through a small orifice in its base at a rate proportional to the square root of the volume remaining at any time. If the cylinder contains 64 gal initially and 15 gal leak out the first day, when will 25 gal remain? How much will remain at the end of 4 days?

10. If interest draws interest at the instant it accrues† and A is the amount at any time t years at r per cent per year, then

$$\Delta A = \left[\frac{r}{100}(A + \epsilon)\right]\Delta t \qquad \text{where } \epsilon \to 0 \text{ as } \Delta t \to 0$$

Show that $A = A_0 e^{rt/100}$. If $r = 6$ show that $A = A_0 2^{t/11.55}$, where A_0 is the amount invested when $t = 0$. When will $A = 2A_0$, and when will $A = 4A_0$?

16 Motion in a straight line

In this section we use the notation t, s, v, a, m, and F for time, distance, velocity, acceleration, mass, and force, respectively. From calculus

* The rate of change of pressure depends on air pressure, temperature of the air, and other conditions. Hence, a formula neglecting all conditions except air pressure will give only rough approximations.

† This kind of interest is said to be compounded continuously. It has no practical importance.

we have

$$v = \frac{ds}{dt} \qquad a = \frac{dv}{dt} = \frac{v\,dv}{ds} \tag{1}$$

If a particle of mass m moves in a straight line under the influence of one or more forces having resultant F, then, in accordance with Newton's laws of motion, we have

$$F = \frac{d}{dt}(mv) \tag{2}$$

According to Einstein's special theory of relativity,

$$m = \frac{m_0}{\sqrt{1 - (v/c)^2}} \tag{3}$$

where m_0 is the rest mass of a body under consideration, $|v|$ the speed of the body, and c the speed of light. However $c = 186{,}000$ mi/sec and v for most purposes is small, so that $m = m_0$ nearly. Therefore we assume that m is constant and from (3) obtain

$$F = m\frac{dv}{dt} = ma \tag{4}$$

In order that (2) and (4) may hold, it is necessary that consistent values for F, s, v, and t be used. In the centimeter-gram-second (cgs) system of units, we use

$$F \text{ (dynes)} = m \text{ (grams) } a \text{ (cm/sec}^2\text{)} \tag{5}$$

In the foot-pound-second (fps) system, we use

$$F \text{ (lb)} = m \text{ (slugs) } a \text{ (ft/sec}^2\text{)} \tag{6}$$

Here we take $m = w/g$, where w represents weight in pounds of a body as measured by a spring scales at some point on the earth,* and g is the acceleration that would be given to a freely falling body at that point. Since $g = 32.2$ ft/sec² nearly on the earth we use the formula

$$F\text{(lb)} = \frac{w\text{(lb)}}{32.2\text{ (ft/sec}^2\text{)}}\frac{dv}{dt} \qquad v \text{ (ft/sec), time } t \text{ (sec)} \tag{7}$$

* The sentence following (6) will apply to any large celestial body if in it, the earth be replaced by the celestial body. For example, if subscript e refers to Earth and j to Jupiter, we know that $g_j = 2.65g_e$ (nearly), and we have

$$m_j = \frac{w_j}{g_j} = \frac{2.65w_e}{2.65g_e} = \frac{w_e}{g_e} = m_e$$

Example. A coasting party weighing 1,000 lb coasts down a 5° incline. The component of gravitational force parallel to the direction of motion is 1,000 sin 5° lb = 87.2 lb. If the force of friction opposing the motion is 40 lb and the air resistance in pounds is numerically equal to 1.5 times the speed in feet per second,* find an expression for the speed after t sec from rest, the speed after 10 sec from rest, and the limiting speed.

Solution. If downhill is chosen as the positive direction, we see from Fig. 1 that $F = 87.2 - 40 - 1.5v$. Therefore, the equation of motion from (7) is

$$47.2 - 1.5v = \frac{1{,}000}{32.2}\frac{dv}{dt} \tag{a}$$

The initial conditions may be written

v	0	v_{10}
t	0	10

(b)

Separating the variables in (a) and integrating, we obtain

$$\int_0^{v_{10}} \frac{-1.5\,dv}{47.2 - 1.5v} = -\int_0^{10} 0.0483\,dt$$
$$\int_0^{v} \frac{-1.5\,dv}{47.2 - 1.5v} = -0.0483\int_0^t dt \tag{c}$$

From the first part of (c),

$$[\ln(47.2 - 1.5v)]_0^{v_{10}} = \ln\frac{47.2 - 1.5v_{10}}{47.2} = -0.483$$

Then

$$\frac{47.2 - 1.5v_{10}}{47.2} = e^{-0.483} = 0.617 \qquad \text{and} \qquad v_{10} = 12.1 \text{ ft/sec} \tag{d}$$

From the second part of (c),

$$\ln\frac{47.2 - 1.5v}{47.2} = -0.0483t \qquad \text{or} \qquad \frac{47.2 - 1.5v}{47.2} = e^{-0.0483t} \tag{e}$$

* The problem of finding the resistance of a fluid on a body moving through it is very complicated. It depends on the speed, the shape of the body, and the properties of the fluid. Any such simple expression as $1.5v$ can represent it reasonably well for only a short period of time in most cases.

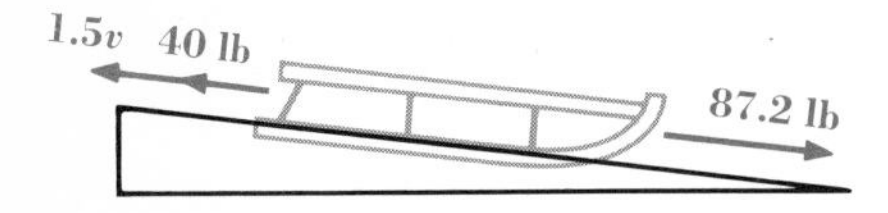

Figure 1

Solving (e) for v, we obtain

$$v = 31.5(1 - e^{-0.0483t}) \tag{f}$$

From (f), it appears that, as t increases without limit, $e^{-0.0483t}$ approaches zero as a limit and v approaches 31.5 ft/sec.

This last result could have been found from the fact that, as v approaches a limiting value, the rate of change of v, or dv/dt, approaches zero. Hence, from equation (a), $47.2 - 1.5v$ approaches zero, and v approaches 31.5 ft/sec.

Exercises

1. A body moves in a straight line with a constant acceleration of $a = 10$ ft/sec². If the speed $v = 5$ ft/sec when $t = 0$ sec, replace a in $a = 10$ by dv/dt, and show that $v = 10t + 5$. Also, replace a by $v\,dv/ds$, and show that $v^2 = 25 + 20s$.

2. Using the equation $a = v\,dv/ds$ for rectilinear motion, prove that if a is constant, $v^2 = v_0^2 + 2as$, where $v = v_0$ when $s = 0$. Also, prove that if a is constant and $v = v_0$ when $t = 0$, then $v = v_0 + at$.

3. If distance is expressed in feet and time in seconds, then a and $-0.3v^2$ for a certain rectilinear motion are expressed by the same number; that is, $a = -0.3v^2$.* If $v = 20$ ft/sec when $t = 0$, find v in terms of t. What does v approach as $t \to \infty$? Also, using $s = 0$ when $v = 20$ ft/sec and $v\,dv/ds$ for a, find v in terms of s.

4. A boat with its load weighs 322 lb (see Fig. 2). If the force exerted upon the boat by the motor in the direction of motion is equivalent to a constant force of 15 lb, if the resistance (in pounds) to motion is equal numerically to twice the speed (in feet per second), that is, $2v$ lb, and if the boat starts from rest, find the speed:

* This equation appears to be incorrect dimensionally; for if L represents distance and T time, a has the dimensions LT^{-2} and v^2 the dimensions L^2T^{-2}. To obtain a balance, we assign to the constant 0.3 the dimension of L^{-1}. In general, we shall assume that the constants in our equations are such that the equations are dimensionally correct.

Figure 2

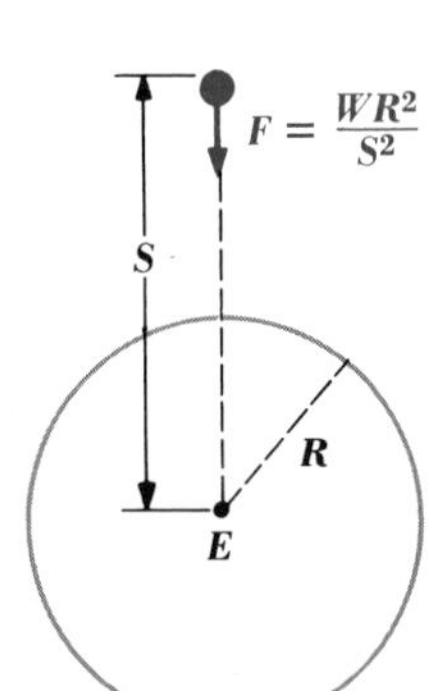

Figure 3

(*a*) after t sec; (*b*) after 10 sec; (*c*) when $t = \infty$, that is, the limiting speed.

5.* Figure 3 represents a ball E, having uniform density, having the weight and radius of the earth, and pulling a w-lb body toward it with a force inversely proportional to the square of the distance s from the center of E. By applying Newton's law of motion, we obtain the equation

$$\frac{w}{32.2}a = \frac{w}{32.2}\frac{v\,dv}{ds} = -\frac{wR^2}{s^2} \qquad (a)$$

where $R = 3{,}960 \times 5{,}280$ ft, s is in feet, and t in seconds. Find the velocity attained by the body in falling from rest at a distance of $4R$ from the center of E to its surface. What velocity would correspond to a fall from an infinite distance?

6. Equation (*a*) of exercise 5 applies to a body thrown directly upward from the earth at a velocity of v_0. (*a*) Show that, if $s = R$ and $v = v_0$ when $t = 0$, then $v^2 = v_0^2 - 2gR + 2gR^2/s$. (*b*) If $v_0 = \sqrt{2gR}$,† show that v will be positive if $s \geqq R$ and will approach zero as $s \to \infty$. If $v_0 > \sqrt{2gR}$, show that v will never be less than $\sqrt{v_0^2 - 2gR}$.

7. In exercise 6 assume that $v_0 = \sqrt{2gR}$, and show that $v = R\sqrt{2g/s}$. In the latter equation replace v by ds/dt, and integrate the result to find t in terms of s. Then find the time t when the body is distant $4R$ from the center of the earth.

* In these exercises only the forces exerted by the body are considered. Air resistance is neglected.

† The velocity $\sqrt{2gR} = \sqrt{2(32.2)(3{,}960)(5{,}280)}/5{,}280 = 6.95$ mi/sec is called the escape velocity, because it is the least value of v_0 for which the body will not fall back to the earth.

8. Show that the equations of exercises 6(*a*) and 6(*b*) apply approximately to any large celestial body, provided that g is the acceleration of a freely falling particle at its surface and R its radius. Taking $g = 5.29$ ft/sec² and $R = 1080$ mi for the moon, find the escape velocity and the velocity a particle would have after falling to the moon from a distance of $4R$ from its center.

9. If $-wR^2/s^2$ in equation (*a*) of exercise 5 is replaced by $-ws/R$, the result applies to a body falling through the earth, assumed uniform in density, in a hole through its center. If the body falls from rest at the surface of the earth through such a hole, find its velocity at the center after showing that $v = -\sqrt{(g/R)(R^2 - s^2)}$. Now replace v by ds/dt, solve the resulting equation, and find the time required to fall from surface to center. *Hint:* One fourth the period of $\cos at$ is $\frac{1}{2}\pi/a$.

★10. The traction resistance of a 3,220-lb car is 60 lb, and the air resistance is $0.08v^2$, where v represents velocity in feet per second. The engine applies a constant force to the car which we will denote by $(0.08H^2 + 60)$ lb. Using (6) and assuming that $v = kH$ when $t = 0$, show that $\ln |(H + v)/(H - v)| = (H/625)t + \ln |(1 + k)/(1 - k)|$. Note the discontinuity at $v = H$ and at $v = -H$. If t may take all values, negative and positive, and if $H > 0$, what ranges of velocity may the car have when: (*a*) $-1 < k < 1$? (*b*) $k > 1$? (*c*) $k < -1$?

11. An iceboat (see Fig. 4) with load weighs 322 lb. It is propelled by a force of $2(v_0 - v)$ lb when moving at a speed of v ft/sec in a v_0-ft/sec tail wind. There is a constant resistance to motion of 10 lb. Using (7), find its speed v at time t sec from rest in a 40-ft/sec wind. Also, find the distance s in terms of t. Find its

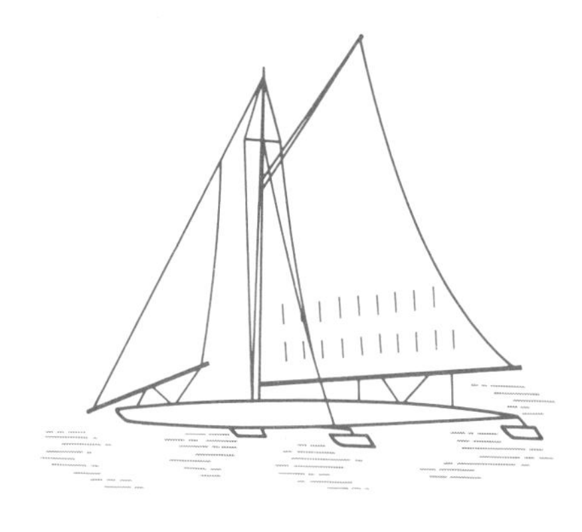

Figure 4

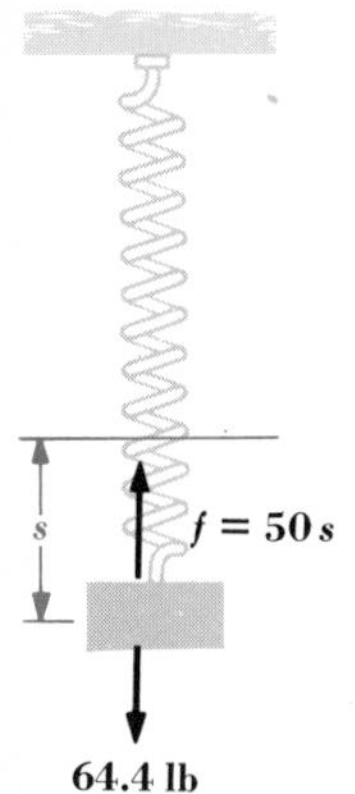

Figure 5

speed after 10 sec from rest and the distance covered during the first 10 sec.

12. A force that increases uniformly at the rate of 6 lb/sec from a value of 0 lb when $t = 0$ acts on a 32.2-lb body initially at rest. Find v in terms of t, then replace v by ds/dt, and integrate again to find s in terms of t. Also, find s in terms of v.

★13. A 64.4-lb weight on a spring attached to a ceiling moves up and down under the action of its weight and a restoring force $f = 50s$ lb (see Fig. 5), where s is the number of feet that the spring is stretched. Using the equation $F = ma$, show that $2a = 64.4 - 50s$. If velocity $v = 0$ when $s = 0$, show that v in feet per second is given by $v = \pm\sqrt{64.4s - 25s^2}$. Replace v in the equation by ds/dt, and show that $s = 1.288 + 1.288 \sin(\pm 5t + \frac{3}{2}\pi) = 1.288(1 - \cos 5t)$, provided that $s = 0$ when $t = 0$. Show that s varies from 0 to 2.576 ft and back to zero again periodically, the period being $\frac{2}{5}\pi$ sec.

14. According to Einstein's special theory of relativity mass increases with velocity. In (2) replace m by $m_0/\sqrt{1 - (v/c)^2}$, and integrate the resulting equation using the condition that $v = 0$ when $t = 0$. Use the formula thus obtained to find the time required to bring a 1-slug mass from speed 0 to a speed $\frac{3}{5}c$ by a constant force of 984 lb, where $c = 9.84 \times 10^8$ ft/sec. Find m when $v = \frac{3}{5}c$.

17 *Other rate problems*

The idea of rate is basic to a great variety of problems. Three more types will be considered in this section.

Consider the flow of heat through a wall of area A which receives or looses heat only through its faces. Assume that the temperature T at any point in the wall at a distance x from a face is a function $T(x)$.

Experiment verifies that for such a wall heat flows from points at one temperature to points at a lower temperature and that the rate of flow Q in units per second is proportional to dT/dx. Hence, we have

$$Q = -kA\frac{dT}{dx} \tag{1}$$

where k, the *conductivity*, is a constant found by experiment. The values used here for k apply for distance measured in centimeters, area A in square centimeters, T in degrees centigrade, and Q in calories per second.

Consider the heat loss per day through a barn wall made of cement, $k = 0.0022$, 3 m (meters) high, 10 m long, and 25 cm thick if the inner temperature is kept at 10°C and the outer temperature is −10°C. Using (1), we get

$$Q = -0.0022(300)(1{,}000)\frac{dT}{dx} \tag{2}$$

The initial conditions are

T	10°C	−10°C
x	0	25 cm

Using these conditions, we get from (2) $Q = 528$ cal/sec. Hence, the loss per day is

$$528 \times 24 \times 3{,}600 = 46 \times 10^6 \text{ cal}$$

A second problem relates to the flow of a liquid through an orifice. The velocity of a substance falling through h ft from rest is $\sqrt{2gh}$ ft/sec. Hence if the surface of the liquid is h ft above the orifice (see Fig. 1) we would expect that the volume per second of liquid issuing from an orifice of area B ft² would be $B\sqrt{2gh}$ ft³. Experiment shows however that we must use $0.6B\sqrt{2gh}$. Hence if Q ft³ is the volume

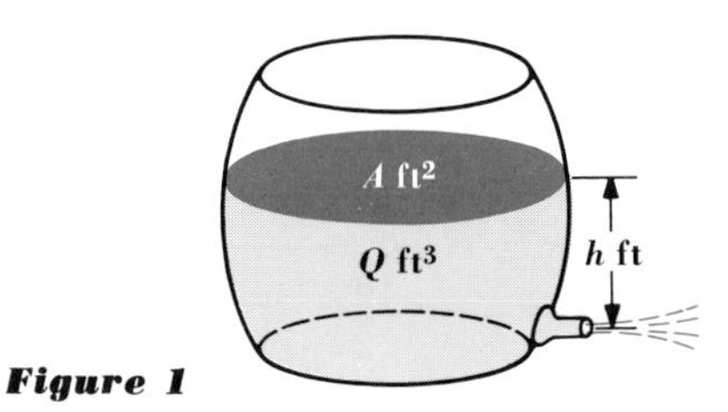

Figure 1

of water in a tank, we get

$$\frac{dQ}{dt} = -0.6B\sqrt{2gh} \tag{3}$$

where B ft^2 is the area of the orifice.

To find the time required to empty a cylindrical can, axis vertical, radius $\frac{1}{2}$ ft, height 2 ft, initially full of water, through an orifice $\frac{1}{3}$ in. in diameter in its bottom, apply (3) to obtain

$$\frac{d[\pi(\frac{1}{2})^2h]}{dt} = -0.6\pi\left(\frac{1}{72}\right)^2\sqrt{2(32)h}$$

The conditions are $h = 2$ when $t = 0$, and $h = 0$ when $t = T$. The solution is $T = 764$ sec nearly, or 12 min 44 sec.

The following example illustrates a third application:

Example. A tank contains initially 100 gal of brine holding 150 lb of dissolved salt in solution. Salt water containing 1 lb of salt per gallon enters the tank at the rate of 2 gal/min, and the brine flows out at the same rate. If the mixture is kept uniform by stirring, find the amount of salt in the tank at the end of 1 hr.

Solution. Evidently 2 lb/min is the rate that salt enters the tank. Let c represent the concentration of salt in the tank at time t, that is c is the number of pounds of salt in each gallon of brine. Hence if n gal/min of fluid is leaving the tank and c is the concentration, then cn lb/min of salt is leaving. Letting Q represent the amount of salt in the tank at time t, we get

$$\frac{dQ}{dt} = \text{rate entering} - \text{rate leaving} \tag{4}$$

Evidently 2 lb/min of salt are entering and $2c = 2Q/100$ lb/min of salt are leaving. We have from (4) (see Fig. 2)

$$\frac{dQ}{dt} = 2 - \frac{2Q}{100} \tag{a}$$

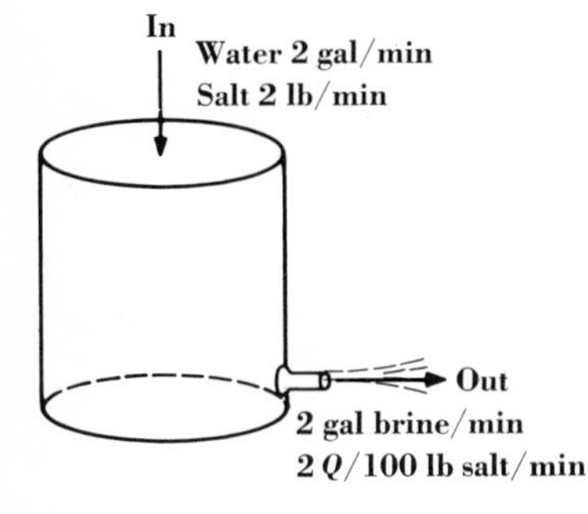

Figure 2

Corresponding values of Q and t are

Q	150	Q_{60}
t	0	60

(*b*)

Solving equation (*a*) under conditions (*b*), we get $Q_{60} = 115.1$ lb

Exercises

1. Use (1) with $k = 0.00023$ to find the number of calories per day passing through the wall of an ice house having an area of 10^7 cm^2, a thickness of 30.5 cm, an inside temperature of 0°C, and an outside temperature of 21.1°C.

2. Find the heat lost per hour through 1 m^2 of furnace wall if the wall is 45.7 cm thick, if k for the masonry is 0.0024, and if the faces of the wall are kept at 1000°C and 120°C, respectively.

3. Use (3) to find the time to empty a cylindrical tank 2 ft in diameter and 3 ft high through a hole 2 in. in diameter in the bottom of the tank. Initially the tank is full of water and its axis is: (*a*) vertical; (*b*) horizontal.

4. Use (3), properly modified, to find the time required to fill a cubical tank of edge 3 ft if there is a round hole 1 in. in diameter in the bottom of the tank and if water is poured into the tank at π ft^3/min.

5. Into a 100-gal tank initially filled with fresh water flow 3 gal/min of salt water containing 2 lb of salt per gallon. The solution, kept uniform by stirring, flows out at the same rate. (*a*) How many pounds of salt will there be in the tank at the end of 1 hr 40 min? (*b*) What is the upper limit for the number of pounds of salt in the tank if the process continues indefinitely? (*c*) How much time will elapse while the quantity of salt in the tank is changing from 100 to 150 lb?

6. Pure water is poured at the rate of 3 gal/min into a tank containing 300 lb of salt dissolved in 100 gal of water, and the solution, kept well stirred, pours out at 2 gal/min. Find the amount of salt in the tank at the end of 100 min.

7. If Q represents the amount of light falling on a sheet of water Δx thick, the amount ΔQ absorbed is given by $\Delta Q = (kQ + \epsilon)\,\Delta x$ and $\epsilon \to 0$ as $\Delta x \to 0$. If two-thirds of the light falling on a surface is absorbed in penetrating 15 ft of water, what part will be absorbed in penetrating 60 ft of water?

8. A 100-gal tank initially filled with fresh water has a mixture of salt and insoluble material at its bottom. If the salt dissolves at a rate per minute equal to one-third of the difference between the concentration (number of pounds of salt per gallon) of the brine and the concentration of a saturated solution (3 lb/gal) and if the concentration is kept uniform by stirring, find the number of pounds of salt dissolved in 1 hr.

9. If a spherical drop of liquid evaporates at a rate proportional to its area, if initially its radius r is $\frac{1}{20}$ in., and if in 15 min r is $\frac{1}{40}$ in., find r in terms of t.

10. A room 30 by 30 by 10 ft contains air at 100°F. A fan blows 900 ft³/min of air at 65°F into the room, and the room air, assumed uniform in temperature, leaves at the same rate. How much time will elapse before the room temperature is 70°F?

11. Air containing 30 per cent oxygen passes slowly into a 3-gal flask initially filled with pure oxygen, and the mixture of air and oxygen, assumed uniform, passes out at the same rate. How much oxygen will the flask contain after 6 gal of air has passed into it?

18 Miscellaneous problems

The following brief outline may be helpful in beginning the solution of a problem based on a differential equation for which the variables are separable (§4).

Geometry:

$$\frac{dy}{dx} = \tan\theta \qquad \text{and Figs. 1, §10, and 1, §11}$$

$$\frac{\rho\, d\theta}{d\rho} = \tan\psi \qquad \text{and Figs. 1 and 2, §12}$$

Isogonal trajectories:

Rectangular coordinates: equations (3) and (5), §11
Polar coordinates: equations (5) and (7), §12

Rate of change of a quantity Q proportional to Q:

$$\frac{dQ}{dt} = kQ \qquad (\S15)$$

Motion caused by forces:

$$F = ma = \frac{w}{g}a \qquad a = \frac{dv}{dt} = \frac{v\,dv}{ds} \qquad (\S16)$$

Heat loss and flow of liquids through an orifice, respectively:

Equations (1) and (3), §17

Rate of change of a quantity Q increasing because of some factors and decreasing because of others:

$$\frac{dQ}{dt} = \text{rate entering} - \text{rate leaving} \qquad (\S 17)$$

Exercises

1. Find the equation of a curve through $(3,-2)$ having at the point (x,y) the slope $(y - 1)/(1 - x)$.

Find the isogonal trajectories with $\alpha = 45°$ and the orthogonal trajectories of each family of curves:

2. $y = x + e^x + c$ **3.** $y = ce^{-3x}$

4. $\rho = c\theta$; $\rho > 0$, $0 < \theta \leqq 2\pi$.

5. The charge of electricity on a body leaks away at a rate proportional to the charge, and half of it leaks away in 10 min. What part of the charge will remain after 50 min?

6. When a gas expands without gain or loss of heat, the rate of change of pressure with volume varies directly as the pressure and inversely as the volume. Find the law connecting pressure and volume in this case.

(**7–10**) Find the equations of the curves having the following properties:

7. The angle between the radius vector and the tangent equals the angle between the radius vector and the initial line. *Hint:* Use $\psi = \theta$ and $\psi = 180° - \theta$; therefore, $\tan \psi = \pm \tan \theta$.

8. The perpendicular from the pole to the tangent is constant.

9. The tangent is equally inclined to the radius vector and to the initial line. *Hint:* $\theta = 180° - 2\psi$, or $\psi = 90° - \frac{1}{2}\theta$.

10. The radius vector is equally inclined to the normal and to the initial line.

11. A freighter of 42,000 tons displacement (1 ton = 2,000 lb) starts from rest. Assuming that the resistance in pounds to motion is $7{,}000v$, where v is the speed in feet per second, and that the force exerted on the ship by the propellers is 120,000 lb, find: (*a*) the speed at any time; (*b*) the limiting speed; (*c*) the time taken to speed up to nine-tenths of the limiting speed.

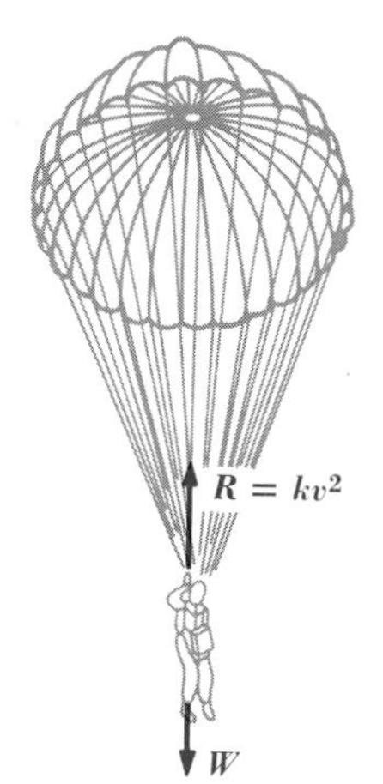

By Newton's law $F = ma$,

$$W - kv^2 = \frac{W}{g}\frac{dv}{dt}$$

Figure 1

12. A man and a parachute are falling (see Fig. 1) with a speed of 173 ft/sec when the parachute opens and the speed is reduced so as to approach the limiting value of 15 ft/sec by an air resistance proportional to the square of the speed. Show that

$$t = \frac{15}{2g} \ln \left(\frac{79}{94} \cdot \frac{v + 15}{v - 15} \right)$$

Hint: When the speed is close to the limiting speed, the velocity is almost constant and the acceleration nearly zero.

★13. Equation (1), §17, applies to a protected cylindrical hot-water or steam pipe if x represents the distance from the axis of the pipe and A the lateral area of a cylinder of radius x and length l equal to the length of the pipe considered, that is, if $A = 2\pi xl$.

Two steam pipes of 20 cm diameter, protected with coverings 10 cm thick of concrete ($k = 0.0022$) and magnesia ($k = 0.00017$), respectively, are run underneath the soil. If the outer surfaces are at 30°C and the pipes themselves are at 160°C, compute the losses per hour per meter length of pipe in the two cases. Also, find the heat lost per hour per meter length of pipe from one of these pipes if it is protected with a covering 5 cm thick of magnesia and, over this, a covering of concrete 5 cm thick.

14. Find the equation of a curve for which the area bounded by the radius vector to any point (ρ,θ) on it, the tangent at the point, and the initial line is proportional to ρ^2. *Hint:* To find the intercept of the tangent on the initial line, apply the law of sines to the triangle bounded by the tangent, the radius vector, and the initial line.

15. Assuming that a man weighing w lb falls from rest, that the resistance of the air is proportional to his speed v, and that his limit

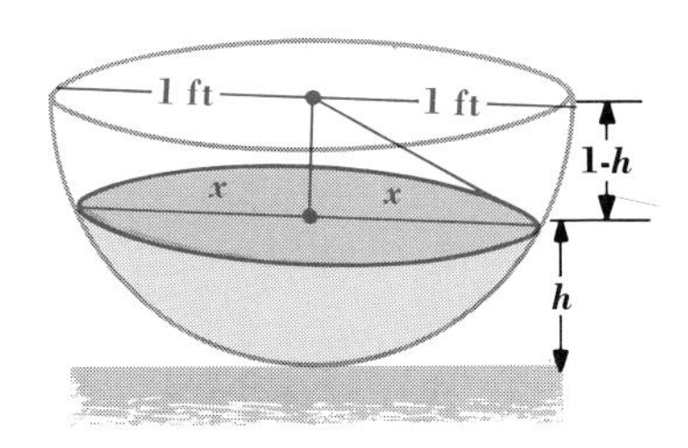

Figure 2

ing speed is 173 ft/sec, find an expression for his speed at any time, and find his speed at the end of the eleventh second.

★**16.** A room 30 by 30 by 10 ft receives 900 ft³/min of fresh air, 0.04 per cent CO_2. If the CO_2 content rises from 0.04 per cent CO_2 to 0.12 per cent CO_2 in $\frac{1}{2}$ hr after a crowd enters the room, what CO_2 content is to be expected 3 hr after the crowd enters? Assume that the crowd breathes CO_2 into the air at a constant rate.

17. A cubical tank of edge 4 ft is full of water which runs out a vertical slit $\frac{1}{8}$ in. wide extending from the top to the bottom of the tank. If the quantity of water per second issuing from a small part of the slit of area a situated at distance x from the surface of the water is $0.6a\sqrt{2gx}$, find the time for the surface of the water to fall 3 ft. *Hint:* First, find the number of cubic feet per minute of water issuing from the slit when the water is h ft deep.

18. A hemisphere having radius 1 ft and base up is full of water which runs out through a hole 1 in. in diameter in its bottom. How long will it take the water to run out? Use (3), §17, and Fig. 2. Note that $dq = -\pi x^2\,dh = -\pi(2h - h^2)\,dh$.

★**19.** A tank 10 by 10 by 10 ft contains 500 ft³ of water and 500 ft³ of air at a pressure of 3,000 lb/ft². How long will it take the water to pass out through an orifice of area 1 in.² in the bottom of the tank? Neglect the pressure due to gravity, and assume that the discharge takes place in a vacuum. Use the modification of equation (3), §17, $dQ/dt = -0.6b\sqrt{2gp/\delta}$,* where δ represents density (62.5 lb/ft³ for water) and p lb/ft² represents pressure. Assume that $pv = c$, a constant, for air.

20. Show that if gravity is taken into account in exercise 19, then $t = (24{,}000/\sqrt{2g}) \int_0^5 (\sqrt{10 - h}/\sqrt{240 + 10h - h^2})\,dh$. *Hint:* Water depth h equivalent to pressure p is $c/(\frac{125}{2}v)$.

21. A uniform rubber band of natural length x in. is stretched kxF in. by a force of F lb. How much will l in. of the string stretch under its own weight of w lb when suspended from one end?

* Note that the pressure at a depth of h feet of water is $p = 62.5h$, or $h = p/62.5$. Hence the velocity of expelled water due to a pressure of p lb/ft² is the same as that for a depth of water $h = p/62.5$ ft.

Differential Equations of the First Order and the First Degree

19 Simple substitutions

Many problems may be reduced to the case of *variables separable* by simple substitutions. Thus, to solve

$$(x + y - 3)\,dx + (x + y + 4)\,dy = 0 \qquad (a)$$

let us try the substitution

$$z = x + y \qquad (b)$$

Then,

$$dz = dx + dy \qquad (c)$$

The next step is to eliminate either y or x from (a) by using (b) and (c). From (b), $y = z - x$; and from (c), $dy = dz - dx$. Substituting these values in (a), we obtain

$$(z - 3)\,dx + (z + 4)(dz - dx) = 0$$

or

$$-7\,dx + (z + 4)\,dz = 0$$

Here the variables are separated, and, solving, we find

$$14x - z^2 - 8z = -c$$

Replacing z by its equal $x + y$, we find

$$14x - (x + y)^2 - 8(x + y) = -c$$

or

$$x^2 + 2xy + y^2 - 6x + 8y = c$$

If the form of an equation indicates that two expressions play a prominent role, it may be well to introduce two new variables. Thus, in considering the equation

$$x(x + y)(dx + dy) = \frac{y}{x}(x\,dy - y\,dx) \tag{d}$$

we note that $x + y$ and y/x stand out. This suggests the substitution

$$z = x + y \qquad w = \frac{y}{x} \tag{e}$$

Taking differentials of equations (e), and also solving (e) for x in terms of z and w, obtain

$$dz = dx + dy \qquad dw = \frac{x\,dy - y\,dx}{x^2} \qquad x = \frac{z}{1 + w} \tag{f}$$

Substituting from (e) and (f) in (d), obtain

$$zx\,dz = x^2w\,dw$$

and from this

$$z\,dz = \frac{zw\,dw}{1 + w} \tag{g}$$

The solution of (g) is

$$z = w - \ln(1 + w) + c \tag{h}$$

Replacing w and z in (h) by their values from (e), obtain

$$x + y = \frac{y}{x} - \ln\left(1 + \frac{y}{x}\right) + c \tag{i}$$

To solve a differential equation by substitution: (*a*) *write the substitution equations;* (*b*) *differentiate the substitution equations;* (*c*) *eliminate all but two of the unknowns from the given differential equation and the results of* (*a*) *and* (*b*); (*d*) *solve the result from* (*c*); (*e*) *replace the new variables in terms of the old in the result of* (*d*).

No general rule for finding effective substitution equations can be given; however, the form of the differential equation may be suggestive. Any outstanding expression may be made the basis of a substitution. Occasionally, substitutions effective for certain types will be given. The following example, based on the fact that $d(xy) = x\,dy + y\,dx$, deals with a class of differential equations and uses the substitution $z = xy$.

Example. Solve $x^m y^n (x\,dy + y\,dx) = \varphi(xy)\,dx$. (*a*)

Solution. Here xy plays a prominent role and we employ the substitution

$$z = xy \qquad y = \frac{z}{x} \qquad dz = x\,dy + y\,dx \tag{b}$$

Making the substitution (*b*) in (*a*) we get

$$x^m \left(\frac{z}{x}\right)^n dz = \varphi(z)\,dx \tag{c}$$

Separating the variables in (*c*) and integrating the result we get

$$x^{n-m}\,dx = \left[\frac{z^n}{\varphi(z)}\right] dz \tag{d}$$

$$\frac{x^{n-m+1}}{n-m+1} = \int_a^{xy} \left[\frac{z^n}{\varphi(z)}\right] dz + c \tag{e}$$

where a is a convenient number.

Exercises

Solve the following differential equations, and determine the constants of integration when initial conditions are given:

1. $2(x - y)\,dx + dy = 0$; let $z = x - y$.
2. $2\,dx + (2x + 3y)\,dy = 0$
3. $(x + y)\,dx + (x + y - 2)\,dy = 0$; let $z = x + y$.
4. $(2x + y + 6)\,dx + (2x + y)\,dy = 0$
5. $(x - 2y + 5)\,dx - [2(x - 2y) + 9]\,dy = 0$
6. $(2x + y)^2\,dx - 2\,dy = 0$
7. $(x^3 + y^3)\,dx + 3xy^2\,dy = 0$; let $y = vx$; when $x = 1$, $y = 1$.
8. $x^m y^n (y\,dx + x\,dy) = \varphi(xy)\,dy$

9. $xy(x\,dy + y\,dx) = 6y^3\,dy$; let $z = xy$; when $y = 1$, $x = 2$.

10. $x^2(x\,dx + y\,dy) = (x^2 + y^2)^2\,dx$; let $z = x^2 + y^2$; when $x = 1$, $y = 2$.

11. $(st + 1)t\,ds + (2st - 1)s\,dt = 0$; let $z = st$.

12. $(x^2 + y^2)\,dx + 2xy\,dy = 0$; let $y = vx$; when $x = 2$, $y = 1$.

13. $3\theta\,d\rho/d\theta + 3\rho = \rho^4\theta^4e^\theta$; let $z = \rho\theta$.

14. $dx + dy = (x + y)(1 + y/x)^2(x\,dy - y\,dx)$; let $z = x + y$, $w = y/x$.

15. $(x^2 + y^2)(x\,dy + y\,dx) - xy(x\,dx + y\,dy) = 0$; let $z = x^2 + y^2$, $w = xy$.

16. $x^my^n(x\,dy - y\,dx) = \varphi\left(\frac{y}{x}\right)dy$

17. $x^my^n(2x\,dy + y\,dx) = \varphi(xy^2)\,dx$

18. $\varphi(2x + 3y + c)\,dx + \psi(2x + 3y + m)\,dy = 0$; let $z = 2x + 3y$.

19. What substitution would you make to solve

$$\frac{dy}{dx} = \varphi\left\{\frac{(ax + by + c)}{[2(ax + by) + \beta]}\right\}$$

Show that your substitution would give an equation coming under the case of *variables separable.*

20 Homogeneous functions

A function $f(x,y)$ is called a homogeneous function of the nth degree if

$$f(tx,ty) = t^nf(x,y) \tag{1}$$

For example $x^3 + xy^2$ is homogeneous of the 3rd degree since $(tx)^3 + (tx)(ty)^2 = t^3(x^3 + xy^2)$. Any function of y/x is homogeneous of the zero-th degree since $f(ty/tx) = t^0f(y/x)$. For example,

$$\frac{1}{(y/x)^2} + \tan\frac{x}{y} + \ln\left(\frac{2x}{y} + 1\right) \tag{2}$$

is homogeneous of the zero-th degree by the test (1).

A useful relation is obtained by letting $t = 1/x$ in the definition expressed by (1). This gives for a homogeneous expression of the nth degree

$$\begin{aligned}\frac{1}{x^n}f(x,y) &= f\left(\frac{x}{x},\frac{y}{x}\right) = \varphi\left(\frac{y}{x}\right)\\ f(x,y) &= x^nf\left(1,\frac{y}{x}\right) = x^n\varphi\left(\frac{y}{x}\right)\end{aligned} \tag{3}$$

A differential equation

$$M\,dx + N\,dy = 0 \tag{4}$$

is homogeneous in x and y if *M and N are homogeneous functions of the same degree in x and y.*

Since, in (3), y/x plays an important role in a homogeneous function, we would expect that the substitution $y/x = v$, or

$$y = vx \qquad dy = v\,dx + x\,dv \tag{5}$$

might be effective in solving a homogeneous equation. We shall prove that *the substitution* (5) *in a homogeneous equation of the first order and first degree leads to an equation of the type variables separable.* Let n be the degree of the homogeneous differential equation (4), and write, in accordance with (3),

$$M\,dx + N\,dy = x^n\varphi_1\left(\frac{y}{x}\right)dx + x^n\varphi_2\left(\frac{y}{x}\right)dy = 0 \tag{6}$$

Now make the substitution (5) in (6) to get

$$x^n\varphi_1(v)\,dx + x^n\varphi_2(v)(v\,dx + x\,dv) = 0 \tag{7}$$

Dividing (7) by x^n and collecting the terms involving dx and those involving dv, we have

$$[\varphi_1(v) + v\varphi_2(v)]\,dx + x\varphi_2(v)\,dv = 0$$

or

$$\frac{dx}{x} + \frac{\varphi_2(v)\,dv}{\varphi_1(v) + v\varphi_2(v)} = 0 \tag{8}$$

and the variables are separated.

The substitution

$$x = vy \qquad dx = v\,dy + y\,dv \tag{9}$$

may be used in place of (5). In solving $M\,dx + N\,dy = 0$, an advantage is sometimes gained by using (5) when N is simpler than M and (9) when M is simpler than N.

Example. Solve $(x^2 + y^2)\,dx - 2xy\,dy = 0$.

Solution. Since the equation is homogeneous, write

$$y = vx \qquad dy = v\,dx + x\,dv$$

Substituting these values for y and dy in the given equation, we get

$$(x^2 + v^2x^2)\,dx - 2vx^2(v\,dx + x\,dv) = 0$$

Collecting coefficients of dx and dv, we obtain

$$(x^2 + v^2x^2 - 2v^2x^2)\,dx - 2vx^3\,dv = 0$$

or

$$x^2(1 - v^2)\,dx - 2vx^3\,dv = 0$$

Division by $x^3(1 - v^2)$ gives

$$\frac{dx}{x} - \frac{2v\,dv}{1 - v^2} = 0$$

Integrating this, we obtain

$$\ln x + \ln(1 - v^2) = \ln c \qquad \text{or} \qquad x(1 - v^2) = c$$

Replacing v by its equal y/x, we have

$$x\left(1 - \frac{y^2}{x^2}\right) = c \qquad \text{or} \qquad x^2 - y^2 = cx$$

Exercises

1. Show that each expression is homogeneous:

(*a*) $8x^2 + 8xy - 10y^2$ (*b*) $x^3 + y^3 - 3x^2y$

(*c*) $x^n + 3x^{n-k}y^k + y^n$ (*d*) $x^2 \sin \frac{y}{x} + y^2 \cos \frac{y}{x} + xy \ln \frac{x+y}{x-y}$

Solve the following differential equations, and determine the constants of integration when initial conditions are given:

2. $(2x - 3y)\,dx - (2y + 3x)\,dy = 0$
3. $(3x + 2y)\,dx + 2x\,dy = 0$
4. $y\,dx + (2x + 3y)\,dy = 0$
5. $(6x^2 - 7y^2)\,dx - 14xy\,dy = 0$
6. $(3\theta + 2\rho)\,d\theta + (2\theta - 4\rho)\,d\rho = 0$
7. $xy^2\,dy - (x^3 + y^3)\,dx = 0$; $y = 0$ when $x = 1$

8. Find the equation of the orthogonal trajectories and of the isogonal trajectories at an angle α of the circles $x^2 + y^2 + 2cx = 0$. Use polar coordinates.

9. $(2xy + y^2)\,dx - 2x^2\,dy = 0$; $y = e$ when $x = e$
10. $y(x^2 + xy - 2y^2)\,dx + x(3y^2 - xy - x^2)\,dy = 0$
11. $x\,dy - y\,dx = \sqrt{x^2 + y^2}\,dx$
12. $x + y \sin \frac{y}{x}\,dx - x \sin \frac{y}{x}\,dy = 0$

13. Show that a straight line through the origin intersects at a constant angle all integral curves of a homogeneous differential equation. *Hint:* $dy/dx = -f_2(y/x)/f_1(y/x)$.

***14.** Find the equation of the orthogonal trajectories and of the isogonal trajectories of: (*a*) the hyperbolas $xy = c$; (*b*) the ellipses $x^2 + 4y^2 = c$.

15. Show that an equation of the type

$$\varphi(y)\,dy + x^n\varphi(x/y)(x\,dy - y\,dx) = 0$$

can be transformed by the substitution $x = vy$ to one of the type *variables separable.*

16. Solve $dy/dx = 4xy/(4x^2 - y^2)$, and state why $y \neq \pm 2x$. How does the solution differ from that of $dx/dy = (4x^2 - y^2)/(4xy)$?

17. Show that $dy/dx = f([ax + by]/[cx + hy])$ is homogeneous.

18. Solve

$$[2(y-1)(x+y-1) + (y-1)^2]\,dx + [4(y-1)(x+y-1) + (x+y-1)^2]\,dy = 0$$

21 Equations of the type $(ax + by + c)\,dx + (\alpha x + \beta y + \gamma)\,dy = 0$

Consider differential equations having the form

$$(ax + by + c)\,dx + (\alpha x + \beta y + \gamma)\,dy = 0 \tag{1}$$

Figure 1 shows two lines

$$ax + by + c = 0 \qquad \alpha x + \beta y + \gamma = 0 \tag{2}$$

meeting at point (h,k). Hence

$$ah + bk + c = 0 \qquad \alpha h + \beta k + \gamma = 0 \tag{3}$$

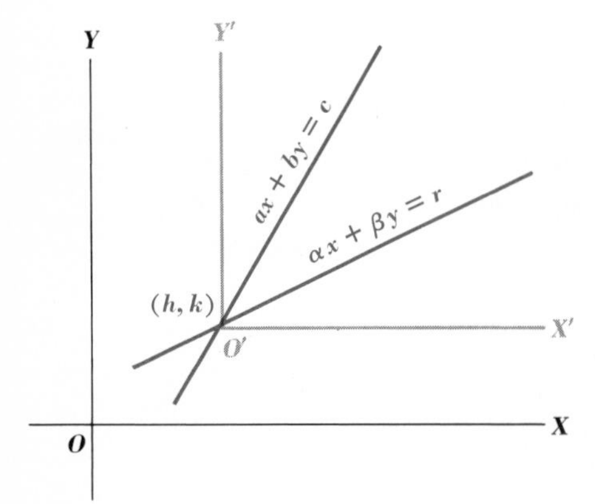

Figure 1

If we refer these two lines to parallel axes with origin (h,k) by the translation

$$x = x' + h \qquad y = y' + k \tag{4}$$

the constant terms must vanish. In fact, applying the translation (4) to (1) we get

$$[ax' + by' + (ah + bk + c)]\,dx' \\ + [\alpha x' + \beta y' + (\alpha h + \beta k + \gamma)]\,dy' = 0 \tag{5}$$

and, because of (3), (5) reduces to

$$(ax' + by')\,dx' + (\alpha x' + \beta y')\,dy' = 0 \tag{6}$$

Equation (6) is homogeneous and we can solve it by the method of §20. Then we must use (4) in the result to replace x' by $x - h$, $y' = y - k$, where h and k are found by solving (3) for h and k. *Observe that the plan consists in writing* (6) *and solving it, then solving* (3) *for* h *and* k, *and finally replacing* x' *by* $x - h$ *and* y' *by* $y - k$ *in the solution of* (6).

If $\alpha/a = \beta/b = m$, then $\alpha x + \beta y = m(ax + by)$ and the lines (2) are parallel. In this case $ax + by$ plays a very prominent role and suggests the substitution

$$z = ax + by \qquad y = \frac{z - ax}{b} \qquad dy = \frac{(dz - a\,dx)}{b} \tag{7}$$

and this reduces the equation to the type *variables separable.*

Example. Solve $(2x - 3y + 4)\,dx + (3x - 2y + 1)\,dy = 0$. (a)

Solution. The process indicated in italics above will not be used. In this case equations (3) are

$$2h - 3k + 4 = 0 \qquad 3h - 2k + 1 = 0 \tag{b}$$

The solution of (b) is $h = 1$, $k = 2$, and the corresponding substitution (4) is

$$x = x' + 1 \qquad y = y' + 2 \tag{c}$$

Making the substitution (c) in (a) we get

$$(2x' - 3y')\,dx' + (3x' - 2y')\,dy' = 0 \tag{d}$$

The solution of homogeneous equation (d) is

$$(y' + x')^5 = c(y' - x') \tag{e}$$

Substituting from (c), $x - 1$ for x' and $y - 2$ for y', in (e) we get

$$(x + y - 3)^5 = c(y - x - 1) \tag{f}$$

Exercises

Solve the following differential equations:

1. $(x - 2y + 4)\,dx + (2x - y + 2)\,dy = 0$
2. $(2x + 3y - 1)\,dx - 4(x + 1)\,dy = 0$
3. $(2x + 3y)\,dx + (y + 2)\,dy = 0$
4. $(2x + y)\,dx - (4x + 2y - 1)\,dy = 0$. *Hint:* $2h + k = 0$ and $4h + 2k - 1 = 0$ have no solution. Hence let $z = 2x + y$.

5. $(2x - 3y + 2)\,dx + 3(4x - 6y - 1)\,dy = 0$
6. $(4x + 3y - 7)\,dx + (3x - 7y + 4)\,dy = 0$
7. $(2x - 2y)\,dx + (y - 1)\,dy = 0$

8. Find the equation of the isogonal trajectories cutting the system $y^2 - (x - 2)^2 = c$ at an angle $\text{Tan}^{-1}\,2$.

9. Write the substitution that would derive from $dy/dx = f([ax + by + c]/[\alpha x + \beta y + \gamma])$ a homogeneous equation.

22 Exact differentials

The formula for the total differential of a function $f(x,y)$ is

$$df\,(x,y) = \frac{\partial f}{\partial x}\,dx + \frac{\partial f}{\partial y}\,dy \tag{1}$$

The right member of (1) *is called an* exact differential, *the right member equated to zero is called an* exact differential equation, *and* $f(x,y)$ *is called an* integral *of the exact differential and, also, of the exact differential equation.* For example,

$$d(x^2 + 8x^2y - 10y^3) = (2x + 16xy)\,dx + (8x^2 - 30y^2)\,dy \tag{2}$$

is an exact differential,

$$(2x + 16xy)\,dx + (8x^2 - 30y^2)\,dy = 0 \tag{3}$$

is an exact differential equation, and $x^2 + 8x^2y - 10y^3 + c$ is an integral of (2).

Example. Find an integral of the exact differential

$$(6xy^2 - 3x^2)\,dx + (6x^2y + 3y^2 - 7)\,dy \tag{a}$$

Solution. Let $f(x,y)$ be the required integral. Then comparison of (a) and (1) shows that

$$\frac{\partial f}{\partial x} = 6xy^2 - 3x^2 \qquad \frac{\partial f}{\partial y} = 6x^2y + 3y^2 - 7 \tag{b}$$

Integrating the first expression with respect to x (treating y as constant) and the second with respect to y, and equating the two values of f thus obtained, we get

$$f(x,y) = 3x^2y^2 - x^3 + \varphi(y) = 3x^2y^2 + y^3 - 7y + \psi(x) \tag{c}$$

where $\varphi(y)$ and $\psi(x)$ are to be found. Inspection of (c) shows that we must have

$$\varphi(y) = y^3 - 7y + g \qquad \psi(x) = -x^3 + g \tag{d}$$

where g is a constant. Hence,

$$f(x,y) = 3x^2y^2 - x^3 + y^3 - 7y + g \tag{e}$$

Alternative method. Expression (a) can be written as the sum of three parts each easily integrable, as follows:

$$(-3x^2\,dx) + (3y^2 - 7)\,dy + 6xy^2\,dx + 6x^2y\,dy$$
$$= d(-x^3) + d(y^3 - 7y) + d(3x^2y^2) \tag{f}$$

Integration of the sum of these differentials gives

$$-x^3 + y^3 - 7y + 3x^2y^2 + c \tag{g}$$

Exercises

Find the total differential of each expression:

1. $x^2 + y^2$ **2.** $\dfrac{y}{x}$ **3.** x^my^n **4.** $\ln(y + y^{-n}e^{mx})$

Form an exact differential equation $M\,dx + N\,dy = 0$ from each expression, and in each case show that $\partial M/\partial y = \partial N/\partial x$:

5. $x^3y - 3y^2$ **6.** y^2e^{ax} **7.** $y^2 + \sin(2xy + x^2)$

Using a method of the example, find an integral of each of the following exact differentials:

8. $(2xy + 4x + 3)\,dx + (x^2 + 2y - 5)\,dy$

9. $(2x - y + 2)\,dx - \left(x - \dfrac{1}{y}\right)dy$

10. $(4x^3 + 8x^3y^3)\,dx + (10y^4 + 6x^4y^2)\,dy$

11. $\frac{1}{x}\,dy - \left(\frac{y}{x^2} + 2\ln x^2\right) dx$

12. $(\sin x + y)\,dx + (x - 2\cos y)\,dy$

13. $\left(\frac{1}{y^2} - \frac{y}{x^2}\right) dx + \left(\frac{1}{x} - \frac{2x}{y^3} - 2\right) dy$

14. $x^{-3}y^{-2}\,dx + (x^{-2}y^{-3} + 2y)\,dy$

15. Show that if $u(x,y)$ and $v(x,y)$ are two integrals of $M(x,y)\,dx + N(x,y)\,dy$ then $u - v = c$, a constant. *Hint:* Show that $du - dv = 0$.

23 Exact differential equations

Assume in §§23 to 25 that the discussion applies to a rectangular region R defined by such inequalities as $|x - p| \leqq a$, $|y - q| \leqq b$, $a > 0$, $b > 0$ and that M, N, $\partial M/\partial x$, $\partial M/\partial y$, $\partial N/\partial x$, and $\partial N/\partial y$ are single-valued and continuous in the region R.

From (1), §22, it appears that an equation

$$M(x,y)\,dx + N(x,y)\,dy = 0 \tag{1}$$

is exact if there exists a function $f(x,y)$ such that

$$M = \frac{\partial f}{\partial x} \qquad N = \frac{\partial f}{\partial y} \tag{2}$$

From (2), we get

$$\frac{\partial M}{\partial y} = \frac{\partial^2 f}{\partial y\,\partial x} \qquad \frac{\partial N}{\partial x} = \frac{\partial^2 f}{\partial x\,\partial y} \tag{3}$$

Since $\partial^2 f/\partial x\,\partial y = \partial^2 f/\partial y\,\partial x$, from (3) we get

$$\frac{\partial M}{\partial y} = \frac{\partial N}{\partial x} \tag{4}$$

Hence, if (1) is an exact differential equation, (4) holds.

Conversely if (4) holds, (1) is an exact differential equation. To prove this, we shall show that, if (4) holds and if fixed point (a,b) and any point (x,y) are in the region R, then (2) is satisfied by

$$f(x,y) = \int_a^x M(t,y)\,dt + \int_b^y N(a,t)\,dt \tag{5*}$$

* Note that (5) is an integral of the exact differential $M\,dx + N\,dy$. It could have been used to solve the exercises of §22.

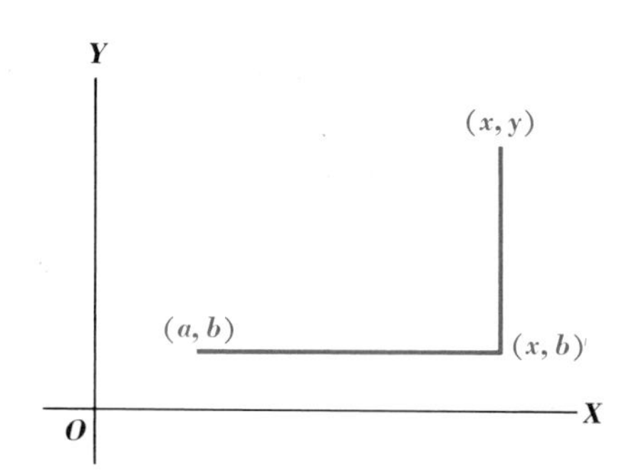

Figure 1

where y in the first integral of (5) is considered as constant for the integration. Note that (5) is the line integral of $M\,dx + N\,dy$. See Fig. 1 for a path parallel to the X axis from (a,b) to (x,b) and then parallel to the Y axis from (x,b) to (x,y). We know from calculus that, for any function $\varphi(x,c)$ fulfilling the conditions assumed for M and N,

$$\frac{\partial}{\partial x}\int_a^x \varphi(x,c)\,dx = \varphi(x,c) \qquad \frac{\partial}{\partial c}\int_a^x \varphi(x,c)\,dx = \int_a^x \frac{\partial\varphi(x,c)}{\partial c}\,dx \tag{6}$$

Therefore, from (4), (5), and (6) we get

$$\frac{\partial f}{\partial x} = M(x,y) + \frac{\partial}{\partial x}\int_b^y N(a,y)\,dy = M(x,y)$$

$$\frac{\partial f}{\partial y} = \int_a^x \frac{\partial M(t,y)}{\partial y}\,dt + \frac{\partial}{\partial y}\int_b^y N(a,t)\,dt = \int_a^x \frac{\partial N(t,y)}{\partial t}\,dt + N(a,y)$$

$$= N(x,y) - N(a,y) + N(a,y) = N(x,y)$$

This completes the proof.

By a similar proof we can show that, when (4) holds, $dF\,(x,y) = M\,dx + N\,dy$, where

$$F(x,y) = \int_b^y N(x,t)\,dt + \int_a^x M(t,b)\,dt \tag{7}$$

where x is considered as constant in the first integral.

We now see that, when (4) holds, $f(x,y)$ from (5) satisfies (2) and, accordingly, that

$$f(x,y) = \int_a^x M(t,y)\,dt + \int_b^y N(a,t)\,dt = c \tag{8}$$ *

satisfies (1) since $d[f(x,y)] = dc = 0$ from (8).

* This formula has been developed with excellent precision and completeness in Walter Leighton, "An Introduction to the Theory of Differential Equations," p. 11, McGraw-Hill Book Company, New York, 1952.

To solve a differential equation $M\,dx + N\,dy = 0$, for which $\partial M/\partial y = \partial N/\partial x$, substitute in (8). Note that any particular numbers for a and b, provided that (a,b) is inside R, may be used, and additive constants arising in the integration may be absorbed in the constant of integration.

If M and N are polynomials for which $\partial M/\partial y = \partial N/\partial x$, use $a = 0$, $b = 0$, since polynomials and their derivatives are continuous everywhere and R may be considered as the whole plane.*

Example 1. Solve

$$(6x^2 + 4xy + y^2)\,dx + (2x^2 + 2xy - 3y^2)\,dy = 0.$$

Solution. Here $M = 6x^2 + 4xy + y^2$, $N = 2x^2 + 2xy - 3y^2$, and we get

$$\frac{\partial M}{\partial y} = \frac{\partial N}{\partial x} = 4x + 2y$$

Now, using (8) with $a = 0$ and $b = 0$, we get

$$\int_0^x (6t^2 + 4ty + y^2)\,dt + \int_0^y [2(0)^2 + 2(0)t - 3t^2]\,dt = c$$
$$2x^3 + 2x^2y + y^2x - y^3 = c$$

Example 2. Solve $(3x^2 + 2y \sin 2x)\,dx + (2 \sin^2 x + 3y^2)\,dy = 0$.

Solution. Here we have $M = 3x^2 + 2y \sin 2x$, $N = 2 \sin^2 x + 3y^2$, and $\partial M/\partial y = \partial N/\partial x = 2 \sin 2x$. Also, M and N are continuous everywhere. Therefore, use (8) with $a = 0$ and $b = 0$, to get

$$\int_0^x (3t^2 + 2y \sin 2t)\,dt + \int_0^y (2 \cdot 0^2 + 3t^2)\,dt = c$$
$$x^3 - y \cos 2x - (0^3 - y \cos 0) + y^3 = c$$
$$x^3 - y \cos 2x + y + y^3 = c$$

Example 3. Solve $(3x^2 + 2y/x)\,dx + (2 \ln 3x + 3/y)\,dy = 0$.

Solution. Here R may be taken as any one of the four quadrants exclusive of the coordinate axes. In R, $\partial M/\partial y = \partial N/\partial x = 2/x$. Taking $a = 1$ and $b = 1$ in (8), we get as the required solution

$$\int_1^x \left(3t^2 + \frac{2y}{t}\right) dt + \int_1^y \left(2 \ln 3 + \frac{3}{t}\right) dt = c$$
$$x^3 + 2y \ln x + 2y \ln 3 + 3 \ln y = c$$
$$x^3 + 2y \ln 3x + 3 \ln y = c$$

* Of course, a solution of $M\,dx + N\,dy = 0$ may be found by equating to c any integral of $M\,dx + N\,dy$ found by a method of §22 or otherwise.

Exercises

Test the differential equations numbered 1 to 19 for exactness by using (4) and solve:

1. $(4x - 2y + 5)\,dx + (2y - 2x)\,dy = 0$
2. $(3x^2 + 3xy^2)\,dx + (3x^2y - 3y^2 + 2y)\,dy = 0$
3. $(a^2 - 2xy - y^2)\,dx - (x + y)^2\,dy = 0$
4. $(2ax + by + g)\,dx + (2ey + bx + h)\,dy = 0$
5. $\dfrac{1}{y}\,dx - \dfrac{x}{y^2}\,dy = 0$
6. $\dfrac{y\,dx - x\,dy}{x^2} = 0$
7. $\dfrac{y}{x}\,dy - \left(\dfrac{y^2}{2x^2} + x\right) dx = 0$
8. $(x - 1)^{-1}y\,dx + \left[\ln(2x - 2) + \dfrac{1}{y}\right] dy = 0$
9. $(x + 3)^{-1}\cos y\,dx - \left[\sin y \ln(5x + 15) - \dfrac{1}{y}\right] dy = 0$
10. $\rho^2 \sec 2\theta \tan 2\theta\,d\theta + \rho(\sec 2\theta + 2)\,d\rho = 0$
11. $(\sin 2\theta - 2\rho\cos 2\theta)\,d\rho + (2\rho\cos 2\theta + 2\rho^2 \sin 2\theta)\,d\theta = 0$
12. $\dfrac{2x}{y}\,dy + \left(2\ln 5y + \dfrac{1}{x}\right) dx = 0$
13. $e^{2x}(dy + 2y\,dx) = x^2\,dx$
14. $e^{x^2}(dy + 2xy\,dx) = 3x^2\,dx$
15. $\dfrac{x\,dy - y\,dx}{y^2} = x^3\,dx$
16. $\dfrac{dx}{\sqrt{x^2 + y^2}} + \left(\dfrac{1}{y} - \dfrac{x}{y\sqrt{x^2 + y^2}}\right) dy = 0$
17. $\dfrac{y^2 - 2x^2}{xy^2 - x^3}\,dx + \dfrac{2y^2 - x^2}{y^3 - x^2y}\,dy = 0$
18. $y^3 \sin 2x\,dx - 3y^2 \cos^2 x\,dy = 0$
19. $\dfrac{3y^2\,dx}{x^2 + 3x} + \left(2y \ln \dfrac{5x}{x + 3} + 3\sin y\right) dy = 0$

24 Integrating factors

If, when a differential equation is multiplied through by an expression, the result is an exact differential equation, the expression is called an *integrating factor* of the equation.

Integrating factors of many differential equations may be found by recognizing certain groups as differentials of known expressions. From $d(y/x) = (x\,dy - y\,dx)/x^2$, it appears that $1/x^2$ is an integrating factor of $x\,dy - y\,dx + f(x)\,dx = 0$, for

$$\frac{x\,dy - y\,dx}{x^2} + \frac{f(x)\,dx}{x^2} = 0$$

is an exact differential equation and its solution is

$$\frac{y}{x} + \int \frac{f(x)}{x^2}\,dx = c$$

Similarly, $1/y^2$ is an integrating factor of

$$x\,dy - y\,dx + f(y)\,dy = 0$$

and its solution is

$$-\frac{x}{y} + \int \frac{f(y)\,dy}{y^2} = c$$

The form $(x\,dy - y\,dx)/(ax^2 + bxy + cy^2)$ is an exact differential,* as may be proved by applying test 4, §23, or by writing

$$\frac{(x\,dy - y\,dx)/x^2}{(ax^2 + bxy + cy^2)/x^2} = \frac{d(y/x)}{a + b(y/x) + c(y/x)^2} \tag{1}$$

Since a, b, and c are any numbers, it appears from (1) that an exact differential is obtained by dividing $x\,dy - y\,dx$ by x^2, y^2, xy, $x^2 + y^2$, $x^2 - y^2$, or any other expression having the form $ax^2 + bxy + cy^2$. To integrate the equation

$$x\,dy - y\,dx = x^2y^3\,dx \tag{2}$$

we derive from it

$$\frac{x}{y}\,\frac{x\,dy - y\,dx}{y^2} = x^3\,dx$$

$$-\int \left(\frac{x}{y}\right) d\left(\frac{x}{y}\right) = \int x^3\,dx \qquad -\frac{\left(\frac{x}{y}\right)^2}{2} = \tfrac{1}{4}x^4 + \tfrac{1}{4}c$$

Also, we could have multiplied both members of (2) by $(x/y)(1/y^2)$ and could have integrated the resulting exact differential equation.

* If $F'(u)$ represents an integrable function, then $F'(y/x)(x\,dy - y\,dx)/x^2$ is an exact differential; for if $u = y/x$,

$$F'\left(\frac{y}{x}\right)\frac{x\,dy - y\,dx}{x^2} = F'(u)\,du = dF(u)$$

Observe that $d(x^p y^q) = px^{p-1}y^q\,dx + qx^p y^{q-1}\,dy$, or

$$d(x^p y^q) = x^{p-1}y^{q-1}(py\,dx + qx\,dy) \tag{3}$$

shows that $py\,dx + qx\,dy$ suggests $x^{p-1}y^{q-1}$ as an integrating factor. For example,

$$\begin{array}{lll} 3y\,dx + 5x\,dy & \text{suggests} & x^{3-1}y^{5-1}(3y\,dx + 5x\,dy) \\ & & \qquad = d(x^3y^5) \\ 3y\,dx - 5x\,dy & \text{suggests} & x^{3-1}y^{-5-1}(3y\,dx - 5x\,dy) \\ & & \qquad = d(x^3y^{-5}) \\ \frac{1}{2}y\,dx - \frac{2}{3}x\,dy & \text{suggests} & x^{-1/2}y^{-5/3}(\frac{1}{2}y\,dx - \frac{2}{3}x\,dy) \\ & & \qquad = d(x^{1/2}y^{-2/3}) \end{array} \tag{4}$$

Also, it is important to note that these simple differential forms, suggesting expressions playing a prominent role, indicate substitutions. In each line of Table 1 is listed, for convenience of reference, an expression of the form $M\,dx + N\,dy$, a corresponding integrating factor u, and a suggested substitution.

Table 1

	$M\,dx + N\,dy$	Integrating factor u	Suggested substitution
I	$py\,dx + qx\,dy$	$x^{p-1}y^{q-1}$	$z = x^p y^q$
I(a)	$y\,dx - x\,dy$	$\begin{cases} 1/x^2,\ 1/y^2,\ 1/(xy) \\ 1/(ax^2 + bxy + cy^2) \end{cases}$	$z = y/x$, x/y, etc.
I(b)	$y\,dx + x\,dy$	1	$z = xy$
II	$px\,dx + qy\,dy$	1	$z = px^2 + qy^2$
III	$dy + yP(x)\,dx$	$e^{\int P(x)\,dx}$	$z = ye^{\int P\,dx}$

The last line in Table 1 needs comment. Observe, from this line, that

$$dz = d(ye^{\int P\,dx}) = e^{\int P\,dx}\,dy + ye^{\int P\,dx}\,d(\int P\,dx) = e^{\int P\,dx}\,(dy + Py\,dx)$$

Note that $e^{\int 2x\,dx} = e^{x^2}$, and $e^{\int 2\,dx/x} = e^{\ln x^2} = x^2$. To solve the equation $dy/dx + y/x = 3x + 2$, we use the substitution $z = ye^{\int 1/x\,dx} = xy$ and obtain the solution $xy = x^3 + x^2 + c$.

Also, see §§25 and 26 for other integrating factors. Observe that the substitution in the third column of Table 1 is suggested by the presence in a differential equation of the corresponding expression in the first column.

The following examples will illustrate methods using integrating factors and substitutions in solving differential equations:

Example 1. Solve $x\,dx + y\,dy = 3\sqrt{x^2 + y^2}\,y^2\,dy$.

Solution. $1/\sqrt{x^2 + y^2}$ is observed to be an integrating factor.

Multiplying the equation through by this, we get

$$\frac{x\,dx + y\,dy}{\sqrt{x^2 + y^2}} = 3y^2\,dy$$

Since this equation is exact, we can solve it as such or we may write it

$$(x^2 + y^2)^{-1/2} \cdot \tfrac{1}{2}d(x^2 + y^2) = 3y^2\,dy$$

and integrate this to get

$$(x^2 + y^2)^{1/2} = y^3 + c$$

Also, the substitution $u = x^2 + y^2$, suggested by $x\,dx + y\,dy$, could have been used.

Example 2. Solve $4y\,dx + x\,dy = xy^2\,dx$. (a)

Solution. The substitution suggested by (3) with $p = 4$ and $q = 1$ or line I of Table 1 is

$$u = x^4y \qquad y = ux^{-4} \qquad dy = x^{-4}\,du - 4x^{-5}u\,dx \tag{b}$$

Replacing y and dy in (a) by their values from (b), we get

$$4ux^{-4}\,dx + x(x^{-4}\,du - 4x^{-5}u\,dx) = xu^2x^{-8}\,dx \tag{c}$$

The solution of this equation, found by the method of separating the variables, is

$$3u^{-1} = x^{-3} + c \tag{d}$$

In this, replace u by x^4y, its value from (b), and simplify slightly to obtain the solution of (a),

$$3 = (x + cx^4)y$$

Example 3. Exercise 12 asks for the proof that if $u(x,y)$ is an integrating factor of $M\,dx + N\,dy = 0$, then

$$\frac{dx}{N} = \frac{dy}{-M} = \frac{du}{u(\partial M/\partial y - \partial N/\partial x)} \tag{5}$$

Assuming (5), prove that $M\,dx + N\,dy = 0$ has an integrating factor:

(a) $u = e^{\int f(x)\,dx}$ if $\dfrac{\partial M}{\partial y} - \dfrac{\partial N}{\partial x} = Nf(x)$

(b) $u = e^{\int -f(y)\,dy}$ if $\dfrac{\partial M}{\partial y} - \dfrac{\partial N}{\partial x} = Mf(y)$

(c) $u = x^k$ if $\dfrac{\partial M}{\partial y} - \dfrac{\partial N}{\partial x} = \dfrac{Nk}{x}$

(d) $u = y^k$ if $\dfrac{\partial M}{\partial y} - \dfrac{\partial N}{\partial x} = -\dfrac{Mk}{y}$

Solution. (*a*) In this case from the equality of the first and third ratios of (5) we get

$$\frac{dx}{N} = \frac{du}{uNf(x)} \qquad \int \frac{du}{u} = \int f(x)\,dx \qquad \ln u = \int f(x)\,dx$$
$$u = e^{\int f(x)\,dx}$$

The proofs of cases (*b*), (*c*), and (*d*) are left for exercise 13.

Exercises

Solve by using integrating factors:

1. $x\,dy + y\,dx = 3x^2\,dx$
2. $x\,dy + y\,dx = xy^3\,dx$. Use the integrating factor $1/(xy)^3$.
3. $x\,dy - y\,dx = (xy)y^2\,dy$
4. $x\,dy - y\,dx = (x^2 + xy - 2y^2)\,dx$
5. $x\,dy - y\,dx = (y^2 - 3)\,dy$
6. $x\,dy - y\,dx = y^3(x^2 + y^2)\,dy$

Use Table 1 to solve the equations numbered **7–11**:

7. $2y\,dx + 3x\,dy = 3x^{-1}\,dy$. Use $p = 2$ and $q = 3$.
8. $x\,dy - 2y\,dx = (xy)^3y\,dy$
9. $x\,dy + 2y\,dx = x^3y^3\,dy$
10. $x\,dy - 3y\,dx = x^4y^{-1}\,dx$
11. $x\,dx + y\,dy = (x^2 + y^2)^3(x\,dy - y\,dx)$

12. Prove that if u is an integrating factor of $M\,dx + N\,dy = 0$, then equation (5) is true. *Hint:* $\partial(uM)/\partial y = \partial(uN)/\partial x$; $\partial M/\partial y - \partial N/\partial x = u^{-1}(N\,\partial u/\partial x - M\,\partial u/\partial y) = u^{-1}N[\partial u/\partial x - (-\,dy/dx)\,\partial u/\partial y] = u^{-1}N\,du/dx$.
13. Using (5) prove parts (*b*), (*c*), and (*d*) of example 3.

Using the integrating factors of example 3, solve:

14. $(3xy^3 + 4y)\,dx + (3x^2y^2 + 2x)\,dy = 0$
15. $(2xy^2 - 2y)\,dx + (3x^2y - 4x)\,dy = 0$

16. Use the substitution of line III, Table 1, to solve:

(*a*) $dy + 2xy\,dx = xe^{-x^2}\,dx$ (*b*) $dy + \frac{3}{x}y\,dx = x^{-3}e^x\,dx$

Solve the following differential equations:

17. $x\,dy - y\,dx = (x^2 - 3)\,dx$
18. $y\,dx + x\,dy = \sqrt{x^2 + y^2}\,(x\,dx + y\,dy)$

19. $y\,dx - x\,dy = (x^2 + y^2)^2(x\,dx + y\,dy)$

20. $x\,dy - y\,dx = \sqrt{4x^2 + 9y^2}\,(4x\,dx + 9y\,dy)$

21. $3y\,dx + 4x\,dy = 5x^2y^{-3}\,dx$ **22.** $4x\,dy - 3y\,dx = xy^{-3}\,dx$

23. $y\,dx + 2x\,dy = x^3y\,dx$

24. $x\,dy - y\,dx = (x^2 - 3axy + 2a^2y^2)(x\,dx + y\,dy)$

25. $x\,dy - y\,dx = (2x^2 + 3y^2)^3(2x\,dx + 3y\,dy)$

26. $\left(\frac{1}{x} - \frac{y^2}{(x - y)^2}\right)dx + \left(\frac{x^2}{(x - y)^2} - \frac{1}{y}\right)dy = 0$

27. Using the integrating factors of example 3, solve $(y^4 + x^3)\,dx + 8xy^3\,dy = 0$.

28. To prove that $1/(xM + yN)$ is an integrating factor of the homogeneous equation $M\,dx + N\,dy = 0$, set $M = x^n\varphi(y/x)$, $N = x^n\psi(y/x)$ in $(M\,dx + N\,dy)/(xM + yN) = 0$, cancel x^n from numerator and denominator of the result, and apply test (4), §23. Let $u = y/x$, denote $d\varphi(u)/du$ by φ', and note that $\partial\varphi/\partial x = \varphi'\,\partial u/\partial x = -(y/x^2)\varphi'$, $\partial\varphi/\partial y = \varphi'\,\partial u/\partial y = (1/x)\varphi'$.

29. Using the integrating factor of exercise 28, solve:

(*a*) $(y^2 - xy)\,dx + x^2\,dy = 0$ (*b*) $(x^3 - y^3)\,dx + xy^2\,dy = 0$

30. If $u(x,y)$ is an integrating factor of $M\,dx + N\,dy = 0$ and if $u\,(M\,dx + N\,dy) = df(x,y)$, show that $u\varphi'(f)$ is an integrating factor and that $\varphi(f) = c$ is a solution of $M\,dx + N\,dy = 0$.

★31. If $dy/dx = f(x,y)$ has an integrating factor of the type $X(x)Y(y)$, show that $f(x,y)$ has the form $[w(x)\psi(y) + \varphi(x)]/\psi'(y)$ where w, φ, and ψ represent arbitrary functions.

32. Using (5) of example 3 show that $e^{\int P(x)\,dx}$ is an integrating factor of $dy/dx + P(x)y = Q(x)$.

25 Linear differential equation

A differential equation of any order is said to be linear *when it is of the first degree in the dependent variable and its derivatives.* Linear differential equations are extremely important because of their wide range of application. In the next chapter we shall find many applications of them, and in Chaps. 6 to 8, dealing with general linear equations, we shall meet with a wide range of problems solved by using them.

The general form of a first order differential equation linear in y is

$$\frac{dy}{dx} + Py = Q \tag{1}$$

where P and Q are functions of x only.

To find an integrating factor of (1), let us solve

$$\frac{dy}{dx} + Py = 0 \qquad \text{or} \qquad \frac{dy}{y} = -P\,dx$$

Here the variables are separated, and the solution is

$$ye^{\int P\,dx} = c \tag{2}$$

The differential of the left member of (2) is $e^{\int P\,dx}(dy + Py\,dx)$. It appears then that, if (1) is multiplied by $e^{\int P\,dx}\,dx$, the left member will be an exact differential and the right member will contain x only. Hence, multiplying (1) by $e^{\int P\,dx}\,dx$, we obtain the exact equation

$$e^{\int P\,dx}(dy + Py\,dx) = Qe^{\int P\,dx}\,dx \tag{3}$$

The solution of (3), and therefore of (1), is

$$ye^{\int P\,dx} = \int Qe^{\int P\,dx}\,dx + c \tag{4}$$

Hence, *to solve an equation having the form* (1), *either substitute in form* (4), *or multiply by* $e^{\int P\,dx}$ *and integrate the result as an exact differential equation.*

As an exercise the student may show that the following differential equation (5) has equation (6) as its solution:

$$\frac{dx}{dy} + G(y)x = H(y) \tag{5}$$

$$xe^{\int G(y)\,dy} = \int e^{\int G(y)\,dy}H(y)\,dy + c \tag{6}$$

Example 1. Solve $x(dy/dx) + 2y = x^3$.

Solution. Division by x gives

$$\frac{dy}{dx} + \frac{2}{x}y = x^2$$

This has the form (1), and

$$Q = x^2 \qquad P = \frac{2}{x} \qquad e^{\int P\,dx} = e^{2\int dx/x} = e^{\ln x^2} = x^2$$

Substituting these values in (4), we obtain

$$yx^2 = \int x^2x^2\,dx + \frac{c}{5} = \frac{x^5}{5} + \frac{c}{5}$$

or

$$5yx^2 = x^5 + c$$

Alternative method. Multiply the given equation through by dx/x to obtain

$$dy + \frac{2}{x}y\,dx = x^2\,dx \tag{a}$$

In accord with the italicized statement, $e^{\int P\,dx} = e^{\int (2/x)\,dx} = x^2$ is an integrating factor of (*a*). Multiplying (*a*) through by x^2, we get

$$x^2\,dy + 2xy\,dx = x^4\,dx \tag{b}$$

This is an exact differential equation, and its solution is

$$x^2y = \tfrac{1}{5}x^5 + c \tag{c}$$

Example 2. Solve $dx/dy - 2xy = 2e^{y^2}y$.

Solution. Substituting $-2y$ for G and $2ye^{y^2}$ for H in (6), we obtain the required solution

$$xe^{\int -2y\,dy} = \int e^{\int -2y\,dy} 2ye^{y^2}\,dy + c$$

or

$$xe^{-y^2} = y^2 + c$$

Also, we could have multiplied (*a*) through by the integrating factor $e^{\int -2y\,dy} = e^{-y^2}$ and then have solved the resulting exact differential equation.

Exercises

State whether each equation is of the form (1), (5), or neither:

1. $\dfrac{dy}{dx} = x^3 - 3x^2y$ **2.** $y\,dx + y^2x\,dy = 7\,dx$

3. $y\,dx + x\,dy = (xy^4 + y^5)\,dy$

Solve each of the following differential equations, and when initial conditions are indicated, find the particular solution satisfied by them:

4. $\dfrac{dy}{dx} + \dfrac{1}{x}y = x^3 - 3$ **5.** $\dfrac{dy}{dx} + \dfrac{2}{x}y = x^2 + 2$

6. $\dfrac{dx}{dy} + \dfrac{3}{y}x = 2y$

7. $y\,dx/dy - 2x = 3y^2 - 2$; $y = 1$ when $x = 1$.

8. $x^2\,dy/dx - 2xy = x^4 + 3$; $y = 2$ when $x = 1$.

9. $x^2\,dy - \sin 2x\,dx + 3xy\,dx = 0$

10. $(a + xy)\,dx = (1 + x^2)\,dy$; $y = 2a$ when $x = 0$.

11. $(\sin 2\theta - 2\rho\cos\theta)\,d\theta = 2\,d\rho$

12. $f(x)\,dy + 2yf'(x)\,dx = f(x)f'(x)\,dx$

13. $x\,dy/dx - 2y = x^2 + x$; $y = 1$ when $x = 1$.
14. $y\,dx - 4x\,dy = y^6\,dy$; $x = 4$ when $y = 1$.
15. $dy(1 + 2x \cot y) = dx$
16. $t\,ds = (3t + 1)s\,dt + t^3e^{3t}\,dt$
17. $(y + 2x)\,dy + dx = 0$; $x = -1$ when $y = 0$.
18. $x\,dy + y\,dx = (5x - 2x^2y)\,dx$; let $z = xy$.
19. $y\,dx - x\,dy + x(2 - y)\,dy = y^4e^y\,dy$; let $z = x/y$.
20. $(1 + x^2)\dfrac{dy}{dx} + xy = x^{-1}\sqrt{1 + x^2}$
21. $[f(y)]^2\dfrac{dx}{dy} + 3f(y)f'(y)x = f'(y)$
22. If $x^2f'(x) + xf(x) = 4x^4 + 3$, find $f(x)$.
23. Solve $(dx/dy) + [d\varphi(y)/dy]x = d\varphi(y)/dy$.

26 Extensions of linear differential equations

The equation

$$\frac{dy}{dx} + Py = Qy^n \tag{1}$$

where P and Q are functions of x only, is named Bernoulli's equation, after James Bernoulli, who studied it in 1695.

From §25, or by differentiation, we can show that

$$e^{\int P\,dx}(dy + Py\,dx) = d(ye^{\int P\,dx}) \tag{2}$$

This suggests for solving (1) the substitution

$$v = ye^{\int P\,dx} \qquad \text{or} \qquad y = ve^{-\int P\,dx} \tag{3}$$

The substitution (3) in (1) produces an equation of the type *variables separable*, and from this the required solution* is found. Similarly, to solve the equation

$$dx + x[G(y)\,dy] = H(y)x^n\,dy \tag{4}$$

make the substitution

$$v = xe^{\int G\,dy} \qquad \text{or} \qquad x = ve^{-\int G\,dy} \tag{5}$$

solve the resulting equation in v and y, and in the result replace v by its value $xe^{\int G\,dy}$ from (5). Two examples will illustrate the method of solution.

* This same subsitution furnishes a solution of an equation of the type $(dy/dx) + P(x)\cdot y = Q(x)\cdot f(ye^{\int P\,dx})$.

Example 1. Solve $dy + 2xy\,dx = xe^{-x^2}y^3\,dx.$ (a)

Solution. In accord with (4) make the substitution

$$v = ye^{\int 2x\,dx} = ye^{x^2} \qquad \text{or} \qquad y = ve^{-x^2} \tag{b}$$

Replacing y in (a) by ve^{-x^2} from (b), we get

$$e^{-x^2}\,dv - 2xe^{-x^2}v\,dx + 2xve^{-x^2}\,dx = xe^{-x^2}v^3e^{-3x^2}\,dx \tag{c}$$

Simplifying (c) and multiplying through by $v^{-3}e^{x^2}$, we get

$$v^{-3}\,dv = xe^{-3x^2}\,dx \tag{d}$$

Solving this equation, replacing v in the result by ye^{x^2} from (b), and simplifying, we get

$$-\tfrac{1}{2}v^{-2} = -\tfrac{1}{6}e^{-3x^2} - \tfrac{1}{6}c$$
$$3y^{-2} = e^{-x^2} + ce^{2x^2}$$

Example 2. Solve $dx - \dfrac{2}{y}x\,dy = x^4\,dy.$ (a)

Solution. In accordance with (5), make the substitution

$$v = xe^{-\int (2/y)\,dy} = xy^{-2} \qquad \text{or} \qquad x = vy^2 \tag{b}$$

to obtain

$$y^2\,dv + 2yv\,dy - \frac{2}{y}vy^2\,dy = v^4y^8\,dy \tag{c}$$

Now solve this equation in v and y, replace v in the result by xy^{-2} from (b), and simplify to get

$$v^{-4}\,dv = y^6\,dy$$
$$-\tfrac{1}{3}(xy^{-2})^{-3} = \tfrac{1}{7}y^7 - \frac{c}{21} \tag{d}$$
$$7x^{-3} = cy^{-6} - 3y \tag{e}$$

Exercises

Use the substitution (3) or (5) in solving:

1. $dy + \dfrac{1}{x}y\,dx = 3x^2y^2\,dx$
2. $dy + y\,dx = 2xy^2e^x\,dx$
3. $dx + \dfrac{2}{y}x\,dy = 2x^2y^2\,dy$
4. $2\dfrac{dy}{dx} - \dfrac{y}{x} = 5x^3y^3$
5. $dx - 2xy\,dy = 6x^3y^2e^{-2y^2}\,dy$
6. $3\dfrac{dy}{dx} + \dfrac{3y}{x} = 2x^4y^4$

7. Show that the substitution $z = y^{1-n}$ in (1) gives a linear equation in z and x. The result may be solved by the methods of §25.

Use the method suggested by exercise 7 to solve:

8. $\dfrac{dy}{dx} + \dfrac{1}{x - 2}y = 5(x - 2)\sqrt{y}$

9. $\dfrac{dx}{dy} + 3y^2x = x^2ye^{y^3}$

The substitution (3) in $(dy/dx) + P(x)y = y^nQ(x)f(ye^{\int P\,dx})$ or (5) in $(dx/dy) + P(y)x = x^nQ(y)f(xe^{\int P\,dy})$ will result in an equation of the type *variables separable.* Solve:

10. $\dfrac{dy}{dx} + \dfrac{2}{x}y = x^2(x^2y)^4$

11. $\dfrac{dx}{dy} - \dfrac{2}{y}x = \sqrt{y}\left(\dfrac{x}{y^2}\right)^{3/2}$

Solve each of the following differential equations, and determine the constant of integration when initial conditions are given:

12. $3\,dy - y\,dx = 3y^3e^{4x/3}\,dx$

13. $(12e^{2x}y^2 - y)\,dx = dy$; $y = 1$ when $x = 0$.

14. $3y^2\,(dy/dx) + [y^3/(x + 1)] - 8(x + 1) = 0$; $y = 0$ when $x = 0$.

15. $dy - y \sin x\,dx = y \ln (ye^{\cos x})\,dx$

16. $x\,dy - 2y\,dx = \frac{1}{3}x^8y^{-2}[3(yx^{-2})^2 + 2yx^{-2}]\,dx$; $x = 1$ when $y = 0$.

17. $\dfrac{dy}{dx} + y = ye^{-2x}(ye^x)^3$

18. $\dfrac{dx}{dy} - 2xy = ye^{-3y^2}[xe^{-y^2} + 3(xe^{-y^2})^2]$

27 Simultaneous equations

Two differential equations in three variables often arise in applications. Only pairs of equations that can be solved by means of the theory already developed will be considered at this time. Two equations

$$\begin{aligned} A_1\,dx + A_2\,dy + A_3\,dt &= 0 \\ B_1\,dx + B_2\,dy + B_3\,dt &= 0 \end{aligned} \tag{1}$$

where the A's and B's represent functions of x, y, and t, have solutions consisting of two relations of the form

$$f_1(x,y,t,c_1,c_2) = 0 \qquad f_2(x,y,t,c_1,c_2) = 0 \tag{2}$$

One relation is generally found by eliminating one of the variables from the given equations and solving the resulting equation in two unknowns by methods already considered. When one relation has been found, it may be used with the given differential equations to find others. Of course, if an equation contains all three variables but separated so that no term contains more than one variable or, more generally, if an equation is exact, it may be integrated directly to obtain one of the required equations. Thus, from $2x\,dx + 2y\,dy + 2t\,dt = 0$, obtain $x^2 + y^2 + t^2 = c$.

Example 1. Solve

$$\frac{dx}{dt} + y = x \qquad \frac{dy}{dt} = 3y \tag{a}$$

Solution. Since there are only two variables in the second equation, we solve it to obtain

$$y = c_1e^{3t} \tag{b}$$

Substituting y from (b) in the first of (a), we obtain

$$\frac{dx}{dt} + c_1e^{3t} = x \tag{c}$$

Since (c) is a linear equation in two variables, we solve it by a method of §25 to find

$$xe^{-t} = -\tfrac{1}{2}c_1e^{2t} + c_2 \tag{d}$$

Rewriting equations (b) and (d) slightly simplified, we have

$$y = c_1e^{3t} \qquad x = -\tfrac{1}{2}c_1e^{3t} + c_2e^t$$

Example 2. Find the particular solution of

$$\frac{dx}{dt} + t\frac{dy}{dt} = 2t \qquad t\frac{dx}{dt} - \frac{dy}{dt} = -x \tag{a}$$

for which $x = 0$ and $y = -3$ when $t = 0$.

Solution. To eliminate y, multiply the second equation of (a) by t, add the result to the first, and obtain

$$(1 + t^2)\frac{dx}{dt} = -tx + 2t \tag{b}$$

Equation (b) may be solved by separating the variables. The solution is

$$\ln(x - 2) = \ln[c_1(1 + t^2)^{-1/2}]$$

or

$$x = 2 + c_1(1 + t^2)^{-1/2} \tag{c}$$

Substituting x from (c) in the first of (a), we get

$$-c_1 t(1 + t^2)^{-3/2} + t\frac{dy}{dt} = 2t \tag{d}$$

Separating the variables in (d) and integrating, we find

$$y = 2t + \frac{c_1 t}{\sqrt{1 + t^2}} + c_2 \tag{e}$$

Equations (c) and (e) constitute the general solution. In (c) and (e) substitute $x = 0$, $y = -3$, and $t = 0$, and obtain

$$0 = 2 + c_1 \qquad -3 = 0 + 0 + c_2$$

or

$$c_1 = -2 \qquad c_2 = -3$$

Hence the required solution is

$$x = 2 - 2(1 + t^2)^{-1/2} \qquad y = 2t - 2t(1 + t^2)^{-1/2} - 3$$

Exercises

Solve the following differential equations and determine the constants of integration when initial conditions are given:

1. $dx/dt - 2t = 0$, $dy/dt - x + t^2 = 0$.
2. $dy/dt + y = e^{-t}$, $dx/dt + y = te^{-t}$.
3. $dx/dt = 1{,}000$, $dy/dt = 0.5\,dx/dt - 16t$.
4. $d\rho/dt + \rho = e^t$, $d\theta/dt = \rho$.
5. $(t - 1)\,dx/dt + dy/dt = 6t^2$, $dx/dt - dy/dt = x$; $x = 18$, $y = 0$ when $t = 1$.
6. $x\,dt + t\,dx = 2t\,dt$, $dx/dt + dy/dt = x - t$; $x = 3$, $y = 2$ when $t = 1$.
7. $(x^2 + t^2)\,dt - xt\,dx = 0$, $t\,dy/dt = x^2t + y$.
8. $t\,dx/dt + dy/dt = 4(t^2 + 1)e^t$, $dx/dt - t\,dy/dt = 4(t^2 + 1)e^{2t}$; $x = -4$, $y = 2$, when $t = 0$.
9. $d\rho/\rho = d\theta/(\rho + \theta + t) = dt/t$.
10. $dx + ay\,dt = 0$, $dy - ax\,dt = 0$.

11. If for a projectile fired from a gun, $dx/dt = 1{,}000$ ft/sec and $dy/dt = (3{,}200 - 32t)$ ft/sec, if $x = 0$ and $y = 0$ when $t = 0$, and if the Y axis is vertical, find the greatest height $y_{\max}$ reached by the projectile.

28 Special differential equations of the second order

The differential equations considered in this section belong to the type represented by

$$\frac{d}{dx}\left[\frac{dy}{dx} + P(x)y\right] = f(x) \tag{1}$$

To solve an equation of type (1), let

$$u = \frac{dy}{dx} + P(x)y \tag{2}$$

so that (1) becomes

$$\frac{du}{dx} = f(x) \tag{3}$$

Solve (3) and in the result replace u by its value from (2), and solve the resulting equation. An example will illustrate the procedure.

Example. Solve $\dfrac{d}{dx}\left(\dfrac{dy}{dx} + x^{-1}y\right) = 8x.$ $\qquad (a)$

Also determine the constants of integration so that $y = 2$ and $dy/dx = 0$ when $x = 1$.

Solution. Let $u = (dy/dx) + x^{-1}y$ in (a) and integrate the result to get

$$\frac{du}{dx} = 8x \qquad u = 4x^2 + c_1 \tag{b}$$

In the right equation of (b) replace u by its equal $dy/dx + x^{-1}y$ and solve the result to obtain

$$\frac{dy}{dx} + x^{-1}y = 4x^2 + c_1 \tag{c}$$

$$y = x^3 + \tfrac{1}{2}c_1x + c_2x^{-1} \tag{d}$$

To get the desired particular solution make the substitution $x = 1$, $y = 2$, $dy/dx = 0$ in (c) and (d), and solve the results for c_1 and c_2 to obtain $c_1 = -2$ and $c_2 = 2$. Replacing c_1 and c_2 in (d) by these values, we get

$$y = x^3 - x + 2x^{-1}$$

Exercises

Solve the following differential equations, and determine the constants of integration when initial conditions are given:

1. $\frac{d}{dx}\left(\frac{dy}{dx} + y\right) = e^{-x}$ **2.** $\frac{d}{dx}\left(\frac{dy}{dx}\right) = 6x + 3$

3. $\frac{d}{dx}\left(x\frac{dy}{dx} + 2y\right) = -2x^{-3}$ **4.** $\frac{d^2y}{dx^2} + \frac{dy}{dx} = 2$

5. $\frac{d}{dx}[x\,dy/dx + (1 + x)y] = 12$; $y = 0$, $dy/dx = 0$ when $x = 1$.

6. $d^2s/dt^2 = 12t$; $s = 0$, $ds/dt = 100$ when $t = 0$.

7. $\frac{d}{dx}(x\,dy/dx + y) = 6x + 2$; $y = 1$, $dy/dx = 4$ when $x = 1$.

8. $\frac{d}{dy}(dx/dy + 2yx) = 9$; $x = 0$, $dx/dy = 0$ when $y = 0$.

9. $\frac{d}{dx}\left[\varphi(x)\frac{dy}{dx} + y\varphi'(x)\right] = \psi''(x)$*

Assuming that $\left[\frac{d}{dx} + f(x)\right]u$ means $du/dx + f(x)u$,† solve the following differential equations:

10. $\left(\frac{d}{dx} + 2x\right)(dy/dx + 2xy) = 2e^{-x^2}$; $y = 0$, $dy/dx = 1$ when $x = 0$.

11. $\left(\frac{d}{dx} + 2\right)(dy/dx - 3y) = 50 \sin x$; $y = 1$, $dy/dx = -7$ when $x = 0$.

12. $\left(\frac{d}{dx} + 1\right)[dy/dx + (\cot x)y] = 4e^x$; $y = e^{\pi/2}$, $dy/dx = 2e^{\pi/2}$ when $x = \frac{1}{2}\pi$.

* Here primes denote derivatives with respect to x.

† Expressions such as $d/dx + f(x)$ indicate operations; they will be taken up in considerable detail in §44.

29 Summary

In solving a differential equation of the form $M\,dx + N\,dy = 0$, the student will often find it helpful to proceed, until a method of solution is found, as follows:

1. Consider whether the equation comes under the case of:
   a. Variables separable (§4)
   b. M and N homogeneous and of the same degree (§20)
   c. Linear equation (§25)
   d. Extensions of linear forms (§26)
   e. Exact differential equation (§23)
   f. M and N linear but not homogeneous (§21)

2. Search for an integrating factor (§24). Note that example 3 and exercise 28 of §24 suggest integrating factors.

3. Make a substitution and consider the result under headings 1 and 2 (§§19 and 23 to 26). Observe that the presence of the form $dy + yP(x)\,dx$ (the left member of the linear equation) suggests the substitution $u = ye^{\int P(x)\,dx}$.

At present, we have studied a few important special types of the differential equation having the form $M\,dx + N\,dy = 0$. It may be of interest to consider what remains to be done with this form. The result of multiplying this equation by $u(x,y)$ will be exact, provided that $\partial(uM)/\partial y = \partial(uN)/\partial x$. In Chap. 14, §127, we shall learn how to solve this partial differential equation for u in terms of x and y. Not only will this furnish a general method of attack, but also it will enable us to make up types of equations that are readily solvable. In Chap. 13, some methods of approximating a particular solution of a differential equation are explained. Also, in Chap. 12 the method of integration in infinite series is considered. This method may be applied to solve a great variety of differential equations.

Exercises

Solve each of the following differential equations, and determine the constant of integration when initial conditions are given:

1. $x^2\,dy + y^2\,dx = x^2y\,dy - xy^2\,dx$
2. $(5x^2 + y^2)\,dx + 2x^2\,dy = 0$
3. $(x^2 + 3)\dfrac{dy}{dx} + 2xy + 5x^2 = 0$

4. $(y + x)^2 \dfrac{dy}{dx} = 2(x + y)^2 - 3$

5. $(3x^2 + 2xy)\,dx + (x^2 + \cos y)\,dy = 0$; $y = \frac{1}{2}\pi$ when $x = 0$.

6. $y\,dx + x\,dy = xy(dx + dy)$ 7. $dy - 4xy\,dx = x^3y^2\,dx$

8. $dx/dt - 3x = e^{3t}$, $dx - dy + x\,dt = 0$.

9. $dx/dt = 2/(x + 2y)$, $dy/dt = (6x - 1)/(x + 2y)$.

10. $2x\,dy + 3y\,dx = (x + 2)(xy^3)\,dx$

11. $x\,dy + y\,dx = x^4y^8\,dy$

12. $(x^2y + y^3)\,dx - 2x^3\,dy = 0$; $y = 3$ when $x = 2$.

13. $(x^3 - x)\,dy/dx = (x^2 + 1)y + 12x(x^2 - 1)^3$; $y = 9$ when $x = 2$.

14. $dy/dx - xy = xy^2 - 2xy^3$; $y = 2$ when $x = 0$.

15. $4y^2\,dx - 3xy\,dy = x(y - 1)\,dy$

16. $\dfrac{d}{dx}\left(x\dfrac{dy}{dx} - 3y\right) = 4x^3 + 2$

17. $\dfrac{d}{dy}\left(\dfrac{dx}{dy} - x\right) = 3(y^2 + 2y)e^y$

18. $\left(\dfrac{1}{x} + \dfrac{2y}{x^2 - 1}\right)dx + \left(\ln\dfrac{x - 1}{x + 1} + \dfrac{1}{y}\right)dy = 0$

19. $4(x^2 - y)^3(2x\,dx - dy) = 3(x^2 - y^2)^{-1/2}(x\,dx - y\,dy)$

20. $(3x^2 + 4y^2 - 5)x\,dx = (6 - 3x^2 - 4y^2)y\,dy$; let $u = x^2$, $v = y^2$.

21. $(x - y^2)\,dx + 2xy\,dy = 0$; $y = -2$ when $x = 1$.

22. $\cos x\,dy + 3y \sin x\,dx - 2\cos^2 x\,dx = 0$

23. $x^2\,dy^2 - y^4\,dx^2 = 0$ 24. $dx + 2xy\,dy = x^2y\,dy$

25. $\dfrac{y + x}{x}\,dx + (2y + \ln 3x)\,dy = 0$

26. $3y\,dx + 5x\,dy = xy^3\,dy$

27. $(2x^2 + y^2 - 3)(x\,dy + y\,dx) = (xy)^3(4x\,dx + 2y\,dy)$

28. $xy(dy/dx)^2 - (2x^2 + y^2)\,dy/dx + 2xy = 0$. *Hint:* Solve for dy/dx.

29. $(xy + 1)(x\,dy - y\,dx) = y^2(x\,dy + y\,dx)$; $y = 2$ when $x = 1$.

30. $y^2\,dx + y\,dy = 2\cos x\,dx$; $y = 0$ when $x = \frac{1}{2}\pi$.

31. $x\,dy = (xy^2 - 3y)\,dx$; $y = 2$ when $x = 2$.

32. $(2x + 3y - 1)\,dx = (5 - 2x - 3y)\,dy$

33. $[6x(x + 2y) + a^2]\,dy + (12xy + 6y^2 + b^2)\,dx = 0$

34. $(x + y - 3)\,dx + (x + y + 5)\,dy = 0$; $y = 0$ when $x = 1$.

35. $(5x + 4y + 4)\,dx + (4x + 3y + 1)\,dy = 0$

36. $(4x - 3y)^2\,dy = 2(4x - 3y)\,dy + 4\,dx$

37. $\cos x \dfrac{dy}{dx} + \sin x = 1 - y$

38. $2x^3y^2\,dx + 2x^2y^3\,dy = x\,dy + y\,dx$; $y = -1$ when $x = 1$.

39. $x(x + y)\,dx + y(y\,dx - x\,dy) = 0$; $y = 2$ when $x = 2$.

40. $(\rho + \sin\theta\cos\theta)\,d\theta + (\theta - \rho^2)\,d\rho = 0$

41. $(x + y)^2(x\,dy - y\,dx) + [y^2 - 2x^2(x + y)^2]\,(dx + dy) = 0$; let $v = x + y$, $w = y/x$.

42. $(x^2 + y^2)(x\,dy + y\,dx) + 2xy(x - y)(dx - dy) = 0$

43. Transform $\varphi(x^my^n)y\,dx + \psi(x^my^n)x\,dy = 0$ to the type *variables separable* by the substitution $z = x^my^n$. Solve $(2 + 4x^2\sqrt{y})y\,dx + x^3\sqrt{y}\,dy = 0$.

44. The equation $(3xy^2 + 7x^3)\,dx + (4x^2y + 3\sqrt{xy})\,dy = 0$ has an integrating factor of the form x^k. Determine k and solve the equation.

45. In the equation $(P + Rx^{k+1})\,dy = (Q + Ryx^k)\,dx$, P, Q, and R are homogeneous functions of x and y, P and Q are of the same degree, and k is a constant. Transform the given equation to the type considered in §26 by the substitution $y = vx$.

Solve:

$$[x^2 + y^2 + (y + 2x)x^{-1}]\,dy = [2(x^2 + y^2) + (y + 2x)x^{-2}y]\,dx$$

46. $dx - dy = dt$, $x\,dt + y\,dx = 0$.

47. $x\,dy + y\,dx = 2t\,dt$, $(xy + t^2)(dx + dy) = 4t\,dt$.

48. $(dx/dt)^2 + (dy/dt)^2 = 25$, $y\,dy/dx = 1$.

49. $xy\,dt + xt\,dy + yt\,dx = 0$, $(x + y)\,dt + (x + t)\,dy + (y + t)\,dx = 0$.

4

Applications Involving Differential Equations of the First Order

30 Miscellaneous elementary applications

The solutions of the following problems involve various types of first-order differential equations. Inasmuch as no new knowledge or new methods are needed in finding and solving the appropriate differential equations, no introductory illustrative examples will be given.

Exercises

(**1–5**) Find the equation of a curve if:

1. It passes through $(3,-2)$, and the slope at any point (x,y) on the curve is $(x^2 + y^2)/(y^3 - 2xy)$.

2. The slope at any point (x,y) on the curve is $\frac{1}{3}\sqrt{2x + 3y}$.

3. The intercept on the X axis of the tangent at any point on the curve divided by the square of the ordinate of the point is constant.

4. The intercept on the X axis of the normal at any point on the curve divided by the nth power of the radius vector to the point is constant. Use rectangular coordinates.

5. The abscissa x of the point of contact of the tangent and a perpendicular from the origin to the tangent have equal lengths. Note that the distance from $(0,0)$ to the tangent at (x,y) on the curve is given by $(y - x\,dy/dx)/\sqrt{1 + (dy/dx)^2}$.

6. A tank contains initially 10 gal of brine with 15 lb of salt in solution. Brine containing 1 lb of salt per gallon enters the tank at 2 gal/min, and the brine, kept well stirred, flows out at 1 gal/min. Find the amount of salt in the tank at the end of 5 min.

7. Find the equation of the orthogonal trajectories (see §11) of all circles $x^2 + y^2 - 2my = 0$ tangent to the X axis at the origin.

In connection with the theory of the rates of chemical reactions, formulas of the following type are used:

$$\frac{dx}{dt} = k(a_1 - x)^{p_1}(a_2 - x)^{p_2} \cdots (a_n - x)^{p_n} \tag{a}$$

where x is the amount of substance transformed and k, a_1, a_2, . . . , a_n, and p_1, p_2, . . . , p_n are constants. Find t in terms of x and also the limiting values of x as t increases without bound if:

8. $dx/dt = k(2 - x)(4 - x)$; $x = 0$ when $t = 0$, $x = \frac{3}{2}$ when $t = 10$.

9. $dx/dt = k(3 - x)^2(4 - x)$; $x = 0$ when $t = 0$, $x = 2$ when $t = 5$.

10. $dx/dt = k(3 - x)(4 - x)(5 - x)$; $x = 0$ when $t = 0$, $x = 2$ when $t = 5$.

11. For a chemical reaction obeying the special case of equation (a), exercise 7

$$\frac{dx}{dt} = k(a - x)(b - x)(c - x)$$

where $0 < a < b < c$ and $x = 0$ when $t = 0$, show that the amount x of substance transformed approaches a but is always less than a.

(**12–14**) Find the equation of a curve if:

12. $ON = OP$, where O is the origin, P is any point (x,y) on the curve, and N is the point where the normal at P meets the X axis.

13. The area bounded by the curve, any two ordinates, and the

X axis is equal to the average of the two ordinates multiplied by the distance between them.

14. The radius vector, length ρ, of a moving point (ρ,θ) sweeps out an area proportional to ρ^n as θ changes from zero to any angle θ in the range $0 \leqq \theta \leqq 2\pi$.

15. A light situated at a point in a plane sends out beams in every direction. The beams in the plane meet a curve and are all reflected parallel to a fixed straight line in the plane. If the angle of incidence with the normal to the curve at the point of incidence is equal to the angle of reflection, find the equation of the curve. Solve by using polar coordinates and also by using rectangular coordinates (see Fig. 1). *Hint:* In Fig. 1, $\psi = 180° - \frac{1}{2}\theta$; also, $\varphi = \frac{1}{2}\theta$, and $\tan \varphi = dy/dx$.

16. Given a point O and a straight line D, find a curve such that the portion of the tangent MN included between the point of contact M and the point of intersection N of the tangent and the line D subtends a constant angle at O (see Fig. 2). *Hint:* Use the law of sines to obtain

$$\frac{ON}{\sin \psi} = \frac{\rho}{\sin (\psi + \alpha)}$$

Also, $ON = a \csc (\theta - \alpha)$.

17. If, in exercise 16, the angle MON, instead of being constant, is equal to angle OMN, show that the differential equation of the curve is either $d\theta = 0$ or $(2a - \rho \sin \theta)\, d\rho + \rho^2 \cos \theta\, d\theta = 0$. Prove that $1/\rho^3$ is an integrating factor of this latter equation, and find its solution.

⋆18. One end of an inextensible string, of length l, is fastened to a weight which rests on a rough horizontal table. The other end is carried slowly along a straight line in the table. Find the path of the weight. Assume that the string is always tangent to the curve described by the weight.

⋆19. Find the isogonal trajectories of the circles $x^2 + y^2 - 2my = 0$ with an angle of intersection α. Use polar coordinates.

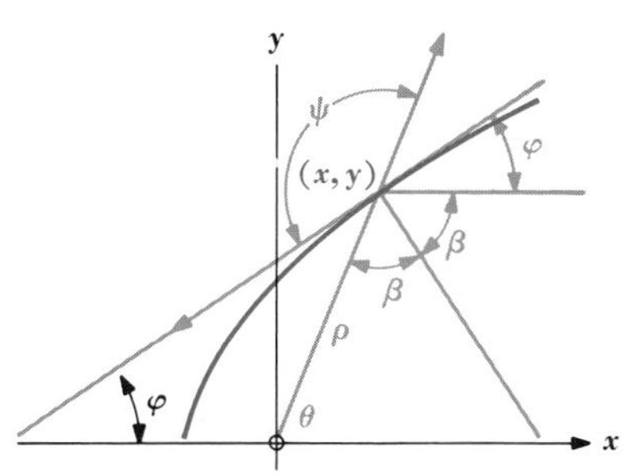

Figure 1

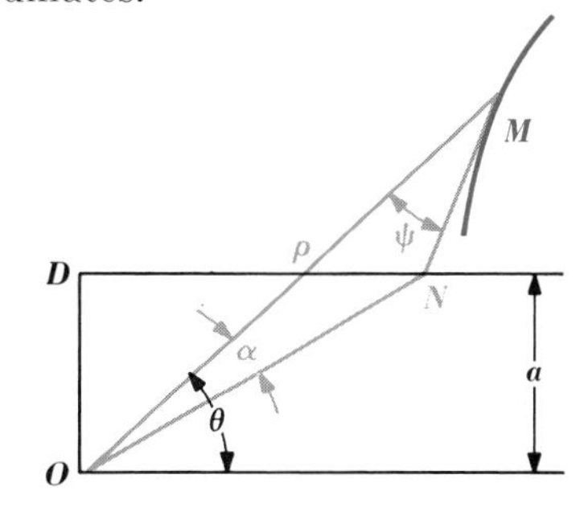

Figure 2

31 Applications involving simultaneous equations

Let x, y, s, v, and t represent abscissa, ordinate, arc length, velocity, and time, respectively, for a point moving in a plane, and let dots denote derivatives with respect to the time. Then the quantities

$$v = \frac{ds}{dt} = \dot{s} \qquad \dot{x} = \frac{dx}{dt} \qquad \dot{y} = \frac{dy}{dt} \tag{1}$$

have the relations indicated in Fig. 1. From the triangle we read, for example

$$\frac{dy}{dx} = \tan\theta \qquad \dot{x} = v\cos\theta \qquad \dot{y} = v\sin\theta \tag{2}$$

$$|v| = \sqrt{\dot{x}^2 + \dot{y}^2} \tag{3}$$

$\dot{x}$ is called the component of velocity along the X axis and $\dot{y}$ the component along the Y axis. A number of the problems in this section refer to plane motion.

Some problems will refer to substances in solution. In these, a pertinent equation will often be obtained by using the expression

Rate of change of substance in a region
= rate of entrance − rate of exit (4)

Example 1. A particle moves on the curve $y = \frac{2}{3}x^{3/2}$ with a constant velocity of $\frac{2}{3}$ unit/sec. Find x and y in terms of t if $\dot{x}$ is positive and $x = 0$ when $t = 1$.

Solution. Two equations for the motion are

$$y = \tfrac{2}{3}x^{3/2} \qquad \dot{x}^2 + \dot{y}^2 = \tfrac{4}{9} \tag{a}$$

Differentiating the first equation of (a) with respect to t and substituting $\dot{y}$ thus obtained in the second equation, we get

$$\dot{y} = \sqrt{x}\,\dot{x} \qquad \dot{x}^2 + x\dot{x}^2 = \tfrac{4}{9} \tag{b}$$

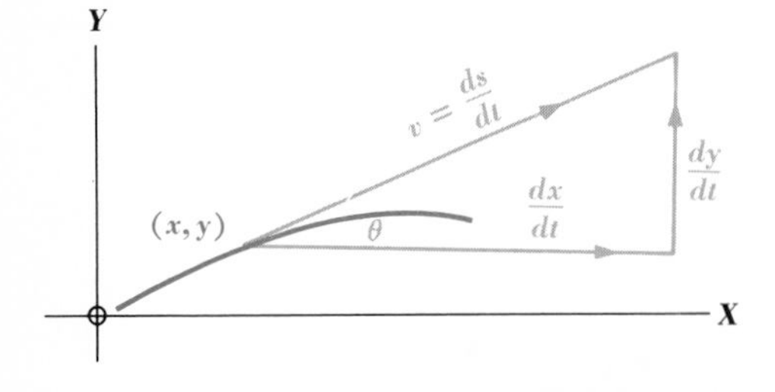

Figure 1

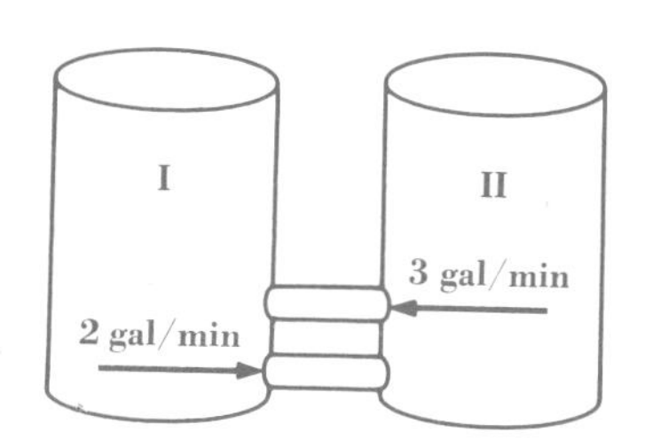

Figure 2

From the second equation of (b), we get

$$(1 + x)^{1/2}\,dx = \tfrac{2}{3}\,dt \qquad \tfrac{2}{3}(1 + x)^{3/2} = \tfrac{2}{3}t + c$$

Since $x = 0$ when $t = 1$, $c = 0$. Hence

$$(1 + x)^{3/2} = t \qquad \text{or} \qquad x = t^{2/3} - 1$$

Since $y = \frac{2}{3}x^{3/2}$ from (a), we have

$$x = t^{2/3} - 1 \qquad y = \tfrac{2}{3}(t^{2/3} - 1)^{3/2}$$

Example 2. Initially tank I and tank II (see Fig. 2) each contain 100 gal of brine, tank I having 200 lb of salt and tank II 50 lb of salt in solution. Brine runs at 2 gal/min from tank I to tank II through one pipe and at 3 gal/min from tank II to tank I through another pipe. The brine is kept well stirred. How much salt will the second tank contain at the end of 50 min?

Solution. Let Q_1 and Q_2 represent the respective number of pounds of salt in tanks I and II (see Fig. 2) at the time t min. Note that tank I gains 1 gal/min of brine and tank II loses 1 gal/min, so that tank I and tank II contain $100 + t$ gal and $100 - t$ gal, respectively, at time t min. Hence the concentration c_1 (number of pounds of salt per gallon) in tank I is $c_1 = Q_1/(100 + t)$ and in tank II $c_2 = Q_2/(100 - t)$. Now, applying equation (4) to each tank, we get

$$\dot{Q}_1 = \frac{3Q_2}{100 - t} - \frac{2Q_1}{100 + t} \tag{a}$$

$$\dot{Q}_2 = \frac{2Q_1}{100 + t} - \frac{3Q_2}{100 - t} \tag{b}$$

Evidently

$$Q_1 + Q_2 = 250 \tag{c}$$

at all times. Substituting Q_1 from (c) in (b), we get

$$\frac{dQ_2}{dt} = \frac{500 - 2Q_2}{100 + t} - \frac{3Q_2}{100 - t} = \frac{-(500 + t)Q_2}{100^2 - t^2} + \frac{500}{100 + t}$$

The general solution of this linear equation is

$$Q_2 = \frac{500}{(100 + t)^2} [100(100 - t) - (100 - t)^2 + c(100 - t)^3] \qquad (d)$$

Using the fact that $Q_2 = 50$ when $t = 0$, we get $c = 0.001$. Then, replacing t by 50, we get $(Q_2)_{t=50} = 58\frac{1}{3}$ lb.

Example 3. A rocket of original mass M g expels gas at the constant rate of k g/sec and at velocity c cm/sec relative to the rocket. Assuming no external forces acting on the rocket, derive a differential equation of its motion.

Solution. The momentum *of a moving body is the product of its mass and velocity.* One of Newton's laws of motion applied to a body moving in a straight line may be written

$$\text{Force} = \text{rate of change of momentum} \qquad (a)^*$$

The mass of the rocket at time t sec is $(M - kt)$ g, and its momentum is $(M - kt)v$ g-cm/sec, where v represents the velocity of the rocket. The change of momentum of $M - kt$ during the Δt sec following time t is composed of two parts, one the momentum of the unburned material $(M - kt - k\,\Delta t)(v + \Delta v)$ and the other the momentum of the material burned during time Δt, or $k\,\Delta t(\bar{v} - c)$, where $\bar{v} = v + \epsilon\,\Delta t$, $|\epsilon| \leqq 1$, and $\epsilon \to 0$ as $\Delta t \to 0$. Hence, since $F = 0$, we have change in momentum during the time Δt is zero, or

$$(M - kt - k\,\Delta t)(v + \Delta v) + k\,\Delta t(\bar{v} - c) - (M - kt)v = F\,\Delta t = 0 \qquad (b)$$

Dividing (b) through by Δt and considering limits as $\Delta t \to 0$, we get

$$(M - kt)\frac{dv}{dt} = kc \qquad (c)\dagger$$

Exercises

1. A particle moves on parabola $y^2 = 4x$ with a velocity such that $\dot{x} = 2t + 2$ at all times. Find x and y in terms of t if the particle passes through (4,4) at time $t = 1$ with a positive y-component.

* This agrees with the law $F = ma = d(mv)/dt$.

† It has been suggested that, since the only change in momentum $(M - kt)v$ during time dt is $k\,dt(\bar{v} - c)$ in the discharged gases and the total change is zero, we have

$$d[(m - kt)v] + k(v - c)\,dt = 0$$

From this we easily derive equation (c).

2. A particle moves on the catenary $y = \cosh x$ with a velocity of constant magnitude 2 ft/sec. It passes through (0,1) at time $t = 0$. Show that $x = \sinh^{-1}(2t)$, $y = \sqrt{1 + 4t^2}$.

3. A particle moves on curve $y = x^2 - \frac{1}{8}\ln x$ with a velocity of constant magnitude 10. If it passes through (1,1) with a positive x-component of velocity at time $t = 0$, show that $8x^2 + \ln x = 80t + 8$.

4. Under certain conditions the motion of a projectile is given approximately by the equations

$$\dot{x} + 0.032x = 1{,}600 \qquad \dot{y} + 0.032y = 1{,}600 - 32t$$

If $x = 0$ and $y = 0$ when $t = 0$, find x and y in terms of t.

5. Brine from a first tank runs into a second tank at 2 gal/min, and brine from the second tank runs into the first at 1 gal/min. Initially, there are 10 gal of brine containing 20 lb of salt in the first tank and 10 gal of fresh water in the second tank. How much salt will the first tank contain after 5 min? Assume that the brine in each tank is kept uniform by stirring.

6. If, for the motion of a particle in the xy plane, $\dot{x} = a\cos pt$ and $\dot{y} = b\sin pt$, show that the particle moves in an ellipse of semi-axes a/p and b/p and that it moves around the ellipse once every $2\pi/p$ units of time. If (m,n) is the center of the ellipse, show that $(x - m)\dot{y} - (y - n)\dot{x}$ is constant for the motion.

7. Using equation (*c*) of example 3 and the initial conditions $v = 0$ when $t = 0$, show that

$$v = -c\ln\frac{M - kt}{M}$$

Why is $t < M/k$? State the physical significance of this fact. Also, find the velocity of the rocket when 0.9 of its mass has been consumed.

8. If distance $x = 0$ when $t = 0$, derive from the equation of exercise 7

$$x = ct + \frac{c}{k}(M - kt)\ln\frac{M - kt}{M}$$

Find x for the rocket when 0.9 of its mass has been consumed.

9. To take account of the earth's pull, or gravity, for the rocket of example 3, show that $g(kt - M)$ dynes should be added to the right member of equation (*c*) of example 3. In this case, find v and x for the rocket at time t and at time $t = 0.9M/k$ if $v = 0$ and $x = 0$ when $t = 0$.

10. A particle moves on a curve in the xy plane. Its motion is defined by

$$\dot{y} = x^3y^2 - xy \qquad \dot{x} = x^2$$

and the fact that it passes through $(1,-1)$ at time $t = 1$. Find the equation of its path, and find x and y in terms of t.

11. The path of a particle moving in the xy plane is the hyperbola $x^2 - y^2 = 25$, and the components $\dot{x}$ and $\dot{y}$ of its velocity satisfy $\dot{x} + \dot{y} = 1$. Find x and y in terms of t if $x = 5$ when $t = 5$.

12. A point moves in a plane curve through $(1,1)$ so that its components $\dot{x}$ and $\dot{y}$ are given by

$$\dot{x} = -2x + 6y \qquad \dot{y} = 2x + 2y$$

Prove that it must move either on the line $x = y$ or on the line $x + 3y = 0$, and find x and y in terms of t for its motion on the line $x = y$. *Hint:* Divide the second equation by the first, member by member.

13. In the chemical process called fractional precipitation, the equations

$$\frac{dx}{dt} = k_1(a - x)(c - z) \qquad \frac{dy}{dt} = k_2(b - y)(c - z)$$

apply. Given that $x = 0$ when $y = 0$, prove that

$$k_1 : k_2 = \ln \frac{a}{a - x} : \ln \frac{b}{b - y}$$

***14.** Brine containing 2 lb of salt per gallon runs into a tank at 2 gal/min, brine from this first tank runs into a second tank at 3 gal/min, and brine runs out of the second tank at 3 gal/min. Initially, the first tank contains 10 gal of brine with 30 lb of salt in solution and the second tank 10 gal of fresh water. Assuming uniform concentration in each tank, find the quantity of salt in the second tank at the end of 5 min.

32 Applications to the flow of electricity

We may think of electricity as a substance which flows through conductors such as wires. *One unit of electricity is the* coulomb. Just as we speak of gallons of water, we speak of coulombs of electricity. The rate of flow of electricity is called current. If I coul of electricity per second are passing a point in a conductor, the current in the conductor is I amp.

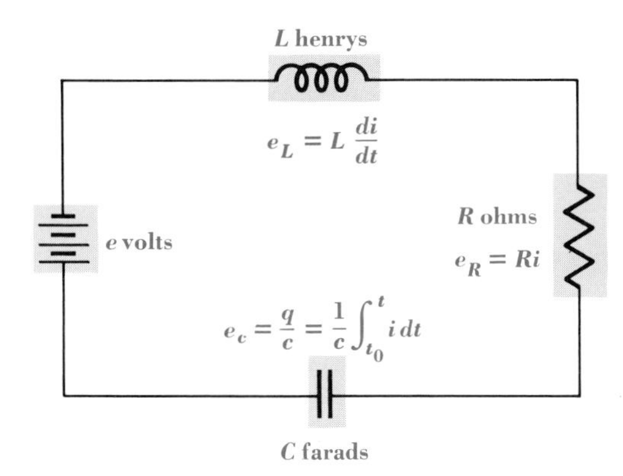

Figure 1

For the circuit indicated in Fig. 1 the following equations hold:

$$L\frac{di}{dt} + Ri + \frac{q}{C} = \epsilon \qquad i = \frac{dq}{dt} \tag{1}$$

where ϵ represents emf, i represents current, q represents the charge, or quantity of electricity, on the capacitor, and L, R, and C are constants. *Electromotive force* ϵ is analogous to *force*, *inductance* L to *mass* or *inertia*, *resistance* R to *friction*, and *capacitance* C to the size of a *storage tank*.

If there is no capacitor in a circuit, the corresponding equation (1) does not contain the term q/C and no consideration of q enters the discussion.

A set of units in common use are *quantity* q in *coulombs*, *current* i in *amperes*, *emf* ϵ in *volts*, *inductance* L in *henrys*, *resistance* R in *ohms*, and *capacitance* C in *farads*.

Example. Discuss the charging of a large capacitor, or battery, in a circuit containing a constant emf E, a resistance R, and no inductance.

Solution. Setting $L = 0$ and $i = dq/dt$ in the first of equations (1), obtain

$$R\frac{dq}{dt} + \frac{q}{C} = E \tag{a}$$

The initial condition may be taken as $q = 0$ when $t = 0$. The solution of (a) subject to this condition is

$$q = CE(1 - e^{-t/(RC)}) \tag{b}$$

From the second equation of (1),

$$i = \frac{dq}{dt} = \frac{E}{R}e^{-t/(RC)} \tag{c}$$

The upper limit of the charge is found from (b) to be CE. When

$t = RC$, $q = CE(1 - e^{-1}) = 0.632CE$, and when $t = 2RC$, $q = CE(1 - e^{-2}) = 0.865CE$. The initial current is E/R when $t = 0$, and it dies away as t increases.

Exercises

1. Solve $L\,di/dt + Ri = E$ for i in terms of t and constants L, R, and E if $i = 0$ when $t = 0$.

(2–4) Using (1) and the result of exercise 1 find indicated unknowns if $i = 0$ when $t = 0$:

2. Find i when $t = 0.1$ sec if $L = 2$ henrys, $R = 20$ ohms, and: (*a*) $E = 10$ volts; (*b*) $E = 100$ volts.

3. Find inductance L if $R = 20$ ohms and $i = 1$ amp when $t = 0.01$ sec, and: (*a*) $E = 40$ volts; (*b*) $E = 200$ volts.

4. Find the constant emf E if $L = 5$ henrys, $R = 15$ ohms, and $i = 1.5$ amp when $t = 1$ sec.

5. Use (1) to find the constant emf E if $C = \frac{1}{20}$ farad, $R = 20$ ohms, $q = 0$ when $t = 0$, and $q = 2$ coul when $t = 1$ sec.

6. Discuss the *discharge of a capacitor* through a resistance R by solving (1) with $L = 0$, $\epsilon = 0$, subject to initial conditions $q = q_0$ when $t = 0$, and by finding q when $t = \infty$, $t = CR$, and $t = 2CR$. Also, find t when $q = 0.01q_0$.

7. Discuss the *decay of a current* of initial value I_0 in a circuit containing neither emf nor capacitor, after solving $L\,di/dt + Ri = 0$ and finding the value of i when $t = L/R$, $2L/R$, and ∞.

8. Discuss the growth of current of zero initial value in a circuit containing no capacitor, a resistance R, an inductance L, and emf (*a*) E; (*b*) $E \sin \omega t$.

9. By setting $L = 0$ and $\epsilon = E \sin \omega t$ in (1), show that the corresponding current i approaches $[EC\omega/(1 + R^2C^2\omega^2)](\cos \omega t + RC\omega \sin \omega t)$ as t increases without bound.

10. By solving (1) for q and i in terms of t with $L = 0$, $R = 10$ ohms, $C = 250 \times 10^{-6}$ farad, and $\epsilon = 110 \sin 300t$, show that q rapidly approaches $11(4 \sin 300t - 3 \cos 300t)/2{,}500$ and that i rapidly approaches $1.32(4 \cos 300t + 3 \sin 300t)$.

11. If there is no capacitor in a circuit of the type shown in Fig. 1 and if $L = 0.1$ henry, $R = 10$ ohms, and $\epsilon = 100 \sin 200t$, show that the current i is given (nearly) by $i = 2 \sin 200t - 4 \cos 200t$ after a very short time.

★12. Use (1) to find resistance R if $L = 0.5$ henry, $\epsilon = 40$ volts, $i = 0$ when $t = 0$, $i = 2$ amp when $t = 0.05$, and there is no capacitor in the circuit.

★13. Use equations (1) with $R = 0$ and $\epsilon = 0$ to find i and q for the discharge of a capacitor through an inductance L. Assume as initial conditions $q = q_0$ and $i = 0$ when $t = 0$. *Hint:* $di/dt = (di/dq)(dq/dt) = i\, di/dq$. Use $i\, di/dq$ for di/dt in (1), solve the result for i in terms of q, then replace i by dq/dt, and solve the result.

33 Air pressure

To obtain an expression for air pressure at height h above the earth, consider a vertical column of air (see Fig. 1) having a small square cross section of area A and extending from the ground upward indefinitely. An element of this column bounded above and below by two horizontal planes at height h ft and $h + \Delta h$ ft, respectively, from the ground is subjected to an upward force on its lower side of p lb/in.^{2}, to a downward force on its upper side of $p + \Delta p$ lb/in.^{2}, to the weight of the element, and to horizontal forces. Since the element is in equilibrium, the vertical forces balance. Equating the algebraic sum of the vertical forces to zero, we have

$$pA - (p + \Delta p)A - \bar{\rho}A\,\Delta h = 0 \tag{1}$$

where $\bar{\rho}$ represents the average density of the element of air. Dividing through by $A\,\Delta h$, and finding the limit approached as Δh approaches zero, we obtain

$$dp + \rho\, dh = 0 \tag{2}$$

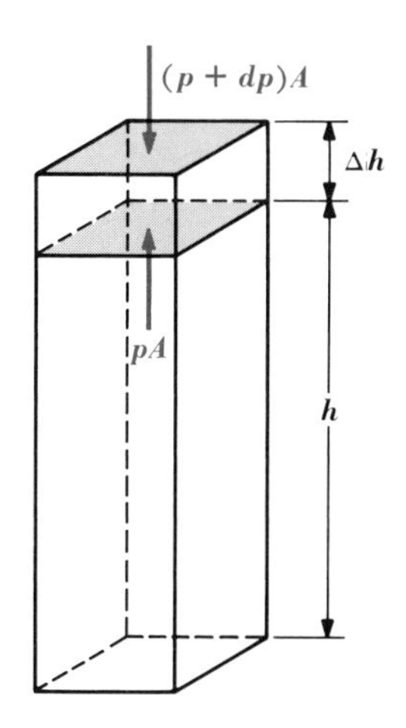

Figure 1

If the air obeys Boyle's law for perfect gases

$$\rho = kp^{*} \qquad (3)$$

Solving (2) and (3) as simultaneous equations, we obtain

$$p = ce^{-kh} \qquad \rho = kce^{-kh}$$

Exercises

1. Assuming that the atmosphere obeys Boyle's law, find the air pressure at a height of 70,000 ft. Assume that the pressure at the surface of the earth is 14.7 lb/in.^{2} and that it is 10.08 lb/in.^{2} at an altitude of 10,000 ft.

2. Find the air pressure at an altitude h if air obeys the adiabatic law $p = k\rho^{1.4}$. Show that in this case the pressure would become zero at a finite height. Find this height in terms of k and the pressure p_0 at the surface of the earth.

3. Compute the theoretical height of an atmosphere which obeys the adiabatic law $p = k\rho^{1.4}$, assuming that the pressure at height zero is 2120 lb/ft^2 and the pressure at a height of 10,000 ft is 1450 lb/ft^2.

★4. Assume that the air pressure on some planet is 2000 lb/ft^2 at the surface, 1500 lb/ft^2 at a height of 10,000 ft, and that the height of the atmosphere is 150,000 ft. Using the units p lb/ft^2, h ft, and ρ lb/ft^3 and assuming that $p = k\rho(1 - nh)$ and (1) holds, show that $p = 2000(1 - \frac{2}{3}h10^{-5})^{4.17}$ approximately.

34 Applications involving forces and velocities

The applications of this section deal with forces acting in a plane and with velocities of particles moving in a plane.

A force may be represented by a directed line segment or vector. Thus, vector PQ in Fig. 1, f units long, represents a force of *magnitude* f units of force and having a direction from P to Q. If PQ makes an angle θ with the directed X axis, the signed numbers

$$X = f\cos\theta \qquad Y = f\sin\theta \qquad (1)$$

* A good approximation to air pressure is not to be expected by assuming Boyle's law or the law of adiabatic expansion because the density of a gas depends on its temperature and many other factors.

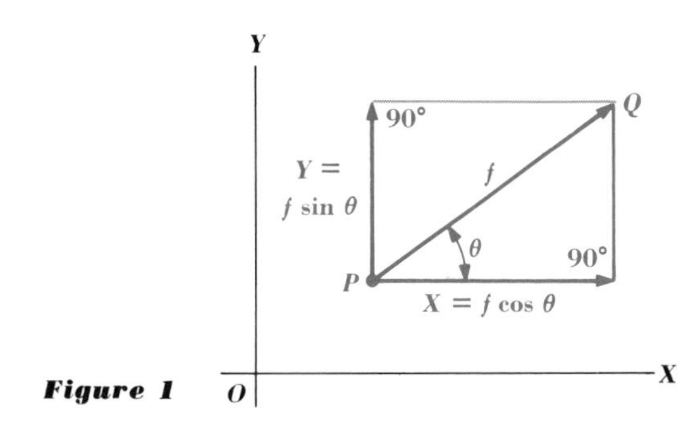

Figure 1

are called, respectively, the x-component and the y-component of the force. X may be considered as a force having the direction of the X axis or the opposite direction according as X is positive or negative, and a similar statement applies to Y. Note that a force is defined in magnitude and direction by its two components X and Y referred to a given system of axes. We shall denote a force with components X and Y by $[X,Y]$ and, in agreement with experiment, define the sum of two forces by

$$[X_1,Y_1] + [X_2,Y_2] = [X_1 + X_2,\ Y_1 + Y_2] \tag{2}$$

Observe that, if the magnitude of the force of Fig. 1 were taken as cf, the components would be $X = cf \cos \theta$, $Y = cf \sin \theta$, and

$$c[X,Y] = [cX,cY] \tag{3}$$

By means of (2) and (3), we can easily derive

$$a[X_1,Y_1] + b[X_2,Y_2] = [aX_1 + bX_2,\ aY_1 + bY_2] \tag{4}$$

Let $X(x,y)$ and $Y(x,y)$ represent functions of x and y, and let $[X(x,y),Y(x,y)]$ represent a force acting at point (x,y). Then, the corresponding set of forces, one for each point (x,y) in a region, is called a field of force. Figure 2 represents the field of force $[0,-w]$. This is approximately the field of force, or weight, in a small part of a vertical plane near the surface of the earth.

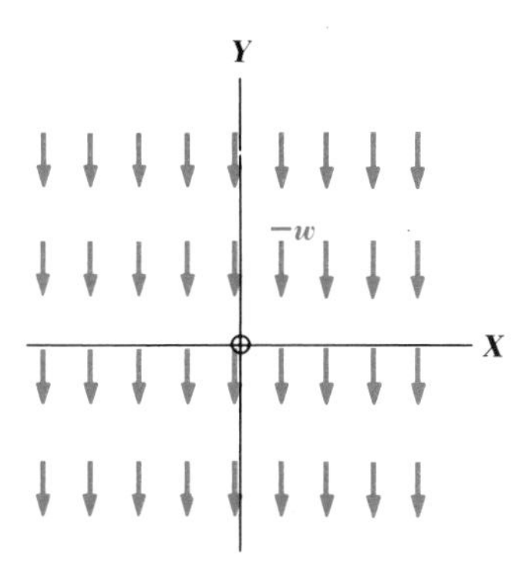

Figure 2

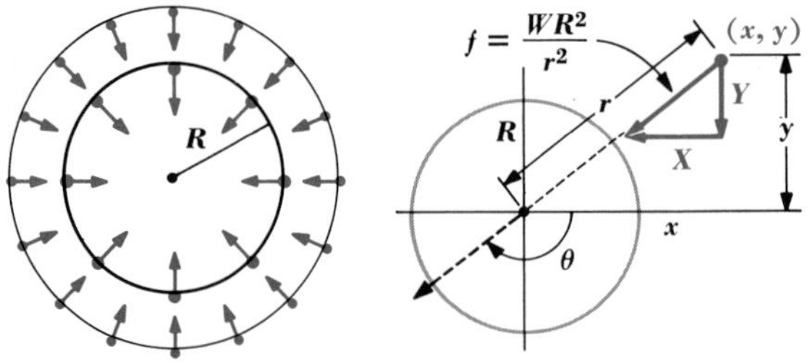

(*a*) (*b*)

Figure 3

Figure 3a indicates approximately the field of force of gravity, or the earth's pull, in a plane through the center of the earth. Figure 3b shows a representative force acting at point (x,y) and having magnitude $f = wR^2/r^2$, where R is the radius of the earth and $r = \sqrt{x^2 + y^2}$. Note that $\cos\theta = -x/r$, $\sin\theta = -y/r$; hence, the field is represented by

$$\left[-\frac{wR^2}{r^2}\frac{x}{r}, -\frac{wR^2}{r^2}\frac{y}{r}\right] \tag{5}$$

Figure 4 represents the field of force defined by $[y, -x]$. The magnitude of the force is $\sqrt{x^2 + y^2}$ units.

A line of force *is a curve in the field of force which has at each of its points a tangent line having the same slope as the vector representing the force through the point.* From Fig. 1, we see that the lines of force of field $[X, Y]$ satisfy the differential equation

$$\tan\theta = \frac{dy}{dx} = \frac{Y}{X} \qquad \text{or} \qquad Y\,dx - X\,dy = 0 \tag{6}$$

The lines of force for the field of Fig. 3 are defined by

$$\frac{dy}{dx} = \frac{-(wR^2/r^2)y}{-(wR^2/r^2)x} = \frac{y}{x} \qquad \text{or} \qquad y = cx \tag{7}$$

This represents all lines through the origin.

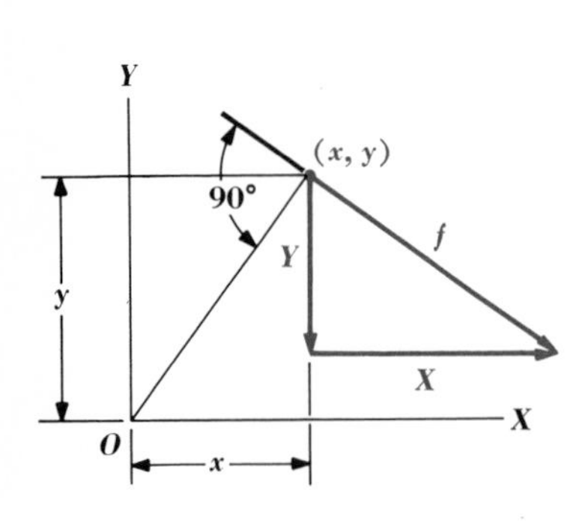

Figure 4

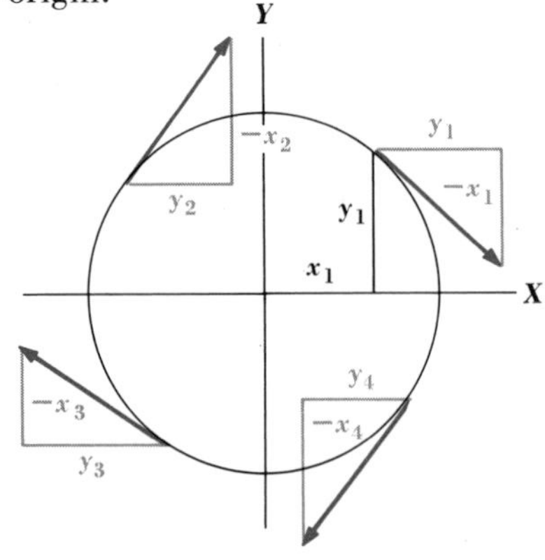

Figure 5

The lines of force for the field $[y, -x]$ satisfy

$$\frac{dy}{dx} = \frac{-x}{y} \qquad \text{or} \qquad x^2 + y^2 = c$$

Figure 5 indicates one line of force and some associated forces.

The energy a body has because of its position in a field of force is called its potential. Thus, a weight on a platform has potential energy because energy was expended in lifting it up to the platform. Let s ft represent distance from a fixed point A on a line. If a force $f(s)$ lb acts along a straight line on a particle while it moves from a point s_0 ft from A to a point s ft from A, the work W done by the force is defined by

$$W = \int_{s_0}^{s} f(r)\, dr \qquad \text{ft-lb} \tag{8}$$

If $f(s) = c$ then from (8) $W = \int_{s_0}^{s} c\, dr = c(s - s_0)$ or *force times distance.* For the force of gravity we may take $s_0 = \infty$* and obtain

$$W = \int_{\infty}^{r} \frac{wR^2}{s^2}\, ds = \frac{-wR^2}{r} \tag{9}$$

It is shown in books on physics and on advanced calculus† that the work W done by a plane field of force $[X(x,y),\ Y(x,y)]$ is given by the line integral

$$W = \int (X\, dx + Y\, dy)$$

evaluated along the path described by the moving mass. Here we consider only the case where $X\, dx + Y\, dy$ is an exact differential, so that there is a function $\varphi(x,y)$ such that

$$X\, dx + Y\, dy = d\varphi = \frac{\partial \varphi}{\partial x}\, dx + \frac{\partial \varphi}{\partial y}\, dy \tag{10}$$

and

$$W = \int (X\, dx + Y\, dy) = \varphi(x,y) + c \tag{11}$$

Here $\varphi(x,y)$ is called the potential‡ of $[X, Y]$. If the potential $\varphi(x,y)$

* Any value may be used for the lower limit. A convenient value is generally selected.

† See R. Courant, "Differential and Integral Calculus," vol. I, pp. 304–306, and vol. II, pp. 343, 414.

‡ A field of force definable as a function of (x,y,z), as in (10), is called a conservative field of force. To gain simplicity in the differential equations, we have used the definitions given above. The function $\varphi(x,y,z)$ is related to the potential energy function $V(x,y,z)$ used in physics by

$$\varphi(x,y,z) = -V(x,y,z)$$

that is, the potential energy function is thought of in terms of the work done on the body to move it against the forces of a conservative field instead of as the work done by the forces of the field. The potential of a body at (x,y,z) as used in physics is the potential energy per unit mass.

expressing the work done by a field of force $[X,Y]$ is taken as zero at point (a,b), then the potential at (x,y) is given, for length in feet and force in pounds, by

$$W(x,y) = \varphi(x,y) - \varphi(a,b) \qquad \text{ft-lb} \tag{12}$$

Also, the energy $U(x,y)$ imparted to the particle is $V(x,y) = -\varphi(x,y) + \varphi(a,b)$.

Observe that when the potential function $\varphi(x,y)$ of $[X,Y]$ is known, from (10) we get

$$X = \frac{\partial\varphi}{\partial x} \qquad Y = \frac{\partial\varphi}{\partial y} \tag{13}$$

Curves along which the potential is constant are called level curves and equipotential curves; they are represented by

$$\varphi(x,y) = c \tag{14}$$

For example, if $\varphi(x,y) = wR^2(x^2 + y^2)^{-1/2}$, the equipotential curves are given by

$$wR^2(x^2 + y^2)^{-1/2} = c_1 \qquad \text{or} \qquad x^2 + y^2 = c \tag{15}$$

Also, in this case from (13) we get

$$X = -wR^2x(x^2 + y^2)^{-3/2} \qquad Y = -wR^2y(x^2 + y^2)^{-3/2} \tag{16}$$

Substituting these values of X and Y in (6) and integrating the result we obtain as an equation of the lines of force

$$y = cx \tag{17}$$

Note that equipotential circles (15) are the orthogonal trajectories of the lines of force (17) as indicated in Fig. 6. This is true generally. For along an equipotential curve $\varphi(x,y) = c$, and from (10),

$$X\,dx + Y\,dy = 0 \tag{18}$$

is the differential equation of the equipotential curves. Comparing $dy/dx = -X/Y$ from (18) with dy/dx from (6) we see that the level curves are the orthogonal trajectories of the lines of force.

A velocity has magnitude and direction and therefore, like force, can be represented by a vector $[u,v]$. To each point in a region, we can associate a velocity by $[u(x,y), v(x,y)]$ and thus define a velocity field. The streamlines corresponding to lines of force are defined by

$$\frac{dy}{dx} = \frac{v}{u} \qquad \text{or} \qquad v\,dx - u\,dy = 0 \tag{19}$$

and the velocity equipotential curves are defined by

$$u\,dx + v\,dy = 0 \tag{20}$$

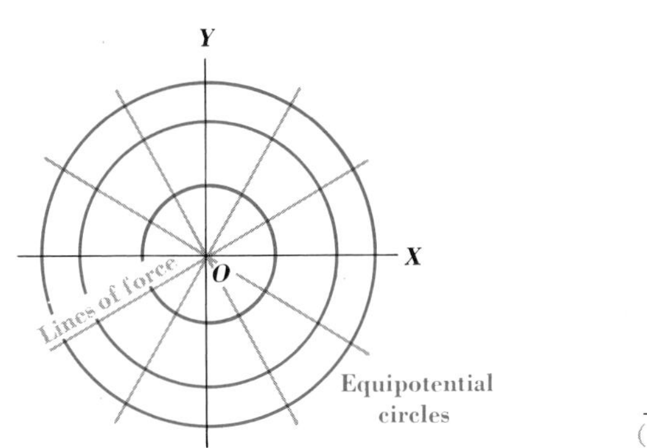

Figure 6

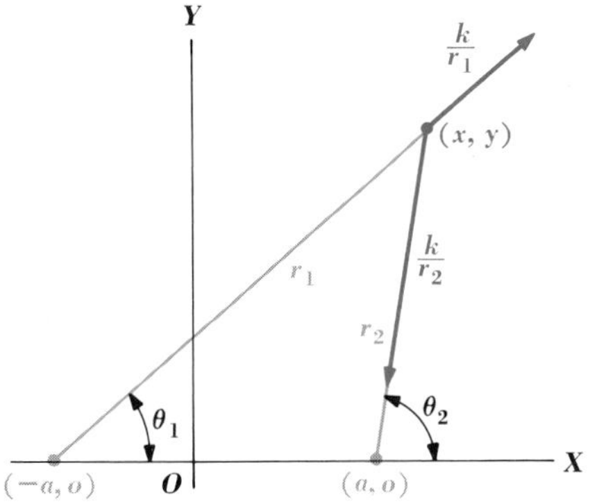

Figure 7

Example. For the field of force indicated in Fig. 7, find the equations of the lines of force and of the equipotential curves.

Solution. To get the equations of the equipotential curves, set the potential $\varphi(x,y)$ equal to a constant and obtain

$$\varphi(x,y) = \int_h^{r_2} \frac{k}{r}\,dr - \int_h^{r_1} \frac{k}{r}\,dr = k \ln r_2 - k \ln r_1 = c_1 \tag{a}$$

Observing that $r_1 = \sqrt{(x+a)^2 + y^2}$ and $r_2 = \sqrt{(x-a)^2 + y^2}$, we change (a) to the form

$$\frac{r_2^2}{r_1^2} = \frac{(x-a)^2 + y^2}{(x+a)^2 + y^2} = c^2 \tag{b}$$

From (b) we easily obtain

$$x^2 + y^2 + a^2 - mx = 0 \qquad \text{where } m = \frac{2a(1+c^2)}{1-c^2} \tag{c}$$

To get the lines of force we may find the orthogonal trajectories of (c), or form the force field $[X,Y]$ from (a) and (13), and use (6).

From (a) and (13) we get

$$X = \frac{\partial\varphi}{\partial x} = \frac{1}{r_2}\frac{\partial r_2}{\partial x} - \frac{1}{r_1}\frac{\partial r_1}{\partial x} \qquad Y = \frac{\partial\varphi}{\partial y} = \frac{1}{r_2}\frac{\partial r_2}{\partial y} - \frac{1}{r_1}\frac{\partial r_1}{\partial y} \tag{d}$$

Using the values of X and Y from (d) in (6) we get for the lines of force

$$Y\,dx - X\,dy = \frac{(x+a)\,dy - y\,dx}{(x+a)^2 + y^2} - \frac{(x-a)\,dy - y\,dx}{(x-a)^2 + y^2} = 0 \tag{e}$$

Each of the two differentials in (e) is exact, and its solution is

$$\tan^{-1}\frac{y}{x+a} - \tan^{-1}\frac{y}{x-a} = \tan^{-1} c_1 \tag{f}$$

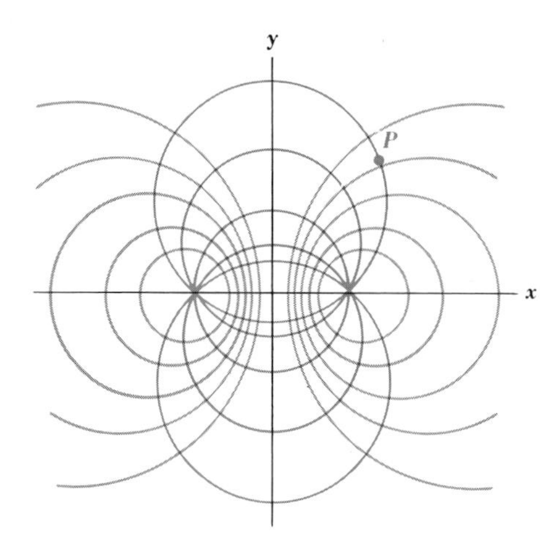

Figure 8

Equating the tangents of the members of (f) and transforming the result, we get

$$x^2 + (y - c)^2 = a^2 + c^2 \tag{g}$$

as the equation of the lines of force. Figure 8 represents the lines of force and the equipotential curves.

Exercises

In all exercises assume that distance is in feet and force is in pounds.

1. The force required to stretch a spring s ft is ks lb. Use (8) to find the work done by a spring when the stretch is changed from a ft to b ft.

2. The potential function $\varphi(x,y) = x^2 + y^2$ relates to the field of force $[2x,2y]$. Use (12) to find the work done by the field on a particle when it moves in the field: (*a*) from (0,0) to (2,4); (*b*) from (3,−2) to (−4,−1).

3. Use (11) to find the potential $\varphi(x,y)$ for the field $[3x,-5y]$, and then use (12) to find the work done when the particle moves in the field: (*a*) from (0,1) to (0,7); (*b*) from (2,−2) to (−4,3).

4. Use (11) to find the potential $\varphi(x,y)$ for the field $[2x + y^3, 3xy^2]$, and then use (12) to find the work done by the field on a particle which moves in it from (−2,3) to (4,−5).

5. Find the potential $\varphi(x,y)$ for the field $[2x,4y]$. Write the equations of the equipotential lines, and use (6) to find the equations of the lines of force. Sketch three lines of force and three equipotential curves.

Using (14), (13), and (6) in succession find the equations of the equipotential curves, the fields of force $[X,Y]$, and the equations of the lines of force from the following potential functions:

6. $\varphi(x,y) = x^2 + y^2$

7. $\varphi(x,y) = xy^{-1} + yx^{-1}$

8. $\varphi(x,y) = x^2y + y^3$

9. Show that two potential functions $\varphi(x,y)$ and $\varphi(x,y) + c$ are associated with the same field of force, the same equipotential curves, and the same lines of force.

10. Using (11) find the potential functions of the fields of force: (*a*) $[x,y]$; (*b*) $[3x^2 - 2xy, 2y - x^2]$; (*c*) $[2x + \sin y, x \cos y]$.

11. Show that a field of force $[X,Y]$ can be derived from a potential function if and only if $\partial X/\partial y = \partial Y/\partial x$. In this case the field of force is called *conservative.*

12. Read exercise 11, and then specify which of the following indicated fields of force have potential functions: (*a*) $[x + y, y - x]$; (*b*) $[4x + y + x^{11}, x + y^{26}]$; (*c*) $[3x + y^2 + 7, -2xy]$.

13. The gravitational field of force for a particle inside the earth obeys approximately the law suggested by Fig. 9. Write a bracket representation of the field, and find the corresponding equations of the equipotential curves and of the lines of force.

14. Find the work done by gravity, or $[-200R^2x(x^2 + y^2)^{-3/2}, -200R^2y(x^2 + y^2)^{-3/2}]$, on a 200-lb mass while it moves from $(0,2R)$ to $(2R,0)$ in its orbit.

15. If a plane field of force contains a particle at every point in the plane moving in the plane so that the particle at any point (x,y) has the vector velocity $[x^2 + 2y^2, 4xy]$, find the equation of the velocity equipotential curves and of the stream lines.

16. If $j^2 = -1$ and if $f(x + jy) = \alpha(x,y) + j\beta(x,y)$, where α and

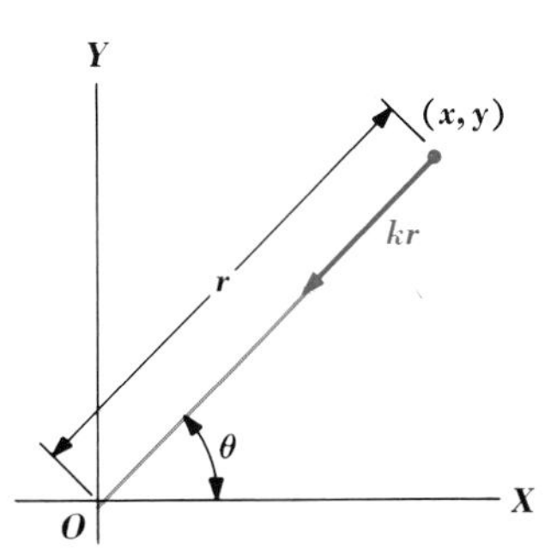

Figure 9

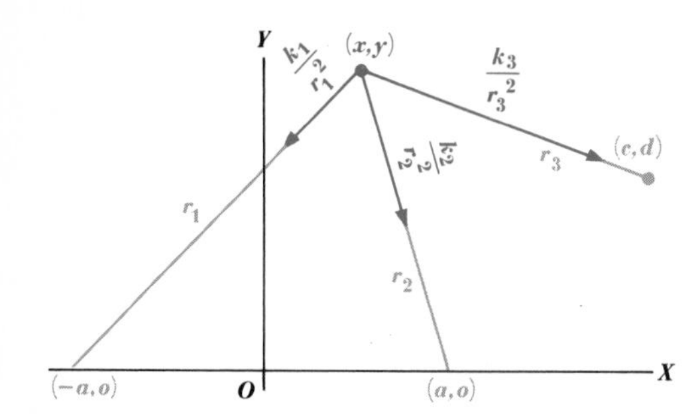

Figure 10

β are real functions of x and y, then the famous Cauchy-Riemann equations, $\partial\alpha/\partial x = \partial\beta/\partial y$ and $\partial\alpha/\partial y = -\partial\beta/\partial x$, hold. Prove in this case that the fields of force $[\beta,\alpha]$ and $[\alpha,-\beta]$ have potential functions (see exercise 11). Illustrate by taking $f(x + jy) = (x + jy)^2$.

17. Figure 10 represents a particle at $P(x,y)$ acted on by three forces as indicated. Show that the potential $\varphi(x,y)$ for the field made by these three forces is $k_1/r_1 + k_2/r_2 + k_3/r_3$. Indicate the field of force by using partial derivatives.

18. If a force $f(x)$ acts on a particle of mass m causing it to move along the X axis, then $f(x) = m\, d^2x/dt^2$. Show that

$$\tfrac{1}{2}mv^2 - \int_{x_0}^{x} f(x)\, dx = c \qquad \text{where } c \text{ is a constant}$$

19. If a plane field of force $[X,Y]$ for which (10) holds acts on a particle of mass m in its plane, then $X = m\, d^2x/dt^2$ and $Y = m\, d^2y/dt^2$. Prove that, in this case

$$\tfrac{1}{2}mv^2 - \varphi(x,y) = c \qquad \text{where } c \text{ is a constant}$$

35 Review problems

The following argument brings out many instructive features and leads to another application of the linear equation.

Figure 1 represents a chain wrapped partially around a cylinder with horizontal axis; PQ represents a small element of the chain, which we assume to be on the point of slipping in the direction from P to Q. T at P and $T + \Delta T$ at Q represent tensions in the chain. The normal reaction of cylinder on PQ is represented by the components $\Delta N(1 + \epsilon_1)$ along OP and $\epsilon_2\, \Delta N$ along the tangent to the cylinder at P. Similarly, friction is represented by components $\mu(\Delta N + \epsilon_3)$ and $\mu\epsilon_4\, \Delta N$. The ϵ's $\to 0$ as $\Delta\theta \to 0$. The nature of these infinitesimal components constitutes an assumption, but it appears justifiable as soon as we consider PQ to be very small. The remaining notation on the figure is self-explanatory. Resolving forces along the line OP and along the tangent at P, we get

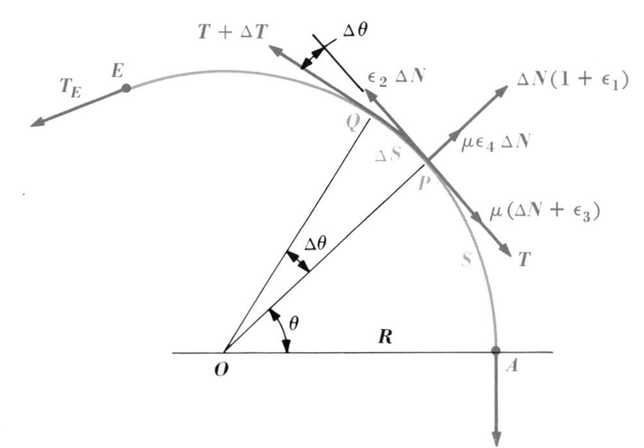

Figure 1

$$\Delta N(1+\epsilon_1) + \mu\epsilon_4\,\Delta N - (T+\Delta T)\sin\Delta\theta = 0 \tag{1}$$
$$(T+\Delta T)\cos\Delta\theta - T + \epsilon_2\,\Delta N - \mu(\Delta N + \epsilon_3) = 0 \tag{2}$$

Note that second-order infinitesimals vanish as $\Delta\theta \to 0$; thus, $\lim_{\Delta\theta\to 0} \epsilon_1(\Delta N/\Delta\theta) = 0 \cdot dN/d\theta = 0$. Also

$$T\,\frac{\cos\Delta\theta - 1}{\Delta\theta} = -T\left(\frac{\sin\frac{1}{2}\Delta\theta}{\Delta\theta}\right) 2\sin\tfrac{1}{2}\Delta\theta \to 0$$

as $\Delta\theta \to 0$. Hence, dividing (1) and (2) through by $\Delta\theta$ and taking limits as $\Delta\theta \to 0$, we get

$$\frac{dN}{d\theta} - T = 0 \qquad \frac{dT}{d\theta} - \mu\,\frac{dN}{d\theta} = 0 \tag{3}$$

From (3), we get

$$\frac{dT}{d\theta} = \mu T \qquad \text{or} \qquad T = ce^{\mu\theta} \tag{4}$$

Inspection of Fig. 1 shows that $T = T_0$ when $\theta = 0$. Hence, from (4) $T_0 = ce^{\mu 0} = c$, and (4) becomes

$$T = T_0 e^{\mu\theta} \tag{5}$$

If $T_0 = 100$ lb, $\mu = 0.3$, and the chain is wrapped halfway round the cylinder so that $\theta = \pi$, from (5) we get

$$T_E = 100e^{0.3\pi}\text{ lb} = 256.6\text{ lb}$$

Exercises

1. The slope of a certain curve at each of its points (x,y) is given by

$$\frac{3ax^2 + 2bxy - ey^2}{3fy^2 + 2exy - bx^2}$$

and it passes through $(m,0)$; find its equation.

2. Find in polar coordinates the equation of a curve passing through (5,2) if $\tan \psi$ (§12) at each point (ρ,θ) on the curve is given by $\tan \psi = \rho\theta/(\rho + \theta^3)$.

3. Find the equation of a curve for which the ordinate y of the point of contact of each tangent and the length of the perpendicular to the tangent from the origin are equal. First show that the distance from (0,0) to the tangent is $[(y - x\,dy/dx)]/\sqrt{1 + (dy/dx)^2}$.

4. Find the equation of the isogonal trajectories (§12) of the system of circles $\rho = c(\cos\theta + 6\sin\theta)$. *Hint:* $c(\cos\theta + 6\sin\theta) = c_1 \sin(\theta + \delta)$, where $\tan\delta = \frac{1}{6}$.

5. A tank initially contains salt in the pores of inert material and 10 gal of fresh water. The salt dissolves at a rate per minute of two times the difference between 3 lb/gal and the concentration of the brine. Two gallons of fresh water per minute enter the tank. How much salt will dissolve in the first 10 min?

6. The motion of a weight hung from a ceiling by a spring is vertical and $a = 16(4 - s)$, where a ft/sec² is acceleration and s is the number of feet the spring is stretched. Also, $s = 0$ and $v = 0$ when $t = 0$. Use $a = v\,dv/ds$ to find v in terms of s and then $v = ds/dt$ to find s in terms of t. Show that $0 \leqq s \leqq 4$ and that the time for a complete motion down and back is $\frac{1}{2}\pi$ sec.

7. A circuit has a resistance of 10 ohms, a capacitor with a capacity of 2×10^{-4} farads, and a battery with an emf of 100 volts. Find an expression for the quantity q of electricity on the capacitor at time t if initially $q = 0$ when $t = 0$. What are the approximate values of q and i after a few seconds?

8. Find the theoretical height of an atmosphere on a planet for which $p = k\rho^2$, where p lb/ft² is pressure, ρ lb/ft³ is density, and k is a constant. Assume that when height h is zero, $p = 2{,}210$ lb/ft² and that at $h = 10{,}000$ ft, $p = 1{,}452$ lb/ft².

(9–11) For each of the following potential functions find the field of force $[X,Y]$, the equations of the lines of force and of the equipotential curves, and the work done on a particle while moving in the field of force from (0,1) to (2,−1):

9. $\varphi(x,y) = x^3 - 3x^2y$

10. $\varphi = \dfrac{k}{(x^2 + y^2)^{5/2}}$

11. $\varphi = \sin \frac{1}{4}\pi x \cos \frac{1}{4}\pi y$

12.* Prove that if $\rho^2\, d\theta/dt = bc$ and $d\rho/dt = (c/\rho)\sqrt{b\rho - b^2}$ for a moving particle and if $\rho = b$ when $\theta = 0$, then the particle moves on the parabola $\rho = 2b/(1 + \cos\theta)$.

13. Find ρ in terms of θ if $\rho^2\, d\theta/dt = 15$, $d\rho/dt = \frac{4}{9}(\sin\theta)\rho^2\, d\theta/dt$, and $\rho = 1$ when $\theta = 0$. Show that the time for the particle to go around its curve (an ellipse) once is 2π time units. Assume that $\int_0^{2\pi} d\theta/(5 + 4\cos\theta)^2 = \frac{10}{27}\pi$.

14. If the acceleration a ft/sec² of a body moving in a straight line is equal numerically to $-s$ ft, where s is the signed distance from a point on the line, and $s = 0$ and velocity $v = 1$ ft/sec when $t = 0$, find v in terms of s and s in terms of t. *Hint:* $a = v\, dv/ds$, and $v = ds/dt$.

15. Find a nondifferential equation connecting i and q if $4L^2\, di/dt + 4LRi + R^2q = 0$ and $i = dq/dt$. *Hint:* $di/dt = (di/dq)(dq/dt) = i\, di/dq$.

16. In exercise 5, assume, in addition to the given conditions, that 1 gal/min of the brine runs out, and then find how much salt dissolves in the first 10 min and how much of the dissolved salt is still in the tank after 10 min.

17. Assume that the chain of Fig. 1 weighs ρ lb/ft. Hence, in Fig. 1 force $\rho R\,\Delta\theta$ acts vertically on element PQ. Assume that OA is horizontal, and, by resolving forces along and perpendicular to the tangent at P and taking limits as $\Delta\theta \to 0$, derive

$$\frac{dT}{d\theta} - \mu T = \rho R(\cos\theta + \mu\sin\theta)$$

Find the general solution of this equation.

18. (*a*) Using the solution of exercise 17 show that $l(1 + \mu^2) = 2\mu R(1 + e^{\pi\mu})$ if $T = 0$ when $\theta = 0$ and $T = \rho l$ when $\theta = \pi$. (*b*) Show that if the chain extends only from $\theta = \frac{1}{2}\pi$ to $\theta = \pi$ and is about to slip, then $e^{\pi\mu/2}(1 - \mu^2) = 2\mu$. (*c*) If the chain extends only from $\theta = \frac{1}{4}\pi$ to $\theta = \pi$ and is about to slip, find the relation that μ must satisfy.

* The motion of exercise 12 is like that of a comet, and the motion of exercise 13 is like that of a planet.

(**19–20**) Find the equations of the lines of force and of the equipotential curves for the fields defined by:

19. $\left[\dfrac{x}{\sqrt{x^2 + y^2}} - 10x, \dfrac{y}{\sqrt{x^2 + y^2}} - 4\right]$

20. $\left[\dfrac{y}{\sqrt{x^2 + y^2}}, \dfrac{-x}{\sqrt{x^2 + y^2}} + 1\right]$

21. For the field of force $[X, Y]$, where $(x + jy)^3 = Y + jX$ and $j^2 = -1$, find the potential function and the equation of the lines of force.

22. In exercise 7, replace the emf by $100 \sin 377t$, and solve the resulting problem. Assume that $\sqrt{500^2 + 377^2} = 626$.

5

First-order Equations of Degree Higher than the First

36 Foreword

This chapter relates mainly to three types of differential equations of the first order and degree higher than the first. Since certain special solutions, called *singular solutions*, involve envelopes of families of curves, a little theory of envelopes will be reviewed and used.

37 Equations solvable for dy/dx

This section deals with a type of first-order differential equation which can be resolved into factors linear in dy/dx. The notation

$$p = \frac{dy}{dx} \tag{1}$$

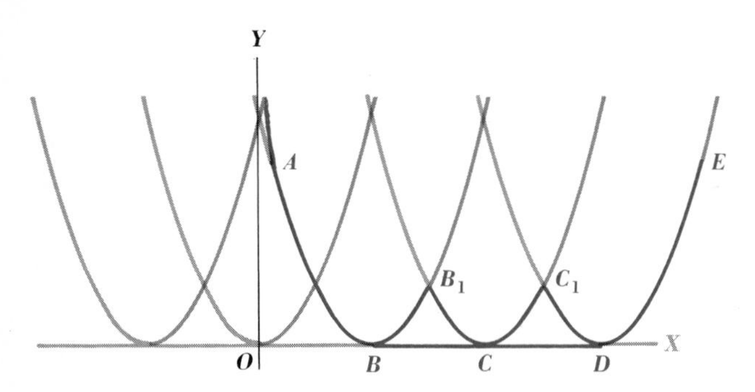

Figure 1

will be used throughout this chapter. An equation which can be reduced to the form

$$(p - A_1)(p - A_2) \cdot \cdot \cdot (p - A_n) = 0 \tag{2}$$

where the $A_1, A_2, \ldots, A_n$ are functions of x and y, may be solved by equating each factor to zero and integrating the resulting equations. The solutions thus obtained are represented by

$$\varphi_1(x,y,c_1) = 0, \ \varphi_2(x,y,c_2) = 0, \ \ldots, \ \varphi_n(x,y,c_n) = 0 \tag{3}$$

If at each of the points on a curve C, at least one of the equations (3) holds and if the functions φ_i, $i = 1, 2, \ldots, n$, are real for all points on C, then C is an integral curve of (2). Instead of (3) we may use

$$\varphi_1(x,y,c)\varphi_2(x,y,c) \cdot \cdot \cdot \varphi_n(x,y,c) = 0 \tag{4}$$

since the entire set of curves represented by (4) is the same as that represented by (3).

Example 1 below will indicate that an equation having form (2) may have solutions other than those indicated in (3), and also solutions consisting of various combinations of integral curves.

Example 1. Find various types of solutions of $p^2 - 4y = 0$.

Solution. Factoring $p^2 - 4y$, equating each factor to zero, and solving the results, we get

$$p + 2\sqrt{y} = 0 \qquad p - 2\sqrt{y} = 0 \tag{a}$$
$$y^{1/2} + x = c_1 \qquad y^{1/2} - x = c_2 \tag{b}$$

Using form (4), we may also employ

$$(x + y^{1/2} - c)(x - y^{1/2} - c) = 0 \qquad \text{or} \qquad (x - c)^2 - y = 0 \tag{c}$$

Note that (*b*) represents halves of parabolas whereas (*c*) represents complete parabolas.

By trial we find that

$$y = 0 \tag{d}$$

also satisfies $p^2 - 4y = 0$. Hence, the curve $ABCDE$ in Fig. 1 is an integral curve of $p^2 - 4y = 0$, and the same is true of curve ABB_1CC_1DE if points B_1 and C_1 are omitted.

Exercises

(**1–8**) Solve the following differential equations:

1. $(p - 1)(p - x)(p - 3x^2) = 0$ **2.** $9p^2 - x^4 = 0$
3. $4p^2 = 25x$ **4.** $p^2 - 4y^2 = 0$
5. $2x^2p^2 + 5xyp + 2y^2 = 0$ **6.** $8p^3 - 27y = 0$
7. $(yp + x)^2 - x^2 - y^2 = 0$ **8.** $p(p^2 - 2xp - 3x^2) = 0$

9. If from the equation $x^2p^2 + 5xyp + 6y^2 = 0$ we form a new equation by replacing p by $-1/p$, what will be the relation between the systems of curves represented by the solutions of the two equations?

10. Solve $(xp + y)^2 = xy$.
11. Solve $(x + 2y)p^3 + 3(x + y)p^2 + (y + 2x)p = 0$.
12. Denote by P any point on a curve C and by T the point where the tangent to C at P meets the Y axis. Find the equation of C if $PT = k$.
13. Find the isogonal trajectories cutting at 45° the integral curves of $p^2 - 2xp - 3x^2 = 0$.

38 Envelopes

Since the equations of envelopes appear as particular solutions of differential equations, a brief review of the pertinent facts of envelopes is in order. *If $f(x,y,c) = 0$ represents a one-parameter family of curves and E is a curve which contacts tangentially (has a common tangent with) every curve of the family $f = 0$, and contacts tangentially one or more curves of $f = 0$ at each of its points, then E is an envelope of f.*

For example (see Fig. 1), the envelope of the circles

$$(x - c)^2 + y^2 = 1 \tag{1}$$

consists of the lines

$$y = \pm 1$$

For convenience denote a one-parameter family of curves by

$$f(x,y,c) = 0 \tag{2}$$

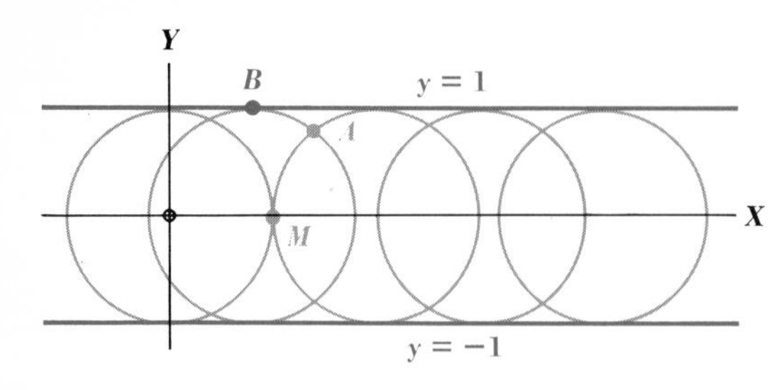

Figure 1

and use subscripts to denote partial derivatives; thus

$$\frac{\partial f(x,y,c)}{\partial x} = f_x(x,y,c) \qquad \frac{\partial^2 f(x,y,c)}{\partial c^2} = f_{cc}(x,y,c) \qquad \text{etc.} \tag{3}$$

Assume that f and its partial derivatives of the first and second orders are continuous in all regions involved. Then we know from calculus* that for fixed values of x and y such that $f_{cc}(x,y,c) \neq 0$, c will be a root of multiplicity 2 or more if and only if

$$f(x,y,c) = 0 \qquad f_c(x,y,c) = 0 \tag{4}$$

We shall now show that (4) is satisfied by any envelope of $f(x,y,c) = 0$. Assume that the envelope E can be represented in parametric form by

$$x = \varphi(c) \qquad y = \psi(c) \tag{5}$$

where φ, ψ, φ' and ψ' are continuous functions of x, y, and c, $(dx/dc)^2 + (dy/dc)^2 \neq 0$, and parameter c has the same value for a point P on the curve that it has for the curve of f contacting E at P. Since x and y from (5) satisfy (2), $f(\varphi,\psi,c) = 0$ identically and

$$f(\varphi,\psi,c) = 0 \qquad f_x\frac{dx}{dc} + f_y\frac{dy}{dc} + f_c = 0 \tag{6}$$

The slope of (2) is given by $dy/dx = -f_x/f_y$, and the slope of (5) by $dy/dx = (d\psi/dc)/(d\varphi/dc)$; these two slopes are equal at a point on (5), and therefore

$$-\frac{f_x}{f_y} = \frac{d\psi/dc}{d\varphi/dc} \qquad \text{or} \qquad f_x\frac{d\varphi}{dc} + f_y\frac{d\psi}{dc} = 0 \tag{7}$$

From (7) and (6), we easily deduce (4) and conclude that *any envelope of* (2) *must satisfy* (4).

The slopes in (7) would be undefined if, at points of (4), $d\varphi/dc$ and $d\psi/dc$ were both zero or if f_x and f_y were both zero. Hence, we assume

* In calculus we proved that the condition that an equation $f(z) = 0$ have a root r of multiplicity greater than 1 is

$$f(r) = 0 \qquad f'(r) = 0$$

that for points on (4)

$$f_x^2 + f_y^2 \neq 0 \qquad (d\varphi/dc)^2 + (d\psi/dc)^2 \neq 0 \tag{8}$$

A curve along which $f_x = 0$ and $f_y = 0$ is a locus of singular points, that is, cusps, nodes, isolated points, and others. The equation of this locus satisfies (4) but is generally not part of the envelope.

Again, a curve defined by (5) and its conditions which satisfies (4), (7), and (8) represents an envelope of $f(x,y,c) = 0$. For, (4) shows that it meets every curve of $f(x,y,c) = 0$, (7) shows that the tangent at each of its points coincides with the tangent to the curve of $f(x,y,c) = 0$ through the point, and (8) shows that it is not a locus of singular points.

The result of eliminating c from (4) *is called the* c-discriminant *of* $f(x,y,c) = 0$. *To find the envelope of a system in the form* (2), *find the c-discriminant of the given equation. All parts of it for which* (7) *and* (8) *hold will belong to the envelope.*

Equation (4) applied for $f = y - cx + c^2 = 0$ gives

$$y - cx + c^2 = 0 \qquad -x + 2c = 0$$

Elimination of c from these gives the equation of the envelope of f, namely

$$y - (\tfrac{1}{2}x)x + (\tfrac{1}{2}x)^2 = 0 \qquad \text{or} \qquad 4y - x^2 = 0$$

Figure 2 represents the lines $y - cx + c^2 = 0$ as tangents to their envelope, the parabola $4y - x^2 = 0$.

Also, (4) applied to f from (1) gives the equation of the envelope of (1), namely

$$(x - c)^2 + y^2 = 1 \qquad 2(x - c)(-1) = 0 \qquad \text{or} \qquad y = \pm 1$$

There may be no envelope. For example, the system of parallel lines

$$5x + 6y - c = 0$$

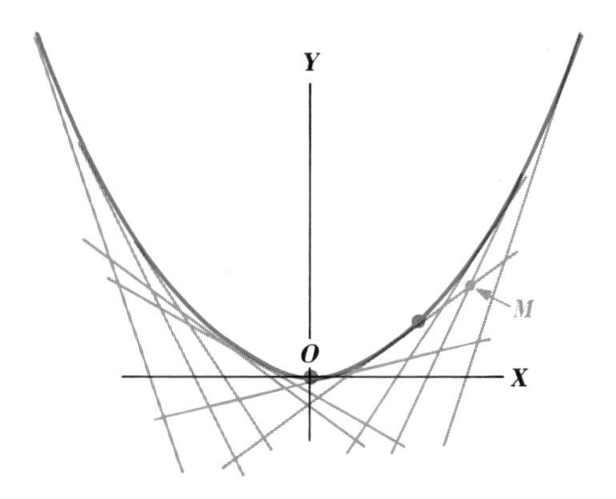

Figure 2

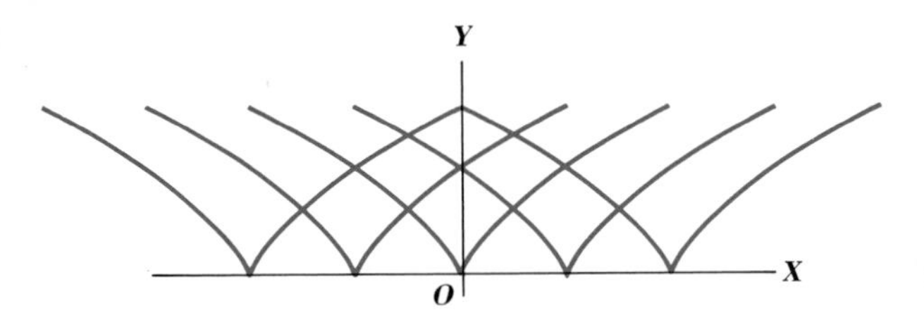

Figure 3

evidently has no envelope. For this system $\partial f/\partial c = -1$, and -1 cannot be zero.

Along a locus of singular points of (1),

$$f_x(x,y,c) = 0 \qquad f_y(x,y,c) = 0 \tag{9}$$

If (5) represents such a locus on $f(x,y,c) = 0$, then (6) is satisfied, and therefore, because of (9), (4) is also satisfied. *Accordingly the c-discriminant contains any locus of singular points that* (2) *may have.*

Example. Find the envelope of the cubical parabolas

$$y^3 - (x - c)^2 = 0 \tag{a}$$

Solution. Here the c-discriminant, from (4), is

$$f = y^3 - (x - c)^2 = 0 \qquad f_c = 2(x - c) = 0 \tag{b}$$

or, in form (5),

$$x = c \qquad y = 0 \tag{c}$$

For the points on locus (c), we find

$$(f_x)_{x=c} = 2(x - c)_{x=c} = 0 \qquad (f_y)_{y=0} = (3y^2)_{y=0} = 0 \tag{d}$$

Hence the c-discriminant (c) is a locus of singular points. Also from (a) and (d) we find that $y' = -f_x/f_y = 0/0$ along (c) and therefore has no value. Hence (c) is not a part of an envelope and the family (a) has no envelope. Figure 3 shows (c), the X axis, as a locus of cusps.*

* A cusp is a double point on a curve at which two tangents to the curve are coincident. Figure 4 shows a cusp at (0,0) on the curve $x^2 = y^3$. Here (0,0) is on both of the branches $x = y^{3/2}$ and $x = -y^{3/2}$, and the Y axis is tangent at (0,0) to both branches.

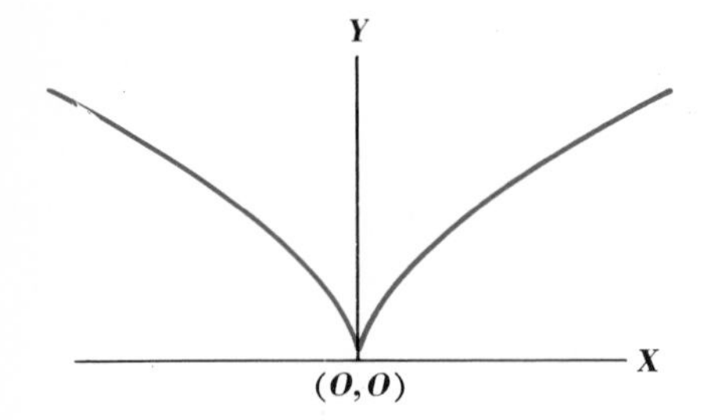

Figure 4

Exercises

1. Using (4) for $f(x,y,c) = y - c^2x + 2c = 0$ show that the envelope is given by $xy = -1$ and in parametric form by $x = 1/c$, $y = -c$. Sketch the equation of the envelope and a few of the lines of the family. Check that $f_x^2 + f_y^2 = [1 + c^4]_{c=x^{-1}} = 1 + x^{-4} \neq 0$ for the c-discriminant.

2. Find the equation of the envelope of the family $(x - c)^2 + (y - c)^2 = 1$ in rectangular form and also in parametric form. Sketch the envelope and a few of the circles of the family.

(3–8) Find the equation of the envelope of:

3. $x^2 + (y - 10c)^2 = 20c^2$. Check the result against Fig. 5.

4. $c^2 + 2cx + y^2 = 0$ **5.** $xc^2 + 2cy - x = 1$

6. $c^3 - 3cy = x$

7. $c^4y^2 + x^2 = 2c^2$ **8.** $x^2 + c^2y = 2c$

9. Show that the c-discriminant of $f(x,y,c) = y^2 - x - (x - c)^2 = 0$ is $x = c$, $y^2 - c = 0$. Show that (7) is satisfied for this locus, and conclude that it is an envelope of $f = 0$.

10. Show that the c-discriminant of the family $y^2 = (x - c)^3$ consists of points $x = c$, $y = 0$ and that $cy^2 = (x - c)^3$ has cusps at $(c,0)$. In the sense that one-sided tangents at the cusps coincide with the X axis, the X axis is an envelope of the family (see Fig. 6).

★11. Show that the family $c^2y^2 = (x - c)^3$ has a set of singular points along $x = c$, $y = 0$ and an envelope $4y^2 = 27x$.

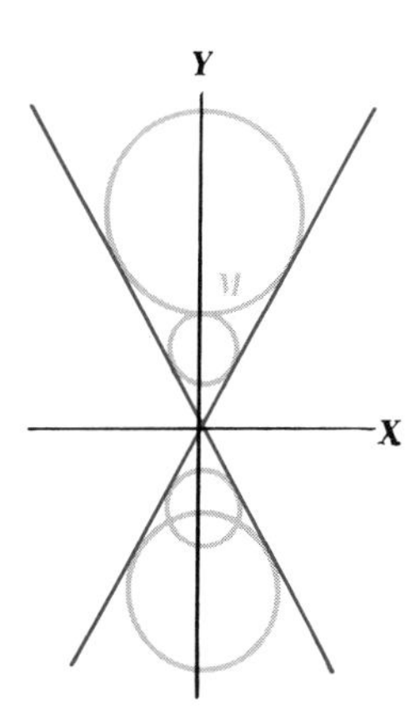

Figure 5

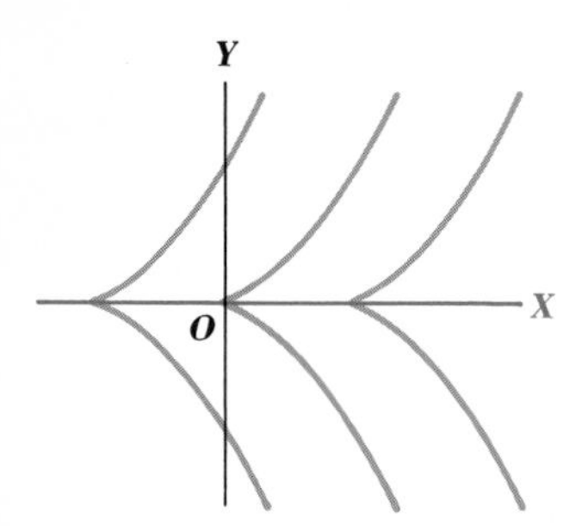

Figure 6

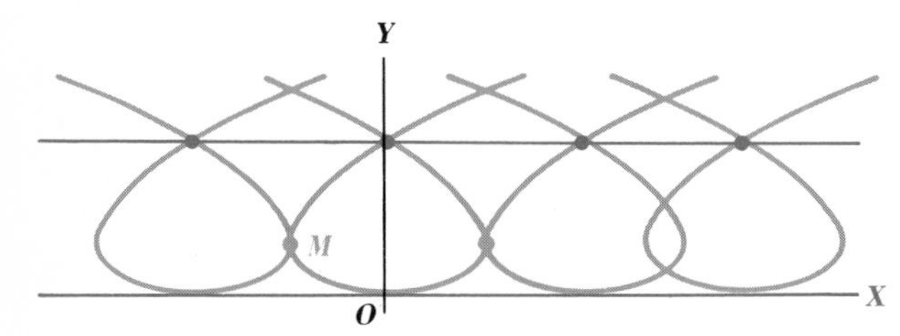

Figure 7

12. Show that $y = 0$, $x = c$ is an envelope of the family of curves $(x - c)^2 = y(y - 3)^2$. Also find a locus of singular points of the family (see Fig. 7).

39 Envelope from differential equation

For a differential equation $f(x,y,p) = 0$, the *p*-discriminant is defined as the triples (x,y,p) satisfying the two equations

$$f(x,y,p) = 0 \qquad \frac{\partial f(x,y,p)}{\partial p} = 0 \tag{1}$$

THEOREM. *The p-discriminant of $f(x,y,p) = 0$ is satisfied at every point on the envelope of the integral curves of $f(x,y,p) = 0$.*

Let $F(x,y,c) = 0$ be the general solution of $f(x,y,p) = 0$, and denote by $F_1(c)$ the set of curves of the family $F(x,y,c) = 0$ which pass through (x,y). Let (x_0,y_0) be any point on the envelope of the curves $F(x,y,c) = 0$. Then, by §38, the c-discriminant of $F(x,y,c) = 0$ holds, and therefore at least two values of c coincide, which implies that at least two curves of $F_1(c)$ at (x_0,y_0) coincide. Hence the slopes p at (x_0,y_0) of these two curves are equal; accordingly (1) is satisfied since it is the condition that $F(x,y,p) = 0$ have at (x,y) a root p of multiplicity greater than 1.

To derive sufficient conditions that locus (1) be an envelope we must express the fact that at any point P on it the slope $dy/dx = p$ where p is the slope of the integral curves of $f(x,y,p) = 0$ through P.

To do this differentiate $f(x,y,p) = 0$ with respect to x, and use (1) on the result to get

$$\frac{\partial f}{\partial x} + \frac{\partial f}{\partial y}\frac{dy}{dx} + \frac{\partial f}{\partial p}\frac{dp}{dx} = 0$$

$$\frac{\partial f}{\partial x} + \frac{\partial f}{\partial y}\,p = 0 \qquad \frac{\partial f}{\partial y} \neq 0 \tag{2}$$

If (2) holds for a curve defined by (1), that curve is an envelope of the integral curves of (1) and is a singular solution of (1).

For example, to consider $y - px - p^n = 0$, write from (1) and (2),

$$y - xp - p^n = 0 \qquad -x - np^{n-1} = 0 \qquad -p + 1(p) = 0 \tag{3}$$

Hence the curve $x = -np^{n-1}$, $y = -(n-1)p^n$ represents an envelope of $y - px - p^n = 0$.

The name tac-locus is given to a curve consisting of points at which two curves of a family have the same tangent. In Fig. 1, §38, $y = 0$ and in Fig. 5, §38, $x = 0$ are tac-loci of points such as M. Since at a point on the tac-locus of the integral curves of $f(x,y,p) = 0$ at least two values of p are equal, the coordinates of the point must satisfy the p-discriminant of $f(x,y,p) = 0$. Therefore, *the p-discriminant of $f(x,y,p) = 0$ must contain the equation of any tac-locus its integral curves may have.* A like argument shows that the p-discriminant of $f(x,y,p) = 0$ contains the equation of any locus of cusps belonging to the integral curves of $f(x,y,p) = 0$. The equation of a tac-locus or a locus of cusps, however, does not satisfy the corresponding differential equation generally. *The equation of the envelope of the integral curves of $f(x,y,p) = 0$ satisfy $f(x,y,p) = 0$.*

If the p-discriminant of $f(x,y,p) = 0$ is written in the form of the product of several factors equated to zero, then we shall call any equation formed by setting one of these factors equal to zero, a part *of the p-discriminant*, and, in like manner, we shall speak of parts of the c-discriminant of $F(x,y,c) = 0$.

SUMMARY. *Any part of the p-discriminant of $f(x,y,p) = 0$ which satisfies (2) represents a part or the whole of the envelope of the integral curves of $f(x,y,p) = 0$ and is called a* singular solution *of $f(x,y,p) = 0$. Any tac-locus or locus of cusps of the integral curves of $f(x,y,p) = 0$ is represented by a part of the p-discriminant.* Also, *if $F(x,y,c) = 0$ is the general solution of $f(x,y,p) = 0$, then any part of the c-discriminant of $F(x,y,c) = 0$ which satisfies (7), §38, is a part or the whole of the envelope of $F(x,y,c) = 0$; any locus of singular points of $F(x,y,c) = 0$ is represented by a part of the c-discriminant of $F(x,y,c) = 0$.*

If (1) involves infinite slopes, it is convenient to use

$$p = \frac{dy}{dx} = \frac{1}{dx/dy} = \frac{1}{q} \qquad \text{or} \qquad p = \frac{1}{q} \tag{4}$$

To get the q-discriminant from $f(x,y,p) = 0$, denote by $f_1(x,y,q) = 0$ the result of clearing $f(x,y,1/q) = 0$ of fractions and use

$$f_1(x,y,q) = 0 \qquad \frac{\partial f_1(x,y,q)}{\partial q} = 0 \tag{5}$$

Thinking of y as the independent variable, we could use the same arguments as we used earlier and prove that (5) is satisfied by envelopes, tac-loci, and cusp loci of the integral curves of $f(x,y,p) = 0$. Example 2 below will illustrate the use of (5).

Example 1. Find the singular solution of

$$4y = 8px + 8x^2 + p^2 \tag{a}$$

Solution. In this case, from (1) we get

$$4y = 8px + 8x^2 + p^2 \qquad 8x + 2p = 0 \tag{b}$$

Replace p in the first equation by $-4x$ from the second, simplify, and get

$$y + 2x^2 = 0 \tag{c}$$

Testing $y = -2x^2$ and $p = dy/dx = -4x$ in (a) we get

$$4(-2x^2) = 8(-4x)x + 8x^2 + (-4x)^2 \qquad \text{or} \qquad -8x^2 = -8x^2$$

Since $y + 2x^2 = 0$ satisfies (a), it is a solution of (a), and, being also the p-discriminant of (a), it represents the envelope of the general solution of (a) and the singular solution of (a). The general solution of (a) is $y = cx - x^2 + \frac{1}{4}c^2$. Observe that (c) is not obtainable by assigning a constant value to c in this general solution.

Example 2. Find the singular solution of

$$(4 - x^2)p^2 = 4 \tag{a}$$

Solution. Here the p-discriminant is

$$(4 - x^2)p^2 = 4 \qquad (4 - x^2)2p = 0 \tag{b}$$

If we take $p = 0$ to satisfy the second equation of (b), we get the impossible result $0 = 4$ from the first. Hence we use (4) and (5) and obtain

$$(4 - x^2) = 4q^2 \qquad 8q = 0 \tag{c}$$

From (*c*) we get $x = \pm 2$ as the q-discriminant. Evidently $x = 2$, $q = dx/dy = 0$ and also $x = -2$, $q = 0$ satisfy $(4 - x^2) = 4q^2$; and these are singular solutions, or envelopes, of the integral curves of $4 - x^2 = 4q^2$ or of $(4 - x^2)p^2 = 4$.

Exercises

1. Show that the p-discriminant of $y = px + p^2$ is $4y = -x^2$, and then show that $4y = -x^2$ satisfies $y = x\, dy/dx + (dy/dx)^2$. Conclude that $4y = -x^2$ is the singular solution of $y = px + p^2$, or the envelope of its integral curves.

(**2–11**) Find the singular solution of each of the following:

2. $y = 2px + 3p^2$ **3.** $y^2(1 + p^2) = 4$

4. $x^2(1 + p^{-2}) = 4$. Use (5). **5.** $y = -x^3p^2 + 2xp$

6. $y = 2px + p^2$ **7.** $y = 2px - yp^2$

8. $2y = p^2 + 4px + 2x^2$ **9.** $y = p^2$

10. $4yx^6 = px^7 + 4p^2$ **11.** $x^2p^2 = 1$

12. Differentiate partially with respect to p the equation

$$(p^2 + 1)(2y - x)^2 = (x + py)^2 \tag{a}$$

to obtain

$$2p(2y - x)^2 = 2(x + py)y \tag{b}$$

Replace $(2y - x)^2$ in (*a*) by its value from (*b*), and simplify to obtain $y = px$; then eliminate p between $y = px$ and (*a*) to obtain the singular solution $y(3y - 4x) = 0$. Prove that this singular solution satisfies (*a*). Also, replace $x + py$ in (*a*) by its value from (*b*), thus showing that $2y - x = 0$ is part of the p-discriminant. Prove that $2y - x = 0$ does not satisfy (*a*). It defines a tac-locus. The solution of (*a*) is

$$(x - 2c)^2 + (y - c)^2 = c^2 \tag{c}$$

Show that the c-discriminant, representing the envelope of (*c*), is $y(3y - 4x) = 0$. Plot the c-discriminant, the tac-locus, and a few of the circles (*c*).

★13. Show that the integral curves of $(x - a)p^2 - (y - b) = 0$ have $y = b$ and also $x = a$ as envelopes. Also show, after solving $(x - a)^2p^2 - (y - b)^2 = 0$, that the corresponding integral curves have no envelope.

40 Equations solvable for y

When a first-order differential equation is solvable for y, it may be written in the form

$$y = f(x,p) \tag{1}$$

Taking the total derivative of this equation with respect to x, we get

$$\frac{dy}{dx} = p = \frac{\partial f}{\partial x} + \frac{\partial f}{\partial p}\frac{dp}{dx} \tag{2}$$

This equation, since y does not appear in it, may be solved as an equation in x and p to get

$$\psi(x,p,c) = 0 \tag{3}$$

Equations (1) and (3) may be thought of as the parametric equations (p being the parameter) of a system of curves and therefore as the general solution of (1).

The result of eliminating p between equations (1) and (3) gives the general solution as an equation in x, y, and an arbitrary constant. If one suspects that the eliminant contains, as it may, extraneous factors that do not represent solutions, one should check by substitution in the differential equation.

Any singular solution may be found by using the p-discriminant as indicated in §39.

Example 1. Solve

$$y = \tfrac{9}{2}xp^{-1} + \tfrac{1}{2}px \tag{a}$$

Solution. Differentiating (a) with respect to x, obtain

$$p = (-\tfrac{9}{2}xp^{-2} + \tfrac{1}{2}x)\frac{dp}{dx} + \tfrac{9}{2}p^{-1} + \tfrac{1}{2}p \tag{b}$$

or, rearranged,

$$(-\tfrac{9}{2}p^{-2} + \tfrac{1}{2})x\frac{dp}{dx} - p(-\tfrac{9}{2}p^{-2} + \tfrac{1}{2}) = 0 \tag{c}$$

Equation (c) will be satisfied if either of the equations

$$x\frac{dp}{dx} - p = 0 \qquad -\tfrac{9}{2}p^{-2} + \tfrac{1}{2} = 0 \tag{d}$$

holds true. From the first of (d), obtain

$$x = cp \tag{e}$$

Parametric equations of the solution, obtained by solving (a) and (e) for x and y in terms of p, are

$$x = cp \qquad y = \tfrac{9}{2}c + \tfrac{1}{2}cp^2 \tag{f}$$

To get the solution in terms of x and y, eliminate p from (f). Replacing p in the second equation of (f) by its value from the first and simplifying, obtain

$$2cy = 9c^2 + x^2 \tag{g}$$

The singular solution is obtained by eliminating p between (a) and the second equation of (d) and simplifying. This gives

$$y = \pm 3x \tag{h}$$

Check by trial that $y = 3x$ and $y = -3x$ both satisfy (a). Also observe that for no value of c does (g) give (h).

Example 2. Solve the equation

$$y = px - \tfrac{4}{27}p^3 \tag{a}$$

Solution. Differentiation of (a) with respect to x gives

$$p = p + (x - \tfrac{4}{9}p^2)\frac{dp}{dx} \qquad \text{or} \qquad (x - \tfrac{4}{9}p^2)\frac{dp}{dx} = 0 \tag{b}$$

Equating dp/dx to zero and solving it, we get

$$p = c \tag{c}$$

Substitution of c for p in (a) gives the general solution

$$y = cx - \tfrac{4}{27}c^3 \tag{d}$$

To get the singular solution, eliminate p from

$$y = px - \tfrac{4}{27}p^3 \qquad \text{and} \qquad x - \tfrac{4}{9}p^2 = 0 \tag{e}$$

to obtain

$$y^2 = x^3 \tag{f}$$

By testing we find that this satisfies (a) and is therefore its singular solution. Also (f) is the envelope of (a) by §39, and the lines (d) are the tangents of the semicubical parabola (f).

Exercises

Find the general solution and the singular solution of:

1. $2y = p^2 + 4px + 2x^2$ **2.** $2yp = 3x + xp^2$
3. $y = 5px + 5x^2 + p^2$ **4.** $p^2x^4 = y + px$

For each equation find a solution in parametric form, and use the p-discriminant to show that there is no singular solution:

5. $y = -px + p^2$ **6.** $y = 2xp - 3p^2$

(7–10) Find the complete solution of:

7. $y = px + p^3$ **8.** $y = px + \frac{1}{2}x^2 + (p + x)^2$

9. $y = px \ln x + p^2x^2$
10. $y = (px + x^2) \ln x + (px + x^2)^2 - \frac{1}{2}x^2$

11. Find in parametric form a solution of $y = (p + p^{-1})x + p^2$.

41 Equations solvable for x

An equation of the first order when written in the form

$$x = \varphi(y,p) \tag{1}$$

can be solved by differentiating with respect to y, replacing dx/dy by $1/p$, and proceeding as in §40. Also, it may be solved by differentiating (1) with respect to y, replacing dx/dy by q and p by $1/q$, and then proceeding as in §40.

A first-order differential equation which is solvable for x may be written in the form

$$x = f(y,p) \tag{2}$$

Taking the total derivative of this equation with respect to y, we get

$$\frac{dx}{dy} = \frac{1}{p} = \frac{\partial f}{\partial y} + \frac{\partial f}{\partial p}\frac{dp}{dy} \tag{3}$$

This equation may be solved as an equation in y and p to obtain

$$\psi(y,p,c) = 0 \tag{4}$$

Equations (2) and (4) may be considered as the general solution in parametric form, or we may eliminate p between (2) and (4) to obtain the solution as a relation between x, y, and a constant of integration.

Example. Solve

$$y = 2px + y^2p^3 \qquad (a)$$

Solution. Solving for x, we get

$$x = \frac{y}{2p} - \frac{y^2p^2}{2} \qquad (b)$$

The derivative of (b) with respect to y is

$$\frac{dx}{dy} = \frac{1}{p} = \frac{1}{2p} - yp^2 - \left(\frac{y}{2p^2} + y^2p\right)\frac{dp}{dy} \qquad (c)$$

or, simplified,

$$\frac{1}{2p} + yp^2 = -\frac{y}{p}\left(\frac{1}{2p} + yp^2\right)\frac{dp}{dy} \qquad (d)$$

Dividing out $(1/2p) + yp^2$ and integrating, we obtain

$$py = c \qquad \text{or} \qquad y = \frac{c}{p} \qquad (e)$$

Substituting y from (e) in (b), we get

$$x = \frac{c}{2p^2} - \frac{c^2}{2} \qquad (f)$$

Equations (e) and (f) give, in parametric form, the general solution required. Eliminating p between (e) and (f), we find the solution in rectangular form to be

$$x = \frac{y^2}{2c} - \frac{c^2}{2} \qquad \text{or} \qquad y^2 = 2cx + c^3$$

Equation (d) is satisfied if $(1/2p) + yp^2 = 0$. Eliminating p between the latter equation and (a), we obtain the singular solution,

$$27y^4 = -32x^3$$

Exercises

1. Use the method of this section to find the general solution of $px = 1 + 4p^3e^{2y}$. Also, find the singular solution.

Solve the following equations completely:

2. $x = y + \ln p$

3. $4px - 2y = p^3y^2$

4. $xp^2 - 2yp + x + 2y = 0$

5. $2px = 2 \tan y + p^3 \cos^2 y$

Find the solution of each of the following equations in parametric form:

6. $x = y + p^2$ **7.** $p^2x = 2yp - 3$

Find the general solution of:

8. $3px = y + 3p\varphi(py^2)$ **9.** $px = 1 + p\varphi(pe^y)$

10. Find a singular solution of the equation in exercise 9, assuming that $\varphi(u) = \frac{1}{3}u^3$.

42 Clairaut's equation

The equation

$$y = px + f(p) \tag{1}$$

known as Clairaut's equation, is named after Alexis Claude Clairaut (1713–1765). He was the first man to differentiate a differential equation, as we have done in §§40 and 41, in order to solve it.

Let us apply the method of §40 to equation (1). Differentiation with respect to x gives

$$\frac{dy}{dx} = p = p + \left(x + \frac{df}{dp}\right)\frac{dp}{dx} \quad \text{or} \quad \left(x + \frac{df}{dp}\right)\frac{dp}{dx} = 0 \tag{2}$$

Then

$$\frac{dp}{dx} = 0 \quad \text{and} \quad p = c$$

Substituting c for p in equation (1), we have

$$y = cx + f(c) \tag{3}$$

as the general solution. Thus, it appears that to solve Clairaut's equation, it is necessary only to replace p by c.

The particular solution arising from equating $x + \dfrac{df}{dp}$ to zero and eliminating p between the result and (1) gives a singular solution. In general, the systems of lines represented by the general solution are the tangent lines of the graph of the singular solution.

Since the equation

$$f(y - px, p) = 0 \tag{4}$$

when solved for $y - px$ takes the form

$$y - px = \psi(p)$$

it is really Clairaut's equation, and its solution is

$$f(y - cx, c) = 0 \tag{5}$$

Note that Clairaut's differential equation (1) represents a family of straight lines $y = cx + f(c)$. The evolute of a curve $Y = \varphi(X)$ is the envelope of the family of normals of $Y = \varphi(X)$. Let (1) represent the family of normals. Then we have for (X,Y), any point on $Y = \varphi(X)$,

$$\text{(a) } Y = \varphi(X) \qquad \text{(b) } Y = pX + f(p)$$
$$\text{(c) } \frac{dY}{dX} = \varphi'(X) = -\frac{1}{p} \tag{6}$$

Hence to find the equation of the evolute of $Y = \varphi(X)$, solve (6)(a) and (6)(c) for X and Y in terms of p, substitute these values in (6)(b), and obtain $f(p)$ from the result. Use this value for $f(p)$ in (1), and then find the corresponding envelope by using (1), §39. The example below illustrates the procedure.

Example. Find the equation of the evolute of

$$Y^2 = 4X \tag{a}$$

Solution. In this case we get from (6),

$$Y^2 = 4X \qquad Y = pX + f(p) \qquad \frac{dY}{dX} = \frac{2}{Y} = -\frac{1}{p} \tag{b}$$

From (b) we obtain

$$Y = -2p \qquad X = \frac{Y^2}{4} = p^2$$
$$f(p) = Y - pX = -2p - p^3 \tag{c}$$

Using $f(p)$ from (c) in (1), we have

$$y = p(x - 2) - p^3 \tag{d}$$

Now using (1), §39, to get the envelope of (d), we write

$$y = p(x - 2) - p^3 \qquad x - 2 - 3p^2 = 0 \tag{e}$$

Substituting $p = \pm\sqrt{(x - 2)/3}$ from the second equation of (e) in the first, we get

$$y = \pm\sqrt{\frac{x - 2}{3}}\,[x - 2 - \tfrac{1}{3}(x - 2)] = \pm\frac{2}{3\sqrt{3}}(x - 2)^{3/2} \tag{f}$$

or

$$y^2 = \tfrac{4}{27}(x - 2)^3 \tag{g}$$

Exercises

1. Using (4) and (5) write solutions of:

(*a*) $(y - px)^2 = \sin(y - px) + p^2$
(*b*) $e^{y-px} = (y - px)^2 - p^3$

Find the singular solution of:

2. $y - px = p^3$ **3.** $y - px = 1 + p^2$ **4.** $(y - px)^2 = 4p$

Find the equation of the evolute of:

5. $Y = X^2$ **6.** $2Y + 2 = X^2$

7. Solve $3p^2e^y - px + 1 = 0$; let $z = e^y$. *Hint:* $dz/dx = P = zp$.

8. Solve $y = y^2p^3 + 2px$; let $z = y^2$.

To find the equation of the family of tangents to $Y = \varphi(X)$, let $y = px + f(p)$ represent the family, and observe that

(*a*) $Y = \varphi(X)$ (*b*) $Y = pX + f(p)$
(*c*) $\dfrac{dY}{dX} = p = \varphi'(X)$

Solve (*a*) and (*c*) for X and Y in terms of p, substitute the results in (*b*), etc. Find the equation of the family of tangents to:

9. $Y = X^2$ **10.** $Y^2 = X$ **11.** $X^2 + Y^2 = 25$

12. Apply the method of §40 to solve $y = px \ln x + \varphi(px)$. *Hint:* $\partial\varphi(u(x,p))/\partial x = [d\varphi(u)/du]\,\partial u/\partial x = \varphi'(u)\,\partial u/\partial x, \partial\varphi(u)/\partial p = \varphi'(u)\,\partial u/\partial p$.

Solve exercise 12 and then solve:

13. $y = px + \frac{1}{2}x^2 + f(p + x)$ **14.** $y = px + 2x^2 + f(p + 4x)$

★15. $y = mpx + \varphi(x^{m-1}p^m)$

16. Find the singular solution of exercise 15 if $\varphi(u) = u$ and $m = \frac{1}{2}$.

43 Review problems

Let p represent dy/dx.

The solution of an equation $f(x,y,p) = 0$ which can be factored is obtained by equating each factor to zero and solving the resulting equations (see §37). The form (4), §37, is sometimes more useful than (3).

The solution of an equation which can be solved for y in terms of x and p is obtained by the method of §40. Use the method of §41 to solve an equation which can be solved for x in terms of y and p.

If $f(x,y,p) = 0$ represents a differential equation and $\varphi(x,y,c) = 0$ its general solution, the p-discriminant, (1), §39, of $f(x,y,p) = 0$ is obtained by eliminating p from

$$f(x,y,p) = 0 \qquad f_p(x,y,p) = 0 \tag{1}$$

and the c-discriminant, §38, by eliminating c from

$$\varphi(x,y,c) = 0 \qquad \varphi_c(x,y,c) = 0 \tag{2}$$

Both the p-discriminant and the c-discriminant contain any singular solution that $f(x,y,p) = 0$ may have. Only parts of the p-discriminant or the c-discriminant which satisfy $f(x,y,p) = 0$ constitute solutions of $f(x,y,p) = 0$ and represent envelopes of $\varphi(x,y,c) = 0$. Parts of the p-discriminant may represent tac-loci, §39, and cuspidal-loci, §§38 and 39. Parts of the c-discriminant, §38, may represent envelopes and loci of singular points. The q-discriminant, (5), §39, is useful in determining whether loci of the form $x = a$ are parts of the singular solution of $f(x,y,p) = 0$.

Clairaut's equation, considered in §42, is $y = px + f(p)$ and its solution is $y = cx + f(c)$. It generally represents a family of lines tangent to an integral curve called its singular solution. Also, §42 considers the problem of finding the evolutes of curves and the differential equation of its tangent lines.

Example. Solve

$$y = px \pm a\sqrt{p^2+1} \qquad a > 0 \tag{a}$$

Solution. The general solution of the Clairaut equation (a) is

$$y = cx + a\sqrt{c^2+1} \tag{b}$$

To get the singular solution, find the p-discriminant of (a), in this case

$$y - px \mp a\sqrt{p^2+1} = 0 \qquad -x \mp \frac{ap}{\sqrt{p^2+1}} = 0 \tag{c}$$

The solution of (c) for p is

$$p = \pm\frac{x}{\sqrt{a^2-x^2}} \tag{d}$$

Substituting x from (c) in (a) we get

$$y = \mp\frac{ap^2}{\sqrt{p^2+1}} \pm \frac{a(p^2+1)}{\sqrt{p^2+1}} = \pm\frac{a}{\sqrt{p^2+1}} \tag{e}$$

From (c) and (e) we obtain

$$x^2 + y^2 = a^2 \tag{f}$$

By trial we find that (f) is a solution of (a) and is therefore its singular solution.

Exercises

1. Solve $(p - 3y)(p - 4x) = 0$.
2. Solve $(p - y)[p^2 + (1 - 2x)p - 2x] = 0$.

(**3–8**) Find the solutions, general and singular, of the equations:

3. $y = px + p^2$ **4.** $x = \dfrac{y}{p} + p^2$
5. $y = px + 3x^2 + \frac{1}{2}(p + 6x)^2$ **6.** $y = 2px + y^2p^3$
7. $4y - 4px \ln x = p^2x^2$ **8.** $3px - 3 = p^3e^{2y}$

9. Find the general solution and the singular solution of $y^2(1 + p^2) = x + yp$. Figure 1 represents the solution. What curves in Fig. 1 are represented by the general solution and what curve by the singular solution?

10. Find the general solution of $f(x,y,p) = 4y - 9(y - 1)^2p^2 = 0$ and also the singular solution. What line in Fig. 2 is defined by the singular solution? The line $y = 1$ satisfies $f_p = 0$. Name this locus relative to the system of integral curves. The c-discriminant is satisfied by the line $x = c$, $y = 3$; name this locus.

11. Find in parametric form the equation of the curve to which the lines defined by $y = px + 2p^4$ are tangents. *Hint:* The envelope of the tangents is the required curve.

12. Find the equation of the evolute of $Y = \frac{1}{2}X^2$. Use the method of the example of §42.

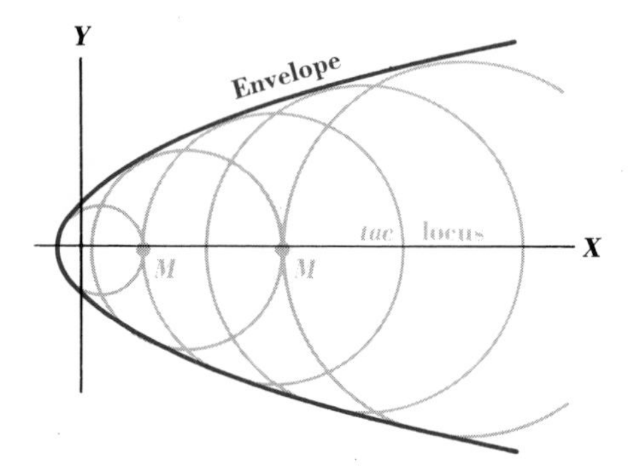

Figure 1

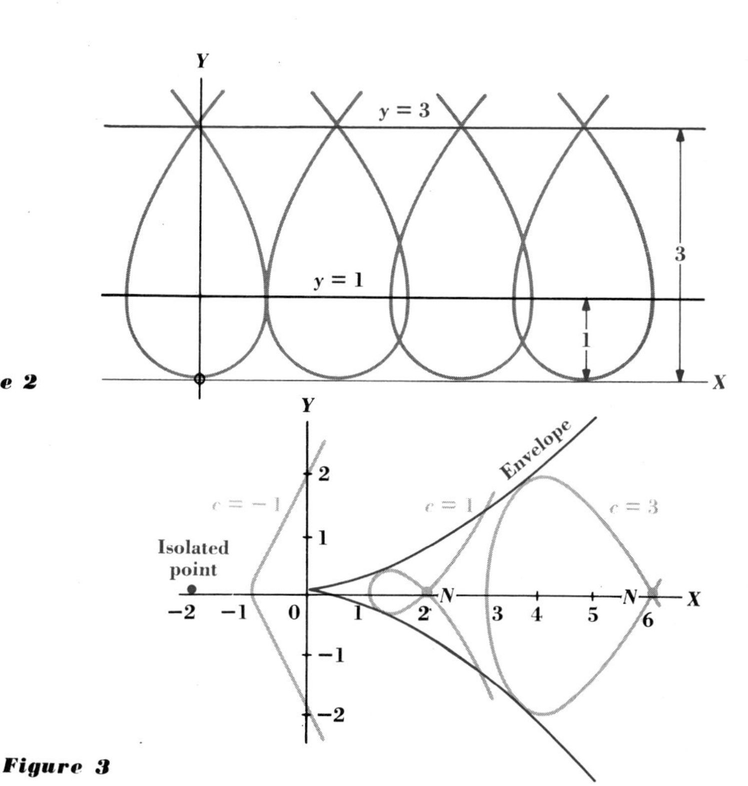

Figure 2

Figure 3

Find the general solution of:

13. $y = 3px + f(x^2p^3)$ **14.** $y = p \tan x \ln(\sin x) + \varphi(p \tan x)$

15. Find the differential equation of the tangents and also the equation of the envelope of the normals for: (*a*) $y^2 = x$; (*b*) $y^2 = 2x + 2$. Use the methods suggested for exercises 11 and 12.

16. Consider $x(x - 1)p^2 - (2x - 1)^2 = 0$. Show by means of the *q*-discriminant that $x = 0$ and $x = 1$ are parts of an envelope. Solve the equation to get

$$y = \pm 2\sqrt{x^2 - x} + c \qquad \text{or} \qquad (y - c)^2 = (2x - 1)^2 - 1$$

Sketch the integral curve with center $(\frac{1}{2}, c)$, and observe that $x = 0$ and $x = 1$ are parts of the envelope but that $(x = \frac{1}{2}, y = c)$ lies on no integral curve.

17. Find the differential equation of the tangent lines of $B^2x^2 + A^2y^2 = A^2B^2$.

18. Show that the *c*-discriminant of $y^2 - (x - c)(x - 2c)^2 = 0$ consists of $y = 0$, $x = 2c$, and $27y^2 - 2x^3 = 0$. Figure 3 represents the three integral curves for which $c = -1$, 1, and 3, respectively. Note that the dotted curve, defined by $27y^2 - 2x^3 = 0$, is the envelope and that $x = 2c$, $y = 0$ is a set of nodes and isolated points.

6

Linear Differential Equations with Constant Coefficients

44 Operators

The operators discussed below furnish a convenient notation and also timesaving methods of solving differential equations.

An operator is a symbol indicating an operation to be performed. We define the operator D to mean *take the derivative with respect to x of*, and generally $D, D^2, \ldots, D^k$ indicate that

$$Du \equiv \frac{du}{dx}, \; D^2u \equiv \frac{d^2u}{dx^2}, \; \ldots, \; D^ku \equiv \frac{d^ku}{dx^k}, \; D^0u \equiv 1u \equiv u \tag{1}$$

where u is a function of x. Also, we denote by $f(x)\,D^k$ an operator such that

$$f(x)\,D^ku = f(x)\,\frac{d^ku}{dx^k} \tag{2}$$

Thus

$$4xD^2x^3 = 4x\frac{d^2x^3}{dx^2} = 4x \cdot 6x = 24x^2$$

Two operators are said to be equal if they produce equal results when applied to any function of x, $u(x)$.

Any expression having the form of a polynomial in D, such as $2D^2 + xD + 5$, will represent an operator defined by (1) and the following definitions of sum and product:

$$\begin{aligned} (aD^k + bD^r)u &= aD^ku + bD^ru \\ aD^k \cdot bD^ru &= aD^k(bD^ru) \end{aligned} \tag{3}$$

where u is a function of x. Thus

$$(3D + 5)u = 3Du + 5u = 3\frac{du}{dx} + 5u$$

and

$$\begin{aligned} [(2D^2) + (3D + 5)]u &= 2D^2u + (3D + 5)u \\ &= 2\frac{d^2u}{dx^2} + 3\frac{du}{dx} + 5u \end{aligned}$$

Note in the product formula that the operator adjacent to u was applied first. Illustrations of (3) are

$$\begin{aligned} [(D^4 + 2D^3) + (3D + 5)]u &= (D^4 + 2D^3)u + (3D + 5)u \\ &= \frac{d^4u}{dx^4} + 2\frac{d^3u}{dx^3} + 3\frac{du}{dx} + 5u \end{aligned}$$

$$\begin{aligned} (D + 2)D(x^2 + 3x) = (D + 2)(2x + 3) &= 2 + (4x + 6) \\ &= 8 + 4x \end{aligned}$$

The elementary operators aD^k, where a is a constant, obey the fundamental laws of algebra; for we see by calculus and definitions (1) to (3) that for a, b, and c constants

$$\begin{aligned} (aD^m + bD^n)u &= (bD^n + aD^m)u \\ (aD^m) \cdot (bD^n)u &= (bD^n) \cdot (aD^m)u \\ [aD^m + (bD^n + cD^r)]u &= [(aD^m + bD^n) + cD^r]u \\ aD^m(bD^n \cdot cD^r)u &= (aD^m \cdot bD^n)(cD^ru) \\ aD^m(bD^n + cD^r)u &= aD^m \cdot bD^nu + aD^m \cdot cD^ru \end{aligned} \tag{4}$$

By applying the definitions (1), (3), and the laws (4), we can build up complicated operators just as we build up complicated expressions in algebra. In particular, we can build the polynomial operators

$$a_0D^n + a_1D^{n-1} + \cdots + a_n \tag{5}$$

where $a_0, a_1, \ldots, a_n$ are constants, and they will indicate that

$$(a_0D^n + a_1D^{n-1} + \cdots + a_n)u = a_0\frac{d^nu}{dx^n} + a_1\frac{d^{n-1}u}{dx^{n-1}} + \cdots + a_nu \quad (6)$$

Because the operators aD^k, a's constant, obey the fundamental laws (4) of algebra, we may obtain from given polynomial operators other equal operators by multiplication, factorization, and introduction or removal of parentheses, just as though the operators were algebraic expressions. Notice particularly that the order of terms in a sum or of factors in a product is immaterial.

Illustrations of operators and operations follow:

$$a^2D^2 + 2aD - 3 = (aD + 3)(aD - 1) = (aD - 1)(aD + 3)$$
$$(2D^2 + 5D + 6)x^3 = 12x + 15x^2 + 6x^3$$
$$(D - a)(D - b)y = [D^2 - (a + b)D + ab]y$$

The following examples illustrate a use of operators in solving differential equations:

Example 1. Solve

$$(D^2 - 5D + 6)y = 6 \quad (a)$$

Solution. The given equation may be written

$$(D - 2)[(D - 3)y] = 6 \quad (b)$$

In this, let

$$(D - 3)y = z \quad (c)$$

and obtain

$$(D - 2)z = 6 \qquad \text{or} \qquad \frac{dz}{dx} - 2z = 6 \quad (d)$$

The solution of (*d*), found by using the method of §25, is $z = -3 + ce^{2x}$. Replacing z in (*c*) by this value $-3 + ce^{2x}$, we get

$$(D - 3)y = -3 + ce^{2x} \qquad \text{or} \qquad \frac{dy}{dx} - 3y = -3 + ce^{2x} \quad (e)$$

The solution of (*e*) by the method of §25 is

$$y = 1 + c_1e^{2x} + c_2e^{3x} \quad (f)$$

where c_1 replaces $-c$.

Example 2. Solve

$$(D - a)(D - a)(D - a)y = 0 \quad (a)$$

Solution. In (a), let

$$(D - a)(D - a)y = z \tag{b}$$

and solve the result, $(D - a)z = 0$, to get

$$z = ce^{ax} \tag{c}$$

Replacing z in (b) by the value ce^{ax} from (c), we get

$$(D - a)(D - a)y = ce^{ax} \tag{d}$$

In (d), make the substitution

$$(D - a)y = w \tag{e}$$

to get $(D - a)w = ce^{ax}$. The solution of this is

$$w = (cx + c_2)e^{ax} \tag{f}$$

Replace w in (e) by its value from (f), and solve the result for y to obtain

$$y = (c_1x^2 + c_2x + c_3)e^{ax}$$

where $c_1 = \frac{1}{2}c$.

Example 3. Solve $(D - 2)(D + 3)y = 1 - 6x$, and determine the constants of integration so that $y = 7$ and $Dy = 0$ when $x = 0$.

Solution. Solving the given differential equation by the method used in the solutions of examples 1 and 2, we get

$$y = x + c_1e^{2x} + c_2e^{-3x} \tag{a}$$

From (a), we get by differentiation

$$Dy = 1 + 2c_1e^{2x} - 3c_2e^{-3x} \tag{b}$$

Replacing x by 0, y by 7, and Dy by 0 in (a) and (b), we get

$$\begin{aligned} 7 &= c_1 + c_2 \\ 0 &= 2c_1 - 3c_2 + 1 \end{aligned} \tag{c}$$

The solution of (c) for c_1 and c_2 is $c_1 = 4$, $c_2 = 3$. Substituting these values of c_1 and c_2 in (a), we get

$$y = x + 4e^{2x} + 3e^{-3x}$$

Exercises

Perform the operations indicated in the following exercises:

1. Dx^2
2. $(D + 1)(8x^2)$
3. $(D - 2)x^3$
4. D^2x^4
5. D^3e^{3x}
6. $(D - 3)e^{5x}$
7. $(D^2 - a^2) \sin ax$
8. $(D^2 + a^2) \cos ax$
9. $(D + 1)^2(xe^{-x})$

Solve the following differential equations by the methods used in the solutions of examples 1 and 2, and determine constants of integration when initial conditions are given:

10. $(D - 2)Dy = 0$
11. $(D - 2)(D + 2)y = 0$
12. $D(D - 2)y = 6x$
13. $(D + 3)D^2y = e^{-x}$. *Hint:* Let $z = D^2y$, solve for z, in the result replace z by DDy, then substitute w for Dy, solve for w, replace w in the result by Dy, and solve for y.
14. $D^2y = 12x^2$
15. $(D - 1)(D - 2)(D - 3)y = 0$
16. $D(D + 3)y = 0$; $y = 5$, $Dy = -9$ when $x = 0$.
17. $(D - 3)(D + 4)y = 12$
18. $D(D - 3)y = e^x$
19. $(D^2 - 4)y = e^{3x}$
20. $(D - 1)(D + 2)y = e^x$
21. $(D - 2)^2y = 2e^{2x}$
22. $(D^2 - 4)y = 0$; $y = 1$, $Dy = -10$ when $x = 0$.
23. $(D - 1)^2y = 2e^x$; $y = 3$, $Dy = 0$ when $x = 0$.

Perform the indicated operations:

24. $(D + 2)(D + 1)^2(x^2e^{-x})$
25. $D^8(D - a)e^{ax}$
26. $(D + 1)^2(D + 2)(x^2e^{-x})$
27. $(D^2 + 4D + 4)(x^2e^{-2x})$
28. $(D^3 + D^2 - 12)e^{2x}$
29. $(D + 1)^4(x^3e^{-x})$

45 Linear independence of functions

The functions $f_i(x)$, $i = 1, 2, \ldots, n$, are linearly dependent in an interval $a \leqq x \leqq b$ if there is a set of constants $a_1, a_2, \ldots, a_n$, not all zero, such that

$$a_1f_1(x) + a_2f_2(x) + \cdots + a_nf_n(x) \equiv 0 \qquad \text{in } a \leqq x \leqq b \tag{1}$$

The left member of (1) is called a linear combination of the functions

$f_1, f_2, \ldots, f_n$. The functions $f_i(x)$, $i = 1, 2, \ldots, n$, are linearly independent if the only set of constants $a_1, a_2, \ldots, a_n$ for which (1) holds is the set $a_1 = a_2 = \cdots = a_n = 0$. For example, $\sqrt{3}\, x^3$ and $7x^3$ are linearly dependent in any interval since

$$7(\sqrt{3}\, x^3) - \sqrt{3}\,(7x^3) = 0$$

in any interval. The functions $1, x, x^2, \ldots, x^{n-1}$ are independent in $0 \leqq x \leqq 1$, for we know from algebra that if

$$a_1 1 + a_2 x + a_3 x^2 + \cdots + a_n x^{n-1} \equiv 0 \qquad 0 \leqq x \leqq 1 \tag{2}$$

is satisfied by more than $n - 1$ values of x, then the only set of values available for the a's is $a_1 = a_2 = \cdots = a_n = 0$.

If the functions of a set are not linearly dependent, then they are linearly independent; for if $f_i(x)$, $i = 1, 2, \ldots, n$, are not linearly dependent, then the only set of values for which (2) holds is $a_1 = a_2 = \cdots = a_n = 0$.

The following theorem is basic in the theory of linear dependence of functions.

THEOREM. *If the functions $f_1(x), f_2(x), \ldots, f_n(x)$ have continuous derivatives of all orders from 1 to n, then a necessary condition that they be dependent in $a \leqq x \leqq b$ is*

$$W^* = \begin{vmatrix} f_1 & f_2 & \cdots & f_n \\ f_1^{(1)} & f_2^{(1)} & \cdots & f_n^{(1)} \\ f_1^{(2)} & f_2^{(2)} & \cdots & f_n^{(2)} \\ \cdots & \cdots & \cdots & \cdots \\ f_1^{(n-1)} & f_2^{(n-1)} & \cdots & f_n^{(n-1)} \end{vmatrix} \equiv 0 \qquad a \leqq x \leqq b \tag{3}$$

Since the functions $f_i(x)$ are dependent in $a \leqq x \leqq b$, there must be constants $a_1, a_2, \ldots, a_n$, not all zero, such that

$$a_1 f_1 + a_2 f_2 + \cdots + a_n f_n = 0 \tag{4}$$

Differentiating (4) $(n - 1)$ times in succession, we get

$$\begin{aligned} a_1 f_1^{(1)} + a_2 f_2^{(1)} + \cdots + a_n f_n^{(1)} &\equiv 0 \\ \cdots\cdots\cdots\cdots\cdots\cdots & \\ a_1 f_1^{(n-1)} + a_2 f_2^{(n-1)} + \cdots + a_n f_n^{(n-1)} &\equiv 0 \end{aligned} \tag{5}$$

From algebra we know that the condition that (4) and (5) hold true in $a \leqq x \leqq b$ for values $a_1, a_2, \ldots, a_n$, not all zero, is that (3) hold. This proves that condition (3) is necessary for the dependence of the functions $f_i(x)$, $i = 1, 2, \ldots, n$.

* Here the superscripts indicate derivatives with respect to x. The determinant in (3) is called the wronskian of the functions $f_i(x)$, $i = 1, 2, \ldots, n$.

If (3) does not hold, then (4) does not hold, and the functions $f_i(x)$, $i = 1, 2, \ldots, n$, are independent. The condition for independence of the functions $f_i(x)$, $i = 1, 2, \ldots, n$, in $a \leqq x \leqq b$ is

$$W \neq 0 \qquad \text{in } a \leqq x \leqq b \tag{6}$$

where W refers to the determinant in (3). For example, $\sin x$ and $\cos x$ are independent in any interval since

$$\begin{vmatrix} \sin x & \cos x \\ \cos x & -\sin x \end{vmatrix} = -\sin^2 x - \cos^2 x = -1$$

The condition that e^{ax}, e^{bx}, e^{cx}, $a < b < c$ be independent is

$$\begin{vmatrix} e^{ax} & e^{bx} & e^{cx} \\ ae^{ax} & be^{bx} & ce^{cx} \\ a^2e^{ax} & b^2e^{bx} & c^2e^{cx} \end{vmatrix} = e^{ax}e^{bx}e^{cx}(a-b)(b-c)(c-a) \neq 0$$

Similarly by factoring the determinant (3) associated with the functions e^{a_ix}, $i = 1, 2, \ldots, n$, we can show that these functions are independent if no two of the a's are equal.

Exercises

1. Use (3) to prove that x^2, $x^2 + 5$, and $2x^2 + 11$ are linearly dependent in all intervals.

2. Use (6) to prove that 1, x, and x^2 are linearly independent in all intervals.

Determine for each set of functions whether its members are linearly dependent or independent in all intervals:

3. e^{2x}, xe^{2x} **4.** $5e^{5x}$, e^{5x}, 5, x **5.** x, $x \sin x$

6. e^{2x}, e^{3x}, $7e^{2x} - 8e^{3x}$ **7.** e^x, $5e^x$, $\tan x$

8. If set A contains 20 linearly dependent functions and if set B contains the 20 functions of set A and 20 others, are the functions of set B necessarily linearly dependent? Why?

9. Are the functions e^x, $5e^x$, and $f(x)$ necessarily linearly dependent? Why?

10. Are the functions $f_1(x)$, $f_2(x)$, $\ldots$, $f_n(x)$, and $f_1(x) + 2f_2(x) + 3f_3(x) + \cdots + nf_n(x)$, necessarily dependent? Why?

11. What condition must $f(x)$ satisfy if $f(x)$ and $xf(x)$ are, in all intervals: (*a*) linearly independent? (*b*) linearly dependent?

12. What is the nature of $f(x)$ if $f(x)$ and $2f(x) + 3$ are linearly dependent?

★13. What conditions must $f(x)$ satisfy if, in all intervals: (*a*) $f(x)$ and $[f(x)]^2$ are linearly independent? (*b*) $f(x)$ and $f'(x)$ are linearly dependent?

14. Are the functions $|x|$ and x dependent in: (*a*) $2 \leqq x \leqq 3$? (*b*) $-2 \leqq x \leqq -1$? (*c*) $-2 \leqq x \leqq 2$?

46 Linear differential equation

linear differential equation *contains the dependent variable and all its derivatives to the first degree only.* Its general form is

$$L(D)y = (a_0D^n + a_1D^{n-1} + \cdots + a_{n-1}D + a_n)y = X \tag{1}$$

where the a's and X are functions of x. If $X = 0$, the equation is said to be homogeneous, since each term is of the first degree in y and its derivatives.

A very important theorem relating to the operator $L(D)$ may be expressed by writing

$$L(D)(y_1 + y_2 + \cdots + y_m) = L(D)y_1 + L(D)y_2 + \cdots + L(D)y_m \tag{2}$$

where $y_1, y_2, \ldots, y_m$ represent functions of x. To prove (2), observe that

$$D^k(y_1 + y_2 + \cdots + y_m) = D^ky_1 + D^ky_2 + \cdots + D^ky_m \tag{3}$$

multiply both members of this by a_0 with $k = n$, by a_1 with $k = n - 1, \ldots$, by a_n with $k = 0$, in succession, and add the results to obtain (2).

If $Y(x)$ is a solution of (1) with $X = 0$, so is $cY(x)$; for

$$L(D)[Y(x)] = 0 \qquad D^k[cY(x)] = cD^k[Y(x)] \qquad k = 0, 1, \ldots n \tag{4}$$

and therefore

$$L(D)[cY(x)] = cL(D)[Y(x)] = c \cdot 0 = 0 \tag{5}$$

If $Y_1(x), Y_2(x), \ldots, Y_m(x)$ are solutions of (1) with $X = 0$, then, in accord with (5), $y_1(x) = c_1Y_1(x)$, $y_2(x) = c_2Y_2(x), \ldots, y_m(x) = c_mY_m(x)$ are solutions. For these functions, the right member of (2), and therefore its left member, is zero; that is, $c_1Y_1(x) + c_2Y_2(x) + \cdots + c_mY_m(x)$ is a solution. Hence, the following theorem is true:

THEOREM. *If* $y = Y_1(x)$, $y = Y_2(x)$, . . . , $y = Y_n(x)$ *are solutions of a linear homogeneous equation, then*

$$y = c_1Y_1(x) + c_2Y_2(x) + \cdots + c_nY_n(x) \tag{6}$$

is also a solution.

Equation (6) will be the general solution of the homogeneous equation (1) with $X = 0$ provided that the solutions $Y_i(x)$, $i = 1, 2, \ldots, n$, are linearly independent. Observe also that if $V(x)$ is a solution of (1) so that $L(V(x)) = X(x)$, then $V(x) + Y_i$ is a solution of (1) since $L(V(x) + Y_i) = X(x) + 0$.

Equation (6) expresses the principle of superposition; that is, any linear combination of solutions of (1) with $X = 0$ is also a solution.

In accordance with a general existence theorem,* an equation having the form (1) with $X = 0$ and with a_i/a_0, $i = 1, \ldots, n$, continuous and single-valued will have n linearly independent solutions $Y_i(x)$, $i = 1, \ldots, n$, each possessing continuous derivatives of order 1 to $n - 1$. Therefore, it will have a solution of the form (6), where the functions $Y_i(x)$, $i = 1, \ldots, n$, satisfy (6), §45. The n arbitrary constants $c_1, c_2, \ldots, c_n$ are used to find a particular solution satisfying given initial conditions, such as

$$y = a_0, \quad Dy = a_1, \quad \ldots, \quad D^{n-1}y = a_{n-1} \qquad \text{when } x = x_0 \tag{7}$$

47 Homogeneous linear differential equation with constant coefficients

We shall first find a method of solving equations of the type

$$(a_0D^n + a_1D^{n-1} + \cdots + a_{n-1}D + a_n)y = 0 \tag{1}$$

where the a's are constants.

A very special case of equation (1) is $Dy + ay = 0$, and its solution is $y = ce^{-ax}$. This suggests that a function of the form

$$y = ce^{mx} \tag{2}$$

might be a solution of (1). Substituting $y = ce^{mx}$, $dy/dx = cme^{mx}$. . . . , $d^ky/dx^k = cm^ke^{mx}$ in (1), we obtain

$$ce^{mx}(a_0m^n + a_1m^{n-1} + \cdots + a_{n-1}m + a_n) = 0$$

This equation will be satisfied if m is a root of the equation

$$a_0m^n + a_1m^{n-1} + \cdots + a_{n-1}m + a_n = 0 \tag{3}$$

* See Theorem II, §98.

Equation (3) is referred to as the auxiliary equation Evidently, if $r_1, r_2, \ldots r_n$ are the roots of (3), the equations

$$y_1 = c_1e^{r_1x},\ y_2 = c_2e^{r_2x},\ \cdots,\ y_n = c_ne^{r_nx} \tag{4}$$

are all solutions of (1). Therefore, in accordance with the theorem of §46,

$$y = c_1e^{r_1x} + c_2e^{r_2x} + \cdots + c_ne^{r_nx} \tag{5}$$

is a solution of (1). Since (5) contains n arbitrary constants, it is the general solution of (1) provided that no two of the roots of (3) are equal.

Example. Solve

$$\frac{d^3y}{dx^3} + 2\frac{d^2y}{dx^2} - 3\frac{dy}{dx} = 0 \tag{a}$$

Solution. The auxiliary equation (3) in this case is

$$m^3 + 2m^2 - 3m = 0 \tag{b}$$

The roots of (b) are 1, -3, 0. Hence the solution of (a), in accordance with (5), is

$$y = c_1e^x + c_2e^{-3x} + c_3$$

Exercises

Solve the following differential equations:

1. $(D^2 - 3D + 2)y = 0$

2. $(D^2 + 4D - 5)y = 0$

3. $(D^2 + 4D + 3)y = 0$

4. $D^2y + 3y = 5\,Dy$

5. $(4D^3 - 5D)y = 0$

6. $D^3y = Dy$

7. $D^2y = k^2y$

8. $(D^3 - 3D^2 - D + 3)y = 0$

9. $(D^3 - 7D + 6)y = 0$

10. $(D^4 - D^3 - 7D^2 + 3D)y = 0$

48 Auxiliary equation has repeated roots

Let $r_1, r_2, \ldots, r_n$ be the roots of the auxiliary equation, and for convenience let us write

$$\begin{aligned} f(m) &= a_0m^n + a_1m^{n-1} + \cdots + a_{n-1}m + a_n \\ &= a_0(m - r_1)(m - r_2)\cdots(m - r_n) \end{aligned} \tag{1}$$

Equation (1) may then be written

$$a_0(D - r_1)(D - r_2) \cdots (D - r_n)y = f(D)y = 0 \tag{2}$$

If $f(m) = 0$ has a double root r, then by §46 the solution of (2) contains the two terms $c_1e^{rx} + c_2e^{rx}$; but since this may be written $(c_1 + c_2)e^{rx} =$ (constant) e^{rx}, only one arbitrary constant is involved. In this case, the solution, got by using (6) of §46, involves fewer than n independent arbitrary constants and consequently is not the general solution.

In example 2, §44, we derived the expression $y = (c_1 + c_2x + c_3x^2)e^{ax}$ as the solution of $(D - a)^3y = 0$. This suggests that *the part of a solution corresponding to a p-fold multiple root a of the auxiliary equation is*

$$(c_1 + c_2x + c_3x^2 + \cdots + c_px^{p-1})e^{ax} \tag{3}$$

Observe that

$$(D - a)(x^me^{ax}) = mx^{m-1}e^{ax} + ax^me^{ax} - ax^me^{ax} = mx^{m-1}e^{ax}$$

Therefore, the result of operating on (3) by $D - a$ would give a result like (3) but with the degree of the polynomial reduced by 1. Hence, $\varphi(D)(D - a)^p$ operating on (3) gives zero, and expression (3) satisfies $\varphi(D)(D - a)^py = 0$. Also, an expression like (3) applies for any other multiple root. For example, the roots of the auxiliary equation of

$$(D - 2)^3(D + 3)^2(D - 4)y = 0 \tag{4}$$

are 2, 2, 2, -3, -3, and 4, and the general solution of (4) is

$$y = (c_1 + c_2x + c_3x^2)e^{2x} + (c_4 + c_5x)e^{-3x} + c_6e^{4x}$$

Example. Find the general solution of $(D^5 - 2D^4 + D^3)y = 0$.

Solution. The auxiliary equation is $m^5 - 2m^4 + m^3 = 0$, and its roots are 0, 0, 0, 1, 1. Hence, in accordance with (3) the general solution is

$$y = c_0 + c_1x + c_2x^2 + e^x(c_3 + c_4x)$$

49 Constants of integration from initial conditions

To determine the constants of integration from the general solution of a differential equation, replace the variables in the solution and the derivatives of the solution by given corresponding values, and solve the resulting equations for the required constants.

Example. Find the particular solution of $(D^3 - 6D^2 + 9D)y = 0$ which satisfies the initial conditions $y = 0$, $Dy = 2$, and $D^2y = -6$ when $x = 0$.

Solution. The auxiliary equation is $m^3 - 6m^2 + 9m = 0$, its roots are 0, 3, 3, and the general solution of the given equation is

$$y = c_1 + e^{3x}(c_2 + c_3x) \tag{a}$$

By differentiation, we obtain from (a)

$$\begin{aligned} Dy &= e^{3x}(3c_2 + c_3 + 3c_3x) \\ D^2y &= e^{3x}(9c_2 + 6c_3 + 9c_3x) \end{aligned} \tag{b}$$

Substitution of the initial conditions in (a) and (b) gives

$$\begin{aligned} 0 &= c_1 + e^0(c_2 + c_3 \cdot 0) = c_1 + c_2 \qquad 2 = 3c_2 + c_3 \\ -6 &= 9c_2 + 6c_3 \end{aligned} \tag{c}$$

Solving these equations for c_1, c_2, and c_3, we get

$$c_1 = -2 \qquad c_2 = 2 \qquad c_3 = -4$$

The required particular solution, obtained by substituting these values of the c's in (a), is

$$y = -2 + e^{3x}(2 - 4x)$$

Exercises

For each of the following equations, find the general solution, and when initial conditions are given, determine the particular solution satisfying them:

1. $(D^2 - 6D + 9)y = 0$
2. $(D^2 + 4D + 4)y = 0$
3. $(D^3 - D^2)y = 0$
4. $(D^5 - 4D^3)y = 0$
5. $(D^3 - 2D^2 + D)y = 0$
6. $D^3y = 0$

7. $(D^2 - 2D + 1)y = 0$; $y = 5$, $Dy = -9$ when $x = 0$.
8. $(D^2 + 2D + 1)y = 0$; $y = 1$, $Dy = -1$ when $x = 0$.

9. $(D^3 - D^2 - D + 1)y = 0$
10. $(D^3 - 3D^2 + 3D - 1)y = 0$
11. $(D^6 - 8D^4 + 16D^2)y = 0$
12. $(D^5 - 12D^3 + 16D^2)y = 0$
13. $(D^2)(D - 1)y = 0$; $y = 2$, $Dy = 3$, $D^2y = 2$ when $x = 0$.
14. $(D^3 + D^2)y = 0$; $y = 4$, $Dy = -2$, $D^2y = 4$ when $x = 0$.

15. $(D^3 - 4D^2 + 4D)y = 0$; $y = 1$, $Dy = 2$, $D^2y = 8$ when $x = 0$.

16. $(D^3 - D^2 - D + 1)y = 0$; $y = 0$, $Dy = 0$, $D^2y = 4$ when $x = 0$.

50 Auxiliary equation has imaginary roots

In the theory of complex numbers* $a + ib$, where $i^2 = -1$ and a and b are real, it is shown that the ordinary processes of algebra and also differentiation and integration apply as well to functions of complex variables as to functions of real variables. Hence, the solutions of differential equations based on real variables may be considered the solutions based on complex variables. For example, the solution $y = x^2 + c$ of $dy/dx = 2x$ may be thought of as

$$y_1 + iy_2 = (x_1 + ix_2)^2 + c_1 + ic_2$$

Evidently, the real solution is obtained from this by setting $y_2 = 0$, $x_2 = 0$, and $c_2 = 0$.

We shall assume that complex exponents obey the ordinary laws of exponents and that

$$e^{i\theta} = \cos\theta + i\sin\theta \tag{1}$$

where, for our present purposes, θ is assumed to be a real number.†

The part of the solution of (1), §47, corresponding to a pair of complex roots $a \pm ib$ of the auxiliary equation, may, in accordance with (4), §47, be written

$$Ae^{(a+ib)x} + Be^{(a-ib)x} \tag{4}$$

where A, B, and x are considered as complex numbers. Transforming

* Most texts on college algebra give an elementary treatment of complex numbers. For a comprehensive treatment, read any book on the theory of functions of a complex variable.

† It is interesting to replace θ by $-\theta$ in (1) and obtain

$$e^{-i\theta} = \cos\theta - i\sin\theta \tag{2}$$

and solve this with (1) for $\cos\theta$ and $\sin\theta$ to obtain

$$\cos\theta = \frac{e^{i\theta} + e^{-i\theta}}{2} \qquad \sin\theta = \frac{e^{i\theta} - e^{-i\theta}}{2i} \tag{3}$$

All the identities connecting the trigonometric functions can be deduced from (3). Also, generalized formulas are deduced from (3) and used in higher mathematics and its applications.

this by (1), we get

$$Ae^{ax}e^{ibx} + Be^{ax}e^{-ibx} \tag{5}$$

$$\begin{aligned} &e^{ax}[A(\cos bx + i \sin bx) + B(\cos bx - i \sin bx)] \\ &e^{ax}[(A + B) \cos bx + (A - B)i \sin bx] \end{aligned} \tag{6}$$

To get the real part of the solution from (6), think of x as real, take $A + B = c_1$ and $(A - B)i = c_2$, and obtain

$$e^{ax}(c_1 \cos bx + c_2 \sin bx) \tag{7}$$

Note that an equation for which $a \pm ib$ are roots of its auxiliary equation may be written in the form

$$f(D)[(D - a)^2 + b^2]y = 0 \tag{8}$$

and direct substitution of (7) in (8) shows that (7) is a solution of (8).

To get a useful alternative form of (7), multiply and divide it by $\sqrt{c_1^2 + c_2^2} = c$, and in the result take

$$c_1 = c \sin \alpha \qquad c_2 = c \cos \alpha \qquad \frac{c_1}{c_2} = \tan \alpha \tag{9}$$

and obtain

$$ce^{ax}[\sin \alpha \cos bx + \cos \alpha \sin b(x)] \tag{10}$$

$$ce^{ax} \sin (bx + \alpha) \tag{11}$$

where c and α are arbitrary constants. *The part of the general solution of an equation in the form* (1), §47, *corresponding to a pair of complex roots* $a \pm ib$ *of the auxiliary equation is generally written in form* (7) *or* (11).

In the case of a double pair of complex roots, the corresponding terms of the general solution are

$$e^{ax}[(A_0 + A_1x) \cos bx + (B_0 + B_1x) \sin bx]^* \tag{12}$$

or

$$e^{ax}[c_0 \sin (bx + \alpha_0) + c_1x \sin (bx + \alpha_1)] \tag{13}$$

and a similar extension applies for a p-fold multiple pair of complex roots.

Example. Solve $(D^3 - 3D^2 + 9D + 13)y = 0$.

Solution. The roots of the auxiliary equation are -1, $2 \pm 3i$. Hence, the general solution is

$$y = c_1e^{-x} + e^{2x}(c_2 \sin 3x + c_3 \cos 3x)$$

or

$$y = c_1e^{-x} + ce^{2x} \sin (3x + \alpha)$$

* This may be proved by substituting expression (12) or (13) for y in

$$[(D - a)^2 + b^2]^2y = 0$$

Exercises

Solve the following equations, and determine the constants of integration where initial conditions are given:

1. $(D^2 - 2D + 2)y = 0$ **2.** $D(D^2 - 4D + 5)y = 0$
3. $(D - 2)(D^2 + 2D + 10)y = 0$ **4.** $(D^2 + 4)y = 0$
5. $(D^4 + 8D^2 + 16)y = 0$ **6.** $(D^4 + D^2)y = 0$

7. $(D^2 + 1)y = 0$; $y = 1$, $Dy = -1$ when $x = \pi$.
8. $(D^2 + 2D + 2)y = 0$; $y = 0$, $Dy = 1$ when $x = 0$.

9. $(D^6 + 6D^4 + 9D^2)y = 0$ **10.** $(D^3 + a^3)y = 0$

11. $(D^2 + 9)y = 0$; $y = 2$, $Dy = 0$ when $x = \frac{1}{6}\pi$.
12. $(D^3 - 2D^2 + 2D)y = 0$; $y = 1$, $Dy = 1$, $D^2y = 2$ when $x = 0$.
13. $(D^3 - 2D + 4)y = 0$; $y = -2$, $Dy = 8$, $D^2y = 0$ when $x = 0$.
14. $(D^2 + 4)y = 0$; $y = 4$, $Dy = 0$ when $x = 1$. *Hint:* Write the solution in the form $y = c_1 \sin 2(x - 1) + c_2 \cos 2(x - 1)$.

51 Right member not zero

This section deals with a special method of solving equations of the type

$$(a_0D^n + a_1D^{n-1} + \cdots + a_{n-1}D + a_n)y = X(x) \tag{1}$$

where the a's are constants and X is a nonzero function of x. We shall be concerned mainly with cases where X consists of combinations of polynomials, sines, cosines, and exponential functions.

To get an idea of the method, consider the special case

$$(D^2 + 3D + 2)y = 40e^{3x} \tag{2}$$

First we attempt to find a particular solution by trial. Assume that

$$y_p = Ae^{3x} \tag{3}$$

To find the value of A substitute Ae^{3x} for y in (2), and get

$$9Ae^{3x} + 9Ae^{3x} + 2Ae^{3x} = 40e^{3x}$$

or

$$20Ae^{3x} = 40e^{3x} \qquad \text{and} \qquad A = 2 \tag{4}$$

Substituting 2 for A in (3) we get the solution

$$y_p = 2e^{3x} \tag{5}$$

The solution of $(D^2 + 3D + 2)y = 0$ is

$$y_c = c_1e^{-x} + c_2e^{-2x} \tag{6}$$

Now for the general solution of (2) we write

$$y = y_c + y_p = c_1e^{-x} + c_2e^{-2x} + 2e^{3x} \tag{7}$$

This, as the student may check, will satisfy (2). For the first two terms when substituted in the left member of (2) will produce zero, and the last term will produce $40e^{3x}$.

This illustration indicates that the general solution of (1) is

$$y = y_c + y_p \tag{8}$$

where $y = y_c$ is the general solution of (1) with X replaced by zero, and y_p is a particular solution of (1). Equation (8) is a solution of (1) involving n independent arbitrary constants and hence is the general solution. y_c is referred to as the complementary function and is found as in §§47 to 50. One method for finding y_p has just been illustrated and is used throughout this section. It is called the method of undetermined coefficients

For suggestive quality, we define like terms as terms linearly dependent in the variable involved, generally x. In other words *two terms are* like *if one may be obtained from the other by multiplying it by a suitable nonzero constant.* Thus $8x$ and $(B + c)x$ are like because $8x = [8/(B + c)](B + c)x$ and $(B + c)x = [(B + c)/8]8x$. Unlike terms *are terms linearly independent in the variable involved.*

The new and highly important procedure is that of finding particular solution y_p. The trial solution y_p for an equation of the type (1) should contain as special cases not only X, but also all the independent terms in its derivatives; for derivatives of the trial solution enter when it is substituted in the left member of (1). The following rule gives sufficient coverage except in cases to be considered in later sections.

RULE. *To obtain a trial form for a particular solution y_p of an equation of the type* (1), *write all unlike (linearly independent) terms in the set consisting of X and its derivatives, replace their coefficients by literal constants A, B, C, . . . , and set y_p equal to the sum of the terms thus obtained.*

To find the values of the literal constants A, B, C, . . . , substitute the trial y_p in the given equation of the type (1), arrange each of the two members of the result in the form of *constants times unlike terms*, equate the coefficient of each term in the left member to that of its like term in the right member, and solve the equations thus obtained for the literal constants. Finally substitute the values of these constants in the trial y_p to obtain the desired particular solution.

To write the trial y_p for $X = 4x \sin 3x$, differentiate X and obtain

$$X = 4x \sin 3x \qquad X' = 12x \cos 3x + 4 \sin 3x$$
$$X'' = -36x \sin 3x + 24 \cos 3x$$

Note that the unlike terms are $\sin 3x$, $\cos 3x$, $x \sin 3x$, and $x \cos 3x$. Hence, in this case,

$$y_p = A \sin 3x + B \cos 3x + Cx \sin 3x + Ex \cos 3x$$

Example. Solve $(D^2 + 2D + 5)y = 12e^x - 34 \sin 2x$ *(a)*

Solution. The solution of $(D^2 + 2D + 5)y = 0$ is

$$y_c = e^{-x}(c_1 \sin 2x + c_2 \cos 2x) \qquad (b)$$

The trial y_p is

$$y_p = Ae^x + B \sin 2x + C \cos 2x \qquad (c)$$

Substituting this value of y_p for y in (a) and transforming slightly, we get

$$8Ae^x + (B - 4C) \sin 2x + (4B + C) \cos 2x = 12e^x - 34 \sin 2x + 0 \cos 2x \qquad (d)$$

Equating coefficients of like terms in the two members of *(d)*, we get

$$8A = 12 \qquad B - 4C = -34 \qquad 4B + C = 0$$

Therefore

$$A = \tfrac{3}{2} \qquad B = -2 \qquad C = 8 \qquad (e)$$

Substituting the values of A, B, and C from *(e)* in *(c)*, we get

$$y_p = \tfrac{3}{2}e^x - 2 \sin 2x + 8 \cos 2x \qquad (f)$$

The required solution is

$$y = y_c + y_p = e^{-x}(c_1 \sin 2x + c_2 \cos 2x) + \tfrac{3}{2}e^x - 2 \sin 2x + 8 \cos 2x \qquad (g)$$

Exercises

Solve the following equations, and determine the constants of integration where initial conditions are given:

1. $(D^2 - 4)y = 12$. *Hint:* Let $y_p = A$.

2. $(D^2 + 2D - 3)y = 42e^{4x}$ **3.** $(D^2 + D - 2)y = 3 - 6x$

4. $(D^2 - D - 2)y = 6e^x$ **5.** $(D^2 + 1)y = 3 + 6e^x$
6. $(D + 2)^2y = x + 8e^{2x}$ **7.** $(D^2 + D)y = 6 \sin 2x$
8. $(D^3 - D^2)y = 2 \cos x$ **9.** $(D^2 + 1)y = 10e^x \sin x$

10. $(D^2 - 9)y = 18 \cos 3x + 9$; $y = -1$, $Dy = 3$ when $x = 0$.

11. $(D^2 - 1)y = 2x^2$; let $y_p = Ax^2 + Bx + C$.
12. $(D^2 + D - 6)y = 2 - 12x$; $y = 3$, $Dy = -7$ when $x = 0$.
13. $(D^2 + D)y = e^x$; $y = 3$, $Dy = 0$ when $x = \ln 2$.
★14. $(D^2 + 4D + 3)y = 8xe^x - 6$; $y = -\frac{11}{4}$, $Dy = \frac{1}{4}$ when $x = 0$.
★15. $(D^3 - 4D)y = 6e^{-x} - 3e^x$; $y = 7$, $Dy = 9$, $D^2y = 19$ when $x = \ln 2$.

52 Special case when the right member is not zero

Let us attempt to solve by the method of §51 the equation

$$(D - a)^2y = 2e^{ax} \tag{1}$$

Here

$$y_c = c_1e^{ax} + c_2xe^{ax} \tag{2}$$

The rule of §51 suggests that we write $y_p = Ae^{ax}$ as trial solution. But this y_p is like the first term in the complementary solution and therefore, when substituted in (1) it will give the impossible equation $0 = 2e^{ax}$. However we may take $y_p = Ax^2e^{ax}$, which is unlike any term in the complementary function (2). Substituting Ax^2e^{ax} for y in (1), we get

$$2Ae^{ax} = 2e^{ax} \tag{3}$$

Hence $A = 1$, $y_p = x^2e^{ax}$, and the solution of (1) is

$$y = y_c + y_p = (c_1 + c_2x)e^{ax} + x^2e^{ax} \tag{4}$$

This solution shows that a term like a term in the complementary function is useless as a part of y_p, but it becomes effective when multiplied by the least power of x that makes it unlike any term in the complementary function.

A little reflection leads us to see that the trial y_p for an equation of the type (1), §51, should contain: (I) *all unlike terms in X and its derivatives,* (II) *no two terms like each other, and* (III) *no term like a term in the complementary function.* Hence the following rule may be used:

RULE. *To obtain a trial form for* y_p: (a) *write the trial form specified by the rule of* §51; (b) *multiply each term from* (a) *which is like a term in the complementary function by the least power of* x *such that conditions* (II) *and* (III) *in the immediately preceding paragraph are satisfied.*

Having written the form for y_p, proceed as in §51.

Example 1. Find y_c and the trial y_p for each of the equations: (*a*) $(D^2 + 4)y = \sin 2x$. (*b*) $(D - 3)^3y = xe^{3x} + e^{4x}$. (*c*) $D^2y = x^2 + x + e^{3x}$.

Solution. (*a*) $y_c = c_1 \sin 2x + c_2 \cos 2x$; $y_p = x(A \sin 2x + B \cos 2x)$. (*b*) $y_c = (c_1 + c_2x + c_3x^2)e^{3x}$; $y_p = (Ax^3 + Bx^4)e^{3x} + Ce^{4x}$. (*c*) $y_c = c_1 + c_2x + c_3x^2$; $y_p = Ax^3 + Bx^4 + Cx^5 + He^{3x}$

Example 2. Solve $(D - 2)^2(D + 3)y = 10e^{2x} + 25e^{-3x}$ (*a*)

Solution. Here $y_c = (c_1 + c_2x)e^{2x} + c_3e^{-3x}$ (*b*)

and the trial

$$y_p = Ax^2e^{2x} + Bxe^{-3x} \qquad (c)$$

Substituting y_p from (*c*) in (*a*) we get

$$10Ae^{2x} + 25Be^{-3x} = 10e^{2x} + 25e^{-3x} \qquad (d)$$

Therefore

$$A = 1 \qquad B = 1 \qquad \text{and} \qquad y_p = x^2e^{2x} + xe^{-3x} \qquad (e)$$

and

$$y = y_c + y_p = (c_1 + c_2x)e^{2x} + c_3e^{-3x} + x^2e^{2x} + xe^{-3x} \qquad (f)$$

Exercises

For each equation write the trial expression for y_p:

1. $(D^2 - 4)y = e^{2x}$
2. $(D^2 - 4)y = xe^{-2x}$
3. $(D^2 + 4)y = x \sin 2x$
4. $D^2y = 10$
5. $D^2y = 10x^2$
6. $(D - 2)^3Dy = e^{2x} + x$

Solve the following differential equations:

7. $D(D + 1)y = 4x$
8. $(D^2 - 1)y = 5e^x$
9. $(D^2 + D)y = e^{-x}$
10. $(D^2 + 1)y = \sin x$
11. $(D - 2)^3y = 12e^{2x}$
★12. $(D^2 - 2D + 2)y = e^x \sin x$

Find a particular solution of each equation:

13. $(D^2 - 2D + 1)y = e^x + 3$ **14.** $(D^2 + 4)y = \cos 2x$
15. $(D^2 + 2D)y = 8x + e^{-2x}$ **16.** $(D^3 + 4D)y = 8\cos 2x + 4$

★17. $(D^3 - 2D^2 + 5D)y = 10 + 15\cos 2x$

53 A basic theorem relating to operators

THEOREM. If $\varphi(y)$ *represents a polynomial, then*

$$\varphi(D)(e^{ax}X) = e^{ax}\varphi(D + a)X \tag{1}$$

where X is any function of x possessing all the derivatives in (1).

Proof. First we shall use mathematical induction in proving that

$$D^n(e^{ax}X) = e^{ax}(D + a)^n X \tag{2}$$

When $n = 0$, each member of (2) is equal to $e^{ax}X$. Hence, (2) is true when $n = 0$. Let k be any nonnegative integer for which (2) holds. Then

$$D^k(e^{ax}X) = e^{ax}(D + a)^k X$$

and, differentiating this, obtain

$$\begin{aligned} D^{k+1}(e^{ax}X) &= D[e^{ax}(D + a)^k X] \\ &= e^{ax}D(D + a)^k X + ae^{ax}(D + a)^k X \\ &= e^{ax}(D + a)^{k+1} X \end{aligned}$$

That is, if (2) holds when $n = k$, it holds when $n = k + 1$. Therefore, by complete induction, (2) is true when $n = 0$ or a positive integer. Consequently, (1) holds, because each term in the left member equals the corresponding term in the right member.

As an illustration of (1), consider that

$$\begin{aligned} (D^2 - 4)(e^{2x}x^2) = e^{2x}[(D + 2)^2 - 4]x^2 &= e^{2x}(D^2 + 4D)x^2 \\ &= e^{2x}(2 + 8x) \end{aligned}$$

Some special cases of (1) are useful.

If in (1) we let $\varphi(D) = \psi(D - a)$ and interchange members, we get

$$e^{ax}\psi(D)X = \psi(D - a)(e^{ax}X) \tag{3}$$

Evidently, $KD^n(e^{ax}) = Ka^n e^{ax}$, and since a polynomial $\varphi(D)$ is a sum of such terms as KD^n, we have

$$\varphi(D)(e^{ax}) = e^{ax}\varphi(a) \tag{4}$$

Thus

$$(D^5 + 5D^2 + 2)e^{-x} = [(-1)^5 + 5(-1)^2 + 2]e^{-x} = 6e^{-x}$$

Example. Solve $(D^2 + 2D + 2)y = e^{-x} \sin x$ (a)

Solution. Multiplying (a) through by e^x we get

$$e^x(D^2 + 2D + 2)y = \sin x \qquad (b)$$

Now use (3) on the left member with $\psi(D) = D^2 + 2D + 2$ and $a = 1$ to get

$$[(D - 1)^2 + 2(D - 1) + 2](e^x y) = (D^2 + 1)z = \sin x \qquad (c)$$

where

$$z = e^x y \qquad (d)$$

Solving (c) for z by the method of §52, we obtain

$$z = (c_1 \sin x + c_2 \cos x) - \tfrac{1}{2}x \cos x \qquad (e)$$

In (e) replace z by ye^x from (d), and solve for y to get

$$y = e^{-x}(c_1 \sin x + c_2 \cos x - \tfrac{1}{2}x \cos x) \qquad (f)$$

Exercises

1. To solve $(D - 2)^3 y = e^{2x}$, multiply through by e^{-2x}, and then use (3) on the left member to obtain $D^3 z = 1$, where $z = ye^{-2x}$. Now find z from $D^3 z = 1$, and then use $y = ze^{2x}$ to get the required solution.

(2–6) Solve the following differential equations:

2. $(D + 1)^2 y = xe^{-x}$

3. $(D + 1)^3 y = 12e^{-x}$

4. $(D - 2)^2 y = e^{2x} \sin x$

5. $(D^2 + 2D)y = e^{-2x} \sin x$

★**6.** $D(D + 1)^2 y = 12e^{-x}$

7. Use (1) to show that: (a) $(D - a)^4(x^4 e^{ax}) = e^{ax} D^4 x^4 = 4!e^{ax}$; (b) $(D - 2)^n(e^{2x}x^n) = n!e^{2x}$; (c) $(D + 2)(D - 2)^3(x^2 e^{2x}) = 0$.

8. Use (3) to show that:

(a) $e^{-3x}(D - 1)(D - 3)X = (D + 2)D(Xe^{-3x})$

(b) $e^{2x}[(D + 2)^4 + (D + 2)^2]x^3 = (D^4 + D^2)(x^3 e^{2x})$

9. Prove that $(D + a)^n(e^{-ax}X) = e^{-ax}D^n(X)$.

Solve the following differential equations:

10. $[(D + 2)^2 + 4]y = e^{-2x} \sin x$

11. $(D + 1)^3 Dy = 12e^{-x}$

12. $[(D - 1)^4 + (D - 1)^2]y = 12x^2 e^x$; $y = 0$, $Dy = 0$, $D^2 y = 0$, $D^3 y = 0$ when $x = 0$.

54 Methods using symbolic operators

In this section, $f(D)$, $\varphi(D)$, and $\psi(D)$ refer to polynomial operators and X refers to a function of x possessing all derivatives and integrals involved.

A solution of the differential equation $\varphi(D)y = X$ is represented by $\varphi^{-1}(D)X$ and also by $[1/\varphi(D)]X$. We write

$$\varphi^{-1}(D)X = y(x) \qquad \text{if and only if} \qquad \varphi(D)y = X \tag{1}$$

Substituting y from the first equation of (1) for y in the second part we get

$$\varphi(D)\varphi^{-1}(D)X = X \tag{2}$$*

From (1) we can see by inspection that $D^{-1}(3x^2) = x^3 + c$, since $D(x^3 + c) = 3x^2$. Any solution of $\varphi(D)y = X$ is a value of $\varphi^{-1}(D)X$. However we generally use a simple value containing no arbitrary constants, especially when the main project is to find a value for y_p of a differential equation under consideration.

To find explicit forms of such expressions as $(D - 2)^{-1}(6x - 6x^2)$, we may solve the corresponding differential equation. We have

$$(D - 2)y = 6x - 6x^2 \qquad y = 3x^2$$
$$(D - 2)^{-1}(6x - 6x^2) = 3x^2$$

The formulas, developed below, will furnish short methods of dealing with inverse operators, mainly to find particular solutions.

Considerable simplification is obtained by using the following definitions:

$$\left[\frac{1}{\varphi(D)} + \frac{1}{\psi(D)}\right] X = [\varphi^{-1}(D) + \psi^{-1}(D)]X$$
$$= \varphi^{-1}(D)X + \psi^{-1}(D)X \tag{3}$$
$$\frac{\varphi(D)}{\psi(D)} X = \varphi(D)\left[\frac{1}{\psi(D)} X\right] \tag{4}$$

Also observe that since the order of the factors in $\varphi(D)$ is immaterial the same thing is true of $1/\varphi(D)$.

The definition (4) justifies the introduction of a factor $\theta(D)/\theta(D)$ in the left member of (4); for, if we denote the left member by $\lambda(x)$, we have $[\theta(D)/\theta(D)]\lambda(x) = [\theta(D)\theta^{-1}(D)]\lambda(x) = \lambda(x)$. Similarly it appears that identical factors in numerator and denominator of an operator in the form of a rational fraction may be canceled.

* $\varphi^{-1}(D)$ is called the inverse operator of $\varphi(D)$ because equation (2) holds true.

In agreement with (1) we define the sign = as used in this section by

$$\frac{\varphi(D)}{\psi(D)} X = y(x) \qquad \text{if and only if} \qquad \psi(D)y = \varphi(D)X \tag{5}$$

From (3), (4), and (5) we get

$$\left[\frac{\varphi_1(D)}{\psi(D)} + \frac{\varphi_2(D)}{\psi(D)}\right] X = \left[\frac{\varphi_1(D) + \varphi_2(D)}{\psi(D)}\right] X \tag{6}$$

for $\psi(D)$ operating on the members of (6) produces the same result $[\varphi_1(D) + \varphi_2(D)]X$. Since (3), (4), and (6) for combining operators formulate the basic procedures for the algebraic combination of rational fractions, *operators involved in an equation may be replaced by an algebraically equal set of simple operators.* For example

$$\begin{aligned} \frac{1}{D(D-2)}(6x - 6x^2) &= \frac{1}{2}\left[\frac{1}{D-2} - \frac{1}{D}\right](6x - 6x^2) \\ &= \tfrac{1}{2}(3x^2 - 3x^2 + 2x^3) = x^3 \end{aligned} \tag{7}$$

Again we have

$$\frac{1}{\varphi(D)}(Xe^{ax}) = e^{ax}\frac{1}{\varphi(D+a)} X \tag{8}$$

For, applying $\varphi(D)$ to both members and taking into account (2) and (1), §53, we get

$$\varphi(D)\left[\frac{1}{\varphi(D)}(Xe^{ax})\right] = Xe^{ax}$$

$$\varphi(D)\left[e^{ax}\frac{X}{\varphi(D+a)}\right] = e^{ax}\varphi(D+a)\frac{X}{\varphi(D+a)} = Xe^{ax}$$

or Xe^{ax} in both cases. For example, by using (8) we get

$$\frac{1}{D-a}(x^2e^{ax}) = e^{ax}\left(\frac{1}{D}x^2\right) = \tfrac{1}{3}x^3e^{ax}$$

and $y_p = \frac{1}{3}x^3e^{ax}$ is a particular solution of $(D - a)y = x^2e^{ax}$.

$$\frac{1}{D}X = \int X\,dx, \frac{1}{D^2}X \cong \int\left(\int X\,dx\right)dx, \ldots \tag{9}$$

For example, to solve the equation $(D^5 + D^4 - 6D^3)y = x^2$, we can find the complementary solution in the ordinary way, and then, in finding a particular solution, operate on both members by $1/D^3$ and obtain

$$(D^2 + D - 6)y = \frac{1}{D^3}x^2 = \tfrac{1}{60}x^5 \tag{10}$$

If X is a polynomial of degree n, the following formula applies to find a particular solution of $(D + a)y = X$:

$$\frac{1}{D+a} X = \frac{(1/a)X}{1 + D/a}$$
$$= \frac{1}{a}\left[1 - \frac{D}{a} + \frac{D^2}{a^2} + \cdots + (-1)^n \frac{D^n}{a^n}\right] X \qquad (11)$$

for, in accord with the italicized statement just above equation (7), we have

$$(D + a)\left\{\frac{1}{a}\left[1 - \frac{D}{a} + \cdots + (-1)^n \frac{D^n}{a^n}\right] X\right\}$$
$$= \left[1 + (-1)^n \left(\frac{D}{a}\right)^{n+1}\right] X$$

and since $D^{n+1}X = 0$, this is equal to X. As an example, let us find $\int x^4e^{2x}\,dx$. By (8) and (11) we have

$$\int x^4e^{2x}\,dx = \frac{1}{D}(x^4e^{2x}) = e^{2x}\frac{1}{D+2}x^4 = \tfrac{1}{2}e^{2x}\frac{1}{1+D/2}x^4$$
$$= \tfrac{1}{2}e^{2x}(1 - \tfrac{1}{2}D + \tfrac{1}{4}D^2 - \tfrac{1}{8}D^3 + \tfrac{1}{16}D^4)x^4$$
$$= \tfrac{1}{2}e^{2x}(x^4 - 2x^3 + 3x^2 - 3x + \tfrac{3}{2})$$

A method of carrying out the indicated operation $[1/\varphi(D)](a_0x^n + a_1x^{n-1} + \cdots + a_n)$ consists in operating on the polynomial in x by the part of the quotient $1/\varphi(D)$ containing terms of degree n or less. Thus

$$\frac{1}{2 + 2D^2 + D^4}x^4 = (\tfrac{1}{2} - \tfrac{1}{2}D^2 + \tfrac{1}{4}D^4 + \cdots)x^4 = \tfrac{1}{2}x^4 - 6x^2 + 6$$

Since $D^2 \sin ax = -a^2 \sin ax$

$$D^{2n+k} \sin ax = (-a^2)^n D^k \sin ax \qquad (12)$$

Hence if $\varphi(D)$ is any polynomial in D, we may replace D^2 by $-a^2$ in $\varphi(D) \sin ax$ or in $\varphi(D) \cos ax$. In symbols

$$\varphi(D) \sin ax = [\varphi(D) \sin ax]_{D^2=-a^2} \qquad (13)$$

Also

$$\frac{1}{\varphi(D)} \sin ax = \left[\frac{1}{\varphi(D)} \sin ax\right]_{D^2=-a^2} \qquad (14)^*$$

for, in each case, $\varphi(D)$ applied to the left member of (14) equals $\varphi(D)$ applied to the right member. In (13) and (14) *sin ax* may be replaced

* The right member of (14) means that in $\varphi(D)$ each even power D^{2n} is to be replaced by $(-a^2)^n$ and each odd power D^{2m+1} is to be replaced by $(-a^2)^m D$.

by *cos ax*. For example, to find a particular solution of $(D^4 - 5D^2 + 4)y = \sin 3x$, write

$$y_p = \frac{1}{D^4 - 5D^2 + 4} \sin 3x = \frac{\sin 3x}{(-9)^2 - 5(-9) + 4} = \tfrac{1}{130} \sin 3x$$

To find a particular solution of $(D^2 + 2D + 2)y = e^x \sin x$, write

$$\begin{aligned} y &= \frac{1}{(D^2 + 2D + 2)} (e^x \sin x) \\ &= e^x \frac{1}{(D + 1)^2 + 2(D + 1) + 2} \sin x \\ &= e^x \frac{1}{D^2 + 4D + 5} \sin x = e^x \frac{D - 1}{(D - 1)4(D + 1)} \sin x \\ &= e^x \frac{(D - 1)}{4(-2)} \sin x = -\tfrac{1}{8} e^x (\cos x - \sin x) \end{aligned}$$

The first step in solving an equation $\varphi(D)y = X$ is to write it in the form $y = \varphi^{-1}(D)X$. Note that the four equations (8), (9), (11), and (14) written in boldface are the basis of short cuts in carrying out operations. Use (8) to remove e^{ax} to a position preceding the operator. Use (14) to operate on $\sin ax$ or $\cos ax$. Use (9) to dispose of factors having the form $1/D^n$. Use (11) to operate on a polynomial in x.

Example 1. Find a particular solution of

$(D - 2)^3(D - 1)y = 6(x^2 + 2x)e^{2x}$

Solution. Using (8), (11), and (9) in that order, we get

$$\begin{aligned} y &= \frac{1}{(D - 2)^3(D - 1)} (6x^2 + 12x)e^{2x} = e^{2x} \frac{6x^2 + 12x}{D^3(D + 1)} \\ &= e^{2x} \frac{1}{D^3} (6x^2 + 12x - 12x - 12 + 12) = \tfrac{1}{10} x^5 e^{2x} \end{aligned}$$

Example 2. Find a particular solution of $(D^2 + 1)y = e^x \sin x$. *Solution.* Using (1) first and then (14), obtain

$$\begin{aligned} y &= \frac{1}{D^2 + 1} (e^x \sin x) = e^x \frac{1}{(D + 1)^2 + 1} \sin x \\ &= e^x \frac{1}{D^2 + 2D + 2} \sin x = e^x \frac{1}{-1 + 2D + 2} \sin x \\ &= e^x \frac{(2D - 1)}{4D^2 - 1} \sin x = \frac{2 \cos x - \sin x}{-5} e^x \end{aligned}$$

Example 3. Find a particular solution of $(D - 2)(D - 1)y = 4x^3e^{2x}$.

Solution. Here

$$y = \frac{4x^3e^{2x}}{(D-2)(D-1)} = e^{2x}\cdot\frac{4x^3}{D(D+1)} = e^{2x}\cdot\frac{x^4}{D+1}$$

Now, using (11), obtain

$$\frac{1}{1+D}x^4 = (1 - D + D^2 - D^3 + D^4)x^4$$
$$= x^4 - 4x^3 + 12x^2 - 24x + 24$$

Therefore

$$y = e^{2x}(x^4 - 4x^3 + 12x^2 - 24x + 24)$$

Example 4. Find a particular solution of $(D^3 + 4D)y = 4\sin 2x$.
Solution. By (9),

$$y = \frac{4\sin 2x}{(D^2+4)D} = \frac{-2\cos 2x}{D^2+4}$$

Here, (14) fails because it involves zero as a divisor. Let Re mean *real part of*. Note that $\cos 2x = \text{Re}\,(e^{i2x}) = \text{Re}\,(\cos 2x + i\sin 2x)$. Hence by means of (1), (9), and (11) we get

$$y = \frac{-2\cos 2x}{D^2+4} = \text{Re}\left(\frac{-2e^{i2x}}{D^2+4}\right) = \text{Re}\, e^{i2x}\frac{(-2)}{(D+2i)^2+4}$$
$$= \text{Re}\, e^{i2x}\frac{-2}{D^2+4iD} = \text{Re}\, e^{i2x}\frac{-2x}{D+4i}$$
$$= \text{Re}\,[(\cos 2x + i\sin 2x)(\tfrac{1}{2}ix - \tfrac{1}{8})] = -\tfrac{1}{8}\cos 2x - \tfrac{1}{2}x\sin 2x$$

Since $-\frac{1}{8}\cos 2x$ satisfies $(D^3 + 4D)y = 0$ it may be discarded. Hence an answer is $y_p = -\frac{1}{2}x\sin 2x$.

Exercises

In each case carry out the operations of the indicated formulas to find an equivalent expression not involving D:

1. $(1/D)x^3$; (9).
2. $(1/D^2)(8e^{2x})$; (9).
3. $[1/(D-1)^2](xe^x)$; (8), (9).
4. $D(D^2+4)^{-1}\sin 3x$; (14).
5. $(D+1)^{-1}x^4$; (11).
6. $(D^2+4)^{-3}\cos 4x$; (14).

7. $[(D+3)D]^{-1}(3x^2)$; (8), (9).
8. $(D-1)^{-2}(e^{2x}\cos x)$; (8), (14).

(9–16) Find particular solutions of:

9. $(D-3)^2y = 48xe^{3x}$. Write $y = [1/(D-3)^2](48xe^{3x})$, and use (8) and (9).

10. $(D+1)^3y = 16(2x+3)^{-3}e^{-x}$

11. $(D^2-2D-3)y = 64xe^{3x}$ **12.** $(D^4+2D^2+1)y = \sin 2x$

13. $(D^2+1)y = \sin 2x + \cos 3x$. Use (14) for each term of the right member.

14. $(1+2D^2+D^3)y = x^3-2x^2$. Divide 1 by $1+2D^2+D^3$ to get the quotient through the term in D^3, and use this quotient to operate on x^3-2x^2.

15. $(D^2+9)y = 36 \sin 3x$. Read the solution of example 4.

16. $(D^2-4)y = 27x^2e^{2x}$. Use $(D^2-4)^{-1} = \frac{1}{4}[(D-2)^{-1} - (D+2)^{-1}]$.

17. Find $\int x^3e^{2x}\,dx$ by using $(1/D)(x^3e^{2x}) = e^{2x}[1/(D+2)]x^3$.

Find particular solutions of the following equations:

18. $(D-1)^3y = e^x \cos x$ **19.** $(D^2+4)y = 9e^{2x}\sin 2x$

20. $(D-1)^2y = e^x \sin x + e^{2x}\cos x$

21. $(D^2-4D+2)y = 8e^x \cos x$

22. $(D^2-D+2)y = 58e^x \cos 3x$

23. $(D-1)(D-2)y = x^2e^{3x}$

24. $(D-3)(D+2)y = e^{2x}(2+6x-4x^2)$

25. $(2+4D-D^2)y = 4x^3$

26. $(D^2+1)y = 4\cos x$. Use the method of example 4.

27. $(D^2+1)y = 16x\cos x$. Use $\cos x = \text{Re}\, e^{ix}$.

28. $(D^2+4)y = 64x \sin 2x + 32 \cos 2x$

29. Show that $[1/\varphi(D)]e^{ax} = e^{ax}/\varphi(a)$. Check that $(1/D^3)e^{ax} = e^{ax}/a^3 + b + cx + gx^2$. Also, check that $(1/D^n)e^{ax} = e^{ax}/a^n$.

30. Use the principle of exercise 29 to write a solution of $(3D^3 + 4D^2 + 1)y = e^{2x}$.

55 Variation of parameters

The methods used in §§51 to 54 generally fail to give a solution of the linear equation when $X, DX, D^2X, \ldots$ contain an infinite number of linearly independent terms. For example, $X = x^{-1}$, $DX = -x^{-2}$, $D^2X = 2x^{-3}, \ldots$ represent an infinite number of linearly independent terms. The powerful method discussed below is called variation of parameters It is used to find a solution of the general linear differential equation when the solution of the corresponding homogeneous equation is known.

Consider the equation

$$(D^3 + D)y = \sec x \tag{a}$$

The solution of $(D^3 + D)y = 0$ is

$$y = A \sin x + B \cos x + C \tag{b}$$

Now we assume that (*b*) is the solution of (*a*), where A, B, C are functions of x to be determined. From (*b*),

$$Dy = A \cos x - B \sin x + A' \sin x + B' \cos x + C' \tag{c}$$

where the primes denote derivatives with respect to x. A, B, and C, being three arbitrary functions, may be subjected to three conditions; hence, let us take

$$A' \sin x + B' \cos x + C' = 0 \tag{d}$$

Then

$$D^2y = -A \sin x - B \cos x + A' \cos x - B' \sin x \tag{e}$$

Take

$$A' \cos x - B' \sin x = 0 \tag{f}$$

Then

$$D^3y = -A \cos x + B \sin x - A' \sin x - B' \cos x \tag{g}$$

Substituting y and its derivatives from (*b*), (*c*), (*e*), and (*g*) in (*a*) while taking account of (*d*) and (*f*), we get, after slight simplification,

$$-A' \sin x - B' \cos x = \sec x \tag{h}$$

The solution of (*d*), (*f*), and (*h*) for A', B', and C' is

$$\begin{aligned} A' &= -\tan x \qquad B' = -1 \\ C' &= \sin x \tan x + \cos x = \sec x \end{aligned} \tag{i}$$

Solving (*i*) for A, B, and C, obtain

$$\begin{aligned} A &= \ln \cos x + c_1 \qquad B = -x + c_2 \\ C &= \ln (\sec x + \tan x) + c_3 \end{aligned} \tag{j}$$

Substituting the values of A, B, and C from (*j*) in (*b*), obtain the solution of (*a*),

$$y = (\ln \cos x + c_1) \sin x + (c_2 - x) \cos x + \ln (\sec x + \tan x) + c_3$$

A consideration of the general case is instructive and suggests short cuts. Take the general equation

$$L(D)y = a_0 D^{(n)}y + a_1 D^{(n-1)}y + \cdots + a_{n-1} Dy + a_n y = X \qquad (1)^*$$

and let the solution of $L(D)y = 0$ be

$$y_c = A_1\varphi_1(x) + A_2\varphi_2(x) + \cdots + A_n\varphi_n(x) \qquad (2)$$

where the A's are considered as constants. Then assume that the A's are functions of x such that (2) is the general solution of (1) and also that

$$A_1'\varphi_1^{(k)}(x) + A_2'\varphi_2^{(k)}(x) + \cdots + A_n'\varphi_n^{(k)}(x) = 0 \qquad k = 0 \cdots (n-2) \qquad (3)^*$$

Substituting (2) in (1) and taking account of (3) and $L(D)y = 0$, if the A's are considered constants, we get

$$a_0[\varphi_1^{(n-1)}A_1' + \varphi_2^{(n-1)}A_2' + \cdots + \varphi_n^{(n-1)}A_n'] = X \qquad (4)^*$$

Solving (3) and (4) for A_i', $i = 1, 2, \ldots, n$, and integrating the results we get values for the A's which when substituted in (2) give the general solution of (1). Note that the wronskian of (3) and (4) must satisfy

$$W = a_0 \begin{vmatrix} \varphi_1 & \varphi_2 & \cdots & \varphi_n \\ \varphi_1' & \varphi_2' & \cdots & \varphi_n' \\ \cdots & \cdots & \cdots & \cdots \\ \varphi_1^{(n-1)} & \varphi_2^{(n-1)} & \cdots & \varphi_n^{(n-1)} \end{vmatrix} \not\equiv 0 \qquad (5)$$

in the interval of x under consideration, since (5) is the condition that it be possible to solve (3) and (4) for the A''s. Also, since (5) is the condition that the functions φ_i be independent, the method fails if the φ's are dependent.

The short cut consists in solving (3) and (4) for the A''s and then proceeding as indicated above. If $n = 1$ use (4) but not (3).

Example. Solve $(x - 1)\, D^2y - x\, Dy + y = (x - 1)^2$ (a)

Solution. By trial we find that $y_1 = \varphi_1 = x$ and $y_2 = \varphi_2 = e^x$ are solutions of

$$(x - 1)\, D^2y - x\, Dy + y = 0 \qquad (b)$$

Check this. Therefore, by the theorem of §46,

$$y = Ax + Be^x \qquad (c)$$

* The superscripts indicate derivatives. Thus $D^{(k)}y = d^ky/dx^k$, $\varphi^{(k)}(x) = d^k\varphi/dx^k$.

is the general solution of (b). In this case the equations (3) and (4) are

$$\begin{aligned} xA' + e^xB' &= 0 \\ (x-1)(A' + e^xB') &= (x-1)^2 \end{aligned} \qquad (d)$$

Solving (d) for A' and B', we get

$$A' = -1 \qquad B' = xe^{-x} \qquad (e)$$

Therefore

$$A = -x + c_1 \qquad B = -(x+1)e^{-x} + c_2 \qquad (f)$$

Substituting the values of A and B from (f) in (c), we get

$$y = c_1x + c_2e^x - x^2 - x - 1 \qquad (g)$$

Note that we could replace $c_1x - x$ by cx. Observe that the wronskian of (d), namely,

$$(x-1)\begin{vmatrix} x & e^x \\ 1 & e^x \end{vmatrix} \equiv (x-1)^2e^x \not\equiv 0$$

holds for all values of x except $x = 1$.

Exercises

1. To solve $(D^2 + 1)y = \tan x$, write

$$y = A \sin x + B \cos x \qquad (a)$$

Find Dy, treating A, B, and x as variables, and take

$$A' \sin x + B' \cos x = 0 \qquad (b)$$

Find D^2y and substitute Dy and D^2y in the given differential equation to obtain, after simplification,

$$A' \cos x - B' \sin x = \tan x \qquad (c)$$

Solve (b) and (c) for A' and B', integrate the results to find A and B, and then replace A and B in (a) by these values.

Solve:

2. $(D^2 + 1)y = \sec^3 x$

3. $(D^2 + a^2)y = \sec^3 ax$

4. $(D^2 + 2D + 2)y = e^{-x} \sec x$

5. $(D^2 + 4)y = 4 \cot 2x$

6. $(D^3 + D)y = \sec^2 x$

7. $(xD^2 - D)y = 4/x$. Two particular solutions of $(xD^2 - D)y = 0$ are $y_1 = 1$ and $y_2 = x^2$.

8. $(D + 2)^2y = x^{-2}e^{-2x}$ **9.** $(D - 2)^2y = x^ne^{2x}$
10. $(D - 2)^2y = 18x^{-2}e^{2x}$ **11.** $(D + 3)^3y = 6x^{-3}e^{-3x}$

Solve each equation after checking the two indicated solutions of the left member equated to zero:

12. $[(x - 1)D^2 - xD + 1]y = (x - 1)^2/x$; $y_1 = x$, $y_2 = e^x$.
13. $(x^4D^2 + 2x^3D - 1)y = 16e^{3/x}$; $y_1 = e^{1/x}$, $y_2 = e^{-1/x}$.

56 Simultaneous differential equations

A solution of n simultaneous equations in $n + 1$ unknowns consists of n independent relations involving one or more of these unknowns but not their derivatives; if these n relations are solved for n of the unknown quantities in terms of the remaining one, and if the results are substituted in the given differential equations, identities must result. *The first object in solving such a system is so to combine the given equations and other equations derived from them as to obtain an equation in two unknowns.* This may be integrated to obtain one relation, and the result may be used to obtain other relations. In the process of eliminating variables, we often find that operators can be used to advantage. Using the facts relating to operators in §§44, 53, and 54, we shall find the process of elimination in the case of linear equations with constant coefficients to be very much like an analogous process of elimination used in algebra.

Example. Solve

$$\begin{aligned} \frac{dx}{dt} + \frac{dy}{dt} + y - x &= e^{2t} \\ \frac{d^2x}{dt^2} + \frac{dy}{dt} &= 3e^{2t} \end{aligned} \tag{a}$$

Solution. Replacing d/dt by D, we may write the equations (a) in the form

$$\begin{aligned} (D - 1)x + (D + 1)y &= e^{2t} \\ D^2x + Dy &= 3e^{2t} \end{aligned} \tag{b}$$

Operating on the first of equations (b) with D and on the second with $(D + 1)$, we obtain

$$\begin{aligned} (D^2 - D)x + D(D + 1)y &= De^{2t} = 2e^{2t} \\ (D^3 + D^2)x + (D + 1)\, Dy &= D(3e^{2t}) + 3e^{2t} = 9e^{2t} \end{aligned} \tag{c}$$

Subtracting the first of equations (c) from the second, we get

$$(D^3 + D)x = 7e^{2t} \tag{d}$$

The solution of this equation is

$$x = c_1 + c_2 \sin t + c_3 \cos t + \tfrac{7}{10}e^{2t} \tag{e}$$

Substituting the value of x from (e) in the second equation of (b) and integrating the resulting equation, we obtain

$$y = \tfrac{1}{10}e^{2t} - c_2 \cos t + c_3 \sin t + c_4 \tag{f}$$

There may be too many constants of integration in the solution given by (e) and (f). It can be shown* that the number of constants of integration to be expected in the general solution of a system of simultaneous linear differential equations with constant coefficients is the same as the degree in D of the determinant of the equations. Thus, the determinant of (b) is

$$\begin{vmatrix} D - 1 & D + 1 \\ D^2 & D \end{vmatrix}$$

its degree is 3, and there should be only three constants of integration in the solution of (a).

The general procedure in finding any relation that may exist between the constants is to substitute the solution in one of the original equations, simplify as much as possible, and equate the coefficients of like terms in the two members of the result. Substituting the value of x from (e) and y from (f) in the first equation of (b), we obtain after simplification

$$c_4 - c_1 = 0 \qquad \text{or} \qquad c_4 = c_1$$

It therefore appears that the solution is

$$\begin{aligned} x &= c_1 + c_2 \sin t + c_3 \cos t + \tfrac{7}{10}e^{2t} \\ y &= c_1 + c_3 \sin t - c_2 \cos t + \tfrac{1}{10}e^{2t} \end{aligned} \tag{g}$$

Remark. Extraneous constants of integration can often be avoided by deriving from the given set of equations an equation of low order to be used in finding the expression for an additional variable after the expressions for several variables have already been found. Thus, after finding (e) subtract the second equation of (b) from the first, solve the result for y, and get

$$y = -2e^{2t} + (D^2 - D + 1)x \tag{h}$$

Now, substitute the right member of (e) for x in (h), simplify, and get the second equation of (g).

* See E. L. Ince, "Ordinary Differential Equations," p. 150.

Exercises

Solve the following systems of differential equations, where $D = d/dt$:

1. $x + Dy = 0$, $(D - 1)x + (D - 1)y = 2t$.
2. $(D - 1)x + Dy = 0$, $Dx + 2\,Dy = 4e^{2t}$.
3. $D^2x = y$, $D^2y = x + 1$.
4. $(D^2 - 3)x - 4y = 0$, $x + (D^2 + 1)y = 0$.
5. $x + y + z = t$, $Dx + z = 0$, $2x - Dy = 0$.
6. $(2D^2 - 4)y - Dx = 4t$, $(4D - 3)x + 2\,Dy = 0$.
7. $(D^2 - 1)x + 8\,Dy = 16e^t$, $Dx + 3(D^2 + 1)y = 0$.
8. Find the particular solution of the system

$$(D^2 - 3)x - 4y + 3 = 0 \qquad (D^2 + 1)y + x + 5 = 0$$

for which $x = y = Dx = Dy = 0$ when $t = 0$.

9. $Dx - 3y = 0$, $Dy + z = 0$, $Dz + y = 0$.
10. $(D^2 - 1)x + 2(D + 1)y + (D + 1)z = e^t$, $(D + 1)^2x + 2(D + 1)y - (D + 1)z = 0$, $(D - 1)x - 2y - z = 0$.

57 Summary and review exercises

An important method of solving a differential equation written in operator form uses factorization of the operator and substitution (see examples 1 and 2, §44).

Referring to the linear differential equation

$$a_0\, D^n y + a_1\, D^{n-1} y + \cdots + a_n y = X(x) \tag{1}$$

by $L(D)y = X$, we found that its general solution is

$$y = y_c + y_p \tag{2}$$

where y_c is the general solution of $L(D)y = 0$ and y_p is a particular solution of (1). To solve $L(D)y = 0$ when the coefficients a are all constants, find the roots r_i, $i = 1, 2, \ldots, n$, of $L(m) = 0$ and write

$$y = c_1 e^{r_1 x} + c_2 e^{r_2 x} + \cdots + c_n e^{r_n x} \tag{3}$$

The shortest methods of finding particular solutions of (1) are based on equation (1), §53, and equations (8), (9), (11), (14), §54. The methods used in §§51 and 52 are instructive but cumbersome. To solve (1) when X, DX, D^2X, . . . without end contains infinitely many linearly independent terms or when some of the a's contain x, use the method of variation of parameters, §55.

To solve a system of differential equations use the procedure of §56.

Exercises

Solve the following differential equations, and determine the constants of integration when initial conditions are given:

1. $(D^2 - 2D - 3)y = 12xe^{-x}$. Use the method of examples 1 and 2, §44.

2. $(D^2 - 9)y = 12x(e^{3x} + e^{-3x})$. Use method of §44.

3. $(D^2 + k^2)y = 0$

4. $(D^4 - a^4)y = 0$

5. $(D^2 - 4)y = 16$

6. $(D^2 + 2D + 1)y = e^{-x}$

7. $(D^3 - D^2)y = 0$

8. $(D - 1)^3(D + 2)^2y = 0$

9. $(D^2 + 2D + 3)y = 0$

10. $D^2(D - 3)y = 54x$

11. $(D^3 + D^2 - 2D)y = 8x$

12. $(D^2 + 2D - 8)y = 16x - 12$

13. $(D^2 - 4)y = 0$; $y = 3$, $Dy = 6$ when $x = 0$.

14. $D^2(D - 3)y = 6$; $y = 1$, $Dy = -1$, $D^2y = -2$ when $x = 1$.

15. $(D - 1)^2y = x^2 - 3x$; $y = 0$, $Dy = 2$ when $x = 0$.

16. $[(D - 1)^2 + 4]^2y = e^x \cos x$

17. $(D^2 + 9)y = 12 \cos 3x$. Use the method of example 4, §54.

18. $(D^2 + 4)y = 8 \tan^2 2x$

19. $(x^2D^2 - 3xD + 3)y = 16x^{-1}$. Check that $y_1 = x$ and $y_2 = x^3$ satisfy $(x^2D^2 - 3xD + 3)y = 0$.

20. $(D - 7)x + y = 0$, $Dy + 3x - 5y = 0$. Here $D = d/dt$.

21. $(D^2 - 1)y = \sin x$; $y = 1$, $Dy = -\frac{3}{2}$ when $x = 0$.

22. $(D^2 - 4)y = 25e^{3x}$; $y = 0$, $Dy = 5$ when $x = 0$.

23. $D(D^2 + 1)y = -6 \cos 2x$; $y = 1$, $Dy = -3$, $D^2y = 0$ when $x = \frac{1}{2}\pi$.

24. $(D - 1)(D^2 + 1)y = 5e^x \sin x$; $y = 0$, $Dy = -2$, $D^2y = -3$ when $x = 0$.

25. $D^2(D + 1)y = 4xe^x$; $y = -4$, $Dy = -4$, $D^2y = 0$ when $x = 0$.

26. $D(D - 1)(D + 1)y = 6 + 130 \cos 5x$; $y = 1$, $Dy = -12$, $D^2y = 1$ when $x = 0$.

27. $(D^2 + 2D + 2)y = e^x(\cos 2x - 8 \sin 2x)$; $y = 0$, $Dy = 2$ when $x = 0$.

28. $(D^3 + 4D)y = 16 \sin 2x$ **29.** $(D - 1)^2 y = x^5 e^x$

30. $(D^2 - 4)y = 64xe^{2x}$
31. $[(D - 1)^2 + 1]y = 24e^x \sin 3x$
32. $(D^3 - 3D^2 + 4D - 2)y = e^x \sec x$
33. $(D^2 - 2D + 2)y = e^x(\tan x + \cot x)$
34. $(2D^2 + D - 3)y = 15e^x + 20e^{-3x/2}$
35. $D(D - 1)^2 y = xe^x$; $y = 0$, $Dy = 0$, $D^2 y = 0$ when $x = 0$.

(36–37) Use the method of example 4, §54, in solving:

36. $(D^2 - 2D + 2)y = 4e^x \sin x$
37. $(D^2 + 4D + 5)y = 12e^{-2x} \cos x$
38. $(D^2 + 2D - 8)y = (6x^{-1} - x^{-2})e^{2x}$
39. $(D^2 + 1)y = -x^{-2} \sin x + 2x^{-1} \cos x$
40. $(D^2 + 1)y = 2 \sec^3 x$
41. $(xD^2 - D - 4x^3)y = 24x^3 e^{2x^2}$. Check that $y_1 = e^{x^2}$ and $y_2 = e^{-x^2}$ both satisfy $(xD^2 - D - 4x^3)y = 0$.
42. $Dx + Dy + 3x = \sin t$, $Dx + y - x = \cos t$. Here $D = d/dt$.
43. $Dx + Dy - x - y = 2e^t$, $Dx - Dz = 0$, $D(x + y + z) = e^t$. Here $D = d/dt$.

7

Laplace Transforms

58 Introduction

Oliver Heaviside, in an effort to solve ordinary linear differential equations with facility, devised a method of operational calculus which led to Laplace transforms. The methods of Laplace transforms are very simple, and they give solutions of differential equations satisfying given boundary conditions directly without the use of the general solution. Since these particular solutions are the ones usually required in physics, mechanics, chemistry, and various fields of practical research, transforms are highly important. Although Laplace transforms have extensive applications, we shall be concerned mainly with their use in solving differential equations. To obtain a thorough knowledge of transforms, the reader should study detailed treatments of them.*

* Consult, for example: Ruel V. Churchill, "Operational Mathematics," 2d ed., McGraw-Hill Book Company, New York, 1958; Murray F. Gardner and John

59 Definition of a Laplace transform

DEFINITION. *If $f(t)$ is defined for all values of t in the interval $t > 0$ and if p is a real number such that the integral $F(p)$ defined by*

$$F(p) = \int_0^\infty e^{-pt}f(t)\,dt \tag{1}$$

converges for some finite value of p and all greater values, then $F(p)$ is called the Laplace transform *of $f(t)$.*

Observe the notation f *for a function and* F *for its transform.* Also, we shall use the notation based on (1),

$$T\{f(t)\} = F(p) \tag{2}$$

An example will give some familiarity with the concept.

Example. Find the transform of: (*a*) $f(t) = 1$; (*b*) $f(t) = t$; (*c*) $f(t) = e^{at}$, $p > a$.

Solution. Using (1) we get

$$(a)\ F(p) = T\{1\} = \int_0^\infty e^{-pt}1\,dt = \left[-\frac{1}{p}e^{-pt}\right]_0^\infty = \frac{1}{p}$$

$$(b)\ F(p) = \int_0^\infty e^{-pt}t\,dt = \left[-\frac{te^{-pt}}{p} - \frac{e^{-pt}}{p^2}\right]_0^\infty = \frac{1}{p^2}$$

$$(c)\ F(p) = \int_0^\infty e^{-pt}e^{at}\,dt = \left[\frac{-e^{-(p-a)t}}{p-a}\right]_0^\infty = \frac{1}{p-a}$$

Exercises

Use (1) to verify the following equations:

1. $T\{at\} = \dfrac{a}{p^2}$
2. $T\{t^2\} = \dfrac{2}{p^3}$
3. $T\{t^3\} = \dfrac{3!}{p^4}$
4. $T\{\sin kt\} = \dfrac{k}{p^2+k^2}$
5. $T\{\cos kt\} = \dfrac{p}{p^2+k^2}$
6. $T\{a+bt\} = \dfrac{a}{p} + \dfrac{b}{p^2}$

Use (1) to find the transform in terms of p indicated by:

7. $T\{\frac{1}{2}(e^{at}+e^{-at})\}$
8. $T\{\frac{1}{2}(e^{at}-e^{-at})\}$
9. $T\{te^{-at}\}$
10. $T\{e^{at}\sin kt\}$

L. Barnes, "Transients in Linear Systems," John Wiley & Sons, Inc., New York, 1949; H. S. Carslaw and J. C. Jaeger, "Operational Methods in Applied Mathematics," Oxford University Press, Fair Lawn, N.J., 1949.

11. Use (1) to verify the Laplace transform of the expression for $f(t)$ in each line of Table 1 §60 beginning with one of the numbers: 2, 4, 6, 7, 8, 12, 15.

60 Some properties of Laplace transforms

DEFINITION I. *A function $f(t)$ is* sectionally continuous *on a finite closed interval if the interval consists of a set of subintervals for each of which $f(t)$ is continuous at every interior point and approaches a finite limit as t approaches either end point from within the interval.*

DEFINITION II. *A function $f(t)$ is of* exponential order as t tends to infinity *if there exist numbers α, M, and L such that*

$$|f(t)| < Me^{\alpha t} \qquad \text{when } t \geqq L \tag{1}$$

For example, t^3e^{at} describes a function in question, for

$$|t^3e^{at}| < e^{(a+1)t} \qquad \text{when } t > 5$$

In fact, the product of e^{at} and any polynomial in t is easily shown to be of exponential order as t tends to infinity. However $e^{t^{3/2}}$ is not of exponential order. The following theorem is of basic importance.

THEOREM. *If $f(t)$ is sectionally continuous for all finite intervals in the domain $t \geqq 0$ and if $f(t)$ is of exponential order as t tends to infinity, then the Laplace transform $F(p)$, defined by*

$$F(p) = \int_0^\infty e^{-pt}f(t)\,dt \qquad p > \alpha \tag{2}$$

exists and converges absolutely.

Proof. To see this, note first that $\int_0^m e^{-pt}f(t)\,dt$ exists if m is finite; for it is the sum of the integrals along the continuous parts of the curve. Therefore, considering (1) we see that, under the conditions mentioned, there is a number M such that

$$|e^{-pt}f(t)| < Me^{-(p-\alpha)t} \qquad p > \alpha$$

Since $\int_0^\infty Me^{-(p-\alpha)t}\,dt$, $p > \alpha$, is $M/(p - \alpha)$,* the integral of the left member exists and $\int_0^\infty e^{-pt}f(t)\,dt$, $p > \alpha$, converges absolutely.

In the material on transforms, *we shall assume, unless otherwise mentioned, that all functions $f(t)$ considered for transformation are sectionally continuous in every finite interval in the domain $t \geqq 0$ and are of exponential order as t tends to infinity.*

* Observe that limit $M/(p - \alpha) \to 0$ as $p \to \infty$, so that a transform $\to 0$ as $p \to \infty$. No such a transform as $(p + 1)/p$ exists.

The relation (2) between $F(p)$ and $f(t)$ is also denoted by

$$T^{-1}\{F(p)\} = f(t) \tag{3}$$

and $T^{-1}\{F(p)\}$ is called the inverse transform of $F(p)$. It can be shown that any two inverse transforms of a function $F(p)$ differ only at the ends of continuous sections of $f(t)$. Such differences may be disregarded for our purposes, and we shall think of the inverse of a function of p as unique.

The usual way of finding transforms and inverse transforms is by means of a table listing standard functions and their transforms. Table 1 is sufficient for our purposes. Other transforms can be found by using the general relations. Reference to any line in the table is made by the first number in the line. Thus from 4, Table 1, we derive

$$\frac{1}{p-a} = T\{e^{at}\} \qquad \text{or} \qquad e^{at} = T^{-1}\left\{\frac{1}{p-a}\right\}$$

Since transforms are integrals, they obey the same laws as regards sums and constant factors; therefore

$$T\{af(t) + bg(t)\} = aT\{f(t)\} + bT\{g(t)\} = aF(p) + bG(p) \tag{4}$$

and, from this and the uniqueness of inverse transforms, we get

$$\begin{aligned} T^{-1}\{aF(p) + bG(p)\} &= aT^{-1}\{F(p)\} + bT^{-1}\{G(p)\} \\ &= af(t) + bg(t) \end{aligned} \tag{5}$$

Thus, using (4) and Table 1 with 3 for a, we get

$$\begin{aligned} T\{2 + 4e^{3t}\} &= 2T\{1\} + 4T\{e^{3t}\} = \frac{2}{p} + \frac{4}{p-3} \\ T\{4\sin 3t - 5\cos 3t\} &= 4T\{\sin 3t\} - 5T\{\cos 3t\} \\ &= \frac{12}{p^2+9} - \frac{5p}{p^2+9} \end{aligned}$$

Similarly, using (5) and Table 1, we get

$$\begin{aligned} T^{-1}\left\{\frac{8p}{p^2+16} - \frac{6}{p^2+16}\right\} &= 8T^{-1}\left\{\frac{p}{p^2+16}\right\} - \tfrac{6}{4}T^{-1}\left\{\frac{4}{p^2+16}\right\} \\ &= 8\cos 4t - \tfrac{3}{2}\sin 4t \\ T^{-1}\left\{\frac{p}{(p^2+9)^2} + \frac{7}{(p^2+9)^2}\right\} &= \frac{t}{6}\sin 3t + \frac{7}{54}(\sin 3t - 3t\cos 3t) \end{aligned}$$

Example. Express in terms of t: $T^{-1}\{(p^3 + p + 3)/(p^2 + 4)^2\}$.
Solution.

$$\frac{p^3 + p + 3}{(p^2 + 4)^2} = \frac{p(p^2 + 4) - 3p + 3}{(p^2 + 4)^2}$$
$$= \frac{p}{p^2 + 4} - \frac{3p}{(p^2 + 4)^2} + \frac{3}{(p^2 + 4)^2}$$

Hence, using the lines of Table 1 beginning with 6, 14, and 16, with $a = 2$, we get

$$T^{-1}\left\{\frac{p^3 + p + 3}{(p^2 + 4)^2}\right\}$$
$$= T^{-1}\left\{\frac{p}{p^2 + 4}\right\} + T^{-1}\left\{\frac{-3p}{(p^2 + 4)^2}\right\} + T^{-1}\left\{\frac{3}{(p^2 + 4)^2}\right\}$$
$$= \cos 2t - \tfrac{3}{4}t \sin 2t + \tfrac{3}{16}(\sin 2t - 2t \cos 2t)$$
$$= \cos 2t(1 - \tfrac{3}{8}t) + \sin 2t(\tfrac{3}{16} - \tfrac{3}{4}t)$$

Exercises

Using Table 1 and (2) to (4) find in terms of p:

1. $T\{1 + t^3\}$

2. $T\{\tfrac{1}{2} \sin 2t + \tfrac{1}{2} \cos 2t\}$

3. $T\{5 \sinh 3t - 4 \cosh 3t\}$

4. $T\{2e^{3t} \sin 2t - 3e^{3t} \cos 2t\}$

Using Table 1 find in terms of t:

5. $T^{-1}\left\{\frac{5}{p^7} + \frac{7}{p - a}\right\}$

6. $T^{-1}\left\{\frac{3}{p - m} + \frac{4}{p + m}\right\}$

7. $T^{-1}\left\{\frac{5 + 5p}{(p^2 + 25)^2}\right\}$

8. $T^{-1}\left\{\frac{2p + 5}{p^2 - 25}\right\}$

9. $T^{-1}\{(p^2 + 5p + 19)/[(p^2 + 4)(p^2 + 9)]\}$. *Hint:* $p^2 + 5p + 19 = (p^2 + 4) + 5p + 15$.

Express directly in terms of p or t:

10. $T\{e^{2t} \sin 4t - 3e^{2t} \cos 4t\}$

11. $T\{t \sin t - 2 \cos t\}$

12. $T\{t^n + n!t\}$

13. $T\{t^n e^t\}$

14. $T^{-1}\left\{\frac{p^3 + p}{(p^2 + 4)^2}\right\}$

15. $T^{-1}\left\{\frac{p^2 + 2p}{(p^2 + 1)(p^2 + 4)}\right\}$

Table 1. Laplace Transforms

	$F(p)$	$f(t)$
1	$\frac{1}{p}$	1
2	$\frac{1}{p^2}$	t
3	$\frac{1}{p^n},\ n = 1, 2, \ldots$	$\frac{t^{n-1}}{(n-1)!}$
4	$\frac{1}{p-a},\ p > a$	e^{at}
5	$\frac{1}{(p-a)^n},\ p > a$	$\frac{e^{at}t^{n-1}}{(n-1)!}$
6	$\frac{p}{p^2+a^2}$	$\cos at$
7	$\frac{a}{p^2+a^2}$	$\sin at$
8	$\frac{p-a}{(p-a)^2+b^2}$	$e^{at}\cos bt$
9	$\frac{1}{(p-a)^2+b^2}$	$\frac{1}{b}e^{at}\sin bt$
10	$\frac{p}{p^2-a^2},\ p > \lvert a\rvert$	$\cosh at$
11	$\frac{a}{p^2-a^2},\ p > \lvert a\rvert$	$\sinh at$
12	$\frac{1}{(p-a)(p-b)},\ a \neq b$	$\frac{1}{a-b}(e^{at}-e^{bt})$
13	$\frac{p}{(p-a)(p-b)},\ a \neq b$	$\frac{1}{a-b}(ae^{at}-be^{bt})$
14	$\frac{p}{(p^2+a^2)^2}$	$\frac{t}{2a}\sin at$
15	$\frac{p^2-a^2}{(p^2+a^2)^2}$	$t\cos at$
16	$\frac{1}{(p^2+a^2)^2}$	$\frac{1}{2a^3}(\sin at - at\cos at)$
17	$\frac{p}{(p^2+a^2)(p^2+b^2)},\ a^2 \neq b^2$	$\frac{\cos at - \cos bt}{b^2-a^2}$
18	$\frac{1}{(p^2+a^2)(p^2+b^2)},\ a^2 \neq b^2$	$\frac{a\sin bt - b\sin at}{ab(a^2-b^2)}$
19	$\frac{1}{p^2+b^2}\cdot\frac{1}{(p+\alpha)^2+a^2}$	$\frac{-2b\alpha\cos bt + N\sin bt}{b(N^2+4\alpha^2b^2)} + \frac{e^{-\alpha t}(2a\alpha\cos at + M\sin at)}{a(M^2+4\alpha^2a^2)}$
	where $M = \alpha^2+b^2-a^2,\ N = \alpha^2+a^2-b^2$	

Table 1. Laplace Transforms (Continued)

	$F(p)$	$f(t)$
20	$\dfrac{p}{p^2+b^2}\cdot\dfrac{1}{(p+\alpha)^2+a^2}$	$\dfrac{N\cos bt+2b\alpha\sin bt}{4b^2\alpha^2+N^2}+\dfrac{e^{-\alpha t}(aP\cos at-\alpha Q\sin at)}{a(4a^2\alpha^2+M^2)}$
	where $M=\alpha^2+b^2-a^2$, $N=\alpha^2+a^2-b^2$, $P=b^2-a^2-\alpha^2$, $Q=b^2+a^2+\alpha^2$	
21	$\dfrac{1}{(p+a)(p^2+c^2)},\ c\neq 0$	$\dfrac{ce^{-at}-c\cos ct+a\sin ct}{c(a^2+c^2)}$
22	$\dfrac{p}{(p+a)(p^2+c^2)},\ c\neq 0$	$\dfrac{a\cos ct+c\sin ct-ae^{-at}}{(a^2+c^2)}$
23	$\dfrac{1}{(p+a)(p+b)(p+c)}$, a, b, c distinct	$\dfrac{(b-c)e^{-at}+(c-a)e^{-bt}+(a-b)e^{-ct}}{(b-a)(a-c)(c-b)}$
24	$\dfrac{p}{(p+a)(p+b)(p+c)}$, a, b, c distinct	$\dfrac{a(b-c)e^{-at}+b(c-a)e^{-bt}+c(a-b)e^{-ct}}{(a-b)(b-c)(c-a)}$
25	$\dfrac{1}{(p-a)(p-b)^2},\ a\neq b$	$\dfrac{e^{at}-[1+(a-b)t]e^{bt}}{(a-b)^2}$
26	$\dfrac{1}{p^n(p-a)},\ a\neq 0$	$a^{-n}e^{at}-\left[\dfrac{t^{n-1}}{a(n-1)!}+\dfrac{t^{n-2}}{a^2(n-2)!}+\cdots+\dfrac{t^0}{a^n(0!)}\right]$
27	$\dfrac{\Gamma(k)}{p^k},\ k>0$	t^{k-1}
28	$\dfrac{\Gamma(\frac{1}{2})}{p^{1/2}}=\dfrac{\sqrt{\pi}}{p^{1/2}}$	$t^{-1/2}$
29	$\dfrac{1}{p^{1/2}}$	$\dfrac{1}{\sqrt{\pi t}}$
30	$\dfrac{1}{p^{3/2}}$	$\dfrac{2}{\sqrt{t/\pi}}$
31	$\dfrac{1}{p\sqrt{p+m^2}}$	$m^{-1}\operatorname{erf}(m\sqrt{t})$
32	$\dfrac{1}{p\sqrt{p-m^2}}$	$m^{-1}\dfrac{2}{\sqrt{\pi}}\displaystyle\int_0^t e^{r^2}\,dr=\dfrac{2m^{-1}}{\sqrt{\pi}}\sum_{n=0}^{\infty}\dfrac{r^{2n+1}}{(2n+1)(n!)}$
33	$\dfrac{1}{(\sqrt{p}+a)}$	$\dfrac{1}{\sqrt{\pi t}}-ae^{a^2t}[1-\operatorname{erf}(a\sqrt{t})]$
34	$\dfrac{\sqrt{p}}{p-a^2}$	$\dfrac{1}{\sqrt{\pi t}}+ae^{a^2t}\operatorname{erf}(a\sqrt{t})$
35	$\dfrac{1}{\sqrt{p}\,(p-a^2)}$	$a^{-1}e^{a^2t}\operatorname{erf}(a\sqrt{t})$
36	$\dfrac{b^2-a^2}{(p-a^2)(b+\sqrt{p})}$	$e^{a^2t}[b-a\operatorname{erf}(a\sqrt{t})]-be^{b^2t}[1-\operatorname{erf}(b\sqrt{t})]$

Table 2. Formulas Involving Laplace Transforms

(I)	$T\{af_1(t) + bf_2(t)\} = aF_1(p) + bF_2(p)$
(II)	$T\{e^{-at}f(t)\} = F(p + a)$
(III)	$e^{at}T^{-1}\{F(p + a)\} = T^{-1}\{F(p)\}$
(IV)	$T\left\{\frac{1}{m}f\left(\frac{t}{m}\right)\right\}, m > 0 = F(mp)$
(V)	$T\{(-t)f(t)\ dt\} = F'(p)$
(VI)	$T\{(-t)^k f(t)\} = F^{(k)}(p)$
(VII)	$T\left\{\int_0^t g(t - r) \cdot f(r)\ dr\right\} = T\left\{\int_0^t f(t - r) \cdot g(r)\ dr\right\} = G(p) \cdot F(p)$
(VIII)	$T\left\{\int_0^t f(r)\ dr\right\} = \frac{1}{p}F(p)$
(IX)	$T\left\{\int_0^t \int_0^r f(\theta)\ d\theta\ dr\right\} = \frac{1}{p^2}F(p)$
(X)	$T\{f^{(n)}(t)\} = p^nF(p) - p^{n-1}f(0) - p^{n-2}f'(0) - \cdots - f^{(n-1)}(0)$
(XI)	$T\left\{\frac{1}{t}f(t)\right\} = \int_p^\infty F(p)\ dp$, where $T\{f(t)\} = F(p)$
(XII)	If $f(x) = 0$ when $x < 0$, $T\{f(t - b)\} = e^{-bp}F(p)$.
(XIII)	When $f(t + a) = f(t)$, $T\{f(t)\} = \left(\int_0^a e^{-pt}f(t)\ dt\right) \Big/ (1 - e^{-ap})$
(XIV)	When $f(t + a) = -f(t)$, $T\{f(t)\} = \left(\int_0^a e^{-pt}f(t)\right) \Big/ (1 + e^{-ap})$
(XV)	$T\left\{\sum_{i=1}^n \frac{N(a_i)e^{a_it}}{(a_i - a_1)(a_i - a_2) \cdots (a_i - a_n)}\right\} = T\left\{\frac{N(p)}{(p - a_1)(p - a_2) \cdots (p - a_n)}\right\}$, where $a_i - a_i$ is omitted from the denominator of the left member and no two a's are equal.
(XVI)	$T\left\{\sum_{i=1}^m \frac{N(a_i)}{M'(a_i)} e^{a_it}\right\} = \frac{N(p)}{M(p)}$
(XVII)	Part of $T^{-1}\{N/[M(p - a)^k]\}$. See (5), §64.
(XVIII)	$T^{-1}\{N/(M[(p - a)^2 + b^2]^k)\}$. See (6), §65.

61 A basic formula of Laplace transforms

The highly important equation (1) below is used to derive many transforms from given ones. Its simplifying properties are closely related to those of the operator equation (1) of §53.

Replacing p by $p + g$ in (1), §59, we get

$$F(p + g) = \int_0^\infty e^{-(p+g)t}f(t)\ dt = \int_0^\infty e^{-pt}[e^{-gt}f(t)]\ dt$$

From this we see that

$$T\{e^{-gt}f(t)\} = F(p + g) \tag{1}$$

For example, from Table 1, in §60, we get $T\{\sin at\} = a/(p^2 + a^2)$; taking $a = 2$ in this and using (1) with $g = 3$ we get

$$T\{e^{-3t}\sin 2t\} = \frac{2}{(p+3)^2+4}$$

Obviously from any given formula of transforms we can obtain any number of related formulas by using (1) with various values of g.

Again, equating the inverse transforms of the members of (1), and observing that $f(t) = T^{-1}F(p)$, we get

$$e^{gt}T^{-1}\{F(p+g)\} = f(t) \tag{2}$$

$$e^{gt}T^{-1}\{F(p+g)\} = T^{-1}\{F(p)\} \tag{3}$$

Using (2) we easily derive from 15, Table 1, that

$$T^{-1}\left\{\frac{(p-3)^2-16}{[(p-3)^2+16]^2}\right\} = e^{3t}(t\cos 4t)$$

The following example will illustrate uses of (1), (2), and (3).

Example. Express in terms of t: $T^{-1}\{(18p+72)/[(3p+9)^2+18]^2\}$.

Solution. Making some algebraic changes and using (3) and (2) together with 14 and 16, Table 1, we get

$$T^{-1}\left\{\frac{18p+72}{[(3p+9)^2+18]^2}\right\} = T^{-1}\left\{\frac{18}{81}\,\frac{(p+3)+1}{[(p+3)^2+2]^2}\right\}$$

$$= \tfrac{2}{9}\,e^{-3t}T^{-1}\left\{\frac{p+1}{(p^2+2)^2}\right\}$$

$$= \tfrac{2}{9}e^{-3t}\left[\frac{t}{2\sqrt{2}}\sin\sqrt{2}\,t + \tfrac{1}{8}\sqrt{2}\,(\sin\sqrt{2}\,t - \sqrt{2}\,t\cos\sqrt{2}\,t)\right]$$

$$= \tfrac{1}{36}\sqrt{2}\,e^{-3t}(2t\sin\sqrt{2}\,t - \sqrt{2}\,t\cos\sqrt{2}\,t + \sin\sqrt{2}\,t)$$

Exercises

Use (1) to derive in Table 1 §60, formula:

1. 4 from 1. **2.** 5 from 3. **3.** 8 from 6. **4.** 9 from 7.

Use (3) to verify each equation, and express each left member in terms of t:

5. $T^{-1}\left\{\dfrac{p+3}{(p+3)^2+1}\right\} = e^{-3t}T^{-1}\left\{\dfrac{p}{p^2+1}\right\}$

6. $e^{-2t}T^{-1}\left\{\dfrac{p}{(p-2)^2+9}\right\} = T^{-1}\left\{\dfrac{p+2}{p^2+9}\right\}$

Use (1) to (3) and 10, 11, 14, and 18 from Table 1 to find in terms of t:

7. $T^{-1}\left\{\dfrac{p-m}{(p-m)^2-a^2}\right\}$ **8.** $T^{-1}\left\{\dfrac{1}{(p-m)^2-a^2}\right\}$

9. $T^{-1}\left\{\dfrac{p+m}{(p+m)^2+a^2}\right\}$

10. $T^{-1}\left\{\dfrac{1}{[(p+m)^2+a^2][(p+m)^2+b^2]}\right\}$

Use Table 1 and equations (1) to (3) to express in terms of p:

11. $T\{e^{-3t}\sin 5t\}$ **12.** $T\{e^{3t}\cos 5t\}$ **13.** $T\{te^{-3t}\sin 5t\}$

Express each transform in terms of p and each inverse transform in terms of t:

14. $T\{t^3e^{2t}\}$ **15.** $T\{e^{-t}\sinh t\}$ **16.** $T\{e^{-2t}(\sin 4t - 4t\cos 4t)\}$

17. $T^{-1}\left\{\dfrac{b(p-a)}{(p-a)^2+k^2}\right\}$ **18.** $T^{-1}\left\{\dfrac{5!}{(p-5)^6}\right\}$

19. $T^{-1}\left\{\dfrac{2p+14}{(p-2)^2+36}\right\}$

20. $T^{-1}\{(3p+23)/((p+2)[(p+3)^2+16])\}$. First use (3) with $g = 3$ and then 21 and 22, Table 1.

21. $T^{-1}\left\{\dfrac{1}{(p-2)[(p-3)^2+9]}\right\}$

22. $T^{-1}\left\{\dfrac{p}{(p-4)[(2p+8)^2+16]}\right\}$

23. $T^{-1}\{(2p+6)/[p^2+8p+17]^2\}$. Note that $p^2+8p+17 = (p+4)^2+1$.

62 Solving differential equations by transforms

Let $Df(t) = df/dt$ satisfy the conditions assumed for $f(t)$ in §60. Then

$$\int_0^\infty e^{-pt}\,Df(t)\,dt \tag{1}$$

exists. Applying integration by parts to (1), we get

$$\int_0^\infty e^{-pt}\,Df(t)\,dt = [e^{-pt}f(t)]_0^\infty + p\int_0^\infty e^{-pt}f(t)\,dt \tag{2}$$

Since $Df(t)$ exists, $f(t)$ is continuous when $t \geqq 0$, and therefore $[e^{-pt}f(t)]_0^\infty = -f(0)$. Hence from (2) we get

$$\int_0^\infty e^{-pt}\, Df(t)\, dt = T\{Df(t)\} = pF(p) - f(0) \tag{3}$$

Similarly if $D^2f(t) = d^2f(t)/dt^2$ satisfies the conditions assumed for $f(t)$ in §60, apply (3) to $D^2f(t)$ and obtain

$$T\{D^2f(t)\} = T\{D[Df(t)]\} = p[F(p) - F(0)] - Df(0)$$

or

$$T\{D^2f(t)\} = p^2F(p) - pf(0) - Df(0) \tag{4}$$

Assuming that $D^nf(t)$ satisfies the conditions placed on $f(t)$ in §60, and using (3) repeatedly, we get

$$T\{D^nf(t)\} = p^nF(p) - p^{n-1}f(0) - p^{n-2}Df(0) - \cdots - D^{n-1}f(0) \tag{5}$$

The following examples will indicate how (3), (4), and (5) are used in solving differential equations.

Example 1. Find the solution of

$$Dy + y = e^{-t} \qquad \text{if } y = 5 \text{ when } t = 0 \tag{a}$$

Solution. Equating the transforms of the members of (a), we get

$$T\{Dy + y\} = T\{Dy\} + T\{y\} = T\{e^{-t}\} = \frac{1}{p+1} \tag{b}$$

Let $T\{y\} = Y$, and from (3) obtain $T\{Dy\} = pY - 5$. Substituting these values for $T(y)$ and $T\{Dy\}$ in (b), obtain

$$pY - 5 + Y = \frac{1}{p+1} \tag{c}$$

The solution of (c) for Y is

$$Y = \frac{1}{(p+1)^2} + \frac{5}{p+1} \tag{d}$$

Equating the inverse transforms of the members of (d), obtain

$$y = te^{-t} + 5e^{-t} \tag{e}$$

Note from (e) that when $t = 0$, $y = 0e^0 + 5e^0 = 5$. Also substitution of y from (e) in (a) gives

$$(-te^{-t} + e^{-t} - 5e^{-t}) + te^{-t} + 5e^{-t} = e^{-t}$$

Hence (e) satisfies (a) not only when $t \geqq 0$, but for all values of t.

Example 2. Solve the equation

$$D^2y - 2\,Dy - 3y = 6e^t \qquad \text{if } y = 1 \text{ and } Dy = 3 \text{ when } t = 0 \qquad (a)$$

Solution. From the given differential equation, we get

$$T\{D^2y - 2\,Dy - 3y\} = T\{6e^t\} = \frac{6}{p - 1} \qquad (b)$$

From (4) and (3) and initial conditions $y(0) = 1$, $Dy(0) = 3$, we get

$$T\{D^2y\} = p^2y - p - 3 \qquad T\{Dy\} = pY - 1 \qquad (c)$$

Using (*c*) in (*b*), we obtain

$$p^2Y - p - 3 - 2pY + 2 - 3Y = \frac{6}{p - 1} \qquad (d)$$

Solving (*d*) for Y, we get

$$Y = \frac{6}{(p - 1)(p + 1)(p - 3)} + \frac{1}{p - 3} \qquad (e)$$

Equating the inverse transforms of the members of (*e*), and using 23 and 4 from Table 1 in §60, we get

$$y = -\tfrac{3}{2}e^t + \tfrac{3}{4}e^{-t} + \tfrac{7}{4}e^{3t} \qquad (f)$$

A check shows that (*f*) satisfies (*a*) not only when $t \geqq 0$ but for all values of t.

Remark. To obtain a general solution of an equation of the nth order, proceed as above using a set of arbitrary constants for y, Dy, . . . , $D^{n-1}y$ when $t = 0$.

Exercises

1. To find the solution of $D^2y + Dy = 2e^t$ for which $y = 5$ and $Dy = -1$ when $t = 0$, show that

$$p^2Y - 5p + 1 + pY - 5 = \frac{2}{p - 1}$$

Solve this for Y, and then use transforms to get $y = 2 + e^t + 2e^{-t}$.

Find by means of transforms the solution of each equation satisfying the indicated initial conditions:

2. $Dy - y = 2$; $y = 0$ when $t = 0$.
3. $(D^2 + 2D - 3)y = 0$; $y = 0$, $Dy = -8$ when $t = 0$.
4. $(D^2 + 4)y = 12$; $y = 6$, $Dy = -3$ when $t = 0$.

5. $(D^3 + 9D)y = 0;\ y = 0,\ Dy = 0,\ D^2y = 5$ when $t = 0$.

6. $(D^2 + 1)^2y = 0;\ y = 0,\ Dy = 0,\ D^2y = 0,\ D^3y = 5$ when $t = 0$.

7. $(D^2 + 9)y = 12 \cos 3t;\ y = 2,\ Dy = 5$ when $t = 0$.

8. $(D^2 - 6D + 13)y = 26;\ y = 0,\ Dy = 2$ when $t = 0$. *Hints:*

$$T^{-1}\left\{\frac{26}{p(p^2 - 6p + 13)}\right\} = T^{-1}\left\{\frac{26}{p[(p-3)^2 + 4]}\right\}$$
$$= 13e^{3t}T^{-1}\left\{\frac{2}{(p+3)(p^2+4)}\right\}$$

9. $(D^2 - 1)(D + 3)y = 0;\ y = 1,\ Dy = 1,\ D^2y = -1$ when $t = 0$.

10. $(D^3 - 6D^2 + 13D)y = 0;\ y = 1,\ Dy = -4,\ D^2y = 2$ when $t = 0$.

11. $D^3(D - 1)y = 12;\ y = Dy = D^2y = D^3y = 0$ when $t = 0$. *Hint:* Use 26, Table 1, §60 with $n = 4$.

12. $(D^2 + 2D + 2)y = 2e^{-t} \cos t;\ y = 2,\ Dy = -2$ when $t = 0$.

13. $(D + 1)^2y = 6te^{-t};\ y = 2,\ Dy = 5$ when $t = 0$.

14. $D^2(D - 1)y = 12;\ y = 1,\ Dy = 1,\ D^2y = -2$ when $t = 0$.

15. $(D + 1)(D^2 - 2D + 2)y = 0;\ y = 2,\ Dy = 0,\ D^2y = -1$ when $t = 0$.

63 Derivatives and products of transforms

The methods treated in this section enable the student to derive formulas involving transforms to solve difficult types of equations, and to provide short cuts and simplifications in using transforms.

Equating the derivatives with respect to p of the sides of (1), §59, we get*

$$\frac{d}{dp}F(p) = \int_0^\infty -te^{-pt}f(t)\,dt = \int_0^\infty e^{-pt}[-tf(t)]\,dt$$

or

$$\frac{d}{dp}F(p) = T\{-tf(t)\} \tag{1}$$

* The conditions under which differentiation with respect to p under the integral sign is valid are satisfied provided that $f(t)$ fulfills the conditions prescribed in §60 for the existence of $T\{f(t)\}$. See R. Courant, "Differential and Integral Calculus," vol. II, p. 312.

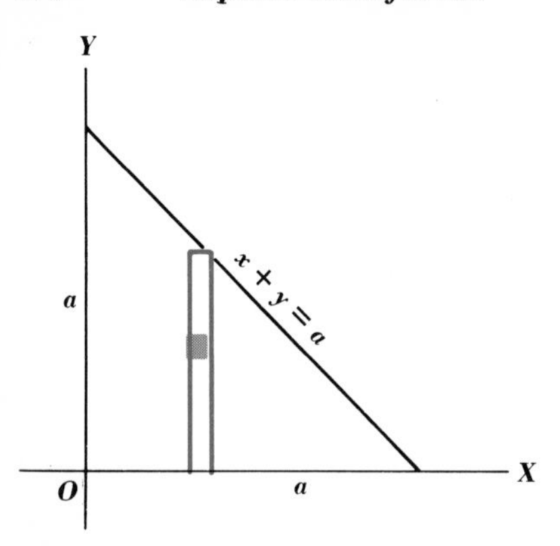

Figure 1

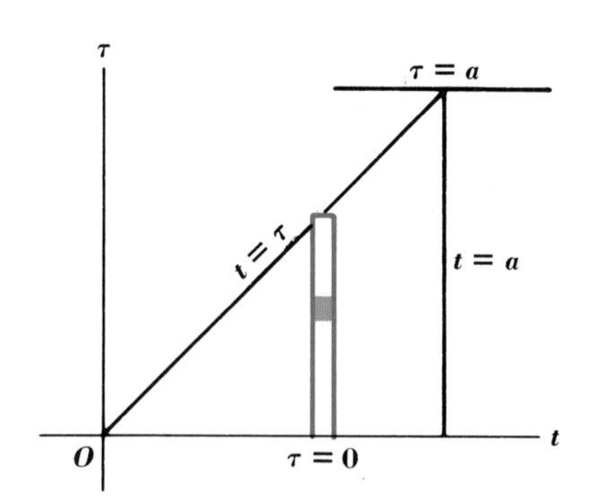

Figure 2

Thus, from $1/p = T\{1\}$ obtain

$$\frac{d}{dp}\left(\frac{1}{p}\right) = T\{-t \cdot 1\} \qquad \text{or} \qquad -\frac{1}{p^2} = T\{-t\}$$

$$\frac{d}{dp}\left(-\frac{1}{p^2}\right) = T\{(-t)^2 \cdot 1\} \qquad \text{or} \qquad \frac{2}{p^3} = T\{t^2\}$$

and so on. Also, from $a/(p^2 + a^2) = T\{\sin at\}$ obtain

$$\frac{d}{dp}\frac{a}{p^2 + a^2} = \frac{-2ap}{(p^2 + a^2)^2} = T\{(-t)\sin at\}$$

$$\frac{d}{dp}\frac{-2ap}{(p^2 + a^2)^2} = \frac{6ap^2 - 2a^3}{(p^2 + a^2)^3} = T\{(-t)^2 \sin at\}$$

Applying (1) repeatedly, we get

$$\frac{d^n}{dp^n}F(p) = T\{(-t)^n f(t)\} \tag{2}$$

The formula relating to the product of transforms is

$$G(p) \cdot F(p) = T\left\{\int_0^t g(t - \tau)f(\tau)\,d\tau\right\} \tag{3}$$

By the definition of a transform, we have

$$G(p) \cdot F(p) = \int_0^\infty e^{-px}g(x)\,dx \cdot \int_0^\infty e^{-py}f(y)\,dy$$

$$= \int_0^\infty \int_0^\infty e^{-p(x+y)}g(x)f(y)\,dx\,dy \tag{4}$$

Reference to the triangles of Figs. 1 and 2 indicates that (4) may be written

$$G(p) \cdot F(p) = \lim_{a \to \infty} \int_0^a \int_0^{a-x} e^{-p(x+y)}g(x)f(y)\,dy\,dx \tag{5}$$

Now, make the change of variables*

$$x = t - \tau \qquad y = \tau \tag{6}$$

in (5), and obtain

$$G(p) \cdot F(p) = \lim_{a \to \infty} \int_0^a e^{-pt} \left[\int_0^t g(t - \tau) f(\tau)\, d\tau \right] dt \tag{7}$$

Figure 2 indicates the reason for the limits of integration in (7). From this we see that (3) follows.

It is easy to show that

$$T^{-1}\{G(p) \cdot F(p)\} = T^{-1}\{F(p) \cdot G(p)\} \tag{8}$$

In (3) make the substitution $u = t - \tau$, $du = -d\tau$, and adjust the limits of integration to get

$$\int_0^t g(t - \tau) f(\tau)\, d\tau = \int_0^t f(t - u) g(u)\, du \tag{9}$$

This equation, in view of (3), implies (8). Equation (8) shows that a *change of order of factors of a transform does not change the inverse transform.*

Evidently, equation (3) could be used repeatedly. Thus

$$H(p) \cdot [G(p) \cdot F(p)] = T\left\{ \int_0^t h(t - \tau) \left[\int_0^\tau g(\theta - \tau) \cdot f(\theta)\, d\theta \right] d\tau \right\} \tag{10}$$

Note that the right member of (3) is zero when $t = 0$, and this may be used as a check. Thus $T^{-1}\{[1/(p - 1)] \cdot [p/(p^2 + 4)]\} = \frac{1}{5}(e^t - \cos 2t) + \frac{2}{5} \sin 2t$, and this is zero when $t = 0$. Apply this check to 17, 18, 19, and 20 of Table 1.

In (3), let $G(p) = 1/p$; then, $g(t) = 1$, and (3) becomes

$$\frac{1}{p} F(p) = T\left\{ \int_0^t f(\tau)\, d\tau \right\} \tag{11}$$

Also, take $H(p) = G(p) = 1/p$ in (9) to get

$$\frac{1}{p^2} F(p) = T\left\{ \int_0^t \left[\int_0^\tau f(\theta)\, d\theta \right] d\tau \right\} \tag{12}$$

By way of application, take in (3), $G(p) = 1/(p - a)$, $F(p) = c/(p^2 + c^2)$ so that $g(t) = e^{at}$, $f(t) = \sin ct$; the result is

$$T^{-1}\left\{ \frac{c}{(p - a)(p^2 + c^2)} \right\} = \int_0^t e^{a(t-\tau)} \sin (c\tau)\, d\tau$$
$$= (a^2 + c^2)^{-1}[ce^{at} - c \cos ct - a \sin ct]$$

* *Ibid.*, vol. II, p. 253.

Table 2 §60 gives a list of formulas involving Laplace transforms.

Example 1. Solve the integrodifferential equation

$$Dy + 2y - 3\int_0^t y\,dt = 5 + 5t \qquad y = 2 \text{ when } t = 0 \tag{a}$$

Solution. By (VIII), Table 2, $T\left\{\int_0^t y\,dt\right\} = (1/p)Y$. Therefore, equate the transforms of the members of (a) to obtain

$$pY - 2 + 2Y - \frac{3}{p}Y = \frac{5p + 5}{p^2} \tag{b}$$

Solving (b) for Y and using inverse transforms, we get

$$\begin{aligned} y &= T^{-1}\left\{\frac{5p + 5}{p(p + 3)(p - 1)} + \frac{2p}{(p + 3)(p - 1)}\right\} \\ &= -\tfrac{5}{3} + 3e^t + \tfrac{2}{3}e^{-3t} \end{aligned} \tag{c}$$

By trial we find that y from (c) satisfies (a) for all values of t.

Example 2. Express $T^{-1}\{F(p)\}$ in terms of t if $F(p) = (p^2 + 2p)/\{(p - 1)(p - 2)[(p - 1)^2 + 1]\}$.
Solution. By (III), Table 2,

$$T^{-1}\{F(p)\} = e^t T^{-1}\left\{\frac{p^2 + 4p + 3}{p(p - 1)(p^2 + 1)}\right\} \tag{a}$$

$$\begin{aligned} \frac{p^2 + 4p + 3}{p(p - 1)(p^2 + 1)} &= \frac{(p^2 + 1) + 4(p - 1) + 6}{p(p - 1)(p^2 + 1)} \\ &= \frac{1}{p(p - 1)} + \frac{4}{p(p^2 + 1)} + \frac{6}{p(p - 1)(p^2 + 1)} \end{aligned} \tag{b}$$

By (11) and 21, Table 1, we get

$$\begin{aligned} T^{-1}\left\{\frac{6}{p(p - 1)(p^2 + 1)}\right\} &= 6\int_0^t \tfrac{1}{2}(e^\tau - \cos\tau - \sin\tau)\,d\tau \\ &= 3(e^t - \sin t + \cos t - 2) \end{aligned} \tag{c}$$

The inverse transforms of the other terms in (b) are easily obtained, and from (b) and (a) we finally get

$$T^{-1}\left\{\frac{p^2 + 2p}{(p - 1)(p - 2)[(p - 1)^2 + 1]}\right\} = 4e^{2t} - (3\sin t + \cos t + 3)e^t$$

Exercises

Read example 1 and then solve and check your solution of:

1. $Dy + \int_0^t y\,dt = 1$; $y = 2$ when $t = 0$.

2. $Dy - y - 6\int_0^t y\,dt = 12e^{3t}$; $y = -3$ when $t = 0$.

3. $y + \int_0^t y\,dt = \sin 2t$

4. Applying (1) successively to $T(1) = p^{-1}$, derive $T(t^3) = 6p^{-4}$.

Apply (1) to 6 and 7, Table 1, to show that:

5. $T^{-1}\left\{\dfrac{a^2 - p^2}{(p^2 + a^2)^2}\right\} = -t\cos at$

6. $T^{-1}\left\{\dfrac{2ap}{(p^2 + a^2)^2}\right\} = t\sin at$

Use (V), Table 2, and then (II), Table 2, to express $T\{-te^{3t}f(t)\}$ in terms of p if:

7. $f(t) = \sin 2t$ **8.** $f(t) = t\sin 2t$ **9.** $f(t) = \cosh t$

10. Use (III) and (VII), Table 2, and 7, Table 1, to show that

$$T^{-1}\left\{\frac{1}{(p-3)[(p-3)^2+4]}\right\} = \tfrac{1}{4}e^{3t}(1 - \cos 2t)$$

Using (3) and Tables 1 and 2, express $f(t)$ in the form of a definite integral if:

11. $F(p) = \dfrac{1}{(p-3)^2}\cdot\dfrac{2}{p^2+4}$ **12.** $F(p) = \dfrac{1}{(p-2)^5}\cdot\dfrac{p}{p^2+4}$

13. $F(p) = \dfrac{1}{p^2+a^2}\cdot\dfrac{p}{p^2+b^2}$ **14.** $F(p) = \dfrac{1}{p^2+a^2}\cdot\dfrac{p}{(p^2+b^2)^2}$

15. $D^2y + Dy - 4y - 4\int_0^t y\,dt = 6e^t - 4t - 6$; $y = 0$, $Dy = 0$ when $t = 0$.

Using (3) and Table 1 derive:

16. 21 from 7. **17.** 22 from 6. **18.** 23 from 12. **19.** 25 from 5.

Using (1) and Table 1, derive:

20. 14 from 7. **21.** 15 from 6.

22. $\dfrac{2p(p^2 - 3a^2)}{(p^2 + a^2)^3} = T\{t^2\cos at\}$

23. To prove (IV), Table 2, substitute mp for p in (1), §59, with t replaced by τ, then make the substitution $t = m\tau$ in the resulting integral, and interpret the result.

24. Use (II), (III), and (IV), Table 2, to show that $F(mp + n) = m^{-1}T\{e^{-nt/m}f(t/m)\}$ if $F(p) = T\{f(t)\}$.

64 Resolving a rational fraction into partial fractions

A rational fraction is one having polynomials as numerator and denominator. Methods using partial fractions are powerful and can be performed by routine procedures. They are not necessary but they give an inverse transform directly without reference to previously derived formulas.

Partial fractions are treated in books* on calculus. Here we shall use a general approach.

We consider only fractions $F(p) = N(p)/M(p)$ where $N(p)$ and $M(p)$ are polynomials having no common factor in p and $N(p)$ is of lower degree than $M(p)$. If

$$F(p) = \frac{N(p)}{M(p)} = \frac{N(p)}{M_1(p)(p-a)^k} \qquad M_1(a) \neq 0 \tag{1}$$

let

$$\varphi(p) = \frac{N(p)}{M_1(p)} \tag{2}$$

expand $\varphi(p)$ in a Taylor's series, and divide the result by $(p - a)^k$ to obtain

$$\begin{aligned} F(p) &= \frac{\varphi(p)}{(p-a)^k} \\ &= \frac{\varphi(a)}{0!(p-a)^k} + \frac{\varphi^{(1)}(a)}{1!(p-a)^{k-1}} \\ &\quad + \frac{\varphi^{(2)}(a)}{2!(p-a)^{k-2}} + \cdots + \frac{\varphi^{(k-1)}(a)}{(k-1)!(p-a)} \\ &\quad + \frac{\varphi^{(k)}(a)}{k!} + \frac{\varphi^{(k+1)}(a)}{(k+1)!}(p-a) + \cdots \end{aligned} \tag{3}$$

where $\varphi^{(s)}(p) = d^s\varphi/dp^s$ and $0! = 1$. The terms in color are the partial fractions of $F(p)$ associated with factor $(p - a)^k$ where $M(p) = M_1(p)(p - a)^k$, $M_1(a) \neq 0$. To get $T^{-1}\{F(p)\}$ we resolve $F(p)$ into partial fractions and find the sum of the transforms of these fractions.

* For elementary approaches consult calculus texts. Rigorous proofs relating to partial fractions are found in books on advanced calculus. Also consult Ruel V. Churchill, "Operational Mathematics," pp. 57–65.

If $k = 1$ in (3) only one partial fraction is indicated, namely $\varphi(a)/(p - a)$, and the corresponding inverse transform is $T^{-1}\{\varphi(a)/(p - a)\} = \varphi(a)e^{at}$. If the zeros of $M(p)$, $a_1, a_2, \ldots, a_n$ are all distinct then

$$T^{-1}\{F(p)\} = \varphi_1(a_1)e^{a_1 t} + \varphi_2(a_2)e^{a_2 t} + \cdots + \varphi_n(a_n)e^{a_n t} \tag{4}$$

where $\varphi_s(p) = N(p)/M_s(p)$ and $M_s(p)$ is the result of deleting factor $p - a_s$ from $M(p)$. For example, if $F(p) = (7p - 1)/[(p - 3)(p + 2)(p - 1)]$, from (4) we get

$$\begin{aligned} T^{-1}\{F(p)\} &= \frac{7(3) - 1}{(3 + 2)(3 - 1)} e^{3t} + \frac{7 \cdot (-2) - 1}{(-2 - 3)(-2 - 1)} e^{-2t} \\ &\qquad + \frac{7(1) - 1}{(1 - 3)(1 + 2)} e^{t} \\ &= 2e^{3t} - e^{-2t} - e^{t} \end{aligned}$$

Similarly

$$T^{-1}\left\{\frac{p + m}{(p - a)(p + b)}\right\} = \frac{a + m}{a + b} e^{at} + \frac{-b + m}{-b - a} e^{-bt}$$

From (3) we obtain as the part of $T^{-1}\{F(p)\}$ associated with the factor $(p - a)^k$ of M,

$$\begin{aligned} T_a^{-1}\{F(p)\} = e^{at}\Bigg[\frac{\varphi(a)}{0!}\frac{t^{k-1}}{(k - 1)!} + \frac{\varphi^{(1)}(a)}{1!}\frac{t^{k-2}}{(k - 2)!} + \cdots \\ + \frac{\varphi^{(k-1)}(a)}{(k - 1)!}\frac{t^0}{0!}\Bigg] \end{aligned} \tag{5}$$

where $\varphi(p)$ is defined by (1) and (2). For example, to find $T_3\{1/[(p - 4)(p - 3)^3]\}$, we note that $\varphi(p) = (p - 4)^{-1}$ and write three terms from (5) with $a = 3$, $k = 3$, and

$$\begin{aligned} T_3^{-1}&\left\{\frac{1}{(p - 4)(p - 3)^3}\right\} \\ &= e^{3t}\left[\frac{(p - 4)^{-1}}{0!}\frac{t^2}{2!} + \frac{-1(p - 4)^{-2}}{1!}\frac{t}{1} + \frac{2(p - 4)^{-3}}{2!}\right]_{p=3} \\ &= e^{3t}[-\tfrac{1}{2}t^2 - t - 1] \end{aligned} \tag{6}$$

Using the first term of (5) for the factor $p - 4$, we get

$$T_4^{-1}\left\{\frac{1}{(p - 4)(p - 3)^3}\right\} = e^{4t}\left[\frac{(p - 3)^{-3}}{0!}\right]_{p=4} = e^{4t} \tag{7}$$

Combining the results of (6) and (7), we have

$$T^{-1}\left\{\frac{1}{(p - 4)(p - 3)^3}\right\} = e^{3t}(-\tfrac{1}{2}t^2 - t - 1) + e^{4t} \tag{8}$$

Note that in finding $T^{-1}\{F(p)\} = T^{-1}\{N(p)/M(p)\}$ we apply (5) for each factor of $M(p)$ and add the results thus obtained.

Exercises

1. Use (4) to obtain $T^{-1}\{(2p+3)/[p(p-3)]\} = -1 + 3e^{3t}$.

2. Use (4) with $n = 3$ to get $T^{-1}\{(2p-3)/[p(p^2+2p-3)]\} = 1 - \frac{1}{4}e^{t} - \frac{3}{4}e^{-3t}$.

3. Use (4) to verify $T^{-1}\{(mp+n)/[(p-a)(p-b)]\} = (ma+n)(a-b)^{-1}e^{at} + (mb+n)(b-a)^{-1}e^{bt}$.

Use (4) to express in terms of t:

4. $T^{-1}\left\{\dfrac{p^2+1}{(p^2-1)(p^2-4)}\right\}$

5. $T^{-1}\left\{\dfrac{2p-3}{p^3+4p^2+3p}\right\}$

6. $T^{-1}\left\{\dfrac{p+1}{(p+2)(2p+3)}\right\}$

7. Write (5) for the case where : (a) $k = 3$; (b) $k = 4$.

8. Take $F(p) = (2p+6)/[(p+2)(p-1)^4]$. First obtain $T_1^{-1}\{F(p)\}$ by substituting $\varphi(p) = (2p+6)/(p+2) = 2[1 + (p+2)^{-1}]$ and $a = 1$ in (5) with $k = 4$. Then find $T_{-2}^{-1}\{F(p)\}$ by taking $\varphi(p) = (2p+6)/(p-1)^4$ in (5) with $k = 1$. Add the two results thus obtained to get $T^{-1}\{F(p)\}$.

Express in terms of t by using (5):

9. $T^{-1}\{(p^3+p^2+2p+5)/(p+1)^5\}$. Use $\varphi(p) = p^3 + p^2 + 2p + 5$ in (5).

10. $T^{-1}\left\{\dfrac{8+10p^2-2p^3}{p(p-2)^5}\right\}$

11. $T^{-1}\left\{\dfrac{4p^2-16}{p^3(p+2)^2}\right\}$

★12. $T^{-1}\left\{\dfrac{5p^2-23p}{(2p-2)(2p+4)^4}\right\}$

Find the solution of each equation subject to the indicated conditions, and check your answer:

13. $D^2y - 4y = 10e^{3t}$; $y = 5$, $Dy = 0$ when $t = 0$.

14. $(D^2 + 2D - 3)y = -4e^{-t}$; $y = 3$, $Dy = -7$ when $t = 0$.

15. $(D^3 + 5D^2 + 4D)y = 20e^{t}$; $y = 1$, $Dy = 6$, $D^2y = -14$ when $t = 0$.

16. $(D^2 - 2D + 1)y = e^{2t}$; $y = 1$, $Dy = 3$ when $t = 0$.

17. $(D^2 - 2D)y = 8$; $y = 3$, $Dy = -2$ when $t = 0$.

18. $(D^2 + D)y = te^{-t}$; $y = 2$, $Dy = -2$ when $t = 0$.

19. $(D^2 + 4D + 4)y = 6te^{-2t}$; $y = 0$, $Dy = 2$ when $t = 0$.

20. $D(D-1)^2y = 2 + 12e^t$; $y = 0$, $Dy = 0$, $D^2y = 14$ when $t = 0$.

21. Use (4) and (5) to derive entries 13, 23, 24, 25, and 26 of Table 1 in §60.

65 Partial fractions. Quadratic factors

It can be shown that the argument used to derive equation (5), §64, can be modified to apply when a is a complex number. Hence we assume that (5), §64, applies when a is complex.

A polynomial with real coefficients must have both factors $p - a - ib$ and $p - a + ib$, $i^2 = -1$, if it has either one. Consider a fraction

$$\frac{N(p)}{M(p)} \qquad \text{where } M(p) = M_1(p)(p-a-ib)^k(p-a+ib)^k \qquad i^2 = -1 \tag{1}$$

where $M_1(a+ib) \neq 0$, $N(a+ib) \neq 0$. Take the $(r+1)$st terms of (5), §64, relating to the two factors $(p - a \pm ib)^k$ of (1). Let

$$[\varphi_1^{(r)}(p)]_{p=a+ib} = \left[\frac{d^r}{dp^r}\frac{N(p)}{M_1(p)(p-a+ib)^k}\right]_{p=a+ib} = A_r + iB_r \tag{2}$$

where A_r and B_r do not contain i; then

$$[\varphi_2^{(r)}(p)]_{p=a-ib} = \left[\frac{d^r}{dp^r}\frac{N(p)}{M_1(p)(p-a-ib)^k}\right]_{p=a-ib} = A_r - iB_r \tag{3}$$

since equation (2) may be obtained from (1) by replacing i by $-i$. Now the sum of the $(r+1)$st terms of (5), §64, indicated, in part, by (2) and (3), is

$$[e^{(a+bi)t}(A_r + iB_r) + e^{(a-ib)t}(A_r - iB_r]\frac{t^{k-r-1}}{r!(k-r-1)!} \tag{4}$$

Replacing $e^{(a\pm ib)t}$ by $e^{at}(\cos bt \pm i \sin bt)$ in (4) and simplifying we get

$$e^{at}[2A_r \cos bt - 2B_r \sin bt]\frac{t^{k-r-1}}{r!(k-r-1)!} \tag{5}$$

Adding the expressions (5) for $r = 0, 1, \ldots, (k-1)$, we get the part of (1) associated with a factor $[(p-a)^2 + b^2]^k$ of $M(p)$, namely

$$\begin{aligned} &2e^{at}\left[\frac{A_0t^{k-1}}{0!(k-1)!} + \frac{A_1t^{k-2}}{1!(k-2)!} + \cdots + \frac{A_{k-1}t^0}{(k-1)!0!}\right]\cos bt \\ &-2e^{at}\left[\frac{B_0t^{k-1}}{0!(k-1)!} + \frac{B_1t^{k-2}}{1!(k-2)!} + \cdots + \frac{B_{k-1}t^0}{(k-1)!0!}\right]\sin bt \end{aligned} \tag{6}$$

where A_r and B_r are defined by (2) or (3).

Two special cases of (6) are

$$2e^{at}(A_0 \cos bt - B_0 \sin bt) \qquad \text{when } k = 1 \tag{7}$$

$$2e^{at}[(A_0t + A_1) \cos bt - (B_0t + B_1) \sin bt] \qquad \text{when } k = 2 \tag{8}$$

where the A's and B's are defined by (2) with $r = 0$ and $r = 1$

Example. Express in terms of t: $T^{-1}\{(p^2 + 2)/(p^2 + 2p + 5)^2\}$.

Solution. Simpler computations will result from using equation (1), §61, to obtain

$$T^{-1}\left\{\frac{p^2 + 2}{[(p + 1)^2 + 4]^2}\right\} = e^{-t}T^{-1}\left\{\frac{(p - 1)^2 + 2}{(p^2 + 4)^2}\right\} \tag{a}$$

Since $p^2 + 4 = (p - 2i)(p + 2i)$, $a = 0$, $b = 2$, and

$$\varphi(p) = \frac{(p - 1)^2 + 2}{(p + 2i)^2} \qquad [\varphi(p)]_{p=2i} = \frac{(2i - 1)^2 + 2}{(4i)^2} = \tfrac{1}{16} + \tfrac{1}{4}i$$

$$\varphi^{(1)}(p) = \frac{2(p - 1)}{(p + 2i)^2} - \frac{2[(p - 1)^2 + 2]}{(p + 2i)^3} \qquad \varphi^{(1)}(2i) = 0 - \tfrac{7}{32}i$$

Hence $A_0 = \frac{1}{16}$, $B_0 = \frac{1}{4}$, $A_1 = 0$, $B_1 = -\frac{7}{32}$, $a = 0$, and $b = 2$. Using these values in (8) we obtain

$$e^{-t}T^{-1}\left\{\frac{(p - 1)^2 + 2}{(p^2 + 4)^2}\right\} = 2e^{-t}[\tfrac{1}{16}t \cos 2t + (-\tfrac{1}{4}t + \tfrac{7}{32}) \sin 2t]$$

Exercises

1. Use (7) to express $T^{-1}\{(p - 6)/(p^2 - 4p + 8)\}$ in terms of t. Note that $p^2 - 4p + 8 = 0$ has factors $p - (2 \pm 2i)$. Also $\varphi(p) = (p - 6)/(p - 2 + 2i)$.

Express in terms of t:

2. $T^{-1}\left\{\dfrac{mp + n}{p^2 - 2ap + a^2 + b^2}\right\}$

3. $T^{-1}\left\{\dfrac{p^2 + 2}{p(p^2 + 2p + 2)}\right\}$

4. $T^{-1}\left\{\dfrac{p^2 + 2}{p(p + 1)(p^2 + 2p + 2)}\right\}$

5. $T^{-1}\left\{\dfrac{p^2}{(p - 1)(p^2 - 2p + 5)^2}\right\}$

6. Write (6) for $k = 3$.

7. $T^{-1}\{1/(p^2 - 2p + 2)^3\}$. Verify that this is equal to $e^tT^{-1}\{1/(p^2 + 1)^3\}$. Therefore use $\varphi(p) = (p + i)^{-3}$ and the answer to exercise 6.

Using partial fractions, solve:

8. $(D^2 - 4D + 8)y = e^{2t}$; $y = 2$, $Dy = -2$ when $t = 0$.

9. $(D^2 + 4)y = 12 \sin 2t$; $y = 0$, $Dy = 0$ when $t = 0$.

10. $(D^2 - 2D + 10)y = 40e^{-t} \sin 3t$; $y = 1$, $Dy = -1$ when $t = 0$.

Express in terms of t:

11. $T^{-1}\left\{\dfrac{p^2}{(p^2+1)^2}\right\}$ **12.** $T^{-1}\left\{\dfrac{p^2+3}{(p^2-2p+5)^2}\right\}$

13. $T^{-1}\left\{\dfrac{p^2+3}{(p+1)(p^2+2p+2)^2}\right\}$

★**14.** $T^{-1}\left\{\dfrac{p+1}{p(p^2+1)^3}\right\}$

★**15.** Solve $(D^2 + 6D + 13)y = 13e^{-3t} \sin 2t$; $y = 0$, $Dy = -2$ when $t = 0$.

16. Use (7) and (8) to derive entries 8, 14, 15, 17, 19, 21, 22 of Table 1.

66 Simultaneous differential equations

To solve two differential equations in x, y, and t express the equations in terms of transforms, solve the result for X and Y in terms of p, and then use inverse transforms to find x and y in terms of t. A similar remark applies to n equations in $n + 1$ unknowns.

Example. Find that solution $x(t)$ and $y(t)$ of

$$\begin{aligned} Dx + x + Dy - y &= 2 \\ D^2x + Dx - Dy &= \cos t \end{aligned} \tag{a}$$

for which $x = 0$, $Dx = 2$, and $y = 1$ when $t = 0$.

Solution. Applying transforms to (a), we get

$$\begin{aligned} T\{Dx + Dy + x - y\} &= T\{2\} = \frac{2}{p} \\ T\{D^2x + Dx - Dy\} &= T\{\cos t\} = \frac{p}{p^2+1} \end{aligned} \tag{b}$$

Since $T\{D^2x\} = p^2X - 2$, $T\{Dx\} = pX$, $T\{Dy\} = pY - 1$, (b) may be written

$$(p + 1)X + (p - 1)Y = 1 + \frac{2}{p}$$
$$(p^2 + p)X - pY = 1 + \frac{p}{p^2 + 1} \qquad (c)$$

Solving equations (c) for X and Y, we get

$$X = \frac{1}{p^2} + \frac{1}{p^2 + 1} \qquad Y = \frac{p}{p^2 + 1} + \frac{1}{p^2} \qquad (d)$$

From (d),

$$x = T^{-1}\{X\} = t + \sin t \qquad y = T^{-1}\{Y\} = t + \cos t \qquad (e)$$

The solution (e) satisfies (a) and the initial conditions.

Exercises

1. To solve the system of equations $x + Dy = 3$, $Dx - y = -2t$, where $x = 1$ and $y = 0$ when $t = 0$, construct transforms from them to get

$$X + pY = \frac{3}{p} \qquad pX - 1 - Y = -\frac{2}{p^2}$$

Solve these for X and Y by algebraic procedures to get

$$X = \frac{1}{p} \qquad Y = \frac{2p^2 + 2}{p^2(p^2 + 1)}$$

From these obtain $x = 1$ and $y = 2t$. Check this result.

Solve the following systems of equations:

2. $x - Dy = t - 2$, $Dx + y = 1 + 2t$; $x = 0$, $y = 2$ when $t = 0$.

3. $Dx + D^2y = e^t$, $2Dx - Dy = 2e^t - 1$; $x = 1$, $y = 0$, $Dy = 1$ when $t = 0$.

4. $D^2x + Dy = \cos t$, $x - Dy = 3t + 2 - \cos t$; $x = 2$, $Dx = 3$, $y = 0$ when $t = 0$.

5. $Dx + Dy = 2t + 2$, $Dx - Dz = 2t + 1$, $Dx + z = t$; $x = y = z = 0$ when $t = 0$.

6. $D^2x + Dy = 2$, $D^2x - D^2y = 0$; $x = 0$, $Dx = 2$, $y = -2$, $Dy = 2$ when $t = 0$.

7. $Dx + (D - 1)y = 2e^t + 1$, $2D^2x - Dy = e^t$; $x = 2$, $Dx = 3$, $y = 3$ when $t = 0$.

8. $(D - 1)x + y = 2e^t$, $3Dx - 3Dy = 3te^t$; $x = 0$, $y = 1$ when $t = 0$.

9. $D^2x - Dy = 0$, $x + Dy = 1$; $x = 1$, $Dx = 1$, $y = 1$ when $t = 0$.

10. $D^2x - Dy = \cos t$, $Dx + D^2y = -\sin t$; $x = 1$, $Dx = 0$, $y = 0$, $Dy = 1$ when $t = 0$.

***11.** The system of equations

$$m\,D^2x + He\,Dy = Ee \qquad m\,D^2y - He\,Dx = 0$$

occurs in the investigation of finding ratio of charge to mass of an electron. Solve them assuming that m, e, E, H are constants and $x = y = Dx = Dy = 0$ when $t = 0$.

67 Equations with variable coefficients

To solve a linear differential equation containing variable coefficients of derivatives we modify the procedure of §62 by using equation (2), §63, namely

$$\frac{d^n}{dp^n}F(p) = T\{(-t)^n f(t)\} \tag{1}$$

For example, when $y(0) = a$, $y'(0) = b$, we get from (5), §62, and (1)

$$T\{t\,D^2y\} = -\frac{d}{dp}(p^2Y - ap - b) = -p^2\frac{dY}{dp} - 2pY + a \tag{2}$$

The following examples will serve to illustrate procedures.

Example 1. Solve $D^2y + t\,Dy - 2y = k$, $y_0 = Dy(0) = 0$. (a)

Solution. Equating the transforms of the two members of (a), from (5), §62, and (1) we get

$$p^2Y - \frac{d}{dp}(pY) - 2Y = +p^2Y - p\frac{dY}{dp} - Y - 2Y = \frac{k}{p} \tag{b}$$

or

$$\frac{dY}{dp} + \frac{3 - p^2}{p}Y = -\frac{k}{p^2} \tag{c}$$

Solving the linear equation (c) by using §25, we get

$$Y = \frac{k}{p^3} + \frac{ce^{p^2/2}}{p^3} \tag{d}$$

Here $c = 0$ must hold since $\lim_{p\to\infty} e^{p^2/2}/p^3 \neq 0$. Hence

$$y = T^{-1}\{Y\} = T^{-1}\left\{\frac{k}{p^3}\right\} = \frac{kt^2}{2} \qquad \text{or} \qquad y = \tfrac{1}{2}kt^2 \tag{e}$$

Example 2. Solve $t\,D^2y + y = 0$, $y(0) = 0$. $\qquad (a)$

Solution. Using the procedure of example 1 on (a), we get

$$-\frac{d}{dp}[p^2Y - y'(0)] + Y = 0 \qquad \text{or} \qquad -p^2\frac{dY}{dp} - 2pY + Y = 0 \tag{b}$$

the solution of (b) is

$$Y = \frac{ce^{-1/p}}{p^2} \tag{c}$$

Using the fact that $e^x = 1 + x + x^2/2! + x^3/3! + \cdots$, from (c) we get

$$Y = c\left(\frac{1}{p^2} - \frac{1}{p^3} + \frac{1}{2!p^4} - \frac{1}{3!p^5} + \cdots\right) \tag{d}$$

or

$$y = T^{-1}\{Y\} = c\left(t - \frac{t^2}{2!} + \frac{1}{2!}\frac{t^3}{3!} - \frac{1}{3!}\frac{t^4}{4!} + \cdots + \frac{(-1)^n t^n}{(n-1)!n!} + \cdots\right) \tag{e}$$

We readily check that (e) is absolutely convergent for all values of t and that y from (e) satisfies (a).

Exercises

Solve:

1. $t\,D^2y + at\,Dy + ay = 0$; $y(0) = 0$.
2. $D^2y + 3t\,Dy - 6y = m$; $y(0) = y'(0) = 0$.
3. $t\,D^2y - (2 + t)\,Dy + 3y = t - 1$; $y(0) = 0$.
4. $D^2y + t\,Dy - y = 0$; $y(0) = 0$, $y'(0) = 5$.
5. $t\,D^2y + (3 - t)\,Dy - 2y = t - 1$; $y(0) = 0$.
6. $t\,D^2y + y = 12t$; $y(0) = 0$.

Find a solution of the equation $t^2\,D^2y + 4t\,Dy + 2y = f(t)$ if $f(t)$ is:

7. t^n 8. $T^{-1}\left\{p^2\frac{d^2}{dp^2}(p^2+1)^{-1}\right\}$ 9. $T^{-1}\left\{p^2\frac{d^2}{dp^2}\varphi(p)\right\}$

68 Translation functions and periodic functions

Multiplying (1), §59, by e^{-bp}, $b > 0$, we get

$$e^{-pb}F(p) = \int_0^\infty e^{-pb}e^{-pt}f(t)\,dt \tag{1}$$

The substitution $\tau = b + t$ in the right member of (1) gives

$$e^{-pb}F(b) = \int_b^\infty e^{-p\tau}f(\tau - b)\,d\tau \tag{2}$$

Taking $f(t)$ as zero when t is negative, we see from (2) that $f(\tau - b) = 0$ when $\tau < b$, and $f(\tau - b) = f(t)$ when $\tau \geqq b$; therefore the graph of $y = f(\tau - b)$ is that of $y = f(t)$ translated rightward the distance b, as indicated in Fig. 1. In this case we use the notation

$$f(\tau - b) = f_b(t) = T^{-1}\{e^{-pb}F(p)\} \tag{3}$$

Thus Fig. 2 represents $y = \cos t$, $0 \leqq t \leqq \pi$ and the corresponding part of $y = \cos_\pi t$, $0 \leqq t \leqq \pi$, obtained by translating $y = \cos t$ rightward π units.

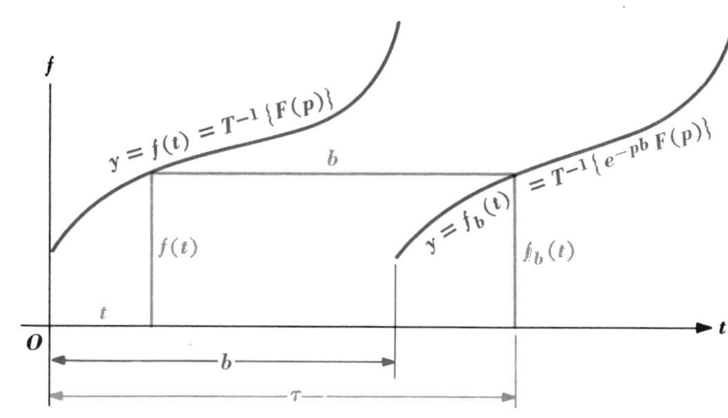

Figure 1

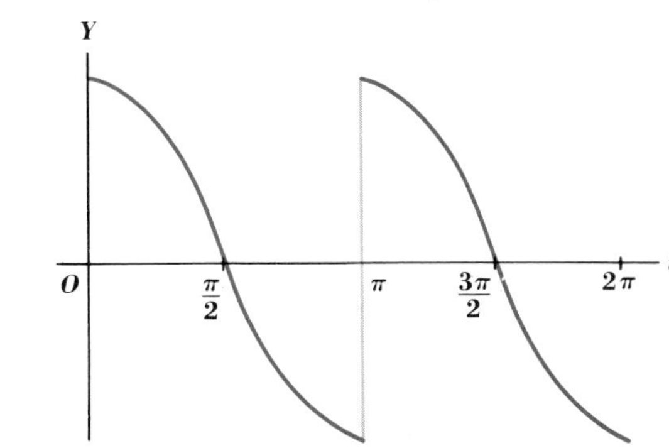

Figure 2

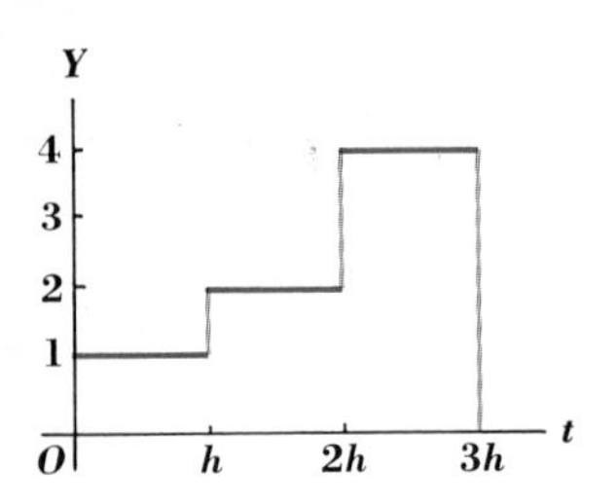

Figure 3

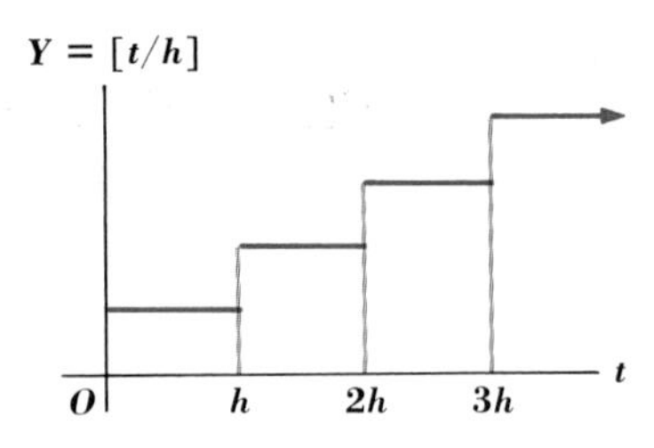

Figure 4

It should be observed that the translation function gives extensive symbolic power. For example the function

$$\begin{aligned} y &= T^{-1}\left\{\frac{1 + e^{-hp} + 2e^{-2hp} - 4e^{-3hp}}{p}\right\} \\ &= 1 + 1_h + 2(1_{2h}) - 4(1_{3h}) \end{aligned} \tag{4}$$

is shown in Fig. 3. Another case is the step function denoted by $[t/h]$, $t \geqq 0$; it is the least integer greater than t/h. Thus $[0.0/h] = 1$, $[0.7h/h] = 1$, $[1h/h] = 2$, $[1.7h/h] = 2$, $[2h/h] = 3$, etc. Figure 4 represents the graph of $[t/h]$. A little reflection verifies that

$$T\left\{\frac{t}{h}\right\} = \frac{1}{p}\,(1 + e^{-hp} + e^{-2hp} + \cdots) \tag{5}$$

if the set of steps is infinite. If there were exactly $n + 1$ steps, the parenthetical sum in (5) would be replaced by $(1 + e^{-hp} + \cdots + e^{-np} - (n + 1)e^{-(n+1)p})$.

Be careful to notice that if $f(t) = \varphi(t)$ when $0 < t \leqq a$ and $f(t) = 0$ when $t > a$, then

$$T\{f(t)\} = \int_0^a e^{-pt}\varphi(t)\,dt + \int_a^\infty e^{-pt}(0)\,dt = \int_0^a e^{-pt}\,\varphi(t)\,dt \tag{6}$$

For example, if $f(t) = \sin t$, $0 \leqq t \leqq \pi$, but $f(t) = 0$, $t > \pi$, then

$$\begin{aligned} T\{f(t)\} &= \int_0^\pi e^{-pt}\sin t\,dt = \left[\frac{e^{-pt}}{p^2 + 1}\,(-p\sin t - \cos t)\right]_0^\pi \\ &= \frac{1 + e^{-\pi p}}{p^2 + 1} \end{aligned} \tag{7}$$

Note from (6) and Fig. 5 that

$$f(t) = T^{-1}\left\{\frac{1 + e^{-\pi p}}{p^2 + 1}\right\} = \sin t + \sin_\pi t = \sin t \qquad 0 \leqq t \leqq \pi$$

The student will gain a clear insight by showing geometrically and also by means of (6) that

$$y = \cos t + \cos_{3\pi/4} t = \cos t \qquad 0 < t \leqq 3\pi/4 \tag{8}$$

A function $f(t)$ is periodic *with period m if*

$$f(x + m) = f(x) \tag{9}$$

for all values of x for which $f(x)$ exists. Generally we speak of *the period of* $f(x)$ as its numerically least period. For example $\sin x = \sin (x + 2\pi)$ and the period of $\sin x$ is 2π. Obviously $4\pi, 6\pi, -8\pi, \ldots$ are also periods. Since $\tan x = \tan (x + \pi)$, $\tan x$ has period π.

Let $f(t + a) = f(t)$ and $T\{f(t)\}$, $0 < t \leqq a$, $= F(p)$. Then

$$T\{f(t)\} = (1 + e^{-ap} + e^{-2ap} + \cdots + e^{-nap} + \ldots)F(p) \tag{10}$$

since $F(p)$ is the transform of $f(t)$ for $0 < t < a$, $e^{-ap}F(p)$ for $a < t < 2a$, $e^{-2ap}F(p)$ for $2a < t < 3a, \ldots$. Denoting the parenthesized operator of (10) by its algebraic equal $1/(1 - e^{-ap})$ we have

$$\begin{aligned} T\{f(t)\} &= \frac{F(p)}{1 - e^{-ap}} \qquad f(t + a) = f(t) \\ F(p) &= \int_0^a e^{-pt} f(t)\, dt \end{aligned} \tag{11}$$

For example, the transform of $f(t)$, where $f(t) = t$, $0 < t < 6$, and $f(t + 6) = f(t)$ (see Fig. 6), derived from (11), is

$$T\{f(t)\} = \frac{\int_0^6 e^{-pt} t\, dt}{1 - e^{-6p}} = \frac{1/p^2 - e^{-6p}/p^2 - 6e^{-6p}/p}{1 - e^{-6p}} \tag{12}$$

The transform of the square function, defined by $f(t) = 1$ when

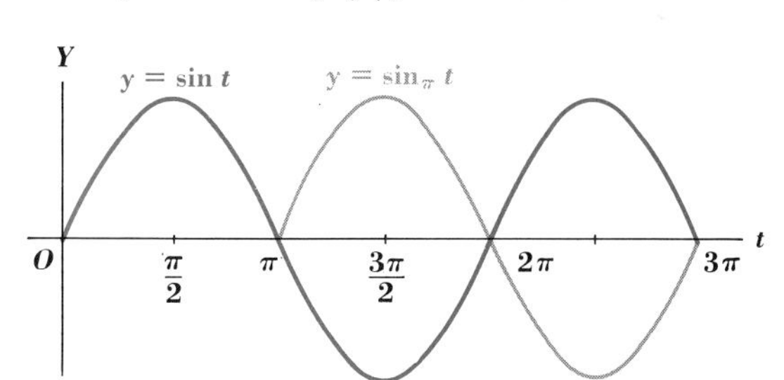

Figure 5

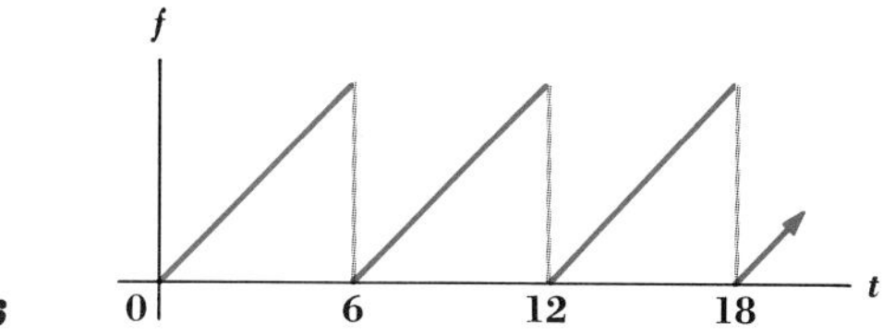

Figure 6

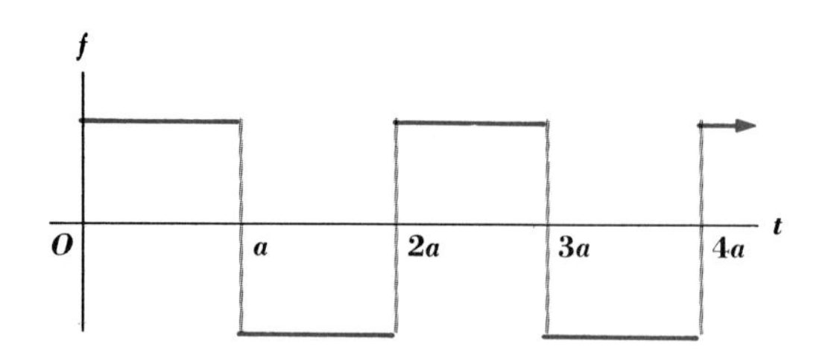

Figure 7

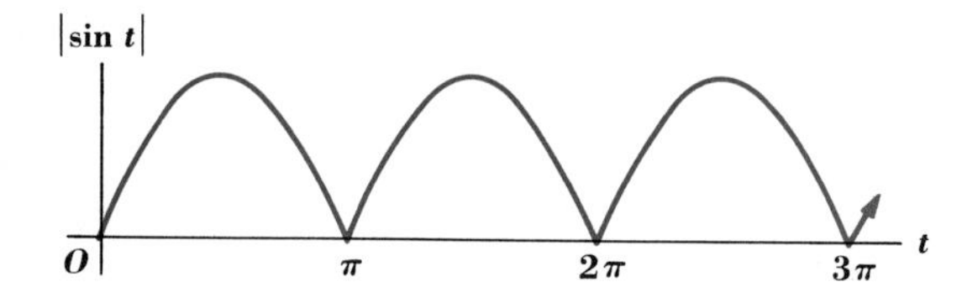

Figure 8

$0 < t < a$, $f(t) = -1$ when $a < t < 2a$, $f(t + 2a) = f(t)$ (see Fig. 7), is given by

$$T\{f(t)\} = \frac{\int_0^a e^{-pt}(1)\,dt + \int_a^{2a} e^{-pt}(-1)\,dt}{1 - e^{-2ap}} = \frac{(1 - e^{-ap})^2/p}{1 - e^{-2ap}}$$
$$= \frac{1 - e^{-ap}}{p(1 + e^{-ap})} = \frac{1}{p}\tanh \tfrac{1}{2}ap \tag{13}$$

Figure 8 represents $y = |\sin t|$, called the rectification of $\sin t$. Using (6) to get $f(p) = (1 + e^{-\pi p})/(p^2 + 1)$ and (11) we obtain

$$T\{|\sin t|\} = \frac{(1 + e^{-\pi p})/(p^2 + 1)}{1 - e^{-\pi p}} = \frac{1}{p}\coth \tfrac{1}{2}\pi p \tag{14}$$

Exercises

1. Draw a geometric representation of $f_3(t)$ where: (*a*) $f(t) = t$; (*b*) $f(t) = 1$ when $0 < t < 2$, and $f(t) = 0$ when $t > 2$.

2. Draw a rough sketch of $y = f(t)$ if: (*a*) $f(t) = T^{-1}\{(1 + e^{-4p})/p\}$; (*b*) $f(t) = T^{-1}\{[1 + e^{-\pi p}]/(p^2 + 1)\}$; (c) $f(t) = T^{-1}\{(1 + e^{-hp} + e^{-2hp} - 3e^{-3hp})/p\}$; (*d*) $f(t) = T^{-1}\{(2 + 3e^{-2hp} + 2e^{-4hp} - 6e^{-7hp})/p\}$.

3. If $T\{f(t)\} = (1 - e^{-hp})/p^2$, define $f(t)$ on the interval $0 < t \leqq h$ and also on the interval $t > h$.

4. Draw on the same axes the graphs of $y_1 = T^{-1}\{1/p^2\}$, $y_2 = T^{-1}\{e^{-2p}/p^2\}$, and $y_3 = T^{-1}\{2e^{-2p}/p\}$; and then show that

$y = y_1 - y_2 - y_3 = t$ when $0 < t < 2$ and $y = 0$ when $t > 2$. Check with (6), by evaluating $y = \int_0^2 e^{-pt}t\,dt$.

5. Define the function $f(t)$ if $T\{f(t)\} = \int_0^k e^{-pt}\,t^3\,dt$. *Hint:* Use (6).

6. If $f(t) = 2$ when $0 < t < 2$ and $f(t) = 0$ when $t > 2$, verify by (11) that $T\{f(t)\}/(1 - e^{-4p}) = T\{\varphi(t)\}$, provided that $\varphi(t) = f(t)$ when $0 < t < 4$ and $\varphi(t + 4) = \varphi(t)$. Define $\psi(t)$ if $T\{f(t)\}/(1 - e^{-10p}) = T\{\psi(t)\}$. Also express $T\{f(t)\}$ in terms of p.

7. Write a definition of $\varphi(t)$ if $T\{\varphi(t)\}$ equals:

(*a*) $\dfrac{5(1 - e^{-p})}{p(1 - e^{-6p})}$ (*b*) $\dfrac{\int_0^4 e^{-pt}t^3\,dt}{1 - e^{-8p}}$

(8, 9) Make rough sketches of:

8. $y = T^{-1}\{e^{-hp}(1 - e^{-hp})/p\}$, and of $y = T^{-1}\{e^{-hp}(1 - e^{-hp})/[p(1 - e^{-2hp})]\}$.

9. $y = T^{-1}\{(1 + e^{-\pi p})p/(1 + p^2)\}$, and of $y = T^{-1}\{[(1 + e^{-\pi p})/(1 - e^{-\pi p})][p/(1 + p^2)]\}$.

10. Show that $T^{-1}\left\{\left(\int_0^{\pi/2} e^{-pt}\cos t\,dt - \int_{\pi/2}^{\pi} e^{-pt}\cos t\,dt\right)\Big/(1 - e^{-\pi p})\right\} = |\cos t|$, the rectification of $\cos t$.

11. If $f(t) = a - t$ when $0 < t < 2a$ and $f(t + 2a) = f(t)$, find $T\{f(t)\}$ in terms of p.

12. Sketch $y = T^{-1}\{(1 + e^{-p} + 3e^{-3p} - 5e^{-5p})/p\}$.

13. Find in terms of p: (a) $T\{|\sin kt|\}$; (b) $T\{|\cos kt|\}$.

14. If $\alpha(t) = t$ when $0 < t \leqq a$, if $\alpha(t) = a$ when $a < t \leqq b$, and if $\alpha(t + a + b) = \alpha(t)$, express $T\{\alpha(t)\}$ in terms of p.

15. Express y in terms of p if $y = \sin 3t$ when $\pi \leqq t \leqq \frac{9}{2}\pi$, and $y = 0$ elsewhere.

16. A function $f(t)$ is called antiperiodic if $f(t + a) = -f(t)$ for all values of $t > 0$ and $a > 0$ for which $f(t)$ exists. (*a*) If $f(t)$ is antiperiodic with period a, is $f(t + 5a)$ equal to $-f(t)$, and is $f(t + 2a)$ equal to $-f(t)$? (*b*) Is $\cos t$ antiperiodic? (*c*) Is $\sin t$ antiperiodic?

17. If $f(t + a) = f(t)$, verify that the rectification of $f(t)$ is given by $T^{-1}\left\{\left(\int_0^a e^{-pt}|f(t)|\,dt\right)\Big/(1 - e^{-ap})\right\}$. If $f(t + a) = -f(t)$, write an expression for $T\{|f(t)|\}$. *Hint:* $f(t + 2a) = f(t)$.

18. Find the transform of the rectification of $\sin 2t$.

69 The gamma function* and the error function

The gamma function , denoted by $\Gamma(n)$ is defined for values of n if $n \geqq 1$ by

$$\Gamma(n) = \int_0^\infty e^{-t}t^{n-1}\,dt \tag{1}$$

Mainly because it is an extension of the factorial function $n!$, it enables us to express many results simply.

Using the substitution $t = p\tau$ in (1) we get

$$\Gamma(n) = \int_0^\infty e^{-p\tau}(p\tau)^{n-1}p\,d\tau = p^n \int_0^\infty e^{-p\tau}\tau^{n-1}\,d\tau$$

or

$$T\{t^{n-1}\} = \frac{\Gamma(n)}{p^n} \qquad t > 0 \tag{2}$$

If n is a positive integer, $T\{t^{n-1}\} = (n-1)!/p^n$; substituting this in the left member of (2), we get, after slight simplification,

$$\Gamma(n) = (n-1)! \qquad n \text{ a positive integer} \tag{3}$$

Also, applying (V), Table 2 in §60 to (2), we obtain

$$T\{(-t)t^{n-1}\} = \frac{d}{dp}\frac{\Gamma(n)}{p^n} = \frac{-n\Gamma(n)}{p^{n+1}} \tag{4}$$

Using $T\{t^n\} = \Gamma(n+1)/p^{n+1}$, got from (2) by replacing n by $n+1$, from (4) we easily obtain

$$\Gamma(n+1) = n\Gamma(n) \tag{5}$$

By repeated application of (5) we obtain

$$\Gamma(n+1) = n(n-1)(n-2)\cdots(n-g)\Gamma(n-g) \tag{6}$$

where $n - g < 1$. The values of $\Gamma(s)$, $0 < s \leqq 1$, are found in tables. To evaluate $\Gamma(3.3)$ we read $\Gamma(0.3) = 2.99$ from the table in §110 and then use (6) to get

$$\Gamma(3.3) = (2.3)(1.3)(0.3)\Gamma(0.3) = (2.3)(1.3)(0.3)(2.99) = 2.68$$

* The reader will find a rather extensive treatment of the gamma function in §110.

approximately. Also from Table 1, §110, we get a useful value

$$\Gamma(0.5) = \sqrt{\pi}^{*} \tag{7}$$

The error function, basic in the theory of probability, is denoted and defined by

$$\operatorname{erf}(t) = \frac{2}{\sqrt{\pi}} \int_0^t e^{-r^2}\, dr \tag{8}$$

We shall derive a formula for $T\{\operatorname{erf}(m\sqrt{t})\}$. From (III), Table 2, in §60 we get

$$T^{-1}\left\{\frac{1}{\sqrt{p+m^2}}\right\} = e^{-m^2 t}\, T^{-1}\left\{\frac{1}{p^{1/2}}\right\} \tag{9}$$

From (2) with $n = \frac{1}{2}$,

$$\frac{1}{p^{1/2}} = \frac{1}{\Gamma(\frac{1}{2})}\, T\{t^{-1/2}\} = \frac{1}{\sqrt{\pi}}\, T\{t^{-1/2}\} \tag{10}$$

Substituting $1/p^{1/2}$ from (10) in (9),

$$T^{-1}\left\{\frac{1}{\sqrt{p+m^2}}\right\} = e^{-m^2 t}\,\frac{t^{-1/2}}{\sqrt{\pi}} \tag{11}$$

From (11) and (VIII), Table 2 §60,

$$T^{-1}\left\{\frac{1}{p}\cdot\frac{1}{\sqrt{p+m^2}}\right\} = \frac{1}{\sqrt{\pi}} \int_0^t e^{-m^2\tau}\tau^{-1/2}\, d\tau \tag{12}$$

In the integral of (12) let $r^2 = m^2\tau$:

$$T^{-1}\left\{\frac{1}{p\sqrt{p+m^2}}\right\} = \frac{2m^{-1}}{\sqrt{\pi}} \int_0^{m\sqrt{t}} e^{-r^2}\, dr \tag{13}$$

* To find $\Gamma(0.5)$ proceed as follows:

$$[\Gamma(\tfrac{1}{2})]^2 = \int_0^\infty e^{-s}s^{-1/2}\, ds \int_0^\infty e^{-r}r^{-1/2}\, dr = \int_0^\infty e^{-(s+r)}r^{-1/2}s^{-1/2}\, ds\, dr$$

In the double integral replace s by x^2 and r by y^2, then express the resulting integral in polar coordinates, and obtain

$$[\Gamma(\tfrac{1}{2})]^2 = \int_0^\infty \int_0^\infty e^{-(x^2+y^2)}4 \cdot dx\, dy = 4\int_0^{\pi/2}\int_0^\infty e^{-\rho^2}\rho d\,\rho\, d\theta = \pi$$

or

$$\Gamma(0.5) = \sqrt{\pi}$$

From (13) and (8),

$$T^{-1}\left\{\frac{1}{p\sqrt{p+m^2}}\right\} = m^{-1}\,\text{erf}\,(m\sqrt{t}) \tag{14}$$

or

$$\frac{1}{p\sqrt{p+m^2}} = m^{-1}T\{\text{erf}\,(m\sqrt{t})\} \tag{15}$$

A great variety of formulas may be derived by using (15) and Table 2 in §60. For example

$$\begin{aligned} T^{-1}\left\{\frac{1}{\sqrt{p}+k}\right\} &= T^{-1}\left\{\frac{\sqrt{p}-k}{p-k^2}\right\} \\ &= -ke^{k^2t} + T^{-1}\left\{\frac{p}{(p-k^2)\sqrt{p}}\right\} \end{aligned} \tag{16}$$

Using (III), Table 2, and (15), we obtain

$$\begin{aligned} T^{-1}\left\{\frac{p}{(p-k^2)\sqrt{p}}\right\} &= e^{k^2t}T^{-1}\left\{\frac{p+k^2}{p\sqrt{p+k^2}}\right\} \\ &= e^{k^2t}\left[\frac{e^{-k^2t}}{\sqrt{\pi t}} + k^2k^{-1}\,\text{erf}\,(k\sqrt{t})\right] \end{aligned}$$

Using this value in (16) we have after simplification

$$T^{-1}\left\{\frac{1}{\sqrt{p}+k}\right\} = \frac{1}{\sqrt{\pi t}} - ke^{k^2t}[1-\text{erf}\,(k\sqrt{t})] \tag{17}$$

Exercises

1. Apply (1) to prove that $\Gamma(1) = 1$.

2. Taking account of (7) show that $\Gamma(2.5) = 0.75\sqrt{\pi}$.

3. If $\Gamma(0.8) = 1.16$, find the value of $\Gamma(3.8)$.

4. Using (2) with $n = \frac{1}{2}$ and (VI), Table 2, express $T^{-1}\{1/(p-a)^{3/2}\}$ in terms of t.

5. Express $T^{-1}\{(p-a)^{-10/3}\}$ in terms of t.

6. Express in the form of an integral $T^{-1}\{p^{-1}(p-a)^{-10/3}\}$.

7. By (1), $\Gamma(0) = \int_0^\infty e^{-t}t^{-1}\,dt$. Prove that this integral has no value. *Hint:* $\int_0^1 e^{-t}t^{-1}\,dt < e^{-1}\int_0^1 t^{-1}\,dt \to \infty$.

8. Using (III) Table 2, (8), and (15) show that $T^{-1}\{1/[\sqrt{p}(p-a^2)]\} = a^{-1}e^{a^2t}\,\text{erf}\,(a\sqrt{t})$.

9. Using the plan of developing (17), show that $T^{-1}\{\sqrt{p}/(p - a^2)\} = 1/\sqrt{\pi t} + ae^{a^2 t}\,\text{erf}\,(a\sqrt{t})$.

10. Solve $(D - a^2)y = 1/\sqrt{\pi t}$ if $y = 0$ when $t = 0$. Give an answer in a form containing an error function. Also expand $Y = \sqrt{\pi}/[p^{3/2}(1 - a^2/p)]$ in an infinite series, integrate it term by term, and check the answer thus obtained.

11. Solve $(D - a^2)(D - b^2)y = t^{-1/2}$ if $y = Dy = 0$ when $t = 0$. *Hint:* $1/[(p - a^2)(p - b^2)] = [1/(b^2 - a^2)][1/(p - b^2) - 1/(p - a^2)]$.

12. Show that $T^{-1}\{1/[p^{1/2}(p - 1)^2]\} = e^t \int_0^t \text{erf}\,(\sqrt{\tau})\,d\tau$.

13. Show that $(1/\sqrt{\pi t})e^{at}(1 + 2at) = T^{-1}\{p(p - a)^{-3/2}\}$.

14. Express in terms of t, $T^{-1}\{1/[p(p - b^2)\sqrt{p + m^2}]\}$.

15. Read the hint to exercise 11, and then state a method of solving an equation having the form $(D - a_1^2)(D - a_2^2) \ldots (D - a_n^2)y = 1/\sqrt{\pi t}$, $y = Dy = \ldots = D^{(n-1)}y = 0$ when $t = 0$.

16. State a method of solving an equation of the type $(D + a_1^2)(D + a_2^2) \ldots (D + a_n^2)y = m^{-1}e^{m^2 t}\,\text{erf}\,(m\sqrt{t})$.

17. Solve $(D + a^2)y = m^{-1}e^{m^2 t}\,\text{erf}\,(m\sqrt{t})$ if $y = 0$ when $t = 0$. Assume that $m^2 > a^2$.

18. Assuming the formulas 1-31 in Table 1, derive formulas 33–36.

19. Express in terms of t: (a) $T^{-1}\{\sqrt{p}/(p - a^2)\}$; (b) $T^{-1}\{1/[\sqrt{p}\,(\sqrt{p} + a)]\}$; ★(c) $T^{-1}\{\sqrt{p}/(p + a^2)\}$.

Applications of Linear Equations with Constant Coefficients

70 Harmonic motion. Damping

For convenience of reference, we shall recall, at this point, a few facts concerning harmonic motion and damping.

If, as a particle moves in a straight line, its motion is defined by

$$y = c \sin (\omega t + \varphi) + a \tag{1}$$

where a is a constant, y is the distance of the particle from a fixed point on the line, and t is the time, its motion is called simple harmonic motion. The number c, representing the greatest value of $y - a$, is called the amplitude of the motion. Because of the periodic nature of $\sin (\omega t + \varphi)$, it is clear that the motion consists of an endless repetition of the movement that takes place while the angle $\omega t + \varphi$ changes by 2π rad; hence, the motion is called periodic. The time T required for the angle $\omega t + \varphi$ to change by 2π rad is called the period of the motion

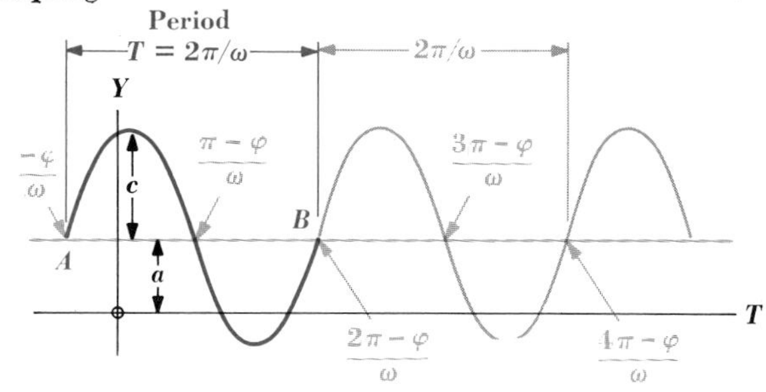

Figure 1

Therefore, we must have

$$\omega(t + T) + \varphi - (\omega t + \varphi) = 2\pi$$

or

$$\text{Period } T = \frac{2\pi}{\omega} \tag{2}$$

The number n of repetitions of the least complete motion, that is, the number of cycles per unit of time, is called the frequency. Hence

$$\text{Frequency } n = \frac{1}{T} = \frac{\omega}{2\pi} \tag{3}$$

The angle φ is often referred to as the angle of epoch, and $\omega t + \varphi$ as the phase. Figure 1 represents the motion. The part of the curve between A and B represents one period of the motion, the length c is the amplitude, and the distance from A to B represents the period $T = (2\pi - \varphi)/\omega - (-\varphi/\omega) = 2\pi/\omega$.

An equation having the form $y - a = c_1 \sin \omega t + c_2 \cos \omega t$ may be written in the form (1). For

$$c_1 \sin \omega t + c_2 \cos \omega t = \sqrt{c_1^2 + c_2^2}\left(\frac{c_1 \sin \omega t}{\sqrt{c_1^2 + c_2^2}} + \frac{c_2 \cos \omega t}{\sqrt{c_1^2 + c_2^2}}\right)$$

and this, in view of Fig. 2, may be written

$$\sqrt{c_1^2 + c_2^2}\,(\sin \omega t \cos \varphi + \cos \omega t \sin \varphi) = \sqrt{c_1^2 + c_2^2}\,\sin(\omega t + \varphi) \tag{4}$$

Hence it appears from Fig. 2 that

$$y - a = c_1 \sin \omega t + c_2 \cos \omega t = c \sin(\omega t + \varphi) \tag{5}$$

where

$$c = \sqrt{c_1^2 + c_2^2} \qquad \varphi = \tan^{-1}\frac{c_2}{c_1} \tag{6}$$

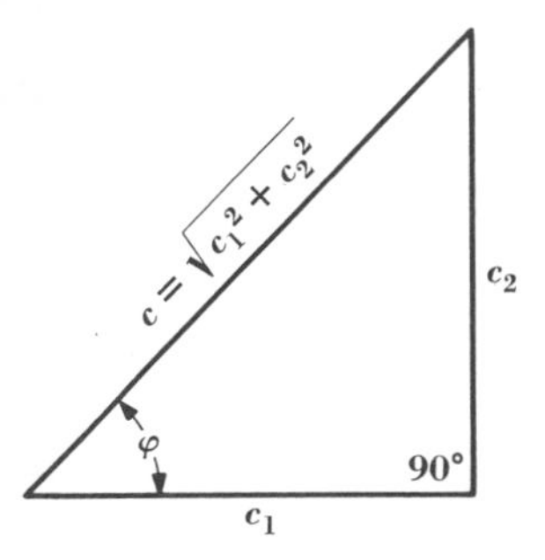

Figure 2

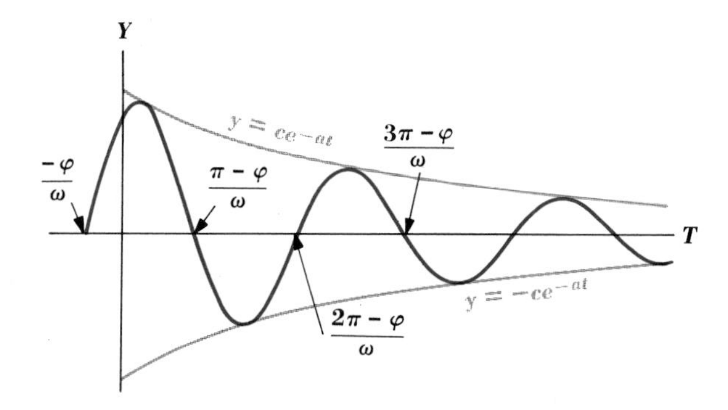

Figure 3

A very important damped oscillatory motion is represented by

$$y = ce^{-at} \sin (\omega t + \varphi) \qquad a > 0$$

Observe that the factor $\sin (\omega t + \varphi)$ describes an oscillatory kind of motion and that e^{-at} becomes smaller and smaller as t increases so that the oscillations become smaller and smaller in magnitude. Figure 3 represents the motion. Observe that the length of time for each wave is the same but that the heights of the waves become smaller and smaller with increasing t; that is, the motion is damped. *The factor* e^{-at} *is called the* damping factor, a *the* damping constant, $2\pi/\omega$ *the* period of oscillation, and $\omega/2\pi$ *the* frequency of y. For example, if

$$y = 10e^{-0.02t} \sin (120\pi t - \tfrac{1}{2}\pi)$$

the period is $2\pi/(120\pi) = \frac{1}{60}$, the frequency is 60 cycles, 0.02 is the damping constant, and $e^{-0.02t}$ is the damping factor. To find the time required by the damping factor to decrease one-half its value, we have $e^{-0.02t} = \frac{1}{2}$; hence, $\ln e^{-0.02t} = \ln \frac{1}{2}$, or $-0.02t = -\ln 2 = -0.6931$, and $t = 34.7$. This shows that the magnitude of the damping factor at the end of a 34.7-unit interval is one-half its magnitude at the beginning.

Exercises

In the following exercises, assume that distance is in feet, that time is in seconds, and that $\dot{x} = dx/dt$, $\ddot{x} = d^2x/dt^2$.

1. For a straight-line motion represented by $y = 5 \sin (12\pi t + \pi/6)$, find the amplitude, the period, the frequency, and two positive values of t for which $y = 0$.

2. For a straight-line motion represented by $y = 5 \sin 32t + 12 \cos 32t$, find the period, the frequency, and the amplitude, and show that the maximum value of dy/dt is 416.

3. The equation $\ddot{y} + 100y = 0$ represents a simple harmonic motion. Find the general solution of the equation, and determine the constants of integration if $y = 10$ and $dy/dt = 50$ when $t = 0$. Tell the frequency, the period, and the amplitude of the motion represented.

4. For a straight-line motion represented by $y = 25e^{-0.035t} \sin (377t + 1)$, find the period and the frequency, and show that the damping factor decreases from 1 when $t = 0$ to $\frac{1}{2}$ when t is approximately 20 units of time.

5. If for the motion of a particle along the X axis, $\ddot{x} + 3\dot{x} + 2x = 0$ and if $x = 0$ and $\dot{x} = 2$ when $t = 0$, find x in terms of t. Show that as t increases without bound from zero, x increases to $\frac{1}{2}$ ft, then decreases, and $x \to 0$ as $t \to \infty$.

6. What must be true of k in the equation $\ddot{s} + k\dot{s} + 30s = 0$ if the motion represented is oscillatory?

7. An oscillatory motion is represented by

$$\ddot{y} + \tfrac{1}{10}\dot{y} + 10y = 0$$

Find the period of oscillation of y, the damping factor, and the time required for the damping factor to decrease 50 per cent.

8. An oscillatory motion represented by an equation of the form

$$\ddot{x} + b\dot{x} + cx = 0$$

has a frequency of oscillation $n = 60$ and a damping constant $a = \frac{1}{10}$. Find b and c.

9. Find b and c in the differential equation of exercise 8 if the period of oscillation of x is $\frac{1}{10}$ sec and the damping factor decreases 50 per cent in 30 sec.

71 Types of Damping. Resonance

Important types of motions of bodies are defined by equations having the form

$$a\frac{d^2x}{dt^2} + b\frac{dx}{dt} + cx = f(t) \tag{1}$$

where a, b, and c are constants and $a > 0, b > 0, c > 0$. Its auxiliary equation is

$$ar^2 + br + c = 0 \tag{2}$$

First, we consider three types of motion indicated by (1) with $f(t) = 0$ and: (*a*) $b^2 - 4ac > 0$; (*b*) $b^2 - 4ac = 0$; (*c*) $b^2 - 4ac < 0$.

For case (*a*), the roots α and β of (2) are real and negative, the solution of (1) has the form $x = ce^{\alpha t} + c_2e^{\beta t}$, there is no oscillation, and the motion is said to be overdamped.

In case (*b*), $b^2 - 4ac = 0$, the roots α, of (2) are equal and negative, the solution of (1) is $x = (c_1 + c_2t)e^{\alpha t}$, the motion is nonoscillatory, and it is said to be critically damped.

In case (*c*), $b^2 - 4ac < 0$, the roots $-k \pm \omega i$ of (2) are imaginary, the solution has the form $x = e^{-kt}(c_1 \sin \omega t + c_2 \cos \omega t)$, and the motion indicated is said to be damped. In all three types of motion, x approaches zero as t becomes infinite.

When $f(t) \neq 0$, we have, in accordance with §51, $x = x_c + x_p$, where x is the solution just considered and x_p is due to $f(t)$ and may or may not be oscillatory. An interesting situation, called resonance, arises when, in (1), $b = 0$ and x_c and x_p have the same period. Consider, for example

$$\frac{d^2x}{dt^2} + 4x = 8 \cos 2t$$

The solution of this equation is

$$x = c_1 \sin 2t + c_2 \cos 2t + 2t \sin 2t \tag{3}$$

Observe that the term $x_p = 2t \sin 2t$ oscillates and that there is no bound to its magnitude. For example, if $t = 10{,}001\pi/4$, $x_p = 10{,}001\pi/2$ units $= 15{,}710$ units nearly.

Exercises

Assume that distance is in feet, that time is in seconds, and that $\dot{x} = dx/dt$ and $\ddot{x} = d^2x/dt^2$.

1. For the motion represented by

$$\ddot{x} + 4\dot{x} + kx = 0$$

find the numerically least integral value of k for the occurrence of: (*a*) critical damping; (*b*) overdamping; (*c*) damping. (*d*) Sketch the curve of part (*b*).

2. Name the type of damping associated with each equation:

$$(a)\ \ddot{x} + \dot{x} + 3x = 0 \quad (b)\ \ddot{x} + 2\dot{x} + x = 0 \quad (c)\ \ddot{x} + \dot{x} + \tfrac{1}{5}x = 0$$

3. (*a*) What is the value of ω at resonance for $\ddot{x} + 64x = 64 \sin \omega t$? (*b*) Find the value of $-4t \cos 8t$ when $t = 10^6\pi$ and when $t = 10^6\pi + \frac{1}{8}\pi$.

4. Show that if α is positive, the maximum value of $te^{-\alpha t}$ for t positive is e^{-1}/α. Sketch the curve $x = c_1te^{-t/3}$ for positive values of t.

5. Show that the solution of $\ddot{x} + \dot{x} = 3e^{-t}$, if $x = \dot{x} = 0$ when $t = 0$, is $x = 3 - 3e^{-t} - 3te^{-t}$ and that as t changes from 0 to ∞, x changes from 0 to 3.

6. Show that every solution for x of (1) with $f(t) = 0$ and a, b, and c positive numbers must approach zero as t becomes infinite.

72 Forces. Accelerations. Moments

Vectors, their components along the coordinate axes, and their sum in magnitude and direction are discussed in §34. Since forces, velocities, accelerations, and other quantities defined by magnitude and direction obey the laws of vectors, the following statements relate directly to these quantities.

A vector AB (see Fig. 1) is the segment AB of a straight line containing an arrowhead pointed toward B to indicate a direction from its initial point A to its terminal point B. The length of the segment indicates the *magnitude* of the vector, and the line with the attached arrowhead indicates direction. If, from the ends A and B of the vector, perpendiculars be dropped to the line of a vector $A'B'$ and meet

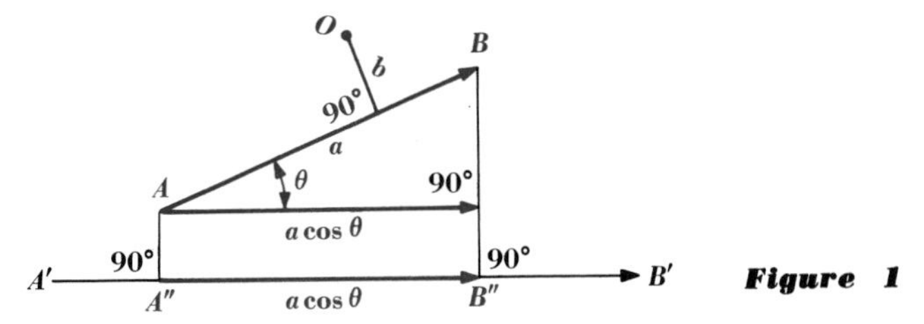

Figure 1

it in the points A'' and B'', respectively, then the vector $A''B''$ directed from A'' to B'' is called the *component* of vector AB in the direction of $A'B'$ (see Fig. 1).

Consider a vector AB of magnitude a, making an angle θ with the positive direction of vector $A'B'$. Then, the absolute value of the quantity $a \cos \theta$ is the magnitude of the component of vector AB in the direction of $A'B'$. If $a \cos \theta$ is positive in sign, the component has the same direction as $A'B'$; if $a \cos \theta$ is negative, the component has the direction opposite to that of $A'B'$. Similarly, if a_i, θ_i, $i = 1, 2, \ldots, n$, represent, respectively, magnitudes and angles made with vector $A'B'$ for n vectors A_iB_i, the algebraic sum $\Sigma a_i \cos \theta_i$, $i = 1, 2, \ldots, n$, will give in magnitude and sense along $A'B'$ the sum of the components of the n vectors in the direction of $A'B'$.

For example, the forces indicated in Fig. 2 have, as the sum of their components along OA, a vector of magnitude $10 \cos 0° + 5 \cos 90° + 7 \cos 240° + 5 \cos (-135°) = 10 - 3.50 - 3.54 = 2.96$ lb directed along OA, and as sum of components along OB a vector of $10 \cos 90° + 5 \cos 0° + 7 \cos 150° + 5 \cos 135° = 5 - (7\sqrt{3}/2) - 5/\sqrt{2} = -4.60$ lb along OB, or 4.60 lb in the direction opposite to that of OB.

Acceleration *a, like force, is a directed quantity which has components a_x and a_y in the directions of the coordinate axes*

$$a_x = \frac{d^2x}{dt^2} \qquad a_y = \frac{d^2y}{dt^2}$$

where t represents time. The statements made about forces apply generally to accelerations.

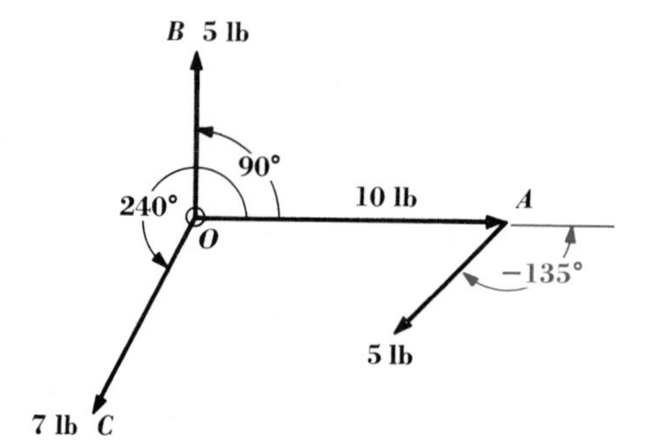

Figure 2

Again, consider a force represented by a vector AB lying in a fixed plane and an axis which is perpendicular to the plane and cuts it in point O (see Fig. 1). *The* moment, *or* torque, *of the force about the axis, also referred to as the* torque about O, *is defined to be the magnitude a of the force multiplied by the distance b from O to the line of action AB of the force.* Moment measures the tendency to cause turning. Two forces, one tending to turn a body about O in one sense and the other tending to turn it about O in the opposite sense, would have moments about O opposite in sign. For example, forces OB and OC in Fig. 2 have moments about A opposite in sign. In general, *if θ is the angle through which a body attached to the axis is turned from some fixed position of reference, the torque of a force is considered as positive when it tends to turn the body in the sense of increasing angle θ and negative when it tends to turn the body in the opposite sense.*

73 Some fundamental equations of motion

Plane motion of a rigid body is a motion such that each point in the body remains at a constant distance from a fixed plane; for example, a wheel on an automobile has plane motion when the car is moving in a straight line.

When all the particles of the body, rigid or not, have plane motion with respect to the same plane, the sum ΣF_d of the components in any direction of all the external forces acting on it is equal to the product of the mass m of the body and the component a_d in the same direction of the acceleration of the center of gravity of the body;* that is

$$\Sigma F_d = ma_d \tag{1}$$

Applying this rule for the direction of the X axis and for the direction of the Y axis, we obtain

$$\sum F_x = ma_x = m\frac{d^2x}{dt^2} \qquad \sum F_y = ma_y = m\frac{d^2y}{dt^2} \tag{2}$$

When a rigid body has plane motion, the moment, or torque T_g, of the external forces acting on it about an axis through the center of gravity of the body and perpendicular to the plane of its motion is equal to the product of the moment of inertia I_g of the body with respect to the same axis and the

* The acceleration g of a body given it by the pull of the earth is nearly 32.2 ft/sec². For many problems in this book, the unit of mass, the slug, will be considered as that of a 32.2-lb body. For such problems the mass m slugs of a w-lb body will be denoted by $w/32.2$.

angular acceleration α *of the body;* that is

$$T_g = I_g\alpha = I_g \frac{d^2\theta}{dt^2} \tag{3}$$

where θ is the angle through which the body is turned from some fixed position of reference. When the motion is pure rotation, we may write

$$T = I\alpha = I \frac{d^2\theta}{dt^2} \tag{4}$$

where T is the torque of the external forces about the axis of rotation and I is the moment of inertia of the body with respect to the same axis. Note that the axis associated with (4) is not necessarily an axis through the center of gravity.

74 Oscillatory motion

Three very important types of motion are referred to as free motion, damped motion, and forced motion. Thus, a weight supported in a vacuum by a spring would tend to move with an oscillatory motion when displaced vertically; if air were admitted, it would tend to slow down, or *damp*, the motion; and if the supporting structure were moved up and down, a motion would be *forced* on the weight. These three types of motion or their counterparts occur in a great many physical phenomena.

Example. The force exerted by a certain spring is proportional to the amount it is stretched, and a force of 8 lb stretches it 3 in. A 16.1-lb weight hanging at rest on the spring is drawn down 6 in. and released. Describe the motion: (*a*) if there is no air resistance; (*b*) if the air resistance in pounds and one-hundredth of the speed in feet per second are equal numerically; (*c*) if, in addition to the air resistance, the supporting structure is given a motion $y = \frac{1}{2}\sin 7t$.

Solution. (*a*) Let s be the number of feet the spring is stretched (see Fig. 1*a*), and let f represent the force exerted by the spring. Then, in accordance with Hooke's law, $f = ks$. Since $f = 8$ lb when $s = \frac{1}{4}$ ft, it appears that $8 = k\frac{1}{4}$ and $k = 32$. Hence

$$f = 32s \tag{a}$$

From the first equation of (2), §73, with downward considered as the positive direction, we get

$$16.1 - f = 16.1 - 32s = \frac{16.1}{32.2}\frac{d^2s}{dt^2} \tag{b}$$

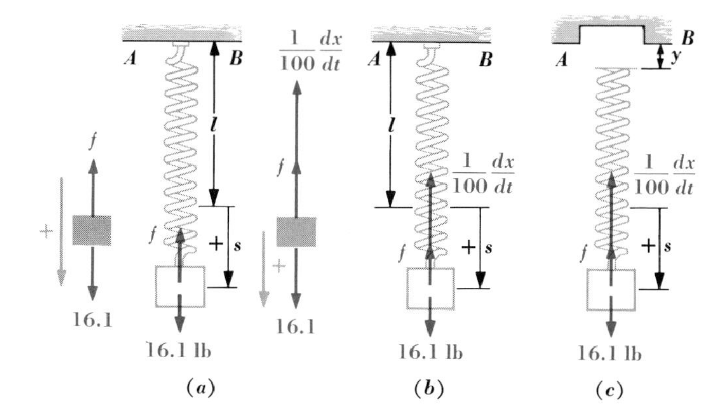

Figure 1 (a) (b) (c)

To simplify this equation, let $s = x + 16.1/32$,* so that

$$16.1 - 32s = -32x \qquad v = \frac{ds}{dt} = \frac{dx}{dt} \qquad a = \frac{d^2s}{dt^2} = \frac{d^2x}{dt^2} \tag{c}$$

In accordance with (*a*), 16.1/32 is the amount the spring is stretched by its weight. Hence, the use of (*c*) amounts to specifying the position of the weight at time t by its distance from the position in which it hangs in equilibrium. The use of (*c*) reduces equation (*b*) to

$$\frac{d^2x}{dt^2} + 64x = 0 \tag{d}$$

Take $t = 0$ just as the weight is released, that is, when it is at rest $\frac{1}{2}$ ft below the equilibrium position. Hence, the initial conditions are

$$x = \tfrac{1}{2} \qquad v = 0 \qquad \textit{when } t = 0 \tag{e}$$

Applying the method of Laplace transforms, §62, to (*d*) under conditions (*e*), we get

$$p^2X - \tfrac{1}{2}p + 64X = 0$$

$$x = T^{-1}\{X\} = T^{-1}\left\{\frac{\frac{1}{2}p}{p^2 + 64}\right\} = \tfrac{1}{2}\cos 8t \tag{f}$$

This represents a harmonic motion with amplitude $\frac{1}{2}$ ft and period $2\pi/8 = 0.785$ sec, or frequency $8/(2\pi) = 1.27$ cycles/sec.

The solution of (*c*) subject to the initial conditions (*e*) could have been obtained by the methods of Chap. 6 instead of the method of transforms.

(*b*) The differential equation of motion in this case is obtained from (*b*) by adding $-0.01\ ds/dt$ to the force members, or by adding 0.02

* It is not necessary to introduce x to replace s as indicated, but a simpler differential equation and a simpler solution result.

dx/dt to the left member of (d). This gives

$$\frac{d^2x}{dt^2} + 0.02\frac{dx}{dt} + 64x = 0 \tag{g}$$

Solving (g) subject to initial conditions (e), we get

$$p^2X - \tfrac{1}{2}p + 0.02pX - 0.02(\tfrac{1}{2}) + 64X = 0$$

$$x = T^{-1}\{X\} = T^{-1}\left\{\frac{\frac{1}{2}p + 0.01}{(p + 0.01)^2 + 64}\right\} \quad \text{nearly}$$

$$= e^{-0.01t}T^{-1}\left\{\frac{\frac{1}{2}(p - 0.01) + 0.01}{p^2 + 64}\right\}$$

$$x = e^{-0.01t}(\tfrac{1}{2}\cos 8t + 0.0006 \sin 8t) \tag{h}$$

Here the period is the same as before, accurate to three figures. The damping factor is $e^{-0.01t}$. To get an idea of the rate of damping, notice that the damping factor is $\frac{1}{2}$ when $e^{-0.01t} = \frac{1}{2}$, or when $-0.01t = -\ln 2 = -0.693$ and $t = 69.3$ sec; that is, the magnitude of the damping factor at the end of a 69.3-sec period is one-half its magnitude at the beginning. Here again we could have used the method of Chap. 6 for solving (g) subject to conditions (e).

(c) In this case (see Fig. 1), $s - y$ is the amount the spring is stretched, and $y = \frac{1}{2}\sin 7t$. Then

$$f = 32(s - y) = 32\left(x + \frac{16.1}{32} - y\right) = 32x + 16.1 - 16 \sin 7t \tag{i}$$

Substituting this value for f in (b), supplying the resisting force $-0.01\ dx/dt$, and simplifying, we get

$$\frac{d^2x}{dt^2} + 0.02\frac{dx}{dt} + 64x = 32 \sin 7t \tag{j}$$

Solving (j) subject to the conditions (e), we get

$$p^2X - 0.5p + 0.02pX - 0.02(\tfrac{1}{2}) + 64X = \frac{224}{p^2 + 49}$$

$$X = \frac{224}{(p^2 + 49)[(p + 0.01)^2 + 64]} + \frac{0.5p + 0.01}{(p + 0.01)^2 + 64} \quad \text{nearly} \tag{k}$$

Using relation 19, Table 1, Chap. 7, and others, and collecting like terms, we get

$$x = T^{-1}\{X\} = e^{-0.01t}(0.520 \cos 8t - 1.86 \sin 8t) + 2.13 \sin 7t - 0.020 \cos 7t \tag{l}$$

approximately. Observe that the motion of the weight is composed of two motions: a damped oscillatory motion and a harmonic motion. As time increases, the damped harmonic motion dies away while the harmonic motion $2.13 \sin 7t - 0.020 \cos 7t$ remains. The flow of electricity in many circuits follows this same plan, being made up of two parts: a transient part, which quickly dies out, and a steady-state part, which remains indefinitely.

Exercises

1. Solve parts (*a*) and (*b*) of the example by the methods of Chap. 6.

2. The force exerted by a spring is proportional to the amount that the spring is stretched and is 200 lb when the spring is stretched 1 ft. A 64-lb weight suspended by the spring as indicated in Fig. 2 is released from rest 0.4 ft below the equilibrium position. Find the equation, the amplitude, the period, and the frequency of the resulting motion. Take the mass as $\frac{64}{32} = 2$ slugs.

3. Assume in excerise 2 an additional vertical resisting force numerically equal to $0.04v$, where velocity v is in feet per second, and solve the resulting problem. Omit the part referring to amplitude. Also, find the damping factor and the time it takes it to decrease to 50 per cent of its initial value.

4. Assume the situation of exercise 2, together with the additional requirement that the upper end of the spring be given the motion $y = 0.64 \sin 6t$, where y ft represents distance below the point of suspension at t sec after motion starts. Find: (*a*) the equation of

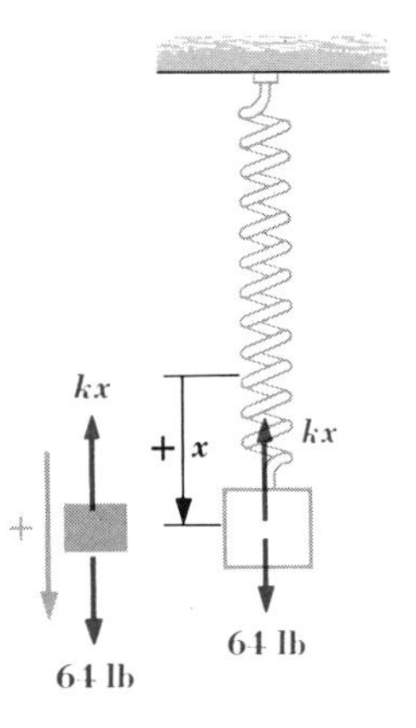

Figure 2

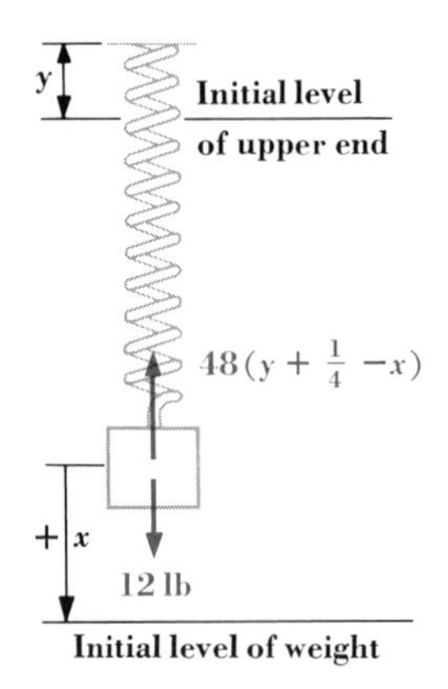

Figure 3

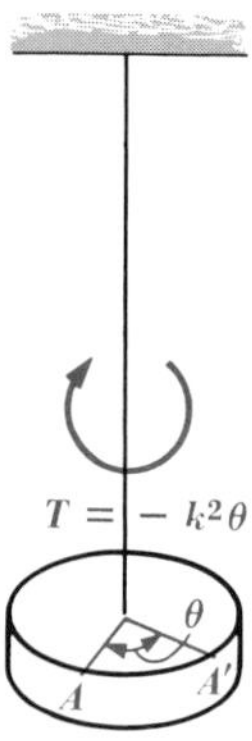

Figure 4

the motion of the 64-lb weight; (b) the period of the complementary solution, that of the particular solution, and that of the motion; and (c) the amount the spring is stretched 1 sec after motion starts.

5. A spring is stretched 1 in. when a 4-lb weight is hung on it. If a 12-lb weight is hanging at rest on the spring when the upper end of the spring is given the motion $y = \sin \sqrt{3g}\, t$, find a differential equation of the motion of the weight, solve this equation, and determine all constants of integration. Find the position of the weight $50\pi/\sqrt{g}$ sec after the motion starts. *Hint:* At time t (see Fig. 3), y is the distance of the upper end above its initial position, and we let x be the distance of the weight above its initial position. Hence, at time t, the spring is stretched $y + \frac{1}{4} - x$, and the upward force on the weight is $48(y + \frac{1}{4} - x) - 12$.

6. A rigid body suspended by a wire (see Fig. 4) has a motion of pure rotation about the line of the wire as an axis. If the only torque acting is a torque in the wire proportional to the angle that the body is turned from the position in which it hangs in equilibrium, find the period of the motion. *Hint:* Let θ be the angle through which the body is turned from equilibrium. Then, use (4)§73 to obtain

$$-k^2\theta = I\frac{d^2\theta}{dt^2}$$

7. A uniform sphere rotates about a supporting wire as an axis. If the number expressing the torque in the wire in pound-feet is equal to the number of radians through which the sphere is turned from the position in which it will hang in equilibrium, and if the sphere makes two complete oscillations per second, find the moment of inertia of the sphere with respect to the line of the wire.

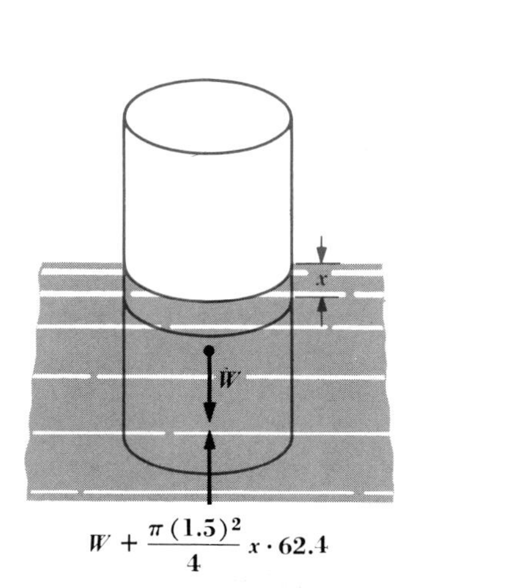

Figure 5

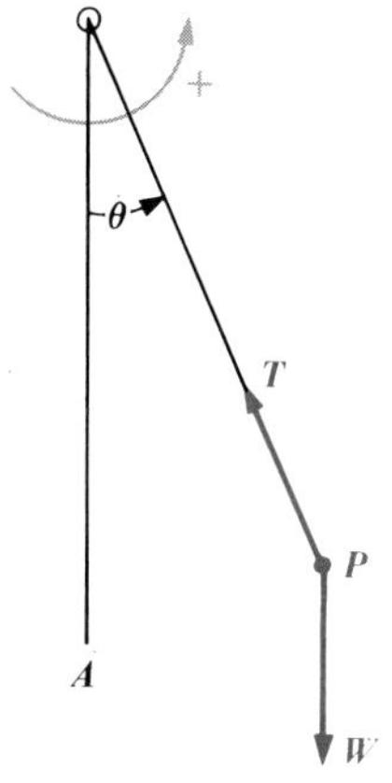

$$T_0 = I_0\alpha = \frac{W}{g} l^2\alpha$$

Figure 6

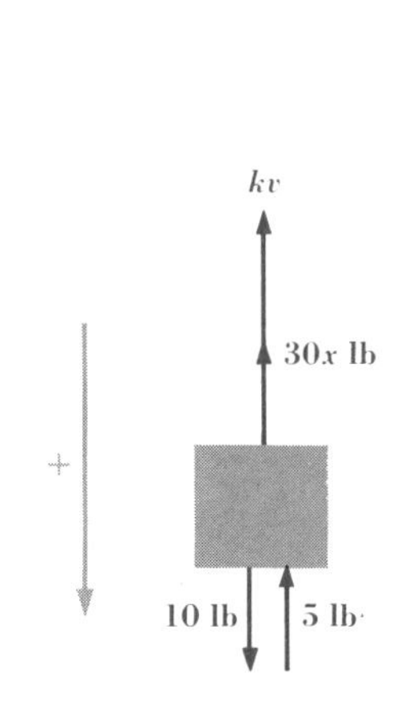

Figure 7

8.* A cylindrical spar buoy 18 in. in diameter stands in fresh water with its axis vertical (see Fig. 5). When depressed slightly and released, its period of vibration is found to be 2.7 sec. Find the weight of the cylinder.

9. Find, approximately, the period of vibration of a simple pendulum l ft long. Assume that the angle θ between the vertical and the cord of the pendulum is always so small that $\sin\theta$ may be replaced by θ without appreciable error. *Hint:* Use Fig. 6 and apply equation (4) §73. Here $I = (w/g)l^2$.

10. A 10-lb weight having specific gravity 2 is immersed in water and supported by a spring which it stretches 2 in. It is drawn down 1 ft from its position of equilibrium and let go. The resistance of the liquid to the motion of the weight is proportional to its velocity. If, at the end of two complete vibrations, the value of the damping factor is 25 per cent of its initial value, find the equation of the motion and its period. Figure 7 indicates the forces acting on the body.

11. A rubber band of natural length $AB = l$ (see Fig. 8) is suspended vertically from a point A, and a weight is attached to it at B. The weight stretches the band to a length $AO = l + h$. The weight is given a displacement $OP = a$, $a < h$, and then released. Find the equation of the motion.

* Archimedes' principle is involved in exercise 8. It states that a body in a liquid is acted upon by an upward force equal to the weight of the liquid it displaces. This force for a body floating at rest in a liquid is the weight of the body.

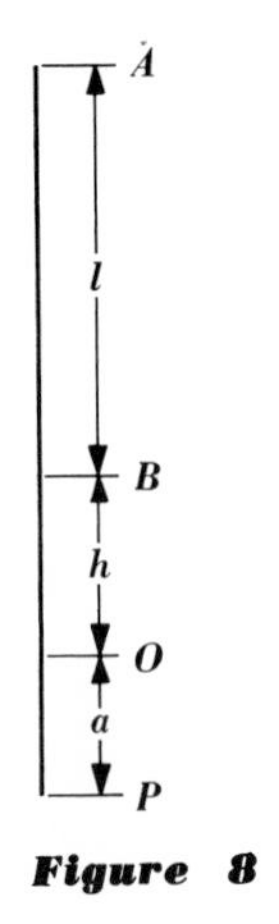

Figure 8

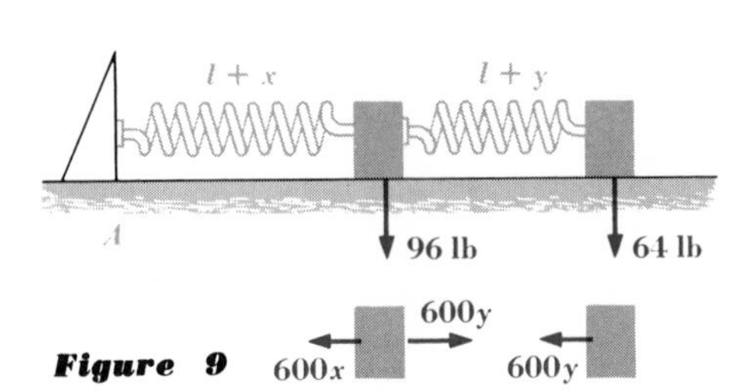

Figure 9

12. Solve exercise 5 with $y = \sin\sqrt{3g}\,t$ replaced by $y = \sin 2\sqrt{g}\,t$. Is there a theoretical upper limit to the distance of the weight from the starting point?

13. If, for the sphere and wire of exercise 7, a change of motion is caused by a frictional torque proportional to the angular velocity together with the torque in the wire and if the corresponding damping factor decreases to 25 per cent of its initial value during the first 20 sec of motion, find the equation and the period of the motion. Use the value of I found in exercise 7.

14. A rectangular block of wood 2 by 2 by 1 ft floats in fresh water with its 1-ft edge vertical. If the block weighs 160 lb, find the time of vibration when it is depressed slightly and released. Find also the time of vibration of the same block in a liquid of specific gravity ρ.

15. If the earth had uniform density, a particle below its surface would be attracted toward its center with a force proportional to the distance of the particle from the center. If a particle were dropped into a smooth straight vacuum passing through this earth's center, how long would it take the particle to reach the center? Assume that the radius of the earth is 3,960 statute miles. *Hint:* $(W/32.2)a = ks$, and $a = -32.2$ ft/sec², when $s = R$.

16. If I ton-ft² is the moment of inertia of a W-ton ship about a longitudinal water-line axis, G is the center of gravity of the ship, and M a fixed point in the ship through which the buoyancy of the water acts, the differential equation

$$I\frac{d^2\theta}{dt^2} = -gW\overline{GM}\theta$$

applies approximately. (*a*) Find I for a 30,000-ton battleship having $\overline{GM} = 7.7$ ft and period 17 sec. (*b*) Find the period of a 1,600-ton destroyer having $\overline{GM} = 2.3$ ft and $I = 2.9 \times 10^5$ ton-ft².

17.* Figure 9 represents a 96-lb weight and a 64-lb weight moving on a smooth, straight, horizontal track subject to the action of springs, as indicated. The force exerted by each spring is $600s$, where s ft represents elongation of the spring. Assume that, when $t = 0$, the springs are unstretched, the 96-lb weight is moving away from A at 600 ft/sec, and the 64-lb weight is at rest. Take 3 slugs and 2 slugs as the respective masses. Find the equations of motion of the weights.

★18.† In Fig. 9, replace 96 lb by P lb, 64 lb by Q lb, and assume an additional force $\alpha \sin \omega t$ acting on the P-lb block. As initial conditions, use $x = x_0$, $dx/dt = \dot{x}_0$, $y = y_0$, and $dy/dt = \dot{y}_0$ when $t = 0$. Set up the differential equations for the motion, and, preferably using transforms, eliminate y from them. Then carry the solution far enough to deduce that, if $gk/Q = \omega^2$, no effect is introduced by the force $\alpha \sin \omega t$ that could not be caused by a change of initial conditions.

75 Plane motions of bodies

In §74 motions in a straight line and simple rotary motions were considered. The equations of §73 will now be applied to the curved-line motion of projectiles and of bodies rotating and translating at the same time. In Fig. 1, the curve ODH is the path, or trajectory, of a projectile fired from a gun at O, φ is the angle of departure, V_0 is the initial velocity, θ is the inclination of the tangent line to the X axis (taken horizontal), OH is the range, vector W represents the force of gravity acting on the projectile, and vector R represents the force due to air resistance assumed to be acting along the tangent to the trajectory in a direction opposite to that in which the projectile is moving.

If $\dot{x} = dx/dt$ and $\dot{y} = dy/dt$, the relations between components $\dot{x}$

* Transforms should be used in solving exercises 17 and 18.

† This exercise illustrates the use of auxiliary masses to eliminate or reduce vibratory disturbances.

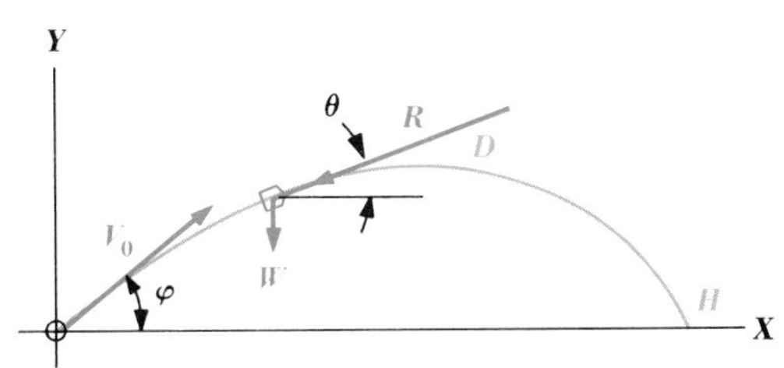

Figure 1

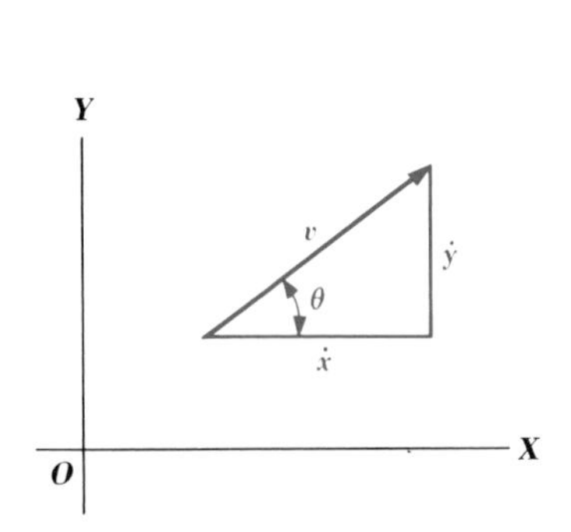

Figure 2

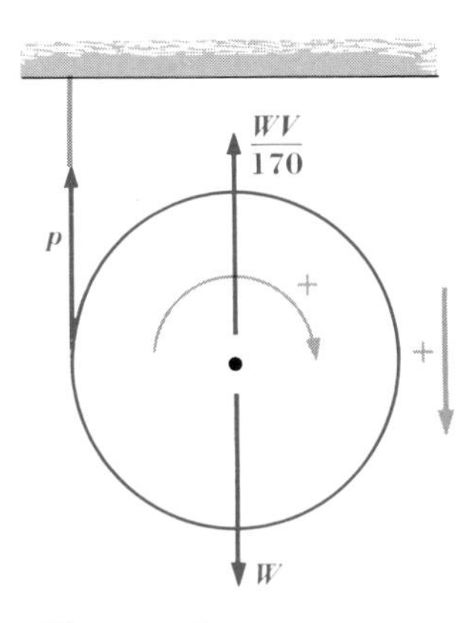

Figure 3

and $\dot{y}$ of the velocity are shown in Fig. 2. From Fig. 2, we read

$$\cos\theta = \frac{\dot{x}}{v} \qquad \sin\theta = \frac{\dot{y}}{v} \tag{1}$$

Applying equations (2) of §73 to the projectile of Fig. 1, we obtain

$$\frac{W}{g}\frac{d\dot{x}}{dt} = -R\frac{\dot{x}}{v} \qquad \frac{W}{g}\frac{d\dot{y}}{dt} = -W - R\frac{\dot{y}}{v} \tag{2}$$

where $\dot{x}$ and $\dot{y}$ represent, respectively, the horizontal and the vertical components of the velocity. The force R is a very complex quantity, so complex, in fact, that only approximations to the solutions of equations (2) can be found. Some rough approximations of special cases will appear as solutions of exercises in the list of this article.

The following example will illustrate a type involving both rotation and translation.

Example. A homogeneous cylinder, having radius $= r$ ft, weight $= W$ lb, and $I_g = \frac{W}{g}\frac{r^2}{2}$ (where $g = 32.2$), has a flexible cord wrapped around its central plane. One end of the cord is attached to a fixed plane as shown in Fig. 3. As the body falls, an air resistance in pounds equal numerically to $W/170$ times its velocity in feet per second retards its motion. If it starts from rest, find the distance y fallen in t sec, the limiting velocity, and the percentage of limiting velocity acquired in 20 sec.

Solution. If y is the distance fallen from rest and θ the angle through which the body has turned, we have

$$y = r\theta \qquad v = \frac{dy}{dt} = r\frac{d\theta}{dt} \qquad a = \frac{d^2y}{dt^2} = r\frac{d^2\theta}{dt^2} = r\alpha \tag{a}$$

Here, downward is considered as the positive direction and clockwise as the positive sense of rotation. Applying equations (2) and (3), §73, to the system represented by Fig. 3, we get

$$-p - \frac{W}{170}v + W = \frac{W}{g}\frac{d^2y}{dt^2} \tag{b}$$

$$pr = \frac{W}{g}\frac{r^2}{2}\alpha = \frac{Wr}{2g}r\alpha = \frac{Wr}{2g}\frac{d^2y}{dt^2} \tag{c}$$

Substituting p from (c) in (b) and simplifying slightly, we obtain

$$\frac{d^2y}{dt^2} + \frac{g}{255}\frac{dy}{dt} = \frac{2g}{3} \tag{d}$$

The initial conditions are

$$y = 0 \qquad \frac{dy}{dt} = 0 \qquad \text{when } t = 0 \tag{e}$$

Solving (d) subject to conditions (e)* by transforms, we get

$$(p^2 - 0p - 0)Y + \frac{g}{255}pY = \frac{2g}{3p}$$

$$Y = \frac{2g/3}{p^2(p + g/255)} = \frac{170}{p^2} - \frac{3(170)^2}{2gp} + \frac{3(170)^2}{2g(p + g/255)}$$

$$y = 170t + \frac{3(170)^2}{2g}(e^{-gt/255} - 1)$$

$$v = \frac{dy}{dt} = 170 - 170e^{-gt/255} \tag{f}$$

Letting t become infinite in (f), we get

$$\lim_{t \to \infty} v = 170 \text{ ft/sec}$$

Again

$$\frac{170(1 - e^{-20(32.2)/255})}{170}(100 \text{ per cent}) = 92 \text{ per cent}$$

Exercises

1. Solve equations (2) for x and y in terms of t if $R = 0$. Determine the constants of integration from the initial conditions $\dot{x} = v_0 \cos\varphi$, $\dot{y} = v_0 \sin\varphi$, $x = 0$, and $y = 0$, all when $t = 0$.

2. Solve equations (2) for the special case where R is numerically equal to $0.02wv/g$ and v is the speed in feet per second. Assume that $v_0 = 3{,}000$ ft/sec, $\varphi = 30°$, and note that when $t = 0$, $x = y = 0$,

* Of course, the solution can be found by the methods of Chap. 6.

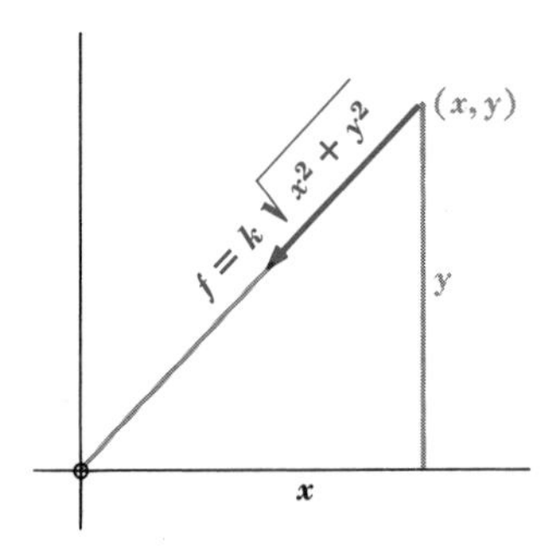

Figure 4

$\dot{x} = 3{,}000 \cos 30°$ ft/sec, and $\dot{y} = 3{,}000 \sin 30°$ ft/sec. Find the greatest height reached by the projectile.

3. When a projectile is fired at a small angle of elevation, the vertical component of its velocity is small, and consequently the vertical component of air resistance is small. A 100-lb shell is fired with initial velocity $v_0 = 2{,}000$ ft/sec and with angle of departure $\varphi = 5°$. Assuming in this case that air resistance is horizontal and equal numerically to $\frac{100}{25g}\frac{dx}{dt}$, derive equations similar to (2); solve these equations for dx/dt, dy/dt, x, and y, and determine the constants of integration.

4. If a projectile is fired at an angle of departure nearly equal to 90°, the horizontal component of air resistance is small. An anti-aircraft gun fires a projectile of weight W with initial velocity $v_0 = 2{,}000$ ft/sec and angle of departure $\varphi = 80°$. Assume air resistance to be vertical and equal numerically to $\frac{W}{1{,}200}\frac{dy}{dt}$. Find the equations of the trajectory, the maximum height attained by the projectile, and the height of the projectile when the vertical component of its velocity is 500 ft/sec.

5. A particle of mass m slugs moves in a plane under the action of a force (see Fig. 4) always directed to a fixed point in the plane and equal in magnitude to k times the distance of the particle from the fixed point. At a certain instant, the particle is moving with velocity v_0 at right angles to a line connecting it with the fixed point and is a units from it. Find the equations of motion and tell the nature of the path. *Hint:* Apply equations (2), §73, to obtain $m\ddot{x} = -kx$, etc.

6. A fixed plane contains a variable point P and two fixed points A and B. P is the position of a particle of unit mass which is acted on by two forces, one equal numerically to the magnitude of BP and exerted in the direction from B toward P and a second equal numeri-

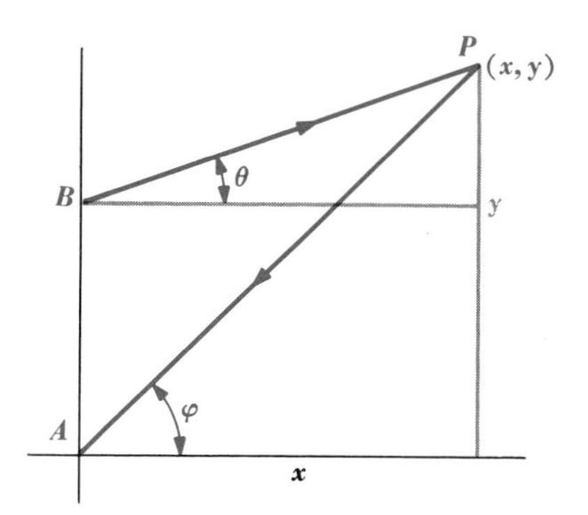

Figure 5

cally to the magnitude of $2PA$ and exerted in the direction from P toward A. Find the equations of motion if $AB = 2$ ft and if the particle is initially at A moving 3 ft/sec in a direction making (*a*) an angle of 90° with AB; (*b*) an angle of 45° with AB. *Hint:* In Fig. 5, $\Sigma F_x = \overline{BP} \cos \theta - 2AP \cos \varphi = \overline{BP}(x/\overline{BP}) - 2\overline{AP}(x/\overline{AP}) = x - 2x = -x$, etc.

7. A circular cylinder having radius r, weight W, and I with respect to its axis $\dfrac{W}{g}\dfrac{r^2}{2}$ is rolling on a rough horizontal plane when two horizontal forces perpendicular to its axis are impressed on it: a constant force equal to $\frac{1}{10}W$ in the direction of motion and an oppositely directed air resistance proportional to the velocity of its axis. If no slipping occurs, if the limiting speed is 60 ft/sec, and if the initial speed is zero, describe the motion and find how far the cylinder rolls during the first minute of motion (see Fig. 6). *Hint:* Apply equations (2) and (3) of §73, eliminate f, and integrate the resulting equations.

8. The cylinder of exercise 7 rolls without slipping and with axis horizontal on a rough plane inclined 30° to the horizontal. If the air resistance is opposite to the direction of motion and is numerically equal to $W/600$ times the magnitude of the velocity of the axis in feet per second, find the limiting speed and the distance traversed during the first minute of motion from rest.

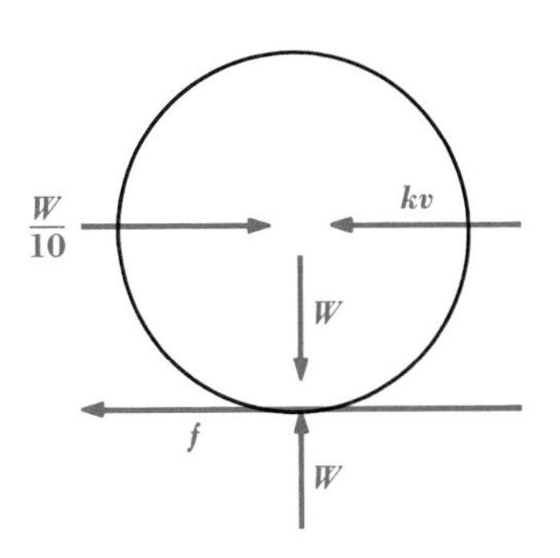

Figure 6

76 Kirchhoff's current law and electromotive-force law*

Electricity is a substance which flows through conductors such as wires. The practical unit of electricity is the coulomb,† and we speak of q coul of electricity just as we would speak of q gal of water. A current of electricity is a rate of flow of this substance. A current of q coul of electricity per second is called q amperes. When a constant current of I amp flows through a wire, $q = It$ coul will pass any cross section of the wire in t sec.

KIRCHHOFF'S CURRENT LAW. *The excess of the current flowing into a given region at a given time over the current flowing out at the same time is the time rate of increase of quantity of electricity within the region at that time.*

If there is no accumulation of electricity within a given region, current flowing into this region equals current flowing out of it; for example, in Fig. 1,

$$i_1 = i_2 + i_3 + i_4 \tag{1}$$

If the region is the positive plate of a capacitor, electricity flows into the region but none, theoretically, flows out. Hence

$$q = \int_c i \, dt \tag{2}$$

where q is the charge (quantity of electricity) on the positive plate of the capacitor, i is the current, t is the time, and the subscript c on the

* Good reference books are G. W. Pierce, "Electric Oscillations and Waves," McGraw-Hill Book Company, New York, 1923; F. W. Sears, "Principles of Physics II," Addison-Wesley Publishing Company, Cambridge, Mass. 1947.

† The electron, an elemental building block of nature, carries a fixed charge of negative electricity. A coulomb consists approximately of the sum of the charges on 6.24×10^{18} electrons.

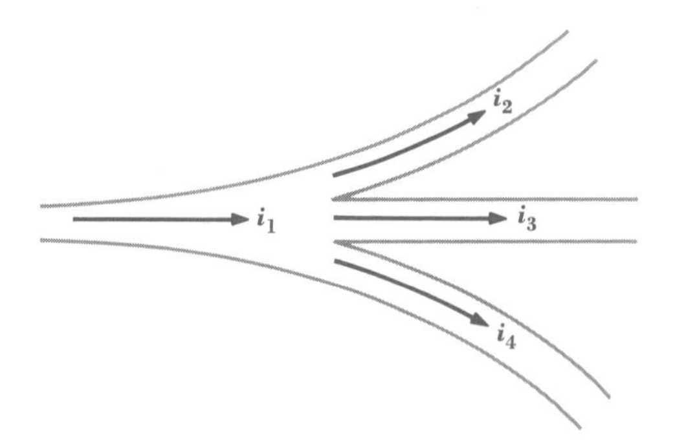

Figure 1

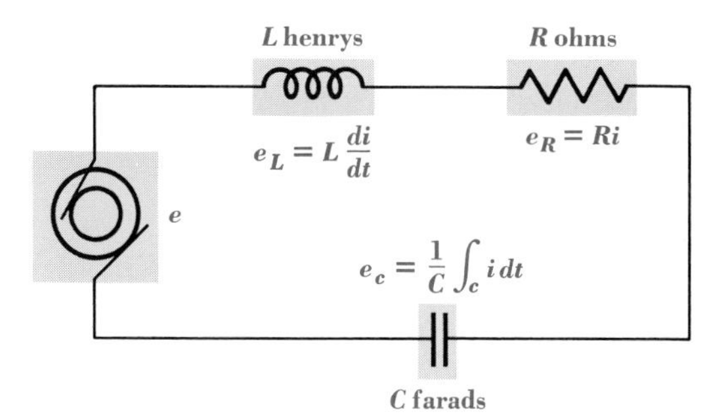

Figure 2

integral sign indicates that limits are to be taken so as to obtain the total quantity of electricity on the capacitor at time t. Differentiating (2) with respect to the time, we obtain

$$\frac{dq}{dt} = i \qquad \frac{d^2q}{dt^2} = \frac{di}{dt} \tag{3}$$

Electromotive force (emf), also called voltage and difference of potential, causes electricity to move just as physical force causes bodies to move. The volt will be used as the unit.

When electricity is flowing through a coil, any change in current sets up a counter-emf opposing the change in current. For this reason, an emf e_L of magnitude $e_L = L(di/dt)$ must act to cause the current to flow. L is a constant, called inductance. It is analogous to the mass of a body. The practical unit of inductance is the henry.

Conductors offer resistance to the flow of electricity through them. The practical unit of resistance is the ohm. Resistance depends upon such things as size and kind of material. If a conductor has a resistance of R *ohms*, an emf force e_R of magnitude $e_R = Ri$ volts is required to cause a current of i amp to flow through it. Resistance is analogous to friction.

A capacitor consists essentially of two plates separated by a nonconducting substance, or insulator. When a current flows to one plate of a capacitor, a charge is deposited there; an equal charge opposite in sign appears at the other plate; and the current away from the capacitor is equal to the current flowing to it. The emf e_C across a capacitor having a charge of q coul and a capacity of C farads is given by

$$e_c = \frac{q}{C} = \frac{1}{C}\int_c i\,dt$$

Figure 2 represents the elements mentioned above, and Kirchoff's electromotive-force law, which follows, gives an equation connecting them:

ELECTROMOTIVE-FORCE LAW. *When several elements, resistance R, inductance L, and capacitance C, all constant, are connected in series, and* *when an instantaneous current i is flowing in them, there is impressed in the direction of i at the terminals of this series from a source of power external to the elements a difference of potential e such that*

$$e = L\frac{di}{dt} + Ri + \frac{1}{C}\int_c i\,dt \tag{4}$$

where $\int_c i\,dt$ represents the total quantity of electricity on the capacitor considered.

It is worthy of note that *Kirchhoff's laws may be applied to any part or the whole of a circuit but that, in such application, fall of potential must be considered positive in the direction of the current and negative in the opposite direction.* Also, L, R, and C do not necessarily apply to concentrated elements in a circuit but are sums of inductances, resistances, or capacitances for the part of the circuit considered.

Substitution of q for $\int_c i\,dt$, dq/dt for i, and d^2q/dt^2 for di/dt in (4) gives

$$L\frac{d^2q}{dt^2} + R\frac{dq}{dt} + \frac{q}{C} = e \tag{5}$$

and differentiation of (4) with respect to the time gives

$$L\frac{d^2i}{dt^2} + R\frac{di}{dt} + \frac{1}{C}i = \frac{de}{dt} \tag{6}$$

Remarks. Typical equations for electrical phenomena and mechanics of moving bodies follow:

$$LD^2q + RDq + \frac{1}{C}q = e(t) \qquad Dq = i$$

$$mD^2s + nDs + Ks = f(t) \qquad Ds = v$$

We see at once the following correspondence:

L	m	Inductance to mass
R	μ	Electrical resistance to a frictional resistance
$1/C$	K	Reciprocal of capacitance to spring constant or potential energy
$e(t)$	$f(t)$	Voltage to external force
q	s	Charge of electricity to distance
i	v	Current of electricity to velocity

Because of this correspondence, numerous solutions of problems of electricity involve the same mathematics as the solution of a problem in the mechanics of moving bodies. This relation is referred to as mechanical analogue or electrical analogue. It is interesting to find that one type of electronic computer solves problems of various kinds by setting up the circuit of the electric analogue and displaying the results thus obtained.

77 Simple circuits containing constant electromotive force

To illustrate the use of equations (5), §76, and (6), §76, consider the charging of a capacitor of capacitance C through a resistance R and an inductance L by a constant emf E. For this case, equation (5), §76, becomes

$$LD^2q + RDq + \frac{1}{C}q = E \tag{1}$$

where $D = d/dt$. Assume as initial conditions $q = 0$ and $i = 0$ when $t = 0$. The roots of the auxiliary equation $Lm^2 + Rm + 1/C = 0$ are

$$m = \frac{-R}{2L} \pm \sqrt{\frac{R^2}{4L^2} - \frac{1}{LC}} \qquad \text{or} \qquad \frac{-R}{2L} \pm \sqrt{\frac{1}{LC} - \frac{R^2}{4L^2}}\,j \tag{2}$$

where $j^2 = -1$. First let us assume that the roots are imaginary, and let

$$a = \frac{R}{2L} \qquad \omega_1 = \sqrt{\frac{1}{LC} - \frac{R^2}{4L^2}} \tag{3}$$

Then we have $m = -a \pm \omega_1 j$, and the solution of (1) is

$$q = \epsilon^{-at}(c_1 \sin \omega_1 t + c_2 \cos \omega_1 t) + CE* \tag{4}$$

Differentiating this, remembering that $i = dq/dt$, and using the initial conditions $q = 0$ and $i = 0$ when $t = 0$, from (4) we obtain

$$q = \frac{-CE}{\omega_1}\epsilon^{-at}(a \sin \omega_1 t + \omega_1 \cos \omega_1 t) + CE$$
$$i = \frac{E}{L\omega_1}\epsilon^{-at} \sin \omega_1 t \tag{5}$$

* Here, ϵ (= 2.7183 approximately) is used to represent the base of natural logarithms

Exercises

1. Carry out the solution required to derive (5) by using (4), the given initial conditions, and $i = dq/dt$.

2. Taking account of (3), find the solution (5) when $R = 0$, and give the period and amplitude of both i and q.

3. If the roots of $Lm^2 + Rm + 1/C = 0$ are the real numbers α_1 and α_2, show from (2) that both are negative and that the corresponding values for q and i shrink toward zero without oscillation as t increases.

4. If $1/(LC) - R^2/(4L^2) = 0$, write the corresponding solution of (2), and show that there is no oscillation of q or i, that q approaches CE, and that i approaches 0 as t increases without bound.

5. Solve equations (5) and (3), §76, to find i and q for a circuit in which $L = 0.1$ henry, $R = 1$ ohm, $C = 250 \times 10^{-6}$ farad, and $e = E = 100$ volts if the initial conditions are $q = 0$ and $i = 0$ when $t = 0$. In what time does the damping factor of the current decrease to one-tenth of its value when $t = 0$, and what is the period of the current? What are the limiting values of q and i?

6. Solve equations (3) and (5), §76, to find i and q in terms of t for a circuit in which $L = 1$ henry, $R = 1$ ohm, $C = 4$ farads, and $e = E = 100$ volts if $q = 0$ and $i = 0$ when $t = 0$. Show that the maximum value of the current is $200\epsilon^{-1}$ (where $\epsilon = 2.718$ nearly) and that the ratio of the current when $t = 10$ sec to this maximum value is $5\epsilon^{-4} < 0.1$.

7. A circuit consists of an impedance coil of inductance L and negligible resistance connected in series with a capacitor of capacitance C. Use equations (3) and (5), §76, to find the charge q on the capacitor and the current i at time t if t is the number of seconds since i was zero and q was q_0. Describe the fluctuation of i and q. *Hint*: $e = 0$.

8. Solve equations (5) and (3), §76, with e replaced by the constant E and R by zero, for i and q in terms of t. Assume that $i = 0$ and $q = 0$ when $t = 0$.

9. A circuit consists of an impedance coil having an inductance L and resistance R connected in series with a capacitor having a capacitance C. Initially the current is zero, and the charge on the capacitor is q_0. Find i and q at any time if $4L > CR^2$.

78 Simple circuits containing a sinusoidal electromotive force

The type of equation to be considered in this article has the form

$$LD^2q + RDq + \frac{1}{C}q = E \sin \omega t \tag{1}$$

The current i is found from the equation $i = dq/dt$. The emf $e = E \sin \omega t$ is generally supplied by a dynamo.

In what follows, it will be convenient to let

$$X = L\omega - \frac{1}{C\omega} \qquad Z = \sqrt{R^2 + X^2} \tag{2}$$

X is called the reactance and Z the impedance.

A particular solution of (1) could be found by using any of the methods discussed in §§47 to 55. That using the symbolic operator (§54) is the shortest. From (1), by using first equation (14) of §54 and then (2) and a slight simplification, obtain

$$\begin{aligned} q_p &= \frac{E \sin \omega t}{(LD^2 + 1/C) + RD} = \frac{LD^2 + 1/C - RD}{(LD^2 + 1/C)^2 - R^2D^2} E \sin \omega t \\ &= \frac{(-L\omega^2 + 1/C)E \sin \omega t - RE\omega \cos \omega t}{(-L\omega^2 + 1/C)^2 + R^2\omega^2} \\ &= \frac{-E}{\omega Z^2}(X \sin \omega t + R \cos \omega t) \end{aligned}$$

Supplying the solution of the auxiliary equation, we obtain the general solution of (1),

$$\begin{aligned} q = q_c + q_p &= \epsilon^{-at}(c_1 \sin \omega_1 t + c_2 \cos \omega_1 t) \\ &\qquad - \frac{E}{\omega Z^2}(X \sin \omega t + R \cos \omega t) \end{aligned} \tag{3}$$

The equation of the current is obtained by differentiating (3) and replacing dq/dt by i.

The part containing the factor ϵ^{-at} generally becomes negligible in a very short time. It is called a transient. Transients are important in the theory of radio and of radar. The other part of the solution is permanent and is called the steady-state solution. Dropping the transient term from (3), we have for the steady state

$$q = \frac{-E}{\omega Z^2}(X \sin \omega t + R \cos \omega t) \tag{4}$$

and since $i = dq/dt$, for the *steady-state value* of i,

$$i = \frac{dq}{dt} = \frac{E}{Z^2}(R \sin \omega t - X \cos \omega t) \tag{5}$$

79 Resonance

Equation (5), §78, may be written in the form

$$i = \frac{E}{Z} \sin\left(\omega t - \tan^{-1}\frac{X}{R}\right) \tag{1}$$

where the quadrant of $\tan^{-1}(X/R)$ is that of point (X,R) when plotted in rectangular coordinates. Observe that the amplitude of i,

$$\frac{E}{Z} = \frac{E}{\sqrt{R^2 + X^2}} = \frac{E}{\sqrt{R^2 + [L\omega - 1/(C\omega)]^2}} \tag{2}$$

will be a maximum for given values of L, R, E, and ω when C is chosen so that $X = 0$, that is, so that

$$L\omega - \frac{1}{C\omega} = 0 \qquad \text{or} \qquad \omega = \frac{1}{\sqrt{LC}} \tag{3}$$

With this condition is associated the name *current resonance*. A person tuning a radio in to a station takes advantage of current resonance.

Exercises

1. Reproduce the solution of (1), §78, to obtain (3), §78, without using the text.

2. Obtain the steady-state equation for i by differentiating (4), §78.

3. If $e = E \sin \omega t$ in the circuit of Fig. 2, §76, derive the steady-state solution for i and q: (*a*) when $R = 0$ and there is no capacitor; (*b*) when $L = 0$ and there is no capacitor; (*c*) when $L = 0$ and $R = 0$; (*d*) when $L = 0$; (*e*) when there is no capacitor; (*f*) when $R = 0$.

4. If $L = 0.1$ henry, $R = 10$ ohms, and $C = 10^{-3}$ farad, use (2) to find the ratio of amplitude of steady-state current when ω in (1) is 50 to that of current resonance when $\omega = 1/\sqrt{LC} = 100$.

5. A sinusoidal emf of frequency $200/\pi$ cycles/sec and maximum value 110 volts is connected in series in a circuit with an inductance of 0.1 henry, a resistance of 10 ohms, and a capacitor of capacity 250×10^{-6} farad. Find the steady-state solution of i and q in terms of t and the maximum values of the steady-state charge on the capacitor and the current.

6. Find the expression of i in terms of t for the circuit of Fig. 2, §76, provided that $e = 20 \sin 500t$, $R = 2$ ohms, $L = 0.2$ henry, $C = 20 \times 10^{-6}$ farad and if i and q are zero when $t = 0$. Find the value of the damping factor of the transient at time $t = 1$ sec.

7. Use (3), §79, to find C at current resonance when the frequency is 10^5 cycles/sec and $L = 6 \times 10^{-6}$ henry. If $R = 100$ ohms, find the ratio of the maximum current E/Z from (1), §79, at current resonance to the current in the same circuit with no capacitor.

8. In (1), §78, take $L = 0.2$ henry, $R = 10$ ohms, $C = \frac{1}{2000}$ farad, $E = 100$ volts, find $i_{\max}$ at current resonance and $i_{\max}$ when $\omega = 1000$.

80 Applications of Kirchhoff's laws to networks

The emf law and the current law stated in §76 may be applied when elements involving inductance, resistance, and capacitance are connected in a more or less complicated network. Equations can be obtained by applying the emf law to complete circuits or the current law at points where two or more conductors meet.

For example, in Fig. 1, apply the current law at A and at B to obtain

$$i = i_1 + i_2 \qquad i_2 = i_3 + i_4$$

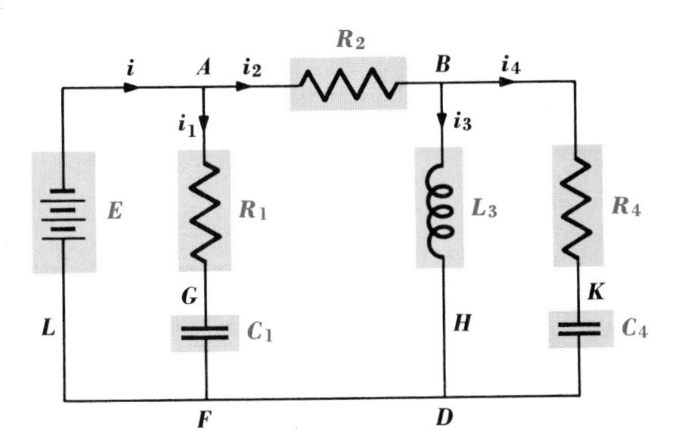

Figure 1

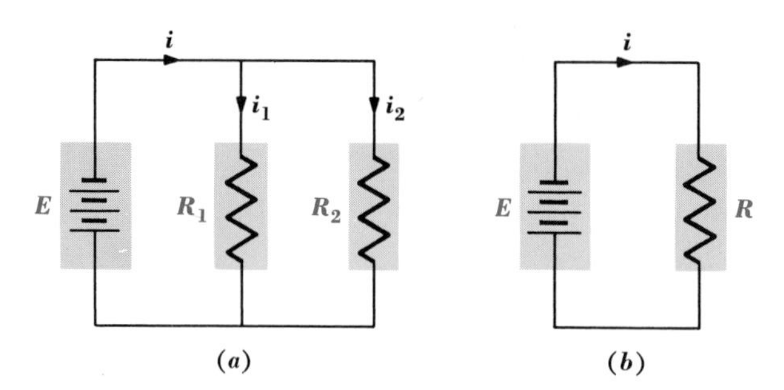

Figure 2

Apply the emf law to circuit $LAGF$ to obtain

$$R_1 i_1 + \frac{q_1}{C_1} = E \qquad i_1 = \frac{dq_1}{dt}$$

Also, apply the emf law to the circuits $LABHDF$ and $BKDH$ to obtain

$$R_2 i_2 + L_3 \frac{di_3}{dt} = E \qquad R_4 i_4 + \frac{q_4}{C_4} - L_3 \frac{di_3}{dt} = 0 \qquad i_4 = \frac{dq_4}{dt}$$

Observe in this last circuit that there was no externally applied emf and that a negative sign was given to $L_3(di_3/dt)$ because, in following the circuit in the direction B to K to D to H, the element of inductance L_3 was traversed opposite to the direction of the assumed current i_3. Solving the equations just derived, we could find the current in each branch of the network and the charge of electricity on each capacitor at time t.

As another application, apply Kirchhoff's laws to the circuits of Fig. 2. We have

$$R_1 i_1 = E \qquad R_2 i_2 = E \qquad i = i_1 + i_2 \tag{1}$$

Substituting i_1 and i_2 from the first two equations of (1) in the third and transforming slightly, we have

$$i = E\left(\frac{1}{R_1} + \frac{1}{R_2}\right) \quad \text{or} \quad \frac{i}{1/R_1 + 1/R_2} = E \tag{2}$$

Comparing this last equation with $Ri = E$ from Fig. 1*b*, it appears that two resistances R_1 and R_2 in parallel are together equivalent to a resistance $R = 1/(1/R_1 + 1/R_2)$.

In general, to find the currents in the branches of a network and the charges on the capacitors, proceed as follows: (1) *Draw a figure representing the elements involved, indicating inductance by* -ꝏ-, *resistance by* -ʌʌʌ-, *and a capacitor by* —||—; (2) *draw arrowheads to indicate the assumed directions of currents through the various branches;* (3) *apply the current law at points where conductors intersect and the emf law to complete circuits to find as many independent equations as are necessary to determine the unknown quantities involved;* (4) *solve these equations for the unknowns.* The following example will illustrate the procedure:

Example. An impedance coil which has a resistance of 14 ohms and an inductance of 0.05 henry and a branch having a noninductive resistance of 15 ohms and a capacitor of capacity 10^{-4} farad in series are connected in parallel across the terminals of a 220-volt source of emf. Find expressions in terms of the time for the charge on the capacitor, the current in the impedance coil, the current in the noninductive resistance, and the total current.

Solution. Figure 3 represents the circuit with indicated elements and currents. The current law applied at point B gives

$$i = i_1 + i_2 \tag{a}$$

The emf law applied to circuit $ABFA$ gives

$$0.05 \frac{di_1}{dt} + 14i_1 = 220 \tag{b}$$

and, applied to circuit $ABDFA$, it gives

$$15i_2 + 10^4 \int_c i_2\, dt = 220 \tag{c}$$

Finally, we have from equation (2) of §76,

$$q = \int_c i_2\, dt \qquad i_2 = \frac{dq}{dt} \tag{d}$$

Elimination of i_2 from (c) by using (d) gives

$$15 \frac{dq}{dt} + 10^4 q = 220 \tag{e}$$

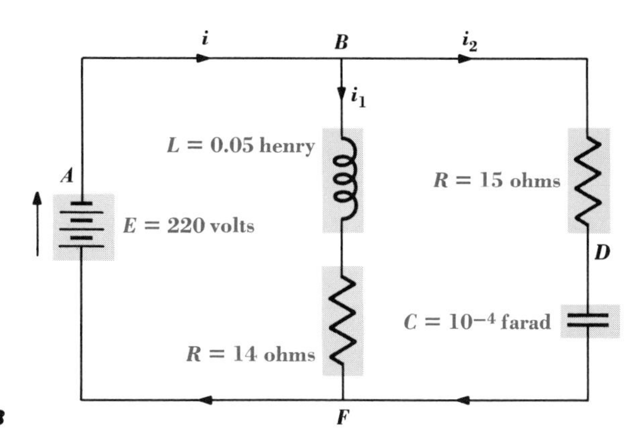

Figure 3

The solution of equation (*e*), subject to the condition $q = 0$ when $t = 0$, is

$$q = 0.022(1 - \epsilon^{-2,000t/3})$$

Therefore,

$$i_2 = \frac{dq}{dt} = \frac{44}{3}\epsilon^{-2,000t/3}$$

The solution of (*b*), subject to the condition $i_1 = 0$ when $t = 0$, is

$$i_1 = 15.71(1 - \epsilon^{-280t})$$

Finally

$$i = i_1 + i_2 = 15.71(1 - \epsilon^{-280t}) + \frac{44}{3}\epsilon^{-2,000t/3}$$

Since the values of ϵ^{-280t} and $\epsilon^{-2,000t/3}$ are practically zero after a fraction of a second, it appears that the capacitor very soon is practically charged and that the inductance in *BF* opposes the current for only a small fraction of a second. Hence, we have practically $q = 0.022$ coul and $i = 15.71$ amp in a very short time.

Exercises

1. For the system represented in Fig. 4, obtain by the current law at point *A*, $i = i_1 + i_2$ and from the emf law applied to circuits *AFHG* and *ABHG*

$$L\frac{di_1}{dt} = E \sin \omega t \qquad Ri_2 = E \sin \omega t$$

Solve these equations for i_1, i_2, and i in terms of t, and determine a constant of integration by using the condition $i = 0$ when $t = 0$.

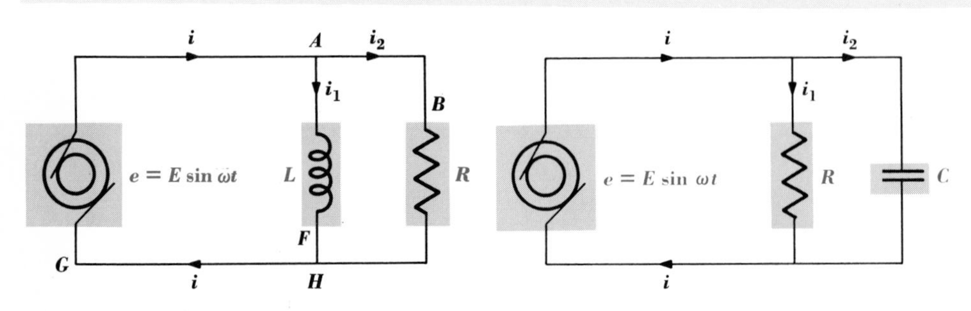

Figure 4 **Figure 5**

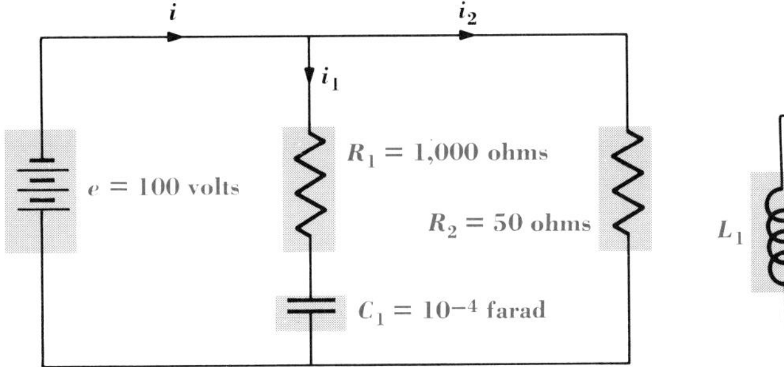

Figure 6

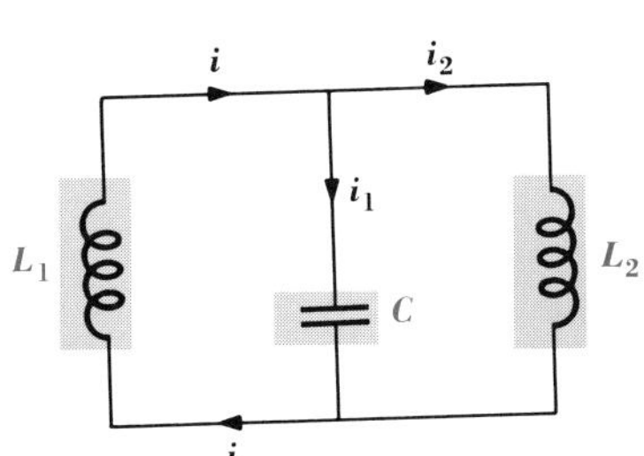

Figure 7

2. From Fig. 5, derive by Kirchhoff's laws

$$i = i_1 + i_2 \qquad Ri_1 = E \sin \omega t \qquad \frac{1}{C}\int_c i_2\,dt = \frac{q}{C} = E \sin \omega t$$

Find q, i_1, i_2, and i in terms of t.

3. For the system indicated in Fig. 6, derive three equations by applying Kirchhoff's laws. Assume that the charge on the capacitor is zero when $t = 0$, and deduce that $i_2 = 2$ amp always and that $i_1 = \frac{1}{10}\epsilon^{-10t}$ and therefore rapidly approaches zero.

4. Replace the 100-volt emf in Fig. 6 by a sinusoidal emf represented by 100 sin $400t$, and then find q, i_1, and i_2 at time t. Assume that the charge on the capacitor is zero when $t = 0$.

5. If initially the charge on the capacitor of Fig. 7 is q_0 and $i_1 = 0$, show that $q = q_0 \cos \sqrt{(L_1 + L_2)/(CL_1L_2)}\; t$. *Hint*: $e = 0$.

6. A hot-wire galvanometer having a resistance of 5 ohms is shunted with a capacitor of capacitance $C = 5 \times 10^{-8}$ farad, as indicated in Fig. 8. If the effective current is $1/\sqrt{2}$ times the maximum current, find the ratio of the effective current i_1 through the galvanometer to the total effective current i when: (*a*) $\omega = 2 \times 10^4$; (*b*) $\omega = 2 \times 10^5$; (*c*) $\omega = 2 \times 10^6$; (*d*) $\omega = 2 \times 10^7$.

7. Find i in terms of the time t in the system represented by Fig. 9 if all initial currents and the initial charge on the capacitor are zero.

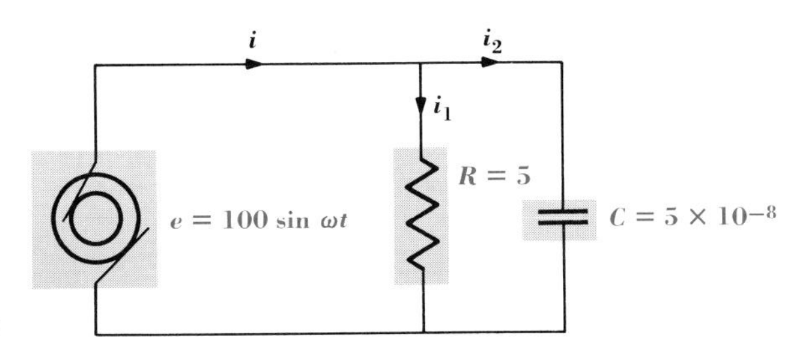

Figure 8

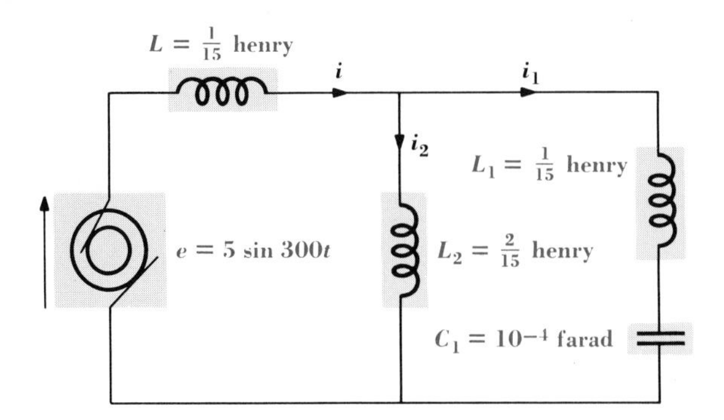

Figure 9

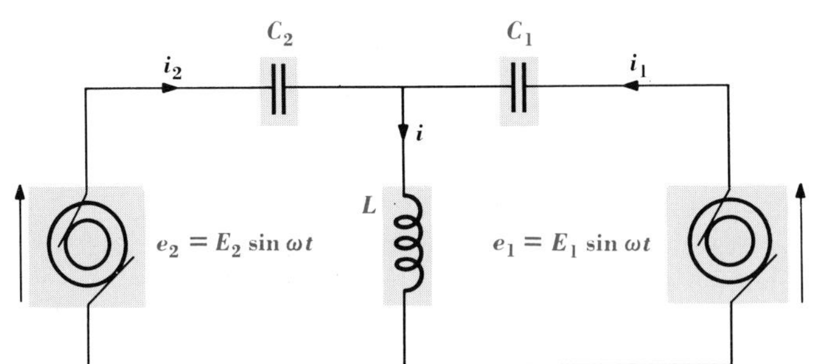

Figure 10

8. Find i in terms of the time t in the system represented in Fig. 10 if the initial currents and the initial charges on the capacitors are zero and $\omega \neq 1/\sqrt{L(C_1 + C_2)}$.

9. Show that the current i indicated in Fig. 11 is the same as it would be if the three inductances were replaced by a single inductance of magnitude $1/(1/L_1 + 1/L_2 + 1/L_3)$ in series with the emf. Also, show that, in the same sense, the three capacitors in parallel in Fig. 12 are equivalent to a single one of capacity $C_1 + C_2 + C_3$.

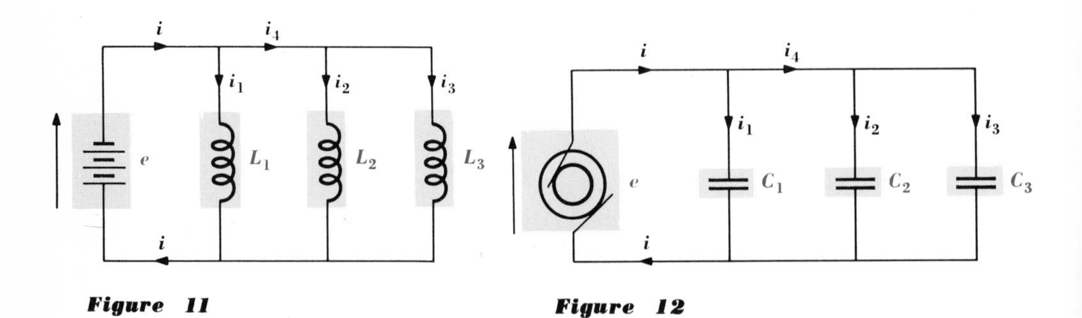

Figure 11 **Figure 12**

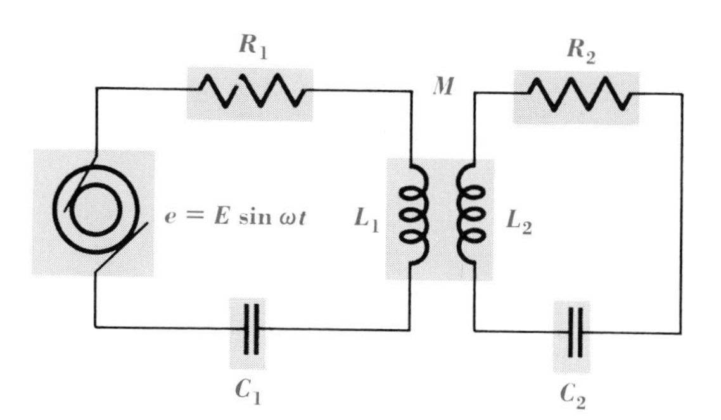

Figure 13

★**10.** The equations for the coupled circuits represented in Fig. 13 are

$$(L_1D + R_1)i_1 + MDi_2 + \frac{1}{C_1}q_1 = E \sin \omega t \qquad Dq_1 = i_1$$

$$MDi_1 + (L_2D + R_2)i_2 + \frac{1}{C_2}q_2 = 0 \qquad Dq_2 = i_2$$

Solve these equations for q_1 and i_2, assuming that $R_1 = R_2 = 0$, $C_2 = \infty$, $L_1L_2 > M^2$, and $q_1 = i_1 = i_2 = 0$ when $t = 0$. Let $L_2/[C_1(L_1L_2 - M^2)] = \omega_1^2$ and $L_2E/(L_1L_2 - M^2) = a$.

81 Review problems

In the following problems D represents d/dt.

1. The equation $D^2y + y = 0$ represents a simple harmonic motion. Find its period, frequency, and amplitude if $y = 5$ and $Dy = 0$ when $t = 0$.

2. Solve the differential equation

$$D^2x + Dx + \tfrac{37}{4}x = 0$$

and determine the constants of integration by using the conditions $x = 0$ and $dx/dt = 6$ when $t = 0$. Find x in terms of the time, the period of oscillation of x, and the magnitude of the damping factor after 3 sec.

3. Solve the differential equation

$$9D^2x + 3aDx + 82x = 0$$

and determine a and the constants of integration if $dx/dt = 6$ and $x = 0$ when $t = 0$ and if the damping factor decreases 50 per cent in 2.08 units of time. Also, find the period.

4. Find b and c in $(D^2 + bD + c)x = 0$ if the corresponding period is $\frac{1}{60}$ sec and the damping factor decreases 50 per cent in 0.1 sec.

5. A body falling from rest in a heavy fluid acquires a velocity which approaches 10 ft/sec as a limit. Assuming the resistance of the medium to be proportional to the velocity and the buoyancy of the fluid to be one-half the weight of the body, find the factor of proportionality and the distance traversed during the first 10 sec.

6. A sphere of uniform density and diameter 5 ft sinks 1 ft in salt water weighing 64 lb/ft³. It is depressed slightly and released. Find the period of the resulting motion. By Archimedes' principle, the weight of the sphere is $\frac{13}{6}\pi 64$ lb.

7. For a simple circuit, $L = 0.1$ henry, $R = 1$ ohm, $C = 250 \times 10^{-6}$ farad, and $e = 0$. If initially $t = 0$, $q = 0.05$ coul, and $i = -0.25$ amp, find i and q in terms of t, the period of the current, and the limiting values of i and q.

8. Starting from $LD^2i + RDi + (1/C)i = E\omega \cos \omega t$, prove that the amplitude of the steady-state current is maximum when $\omega = 1/\sqrt{LC}$, that is, at current resonance.

9. In the equation $(D^2 + 0.01D + c)x = 0$, find c if the corresponding damping is critical.

10. Find the relation satisfied by a, b, and c for a motion defined by $(aD^2 + bD + c)x = 0$ if the motion is: (*a*) overdamped; (*b*) critically damped; (*c*) damped oscillatory; (*d*) undamped.

11. A particle of mass 1 slug moves toward a fixed center of force which repels it with a magnitude in pounds equal to k times the distance of the particle from the center. Initially, the particle is at a distance a from the center and is moving toward it with a velocity equal in magnitude to $\sqrt{ka^2}$. Prove that the particle will continually approach but never reach the center.

12. A body of weight w lb moves vertically under the force of gravity and under the action of a force opposite to the direction of motion and equal numerically to $0.4wv/32.2$, where v is the speed in feet per second. Taking y as the distance above the ground, study the motion under the conditions $y = 0$ and $v = 100$ when $t = 0$.

13. Find the charge on the capacitor of Fig. 1 at time t if initially there is no charge on the capacitor and no current through the inductance.

14. If, in the system represented by Fig. 2, there is initially no charge on the condenser and no current flowing, find i in terms of the time, and describe its fluctuation.

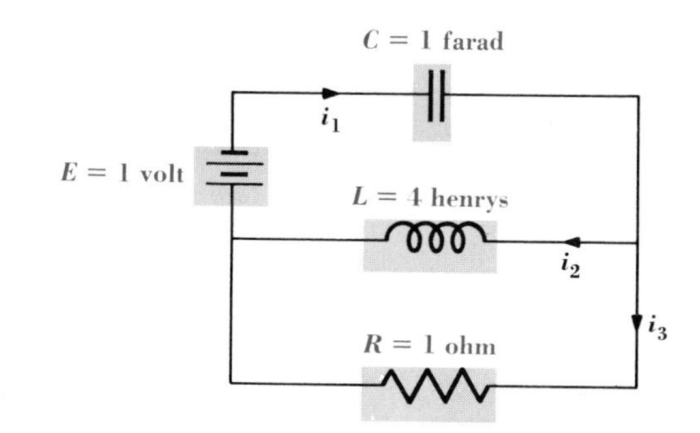

Figure 1

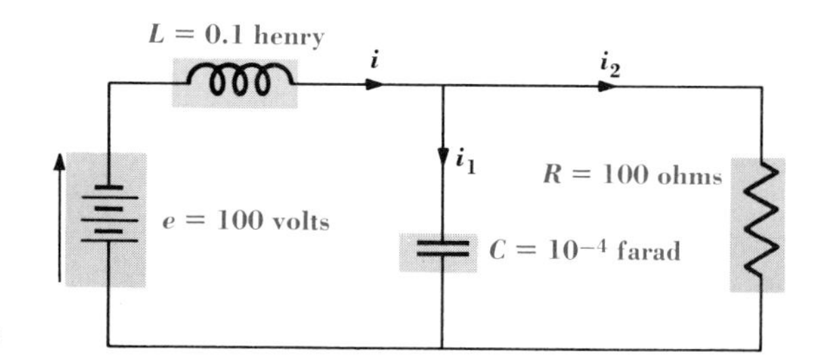

Figure 2

15. Find the solution of the differential equation

$$(D^2 + 2D + 5)x = 34 \sin 2t$$

for which $x = 0$ and $Dx = -8$ when $t = 0$. Use transforms and relation 19 of Table 1, §60. Give the amplitude and period of the permanent part of the solution, and show that the part having the factor e^{-t} is less than 0.005 when t is 9 sec or more.

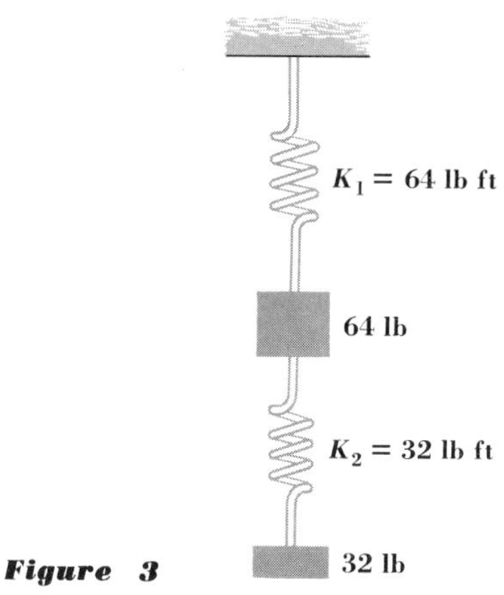

Figure 3

16. If the capacitors of exercise 10, §80, are omitted, the corresponding equations are

$$\begin{aligned} (L_1D + R_1)i_1 + MDi_2 &= E \sin \omega t \\ MDi_1 + (L_2D + R_2)i_2 &= 0 \end{aligned}$$

Find i_1 and i_2 for these, assuming that $R_1 = 0$, $L_1L_2 = M^2$, and that $i_1 = i_2 = 0$ when $t = 0$.

★**17.** Figure 3 represents a 64-lb weight and a 32-lb weight hung from a ceiling and connected by springs having the indicated spring constants. When $t = 0$, the weights are at rest and the springs have natural unstretched lengths of 1 ft each. Find the lengths x ft and y ft of the springs at time t. Take $g = 32$ ft/sec^2.

18. In (1), §78, take $L = 0.3$, $R = 10$, and $1/C = h\omega^2$, and find the ratio of the amplitude of steady-state current when $h \neq 0.3$ to that when $h = 0.3$. What name is associated with the current when $h = 0.3$?

Miscellaneous Differential Equations of Order Higher than the First

82 Reduction of order by substitution

In Chap. 6, we found general methods for solving linear equations with constant coefficients. This was unusual, for we cannot solve most equations in finite form. Certain types, however, are readily solvable. One method of attack is to make a substitution which will reduce the order, and then try to solve the result. This procedure was used in §44. Also consider the equation

$$\frac{d^n y}{dx^n} = f(x) \tag{1}$$

To solve this, substitute q for $d^{n-1}y/dx^{n-1}$, and obtain

$$\frac{d^n y}{dx^n} = \frac{d}{dx}\left(\frac{d^{n-1}y}{dx^{n-1}}\right) = \frac{dq}{dx} = f(x) \tag{2}$$

Therefore

$$q = \int f(x)\,dx + c_1 \qquad \text{or} \qquad \frac{d^{n-1}y}{dx^{n-1}} = \int f(x)\,dx + c_1 \tag{3}$$

Clearly, we can treat equation (3) in a similar manner and obtain

$$\frac{d^{n-2}y}{dx^{n-2}} = \int \left[\int f(x)\,dx \right] dx + c_1 x + c_2 \tag{4}$$

and it appears that we may continue this process until we find y in terms of x by n successive integrations.

Various types of differential equations with appropriate substitutions will be considered in the following sections.

83 Dependent variable absent

If an equation contains derivatives of the dependent variable y but does not contain y directly, then the substitution

$$p = \frac{dy}{dx}, \frac{dp}{dx} = \frac{d^2y}{dx^2}, \quad . \; . \; . \; , \frac{d^{n-1}p}{dx^{n-1}} = \frac{d^ny}{dx^n} \tag{1}$$

will reduce the order of the equation by unity; if the result can be solved for p in terms of x,

$$p = \frac{dy}{dx} = f(x) \tag{2}$$

a single integration will give y in terms of x. In fact, if d^ky/dx^k is the derivative of lowest order in an equation which does not contain y directly, then the substitution

$$p = \frac{d^ky}{dx^k}, \frac{dp}{dx} = \frac{d^{k+1}y}{dx^{k+1}}, \quad . \; . \; . \; , \frac{d^{n-k}p}{dx^{n-k}} = \frac{d^ny}{dx^n} \tag{3}$$

will reduce the order by k; if the result can be solved for p in terms of x,

$$p = \frac{d^ky}{dx^k} = f_1(x) \tag{4}$$

y may be found in terms of x by k successive integrations, as indicated above.

Example. Solve

$$(1 + x^2)\frac{d^2y}{dx^2} + x\frac{dy}{dx} + ax = 0 \tag{a}$$

Solution. In equation (a), substitute

$$p = \frac{dy}{dx} \qquad \frac{dp}{dx} = \frac{d^2y}{dx^2} \tag{b}$$

and obtain

$$(1 + x^2)\frac{dp}{dx} + px + ax = 0 \tag{c}$$

Separating the variables in equation (c) and integrating, we have

$$p + a = c_1(1 + x^2)^{-1/2}$$

Replacing p by dy/dx and integrating, we obtain

$$y = -ax + c_1 \sinh^{-1} x + c_2$$

Exercises

Solve the following differential equations, and determine the constants of integration where sufficient conditions are given:

1. $\dfrac{d^2y}{dx^2} = 12x$

2. $x^3 \dfrac{d^3y}{dx^3} = 12$

3. $x \dfrac{d^2y}{dx^2} + \dfrac{dy}{dx} = 0$

4. $x^2 \dfrac{d^2y}{dx^2} + \left(\dfrac{dy}{dx}\right)^2 = 0$

5. $d^2y/dx^2 + 24x = 0$; $y = -28$, $dy/dx = -10$ when $x = 1$.

6. $(d/dx + 1)(x\, dy/dx + y) = 12e^{-x}$; let $z = x\, dy/dx + y$.

7. $x\, d^3y/dx^3 - 2\, d^2y/dx^2 = 12x^3$; $y = 0$, $dy/dx = 1$, $d^2y/dx^2 = 0$ when $x = 1$.

8. $a\, d^3y/dx^3 = \sqrt{a^2 + (d^2y/dx^2)^2}$; $y = 0$, $dy/dx = -a^2$, $d^2y/dx^2 = 0$ when $x = 0$, $a > 0$.

9. $\left(x^2 \dfrac{d}{dx} + x\right)\left(x \dfrac{dy}{dx} + 2y\right) = x^2e^x$

84 Independent variable absent

If p is substituted for dy/dx, we have

$$\frac{dy}{dx} = p \qquad \frac{d^2y}{dx^2} = \frac{dp}{dy}\frac{dy}{dx} = \frac{p\,dp}{dy}$$
$$\frac{d^3y}{dx^3} = \frac{d}{dy}\left(p\frac{dp}{dy}\right)\frac{dy}{dx} = p^2\frac{d^2p}{dy^2} + p\left(\frac{dp}{dy}\right)^2 \qquad \text{etc.} \tag{1}$$

Therefore, if a differential equation does not contain x directly, the substitution (1) will give a new differential equation in p and y of order one less than that of the original equation. If this new equation can be solved for p in terms of y to get

$$p = \frac{dy}{dx} = f(y) \tag{2}$$

then x may be found in terms of y from

$$x = \int \frac{dy}{f(y)} + c \tag{3}$$

Example. Solve

$$y\frac{d^2y}{dx^2} + \left(\frac{dy}{dx}\right)^2 = \frac{dy}{dx} \tag{a}$$

Solution. Substitition of

$$p = \frac{dy}{dx} \qquad \frac{p\,dp}{dy} = \frac{d^2y}{dx^2} \tag{b}$$

in (a) gives

$$yp\frac{dp}{dy} + p^2 = p \qquad \text{or} \qquad p\left(y\frac{dp}{dy} + p - 1\right) = 0 \tag{c}$$

From (c),

$$y\frac{dp}{dy} + p - 1 = 0 \qquad p = 0 \tag{d}$$

Hence,

$$\frac{dp}{p - 1} + \frac{dy}{y} = 0 \tag{e}$$

The solution of (e) is

$$p = 1 + \frac{c_1}{y} \tag{f}$$

Replacing p by dy/dx and separating the variables, we obtain

$$\frac{y\,dy}{y + c_1} = dx \tag{g}$$

The solution of (g) is

$$x = y - c_1 \ln (y + c_1) + c_2$$

From the second equation of (d) or by inspection, it appears that $y = c$ satisfies equation (a).

Exercises

Solve the following differential equations, and determine constants of integration when initial conditions are given:

1. $y\dfrac{d^2y}{dx^2} + \left(\dfrac{dy}{dx}\right)^2 = 0$

2. $y^2\dfrac{d^2y}{dx^2} + \left(\dfrac{dy}{dx}\right)^3 = 0$

3. $y\dfrac{d^2y}{dx^2} + 2\left(\dfrac{dy}{dx}\right)^2 = 0$

4. $y\dfrac{d^2y}{dx^2} + (1 + y)\left(\dfrac{dy}{dx}\right)^2 = 0$

5. $\dfrac{d^2s}{dt^2} = \dfrac{1}{s^3}$

6. $\dfrac{d^2s}{dt^2} = 64 - \left(\dfrac{ds}{dt}\right)^2$

7. $y\, d^2y/dx^2 + 4y^2 - \frac{1}{2}(dy/dx)^2 = 0; y = 1, dy/dx = \sqrt{8}$ when $x = 0$.

★8. $d^2s/dt^2 = 100 - (ds/dt)^2$; $s = 0$, $ds/dt = 26$ when $t = 0$.

★9. $2\, d^2y/dx^2 = e^y$; $y = 0$, $dy/dx = 0$ when $x = 0$.

10. A ring slides from a point A to a lower point B under the influence of gravity. The ring is smooth, and the time required is minimum.* By higher mathematics, it is proved that the path followed satisfies the equation $1 + (dy/dx)^2 + 2y\, d^2y/dx^2 = 0$. First prove that $(dy/dx)^2 = c_1/y - 1$. Let $y = c_1 \sin^2 \frac{1}{2}\theta$, and show that $x = \frac{1}{2}c_1(\theta - \sin\theta) + c_2$.

11. A certain curve connecting two points A and B passes through (0,1) with slope zero. When revolved about the X axis, it generates a surface of minimum area (see Fig. 1). This will be

* The curve, called the brachistochrone, has the shape of a cycloid.

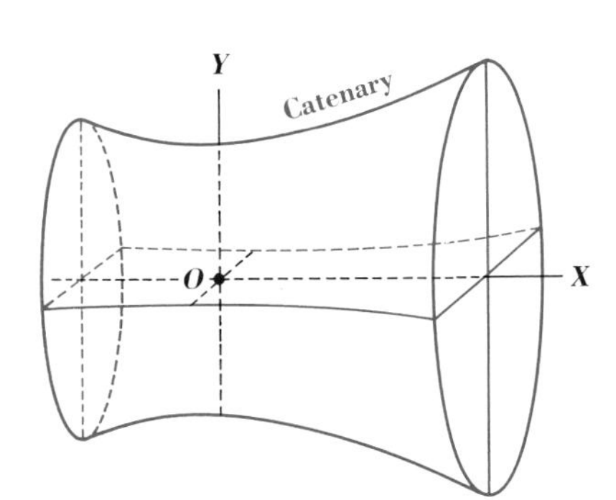

Figure 1

true provided that the equation of the curve satisfies $1 + (dy/dx)^2 = y\, d^2y/dx^2$. Find the equation of the curve.*

85 Euler's linear equation

The equation

$$x^n \frac{d^ny}{dx^n} + A_1x^{n-1}\frac{d^{n-1}y}{dx^{n-1}} + \cdots + A_{n-1}x\frac{dy}{dx} + A_ny = X \tag{1}$$

where the A's are constant and X represents a function of x, is often referred to as Euler's differential equation. The substitution

$$z = \ln x \qquad \frac{dz}{dx} = \frac{1}{x} = e^{-z} \tag{2}$$

reduces this to the linear equation with constant coefficients. Using the notation

$$\frac{d^ky}{dz^k} = D^ky \tag{3}$$

we have

$$\begin{aligned} \frac{dy}{dx} &= \frac{dy}{dz}\frac{dz}{dx} = \frac{1}{x}\frac{dy}{dz} = \frac{1}{x}Dy \\ \frac{d^2y}{dx^2} &= \frac{d(dy/dx)}{dz}\frac{dz}{dx} = \frac{1}{x^2}\frac{d^2y}{dz^2} - \frac{1}{x^2}\frac{dy}{dz} = \frac{1}{x^2}D(D-1)y \\ \frac{d^3y}{dx^3} &= \frac{1}{x^3}D^2(D-1)y - \frac{2}{x^3}D(D-1)y \\ &= \frac{1}{x^3}D(D-1)(D-2)y \\ &\cdots\cdots\cdots\cdots\cdots \\ \frac{d^ny}{dx^n} &= \frac{1}{x^n}D(D-1)\cdots(D-n+1)y \end{aligned} \tag{4}$$

Substitution of the values of d^ky/dx^k from (4) in (1) evidently gives a linear equation with constant coefficients. Solving this equation by the methods of Chap. 6, and replacing z by $\ln x$ in the result, obtain the solution of (1).

Example. Solve

$$x^3\frac{d^3y}{dx^3} + 6x^2\frac{d^2y}{dx^2} + 8x\frac{dy}{dx} - 8y = x^2 \tag{a}$$

* The curve is that of a uniform, perfectly flexible string hanging under its own weight. It is called a catenary

Solution. Using (2) to (4), we have

$$\frac{x^3}{x^3} D(D-1)(D-2)y + 6\frac{x^2}{x^2} D(D-1)y + 8\frac{x}{x} Dy - 8y = e^{2z} \tag{b}$$

$$(D^3 + 3D^2 + 4D - 8)y = e^{2z} \tag{c}$$

The solution of equation (c) is

$$y = c_1 e^z + e^{-2z}(c_2 \sin 2z + c_3 \cos 2z) + \tfrac{1}{20}e^{2z} \tag{d}$$

Replacing z in (d) by $\ln x$ from (2), we have

$$y = c_1 x + \frac{1}{x^2}[c_2 \sin(2\ln x) + c_3 \cos(2\ln x)] + \tfrac{1}{20}x^2$$

Exercises

Solve the following differential equations:

1. $x^2 \dfrac{d^2y}{dx^2} + 2x\dfrac{dy}{dx} - 2y = x^2$

2. $x\dfrac{dy}{dx} + 2y = x^5$

3. $x^3 \dfrac{d^3y}{dx^3} + 3x^2\dfrac{d^2y}{dx^2} + x\dfrac{dy}{dx} = x^3$

4. $x^3 \dfrac{d^3y}{dx^3} + 3x^2\dfrac{d^2y}{dx^2} - 6x\dfrac{dy}{dx} - 6y = 0$

5. $x^2\, d^2y/dx^2 + x\, dy/dx - 9y = x^n$; $n \neq \pm 3$.

6. $(x-1)^3\, d^3y/dx^3 + 2(x-1)^2\, d^2y/dx^2 - 4(x-1)\, dy/dx + 4y = 4\ln(x-1)$; let $z = \ln(x-1)$.

7. $x\dfrac{d^3y}{dx^3} + 2\dfrac{d^2y}{dx^2} = 0$

86 Second-order linear equation

This section presents a method of deriving from a linear differential equation $L(D)y = X(x)$ a new equation of lower degree when a solution of $L(D)y = 0$ is known. Consider the equation of the second order

$$\frac{d^2y}{dx^2} + f_1(x)\frac{dy}{dx} + f_2(x)y = f_3(x) \tag{1}$$

To get an idea of how this may be solved, let us try the substitution

$$y = v(x) \cdot \varphi(x) \tag{2}$$

Substituting y from (2) in (1) and rearranging the terms, we find

$$\varphi \frac{d^2v}{dx^2} + \left(2\frac{d\varphi}{dx} + f_1\varphi\right)\frac{dv}{dx} + \left(\frac{d^2\varphi}{dx^2} + f_1\frac{d\varphi}{dx} + f_2\varphi\right)v = f_3 \tag{3}$$

If now $\varphi(x)$ is chosen so that the coefficient of v in (3) is zero, that is, if

$$\frac{d^2\varphi}{dx^2} + f_1\frac{d\varphi}{dx} + f_2\varphi = 0 \tag{4}$$

equation (3) does not contain v.* Hence, the method of §83 may be applied to solve it. In other words, if $y = \varphi(x)$ *is any particular solution of the equation obtained by setting the left member of* (1) *equal to zero, then the substitution* (2) *applied to* (1) *reduces it to an equation that can be solved by methods already considered.*

Example. Solve

$$(1 + x)\frac{d^2y}{dx^2} + (4x + 5)\frac{dy}{dx} + (4x + 6)y = e^{-2x} \tag{a}$$

Solution. First we try to find a particular solution of

$$(1 + x)\frac{d^2y}{dx^2} + (4x + 5)\frac{dy}{dx} + (4x + 6)y = 0 \tag{b}$$

Often, particular solutions are obtained by trial. The usual plan is to substitute simple expressions, such as $y = e^{ax}$, $y = x^a$, $y = x + a$, y = polynomial in x, in the equation to be solved, and then try to determine the arbitrary constants so that the equation will be satisfied. Substituting $y = e^{ax}$ in (b), we find, after slight simplification

$$[(a^2 + 5a + 6) + x(a^2 + 4a + 4)]e^{ax} = 0 \tag{c}$$

This will be true, if

$$a^2 + 5a + 6 = 0 \qquad \text{and} \qquad a^2 + 4a + 4 = 0$$

Both of these equations have a root -2. Hence, $y = e^{-2x}$ is a particular solution. Therefore, in accordance with (2), we substitute

$$y = ve^{-2x} \tag{d}$$

* Note that if (4) holds and we set $dv/dx = u$ in (3), an equation of the first order results. The method used for (1) will apply to obtain from any linear differential equation a new equation of lower order.

in (*a*) to obtain, after considerable simplification

$$(1 + x)\frac{d^2v}{dx^2} + \frac{dv}{dx} = 1 \tag{e}$$

The solution of this equation, found by the method of §83 is

$$v = x + c_1 \ln (x + 1) + c_2 \tag{f}$$

Substitution of this value of v in (*d*) gives the required solution of (*a*):

$$y = e^{-2x}[x + c_1 \ln (x + 1) + c_2]$$

Exercises

Solve the following differential equations:

1. $(D + 1)y = 3x^2e^{-x}$; let $y = ve^{-x}$, because $y = e^{-x}$ is a solution of $(D + 1)y = 0$.
2. $(D^2 - 1)y = 2xe^x$; let $y = ve^x$.
3. $(D^2 - 4D + 4)y = x^ne^{2x}$
4. $(x^2 - 1)\, D^2y + x\, Dy - y = 0$. First show that $y = x$ is a particular solution.
5. $[x^2D^2 + x^2\, D + (x - 2)]y = 0$. *Hint:* Try $y = x^n$.
6. $(D^2 + 1)y = 2 \cos x$
7. $[(x^2 + 1)D^2 - 2x\, D + 2]y = 6(1 + x^2)^2$
8. $[x\, D^2 - (x + 3)D + 3]y = 4x^4e^x$
9. $\sin^2 x\, D^2y - \sin x \cos x\, Dy + y = -\sin^3 x$

87 Method based on factorization of the operator

Operators in the form of polynomials in D with variable coefficients are defined by relations (1), (2), (3), and (4), §44. If P, M, R, N, and u are functions of x, we get from (3) and (4), §44,

$$(PD + M)(RD + N)y = (PD + M)u \qquad \text{where } u = (RD + N)y$$

Also, the student may verify that $(PD + M)(RD + N)u$ is equal to

$$P\frac{dR}{dx}\frac{du}{dx} + PR\frac{d^2u}{dx^2} + P\left(N\frac{du}{dx} + u\frac{dN}{dx}\right) + MR\frac{du}{dx} + MNu \tag{1}$$

Note that, if O_1 and O_2 are operators, O_1O_2y may not equal O_2O_1y. Note, for example, that

$$(D - x)(D - x^2)y = D^2y - x^2Dy - 2xy - x\,Dy + x^3y \tag{2}$$
$$(D - x^2)(D - x)y = D^2y - x\,Dy - y - x^2Dy + x^3y \tag{3}$$

To solve a differential equation having the form

$$(PD + M)(RD + N)y = f(x) \tag{4}$$

set $(RD + N)y = u$, solve the result to find $u = \varphi(x)$, and then solve $(RD + N)y = \varphi(x)$. The difficulty comes in transforming an operator to the factored form. The following examples illustrate methods of solution using operators:

Example 1. Solve

$$x^2D^2y + (5x - x^2)Dy + (3 - 2x)y = 4e^x \tag{a}$$

Solution. Assume the left member to be

$$(xD + M)(xD + N)y$$

or

$$x^2D^2y + x(M + N + 1)Dy + (MN + xN')y \tag{b}$$

where M and N are functions of x. If expression (b) equals the left side of (a), we must have

$$x(M + N + 1) = x(5 - x) \qquad MN + xN' = 3 - 2x \tag{c}$$

Substituting $M = 4 - x - N$ from the first of (c) in the second, we get

$$(4 - x)N - N^2 + xN' = 3 - 2x \tag{d}$$

We attempt to find a solution of (d) by letting

$$N = ax + b \tag{e}$$

in (d), equating coefficients of like powers of x, and determining a and b. This gives

$$(4 - x)(ax + b) - (a^2x^2 + 2axb + b^2) + xa = 3 - 2x$$
$$-a - a^2 = 0 \qquad 5a - b - 2ab = -2 \qquad 4b - b^2 = 3$$

These are satisfied by $a = -1$ and $b = 3$. Therefore

$$N = -x + 3 \qquad M = 4 - x + x - 3 = 1 \tag{f}$$

and the given equation may be written

$$(xD + 1)(xD + 3 - x)y = 4e^x \tag{g}$$

Now, let $z = (xD + 3 - x)y$ in (*g*), and solve the resulting linear equation by the method of §25 to get

$$(xD + 1)z = 4e^x$$
$$xz = 4e^x + c \tag{h}$$

Now, replacing z in (*h*) by its equal $(xD + 3 - x)y$ and solving the resulting linear equation for y, we get

$$x(xD + 3 - x)y = 4e^x + c$$
$$x^3y = c_1(x + 1) + (c_2 + 2x^2)e^x$$

Example 2.* (*a*) Show that, if M and N are functions of x and c is a constant, then

$$[MD^2 + (Mc + N)D + cN]y = (MD + N)(D + c)y$$

(*b*) Solve $[x^3D^2 + (x^3 - x^2)D - x^2]y = 2x^4$.

Solution. (*a*) This part is left as an exercise.

(*b*) Taking $M = x^3$, $N = -x^2$, $c = 1$ in part (*a*) of this example, we get

$$[x^3D^2 + (x^3 - x^2)D - x^2]y = (x^3D - x^2)(D + 1)y$$

Therefore, the given equation may be written

$$(x^3D - x^2)(D + 1)y = 2x^4$$

In this, let $(D + 1)y = z$, and solve for z to get

$$z = 2x^2 + cx$$

In this, replace z by its equal $(D + 1)y$, solve the resulting equation, and obtain

$$y = c_2e^{-x} + cx - c + 2x^2 - 4x + 4$$

Exercises

1. Show that $(PD + M)(RD + N)u$ is equal to (1) when P, R, M, N, and u are functions of x. Note that $PD[R\,D(u)] = P[R\,D^2u + DR \cdot Du]$.

2. Show that $(xD - 1)(D - x^2)y = [xD^2 - (x^3 + 1)D - x^2]y$.

3. Show that $(D + x^2)(xD - 4)y - (xD - 4)(D + x^2)y = (D - 2x^2)y$.

4. Solve $(xD - 2)(xD + x + 1)y = 0$.

* Numerous simple special cases of solutions by operators can be devised. Example 2 furnishes a case for which factorization of the operator by algebraic processes applies.

Solve the following differential equations:

5. $(xD^2 - D)y = 12$

6. $(x^2D^2 - xD + 1)y = 12x^2$

7. $[x^2D^2 + (4x + x^2)D + 2 + 2x]y = 12x$. Use the trial form $(D + M)(x^2D + N)y = 0$, and determine M and N.

(**8–9**) Write $(MD + N)(MD + N)y$, where M and N are functions of x, in expanded form, and use the result in solving:

8. $[D^2 - 6x^2D + (9x^4 - 6x)]y = 4e^{x^3}$
9. $[x^2D^2 + (4x^3 + x)D + (4x^4 + 4x^2)]y = 12xe^{-x^2}$

(**10–11**) Write $(D + M)(RD + N)y$, where M, R, and N are functions of x, in expanded form, and solve:

10. $[xD^2 + (3x^3 + 4)D + 9x^2]y = 3x^2$
11. $[x^2D^2 + (5x + 2x^3)D + (6x^2 + 3)]y = 12x$

Solve the following differential equations:

12. $[x^2D^2 + (x^2 + 4x)D + (2 + 2x)]y = 1$
13. $[xD^2 + (3 - 2x^2)D - 4x]y = 4xe^{x^2}$

Exercises 14–16 belong to the type of example 2. Solve:

14. $[xD^2 + (2x - 1)D - 2]y = 4x^2$
15. $[x^2D^2 + (2x^2 - x)D - 2x]y = 0$
16. $[x^3D^2 + (5x^3 - x^2)D + (6x^3 - 2x^2)]y = 0$
17. $[D^2 + (1 + \cot x)D + (\cot x - \csc^2 x)]y = 2e^x$

88 Review exercises

In the following exercises D means d/dx.

Solve the following differential equations, and determine the constants when initial conditions are given:

1. $x\,D^2y + Dy = 16x^3$

2. $(x + 1)\,D^2y - (x + 2)\,Dy = 0$

3. $D^2y = -100 - (Dy)^2$; $y = 0$, $Dy = 24$ when $x = 0$.
4. $D^2y + y\,Dy + (Dy)^2 = 0$; $y = 0$, $Dy = 1$ when $x = 0$.
5. $x\,D^3y + D^2y = 0$
6. $(D^2 - 4xD - 2 + 4x^2)y = 2e^{x^2}$
7. $[(2x^2 + x)D^2 - D - 4]y = 0$. *Hint:* Try $y = x^n$.

8. At any point a distance r from the common center of two concentric spheres, the differential equation for the potential v due to an electric charge on the inner sphere is

$$\frac{d^2v}{dr^2} + \frac{2}{r}\frac{dv}{dr} = 0$$

Solve for v in terms of r, given that $v = v_1$ when $r = r_1$ and that $v = v_0$ when $r = r_0$.

9. $x^3 D^3y + 2x^2 D^2y + x\, Dy - y = 5x^2$

10. $D^2y + \cos x (Dy)^2 = 0$; $y = 0$, $Dy = \frac{5}{8}$ when $x = \frac{1}{2}\pi$.

11. $\left[(xD - 2)\left(D + 2x - \frac{1}{x}\right)\right] y = 4x^4 e^{-x^2}$

12. $[D^2 + (1 - 2x)D - 1 - x + x^2]y = e^{x^2/2}$

13. $[x^2D^2 + x^2D - (x + 2)]y = 0$

14. $(2x^2D^2 + xD - 3)y = 12 \ln x$

★**15.** $a\, D^3y = \sqrt{1 + (D^2y)^2}$; $y = 0$, $Dy = -a$, $D^2y = 0$ when $x = 0$.

16. $x^3 D^3y + 3x^2 D^2y + 4x\, Dy = \sin(\sqrt{3} \ln x)$

17. $x\, D^2y + (x - 1)\, Dy + (3 - 12x)y = 0$

18. $x^3 D^3y + 3x^2 D^2y + x\, Dy = x^3$

10

Applications

89 Radius of curvature

Some problems relating to the radius of curvature are rather interesting. As an illustration, consider the following:

Example. Find the equation of the curve whose radius of curvature is double the normal and oppositely directed.

Solution. Equating the expression for the radius of curvature and twice the expression for the normal (see Fig. 1), obtain

$$\frac{[1 + (dy/dx)^2]^{3/2}}{d^2y/dx^2} = \pm 2y\sqrt{1 + \left(\frac{dy}{dx}\right)^2} \tag{a}$$

Assuming that the radius of curvature R is positive and is directed from the curve toward the center of curvature and that the normal N is directed from the curve toward the X axis, we see that y and d^2y/dx^2

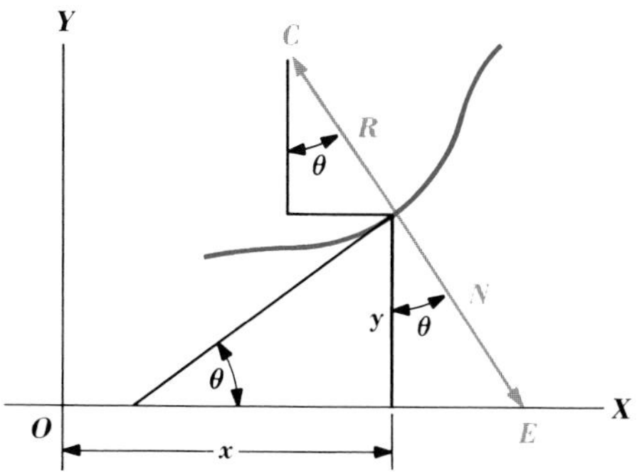

Figure 1

will have the same sign when R and N are directed oppositely. Hence, the plus sign in (a) must be used. Substitution of p for dy/dx and $p\,dp/dy$ for d^2y/dx^2 in (a) gives

$$\frac{(1 + p^2)^{3/2}}{p\,dp/dy} = 2y(1 + p^2)^{1/2} \tag{b}$$

Separating the variables and integrating (b), we obtain

$$\ln (1 + p^2) = \ln \frac{y}{c} \qquad \text{or} \qquad \frac{dy}{dx} = \pm \sqrt{\frac{y}{c} - 1} \tag{c}$$

Integrating (c) and simplifying, we obtain

$$(x - c_1)^2 = 4cy - 4c^2 \tag{d}$$

This represents a system of parabolas with their axes parallel to the Y axis.

Exercises

1. Find the equation of all plane curves which have a constant radius of curvature.

Determine the curves for which the radius of curvature:

2. Is equal to the normal and in the same direction.
3. Is equal to the normal and in the opposite direction.
4. Projected on the X axis equals the abscissa.
5. Varies as the cube of the normal.
6. Projected on the X axis is the negative of the abscissa.
7. Projected on the X axis is twice the abscissa.
8. Integrate completely the differential equation of the curve for which the projection of the radius of curvature upon the X axis is constant.

90 Cables. The catenary

Figure 1 represents a loaded cable. Let us assume that it is perfectly flexible, inextensible, homogeneous, and hanging from two points under the action of gravity on a load distributed along it in a continuous way. Let H be the tension in the cable at its lowest point A, T the tension at any point B of the cable, and L the resultant force of gravity exerted between A and B. Since the forces T, H, and L are in equilibrium

$$T \cos \theta = H \tag{1}$$
$$T \sin \theta = L \tag{2}$$

where θ is the angle between the direction of force T and the horizontal. Evidently, the tension H at A is horizontal, and the tension T at B is directed along the tangent to the curve of the cable; hence, if we take the X axis horizontal and the Y axis vertical, we have $\tan \theta = dy/dx$. Then division of (2) by (1), member by member, gives

$$\tan \theta = \frac{dy}{dx} = \frac{L}{H} \tag{3}$$

Differentiating both sides, we obtain

$$\frac{d^2y}{dx^2} = \frac{1}{H}\frac{dL}{dx} \tag{4}$$

as the differential equation of the curve of the cable.

The curve in which a uniform chain hangs under its own weight is called the catenary In this case, we have

$$L = ws \tag{5}$$

where s represents the arc length AB in Fig. 1 and w represents weight

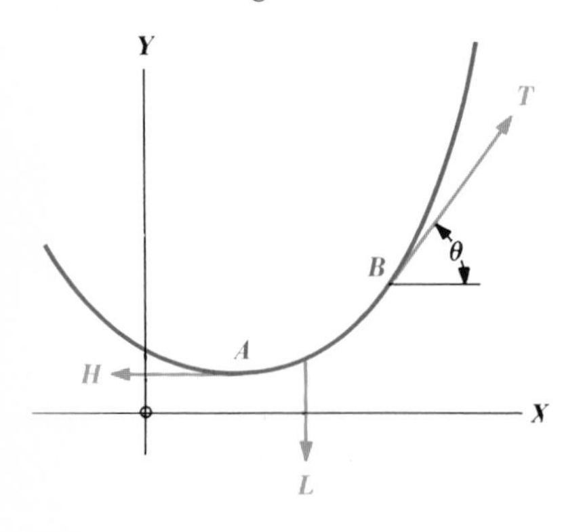

Figure 1

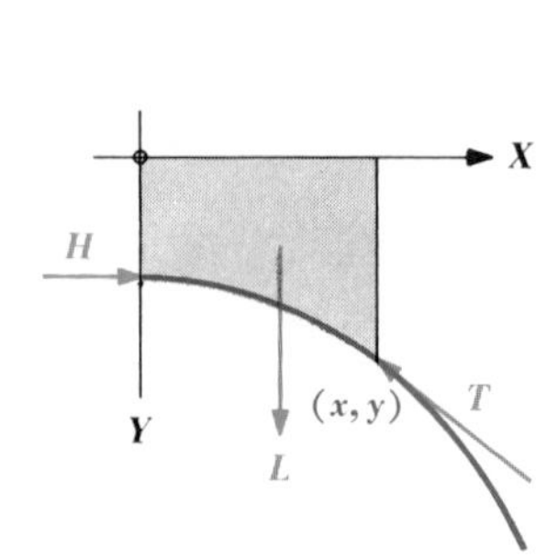

Figure 2

per unit length of the chain. Substitution of L from (5) in (4) gives

$$\frac{d^2y}{dx^2} = \frac{w}{H}\frac{ds}{dx} = \frac{w}{H}\sqrt{1 + \left(\frac{dy}{dx}\right)^2} \tag{6}$$

Substituting in (6) p for dy/dx and dp/dx for d^2y/dx^2, we obtain

$$\frac{dp}{dx} = \frac{w}{H}\sqrt{1 + p^2} \tag{7}$$

Integrating (7), we obtain

$$\sinh^{-1} p = \frac{w}{H}x + c_1 \qquad \text{or} \qquad p = \sinh\left(\frac{w}{H}x + c_1\right) \tag{8}$$

Replacing p by dy/dx in (8) and integrating, we find

$$y = \frac{H}{w}\cosh\left(\frac{w}{H}x + c_1\right) + c_2 \tag{9}$$

Taking the origin H/w units below the lowest point on the curve, we have $y = H/w$, $dy/dx = 0$ when $x = 0$. Using these values in (8) and (9) and remembering that $\cosh^2\theta - \sinh^2\theta = 1$, we get

$$y = \frac{H}{w}\cosh\frac{w}{H}x \tag{10}$$

Exercises

1. Find the curve in which the cable of a suspension bridge hangs, if it carries a uniform horizontal load of w lb/ft run; neglect the weight of the cable. *Hint:* In equation (3), $L = wx$.

2. Set up the differential equation for the cable of exercise 1 if its weight per running foot is assumed constant and taken into account.

3. Find the shape of the arch of a stone-arch bridge if the resultant stress at any point of the arch due to the weight of the masonry above is directed along the tangent to the arch at the point. Assume that the masonry is uniform in density and that the surface of the road is horizontal (see Fig. 2).

4. If an arch carries, in addition to the load of exercise 3, a layer of material spread uniformly over the horizontal road surface, the density of the layer being k times that of the masonry, what would be the shape of the arch?

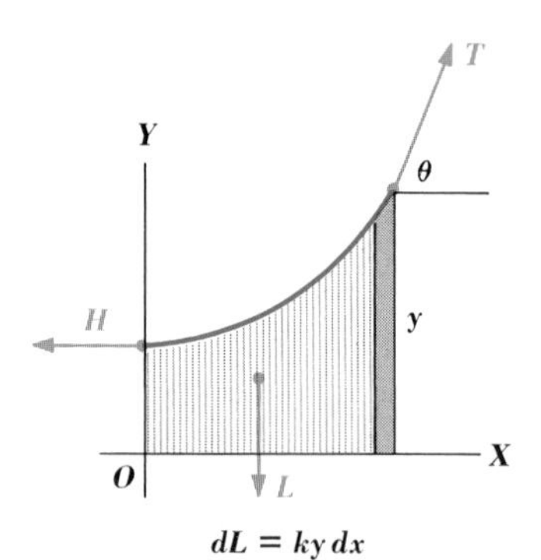

Figure 3

5. Slender uniform rods all of the same diameter are suspended from a string to which each is knotted; two consecutive rods just touch, and they hang so that their ends are in a straight horizontal line. Neglecting the diameter of the rods, find the equation of the curve of the string. *Hint:* dL in equation (4) $= ky\,dx$ (see Fig. 3).

6. What would be the equation of the curve of the string in exercise 5 if all the rods were of the same length instead of having their ends in a straight line?

7. Set up the differential equation for the string of exercise 5 if its weight per running foot is assumed constant and taken into account.

★8. An arch is the bottom of a canal carrying a water load. Find the equation of its curve, assuming that the stress at each point due to the weight of the water is directed along the tangent. Neglect the weight of the arch, and note that the water pressure at each point is directed along the normal to the arch through the point (see Fig. 4).

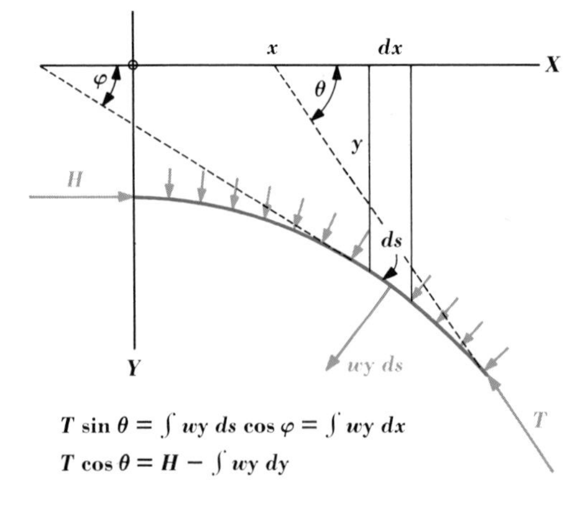

Figure 4

★**9.** Solve exercise 8 if the water is covered with a uniform layer of fluid having a specific gravity k.

★**10.** A uniform cable $2l$ units long has its ends attached at two points, A and B, and hangs under its own weight. A horizontal line through B meets a vertical line through A at point C. If $AC = 2b$ units and $BC = 2c$ units, find the coordinates of the midpoint of line AB referred to a set of rectangular axes with origin at the lowest point of the cable and X axis horizontal.

91 Equation of elastic curve. Beams

Consider a horizontal beam acted upon by vertical loads, and assume that the forces due to these loads lie in a vertical plane containing the centroidal axis (central longitudinal axis) of the beam and that they are such that no part of the beam is stressed beyond its elastic limit. These stresses cause the beam to bend, as indicated in Fig. 1, and the curve of its centroidal axis is called the elastic curve of the stressed beam. An important problem in the consideration of strength of materials is to find the equation of this elastic curve. If a beam is loaded as just indicated, is made of uniform material satisfying Hooke's law, and fulfills certain other conditions relating to shape and to properties of materials, it can be shown that its elastic curve satisfies approximately the differential equation

$$EI\frac{d^2y}{dx^2} = M \tag{1}$$

where the X axis is horizontal along the beam, the Y axis is vertical, E is the modulus of elasticity of the material of the beam, I is the moment of inertia of the cross section of the beam perpendicular to its axis with respect to a horizontal line in the cross section passing through its centroid, and M is the bending moment at the cross section. Since the material of the beam is uniform, E *is constant*, and if the beam has a uniform cross section, I *is constant*.

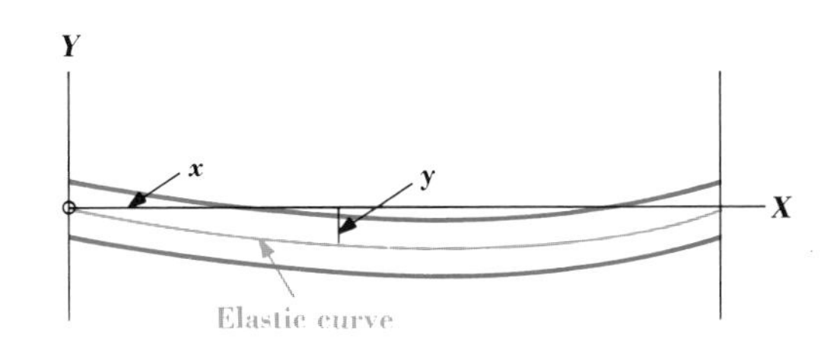

Figure 1

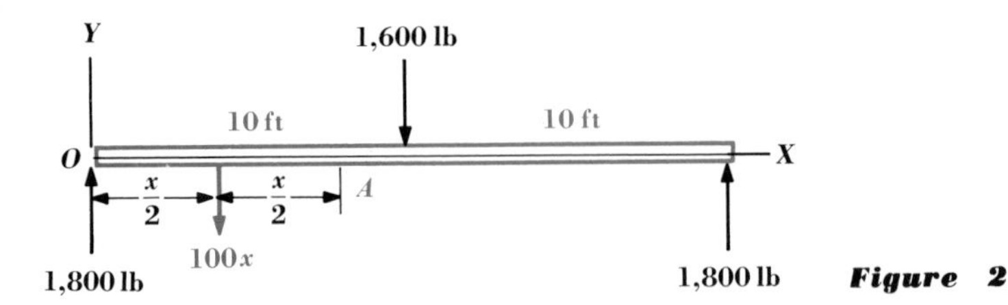

Figure 2

The bending moment M at any cross section may be found by taking the algebraic sum of the moments of the external forces on the part of the beam on one side of the cross section about a horizontal line in this cross section. In finding the bending moment, *consider upward forces as giving positive moments and downward forces negative moments* about the horizontal line in the cross section. Consider, for example, the beam of Fig. 2 loaded as indicated in addition to a uniform running load of 100 lb/ft. To find the bending moment at point A, consider the forces to the left of A: 1,800 lb at O with arm x ft, and $100x$ lb of running load downward and thought of as concentrated at the center of OA. Taking moments about A, we have

$$M = 1{,}800x - 100x\frac{x}{2} = 1{,}800x - 50x^2$$

Example. A uniform beam (see Fig. 3) l ft long is fixed at both ends and carries a uniformly distributed load of w lb/ft length. Find the equation of its elastic curve and its maximum deflection.

Solution. Since the beam is fixed at the ends, the elastic curve is horizontal at both ends. Hence, taking the origin at the left end, we have

$$\begin{aligned} y &= 0 \qquad \frac{dy}{dx} = 0 \qquad \text{when } x = 0 \\ y &= 0 \qquad \frac{dy}{dx} = 0 \qquad \text{when } x = l \end{aligned} \tag{a}$$

The forces acting on the beam to the left of the cross section x ft from the left end are: (1) wx lb due to the running load, (2) a supporting force at the left end, and (3) a couple exerted by the masonry. Writing

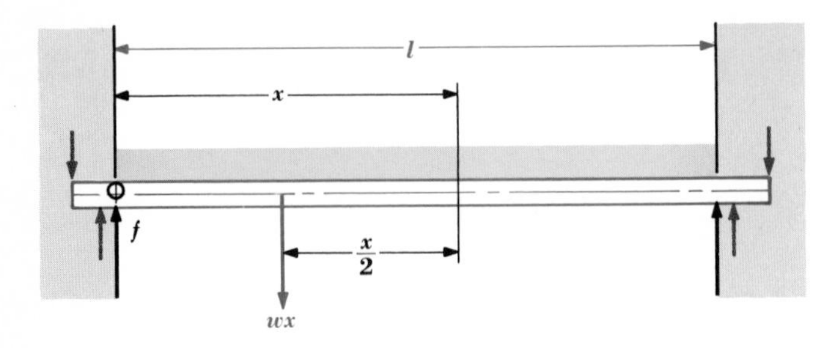

Figure 3

f for the supporting force and A for the moment of the couple, we have for the bending moment x ft from the left end (see Fig. 3)

$$M = A + fx - wx\frac{x}{2} \qquad (b)$$

Substitution of M from (b) in (1) gives

$$EI\frac{d^2y}{dx^2} = A + fx - w\frac{x^2}{2} \qquad (c)$$

Integrating (c), we obtain

$$EI\frac{dy}{dx} = Ax + f\frac{x^2}{2} - \frac{wx^3}{6} + c_1 \qquad (d)$$

$$EIy = \frac{Ax^2}{2} + f\frac{x^3}{6} - w\frac{x^4}{24} + c_1x + c_2 \qquad (e)$$

Substituting the conditions (a) in (d) and (e), we get

$$c_1 = 0 \qquad c_2 = 0 \qquad Al + f\frac{l^2}{2} - w\frac{l^3}{6} = 0$$

$$\frac{Al^2}{2} + \frac{fl^3}{6} - \frac{wl^4}{24} = 0$$

or

$$c_1 = 0 \qquad c_2 = 0 \qquad f = \frac{wl}{2} \qquad A = \frac{-wl^2}{12} \qquad (f)$$

Substituting in (e) the values of c_1, c_2, f, and A from (f) and simplifying, we have the equation of the elastic curve

$$y = \frac{-w}{24EI}(x^2l^2 - 2x^3l + x^4) \qquad (g)$$

Here, $-y$ represents the deflection of the beam x ft from the left end, and since the maximum deflection (d_{max}) will evidently be at the center of the beam where $x = l/2$, we have, from (g),

$$d_{max} = (-y)_{x=l/2} = \frac{w}{24EI}\left(\frac{l^4}{4} - 2\frac{l^4}{8} + \frac{l^4}{16}\right) = \frac{wl^4}{384EI}$$

Exercises

1. A beam (see Fig. 4) l ft long is simply supported at its ends and carries a uniform load of w lb/ft run. Show that the bending moment at a point x ft from the left end is $\frac{1}{2}wlx - \frac{1}{2}wx^2$, and use this in equation (1) to find the deflection y at this point. Also, find the maximum deflection.

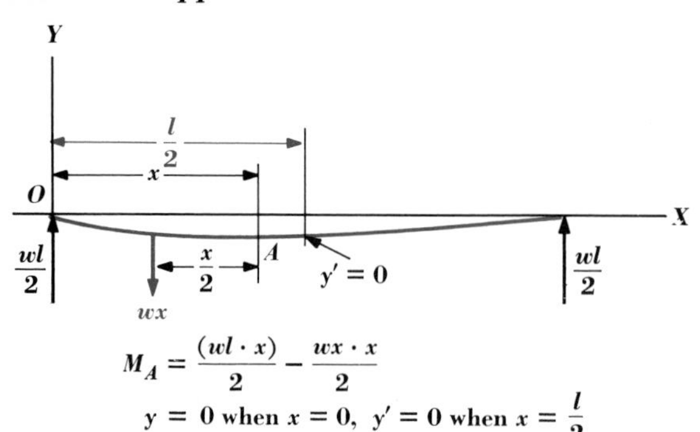

Figure 4

2. A beam l ft long and simply supported carries a load of P lb at its center. Find its deflection x ft from the left end and its maximum deflection. *Hint:* Draw a figure of the beam like Fig. 4, and observe that $M = \frac{1}{2}Px$ if $x < \frac{1}{2}l$.

3. Using the answers to exercise 1 and 2, write a formula for the maximum deflection of a beam simply supported at its ends and carrying a uniformly distributed load of w lb/ft and P lb at its center.

4. Using the formulas from exercises 1 to 3, find the maximum deflection of a simply supported steel beam 20 ft long, having $E = 30 \times 10^6$ lb/in.^{2} and $I = 54$ in.^{4}, if the loading is: (*a*) 5,000 lb at center; (*b*) 40 lb/in. uniformly distributed; (*c*) the combined loading just mentioned.

5. A beam fixed at one end and unsupported at the other (see Fig. 5) is called a cantilever beam. Find the deflection x ft from the fixed end of a cantilever beam l ft long if it carries a load of: (*a*) P lb at the free end ($w = 0$ in Fig. 5); (*b*) w lb/ft run ($P = 0$ in Fig. 5); (*c*) w lb/ft run and P lb at its free end.

★6. Figure 6 represents a cantilever beam carrying a uniformly distributed load of w lb/ft, fixed at one end and simply supported at

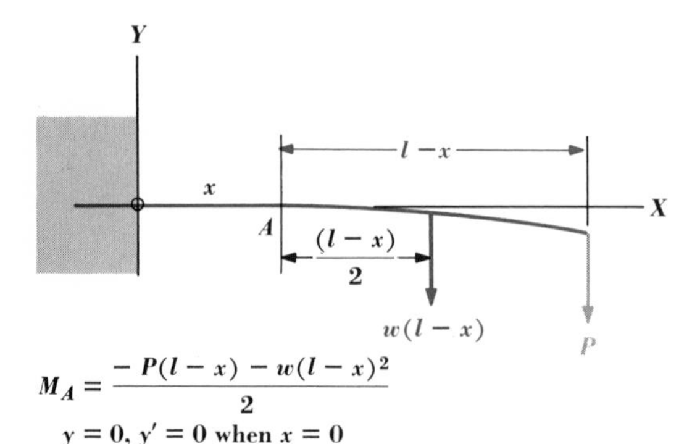

Figure 5

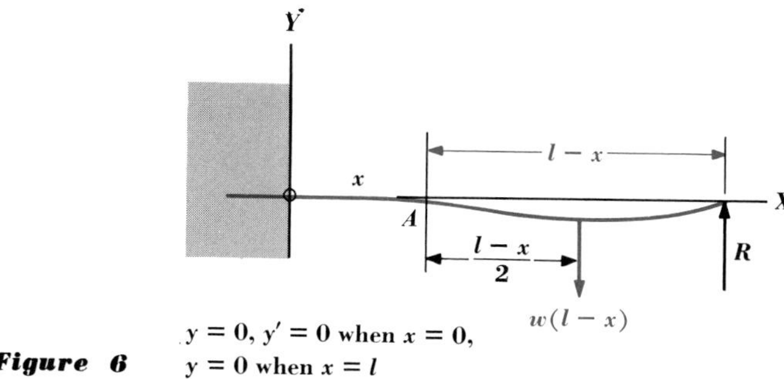

Figure 6 $y = 0$, $y' = 0$ when $x = 0$,
$y = 0$ when $x = l$

the other so that the right end is on a level with the left end. Since force R is unknown, it must be determined like a constant of integration by using an initial condition. Find the equation of the elastic curve of the beam. Also, find the value of x/l at the point where the deflection is maximum.

7. Find the equation of the elastic curve of a uniform beam fixed at both ends and carrying a load P at its center.

★8. Figure 7 represents a uniform beam AB with a concentrated load P at C twice as far from A as from B. The supporting forces $\frac{1}{3}P$ and $\frac{2}{3}P$ are indicated. If y_1 is the deflection for part AC and y_2 for part CB, we have

$$EI\frac{d^2y_1}{dx^2} = \frac{1}{3}Px \qquad EI\frac{d^2y_2}{dx^2} = \frac{1}{3}Px - P(x - 2a)$$

with the initial conditions $y_1 = 0$ when $x = 0$, $y_2 = 0$ when $x = 3a$, $y_1 = y_2$ and $dy_1/dx = dy_2/dx$ when $x = 2a$. Integrate the two equations, using the initial conditions to determine constants, and thus find the equations of the elastic curves of the parts of the beam. Also, find the maximum deflection.

★9. Solve exercise 8, and then find the maximum deflection of the beam of Fig. 7 with the same loading, assuming that it is fixed at both ends. Observe in this case that the supporting force and the couple at A must be found by using initial conditions.

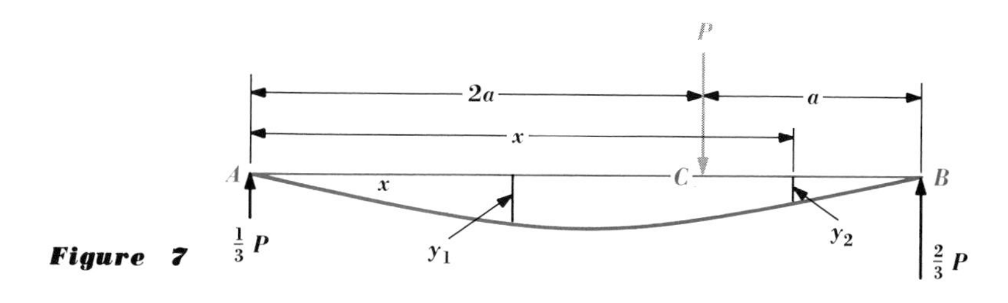

Figure 7

★**10.** A cantilever beam l ft long carries a load of material whose width and density are uniform and whose depth is directly proportional to its distance from the free end. Find the maximum deflection of the beam.

★**11.** Find the equation of the elastic curve of a beam fixed at both ends and carrying a concentrated load P, at a distance a ft from the left end and b ft from the right end.

92 Columns

Beams placed vertically to support vertical loads are often called columns. This section will deal mainly with the four types represented in Figs. 1 to 4. The plane of the elastic curve of a column contains the axis of each cross section about which the moment of inertia is greatest. If now we think of the column as placed horizontally without relative change of forces acting on it and with the plane of the elastic curve vertical (see Fig. 5), we may apply equation (1) of §91 to obtain an approximation to the equation of the elastic curve of the column.

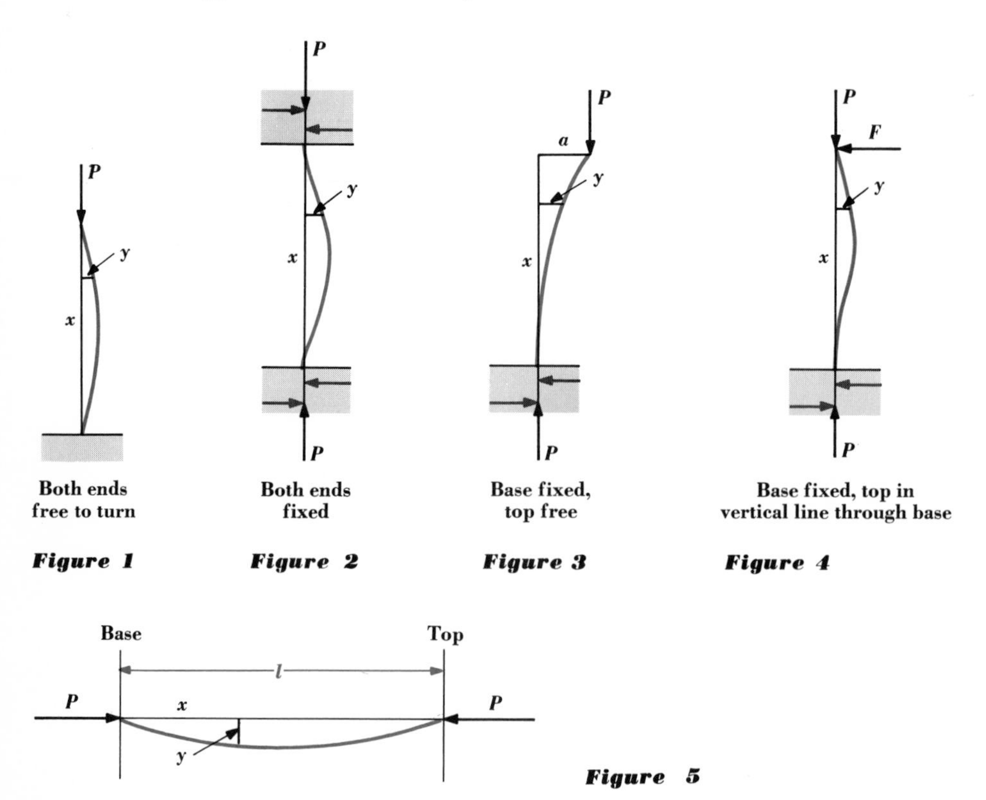

Both ends free to turn

Figure 1

Both ends fixed

Figure 2

Base fixed, top free

Figure 3

Base fixed, top in vertical line through base

Figure 4

Figure 5

Example. Approximate the equation of the elastic curve of a column (Fig. 5) with both ends free to turn, and find the buckling* stress.

Solution. Neglecting the weight of the beam and applying (1) of §91, obtain (see Fig. 5)

$$EI\frac{d^2y}{dx^2} = -Py$$

The general solution of this equation is

$$y = c_1 \sin \omega x + c_2 \cos \omega x \qquad \omega = \sqrt{\frac{P}{EI}} \tag{a}$$

As initial conditions, we have

$$y = 0 \qquad \text{when } x = 0 \text{ and when } x = l \tag{b}$$

Since $y = 0$ when $x = 0$, we have from (*a*) $c_2 = 0$, and

$$y = c_1 \sin \omega x \tag{c}$$

The second condition of (*b*), $y = 0$ when $x = l$, will be satisfied if $c_1 = 0$. In this case, $y = 0$ for all values of x, and the beam is straight. Also, $y_l = c_1 \sin \omega l = 0$ if $\omega l = \pi$; or since $\omega = \sqrt{P/(EI)}$, $P = \pi^2 EI/l^2$. Hence, if A is the area of the horizontal cross section of a column, the stress p, or force per square unit for buckling, is

$$p = \frac{P}{A} = \frac{\pi^2 EI}{Al^2} \tag{d}$$

Since we cannot have much deflection without disagreement with the conditions of §91, a column should not be loaded so that buckling impends.† Also, note that, if l is very small, p from (*d*) is very large and has no design value. Actually, equation (*d*) is useful only for a limited range of $I/(Al^2)$.

* If a rod or column is subjected to sufficiently small axial forces, it remains straight. If the axial forces P are gradually increased, however, a value P_1 will be reached at which the rod suddenly begins to bend or buckle and its shape varies greatly with a small increase in P. The critical force P_1, at which the rod begins to buckle, is called the critical value, and the corresponding stress is the one required in the example. To obtain a detailed treatment of buckling, the student may consult books on the strength of materials.

† Since amplitude c_1 in (*c*) is independent of p, it is not determined, and the maximum deflection of a buckling column seems to be arbitrary. However, Prof. R. P. Bailey, using a more accurate formula than (1), §91, has shown that amplitude of deflection depends on stress p.

Exercises

1. Show for the column represented by Fig. 2 that $EI\, d^2y/dx^2 = -Py + G$, where G is a constant. Check that the initial conditions are $x = 0$, $y = 0$, and $dy/dx = 0$, both when $x = 0$ and when $x = l$. Show that $y = (G/P)(1 - \cos \omega x)$, where $\omega = \sqrt{P/(EI)}$, and that for buckling $P = 4\pi^2 EI/l^2$.

2. In the solution of exercise 1, you took $\omega l = 2\pi$ to provide that $1 - \cos \omega l = 0$. Show that, if you had used $\omega l = 4\pi$, the corresponding P would have been four times as large. Sketch the corresponding elastic curve of the beam, and thus show that it consists essentially of two beams each having one-half the length of the original one.

3. Figure 3 represents a column with its base fixed and the top free to move. Show that for buckling $P = \frac{1}{4}\pi^2 EI/l^2$.

4. For exercise 3, discuss the elastic curve and the strength of the beam corresponding to a value of $\frac{3}{2}\pi$ instead of $\frac{1}{2}\pi$ for ωl to provide that $\cos \omega l = 0$.

★5. Figure 4 represents a column with its base fixed and its top free to turn but held on a vertical line through the base. Prove that for buckling $\tan \omega l = \omega l$, where $\omega = \sqrt{P/(EI)}$. Then show that, if A is the cross-sectional area of the column, assumed uniform, the stress $p = P/A = (4.4934)^2 EI/(Al^2)$, approximately.

6. For a round steel column supported as indicated in Fig. 4 and having $E = 30 \times 10^6$ lb/in.^{2} and a radius of 2 in., show that $p = 6.057 \times 10^8/l^2$ lb/in.^{2}, where l is length in inches. Show that this is greater than 30,000 lb/in.^{2} (the elastic limit of steel) when $l < 142.1$ in.

93 Motion of a particle in a plane

Polar coordinates lend themselves to the solution of certain problems more readily than rectangular coordinates. Accordingly, we shall find an expression for the component a_ρ of the acceleration in the direction of the radius vector and an expression for the component a_θ at right angles to the radius vector for the motion of a particle in a plane. The component a_ρ is the projection of the acceleration vector a on the radius vector, or, what amounts to the same thing, the sum of the projections of the components of a on the radius vector. Hence, denoting derivatives with respect to t by dots thus

$$\frac{dx}{dt} = \dot{x} \qquad \frac{d^2x}{dt^2} = \ddot{x} \qquad \frac{d\rho}{dt} = \dot{\rho} \qquad \frac{d\theta}{dt} = \dot{\theta} \qquad \text{etc.} \tag{1}$$

we see from Fig. 1 that

$$a_\rho = \ddot{x} \cos \theta + \ddot{y} \sin \theta \tag{2}$$

and

$$a_\theta = -\ddot{x} \sin \theta + \ddot{y} \cos \theta \tag{3}$$

Repeated differentiation of each of the equations

$$x = \rho \cos \theta \qquad y = \rho \sin \theta \tag{4}$$

with respect to the time gives

$$\dot{x} = \dot{\rho} \cos \theta - \rho\dot{\theta} \sin \theta \qquad \dot{y} = \dot{\rho} \sin \theta + \rho\dot{\theta} \cos \theta \tag{5}$$

$$\ddot{x} = \ddot{\rho} \cos \theta - 2\dot{\rho}\dot{\theta} \sin \theta - \rho\dot{\theta}^2 \cos \theta - \rho\ddot{\theta} \sin \theta \tag{6}$$

$$\ddot{y} = \ddot{\rho} \sin \theta + 2\dot{\rho}\dot{\theta} \cos \theta - \rho\dot{\theta}^2 \sin \theta + \rho\ddot{\theta} \cos \theta \tag{7}$$

Substituting the values of $\ddot{x}$ and $\ddot{y}$ from (6) and (7) in (2) and (3) and simplifying, we have

$$a_\rho = \ddot{\rho} - \rho\dot{\theta}^2 \tag{8}$$

$$a_\theta = \rho\ddot{\theta} + 2\dot{\rho}\dot{\theta} = \frac{1}{\rho}\frac{d}{dt}(\rho^2\dot{\theta}) \tag{9}$$

Similarly, it is easy to show that the component v_ρ of the velocity along the radius vector and the component v_θ at right angles to it are

$$v_\rho = \dot{\rho} \qquad v_\theta = \rho\dot{\theta} \tag{10}$$

The component a_t of acceleration along the tangent of the path of a moving particle and the component a_n along the normal to the path are

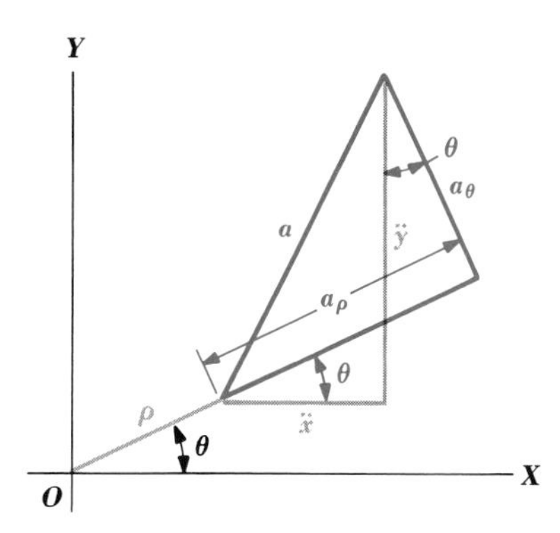

Figure 1

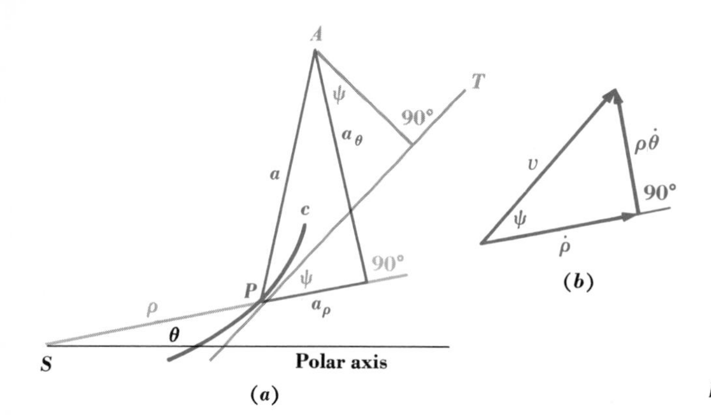

Figure 2

used frequently. Figure 2 represents the path C of a particle, a_ρ, a_θ, a_t, and a_n, and the angle ψ for which $\tan \psi = \rho\dot{\theta}/\dot{\rho}$. From Fig. 2$a$ we read

$$a_t = a_\rho \cos \psi + a_\theta \sin \psi \tag{11}$$
$$a_n = a_\rho \cos (90° + \psi) + a_\theta \cos \psi \tag{12}$$

In (11) and (12) substitute a_ρ from (8), a_θ from (9), $\sin \psi$ and $\cos \psi$ from Fig. 2b, and using the fact that dv/dt and radius of curvature R are given by

$$\frac{dv}{dt} = \frac{\dot{\rho}\ddot{\rho} + \rho\dot{\rho}\dot{\theta}^2 + \rho^2\dot{\theta}\ddot{\theta}}{v} \qquad R = \frac{v^3}{\rho^2\dot{\theta}^3 + 2\dot{\rho}^2\dot{\theta} - \rho\ddot{\rho}\dot{\theta} + \rho\dot{\rho}\ddot{\theta}} \tag{13}$$

derive as an exercise

$$a_t = \frac{dv}{dt} \qquad a_n = \frac{v^2}{R} \tag{14}$$

Example. A particle A fixed at pole $\rho = 0$ attracts a 1-slug mass B at (ρ ft, θ rad) with a force of $64/\rho^2$ lb; and $\rho = 2$ ft, $\theta = 0$, $\dot{\rho}_0 = 0$, and $v_0 = 4$ ft/sec at time $t = 0$. Find the equation of the path of B and the time for B to make one revolution about A.

Solution. Using (8) and (9) and applying $ma = F$ in the direction along and at right angles to the radius vector, we get

$$1(\ddot{\rho} - \rho\dot{\theta}^2) = \frac{-64}{\rho^2} \qquad \frac{1}{\rho}\frac{d}{dt}(\rho^2\dot{\theta}) = 0 \tag{a}$$

The initial conditions are

$$\theta = 0 \qquad \rho = 2 \text{ ft} \qquad \dot{\rho}_0 = 0 \qquad \dot{\theta}_0 = \frac{v_0}{2} = 2 \text{ rad/sec} \tag{b}$$

Integrating the second equation of (*a*) and using conditions (*b*), we get

$$\rho^2\dot{\theta} = c \qquad \rho^2\dot{\theta} = 8 \tag{c}$$

Replacing θ in the first equation of (*a*) by its value $8/\rho^2$ from (*c*) and transforming the result slightly, we get

$$\ddot{\rho} = -\frac{64}{\rho^2} + \frac{64}{\rho^3} \tag{d}$$

Replacing $\ddot{\rho}$ in (*d*) by $\dot{\rho}\,d\dot{\rho}/d\rho$, solving the resulting equation, using $\dot{\rho}_0 = 0$, $\rho = 2$ from (*b*) to determine the constant of integration, and completing the square in the right member, we get

$$\frac{\dot{\rho}^2}{2} = +\frac{64}{\rho} - \frac{64}{2\rho^2} - \frac{64}{2} + \frac{64}{8} \tag{e}$$

$$\dot{\rho}^2 = -\left(\frac{64}{\rho^2} - \frac{128}{\rho} + 64\right) + 16 \tag{f}$$

Equating the square roots of the members of (*f*), separating the variables in the result, and replacing dt by $\rho^2 d\theta/8$ from (*c*), we get

$$\frac{d\rho}{\sqrt{16 - (8/\rho - 8)^2}} = \pm\frac{\rho^2}{8}\,d\theta \tag{g}$$

From (*g*) we easily obtain

$$\sin^{-1}\frac{8 - 8/\rho}{4} = \pm\theta + c_3 \tag{h}$$

Substituting 2 for ρ and 0 for θ from (*b*) in (*h*), we get $c_3 = \frac{1}{2}\pi$. Equating the sines of the members of (*h*) with $c_3 = \frac{1}{2}\pi$ and solving the result for ρ, we get

$$\frac{8 - 8/\rho}{4} = \cos\theta \qquad \text{or} \qquad \rho = \frac{1}{1 - \frac{1}{2}\cos\theta} \tag{i}$$

From analytic geometry, we know that (*i*) represents an ellipse with eccentricity $e = \frac{1}{2}$. The semimajor axis is $\frac{1}{2}(\rho_{\max} + \rho_{\min}) = \frac{1}{2}(2 + \frac{2}{3}) = 4/3 = L$, and the semiminor axis is $L\sqrt{1 - e^2} = \frac{4}{3}(\sqrt{3}/2) = \frac{2}{3}\sqrt{3}$. To get the period use the fact that the area A of the ellipse $= \pi(\frac{4}{3})(\frac{2}{3}\sqrt{3}) = \frac{8}{9}\pi\sqrt{3}$, that $\rho^2\dot{\theta} = 8$, and that

$$A = \frac{8}{9}\pi\sqrt{3} = \frac{1}{2}\int_0^{2\pi}\rho^2\,d\theta = \frac{1}{2}\int_0^{T}\rho^2\dot{\theta}\,dt = \frac{1}{2}8T$$

where $T = \frac{2}{9}\pi\sqrt{3} = 1.21$ sec is the time required for each revolution.

Exercises

1. For the plane curve $\rho = t^{3/2}$, $\theta = t^{1/2}$, use equations (8) to (14) to find at time $t = 1$: $\dot{\theta}$, $\ddot{\theta}$, $\dot{\rho}$, $\ddot{\rho}$, v, $[a_\rho, a_\theta]$, and $[a_t, a_n]$. Check your work by showing that the magnitudes of the two accelerations $[a_\rho, a_\theta]$ and $[a_t, a_n]$ are equal.

2. A particle of weight w lb moves in a straight line from a distance a toward a charged point which attracts with a force whose magnitude in pounds is equal numerically to $w/(2gr^2)$, r denoting the distance of the particle from the point. If the particle had an initial velocity toward the point of $1/\sqrt{a}$, how long will it take to traverse half the distance to the point? Use $F = ma$.

3. A particle moves in a straight line from rest at a distance a toward a center of attraction, the attraction varying inversely as the cube of the distance. How long will it take the particle to reach the center? Use $F = ma$.

4. The areal velocity of a moving particle P with reference to a fixed point A is the rate at which the line AP generates area. Prove that a particle moving in a plane through a point A under the action of a force always directed toward point A has a constant areal velocity with respect to A. *Hint:* First show that a_θ in equation (9) of this section is zero.

5. A particle moves so that its radial acceleration is always 1 ft/sec² and the angular velocity of its radius is always 2 radians/sec. If $\rho = \frac{3}{4}$, $\theta = 0$, and $\dot{\rho} = 6$ when $t = 0$, find ρ and θ in terms of t. Also, find a_θ in terms of t.

6. Use (14) to show that a particle B of mass m moving in a circle of radius r at speed h will have an acceleration h^2/r towards the center of the circle and that the force acting on B will be mh^2/r. Find the tension in a taut cord 5 ft long if one end is attached to a mass of 1 slug and the other to a point in a smooth horizontal table upon which the mass is moving at 20 ft/sec.

7. A particle moves so that its radial velocity $\dot{\rho}$ is always 1 ft/sec, and the force on the particle is directed along the radius vector. If initially $\rho = 1$, $\theta = 0$, and $\dot{\theta} = 2$ when $t = 0$, find ρ, θ, and the magnitude of the force in terms of t.

8. If a body B of mass m slugs is moving in a plane through point A ($\rho = a$, $\theta = 0$) with velocity v_0 directed at angle α to the polar axis SA (see Fig. 3) and if a force of K lb acts on B always directed at right angles to its direction of motion, show that B will move with unchanged speed in a circle of radius mv_0^2/K and will

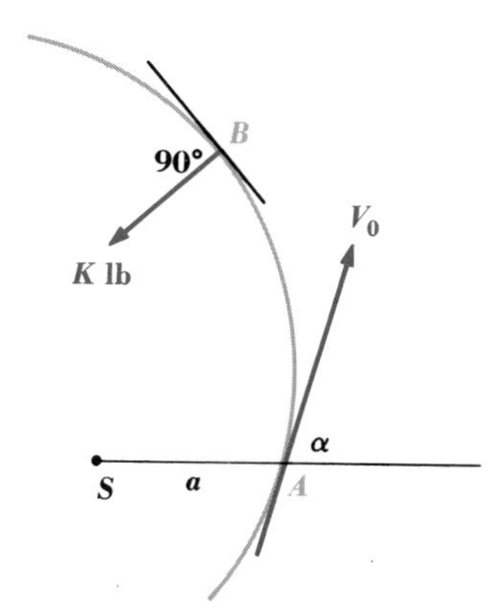

Figure 3

pass through A in the direction of the initial velocity.* *Hint:* Use (14).

9. In the example above replace the force $64/\rho^2$ acting on B by $16/\rho^2$, carry out the solution to show that, in this case, the path of B will be a parabola, and find its equation.

94 Satellites

This section applies the procedure of the illustrative example of §93 to the motion of a satellite orbiting about the earth. The basic assumption, conceived by Isaac Newton, is that two particles attract each other with a force directly proportional to their masses, inversely proportional to the square of the distance between them, and directed along a straight line connecting them. Hence assuming that the earth and the satellite act as particles† and that the earth is so large relative to the satellite that we may think of the earth as fixed at the pole E of a system of polar coordinates (see Fig. 1), consider the following basic example.

Example. A satellite S of mass m slugs moves about the center E of the earth under the attraction mgR^2/ρ^2 lb always directed toward the center E of the earth. Find the equation of the path of S if R is

* This idea can be used in the theory of bringing one satellite alongside of another.

† This involves the assumption that the earth is a perfect sphere having as density at any point P a function of the distance r of P from the center of the earth. It does not take account of the gravitational forces due to other bodies, or the magnetic field of the earth.

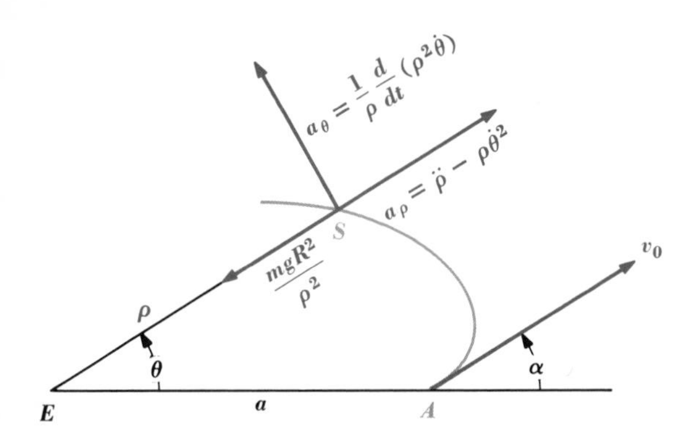

Figure 1

the radius of the earth and (see Fig. 1) $\rho_0 = a$, $\theta_0 = 0$, $v = v_0$, $\dot\theta_0 = (v_0 \sin \alpha)/a$, and $\dot\rho_0 = v_0 \cos \alpha$, all when $t = 0$. Also find the time required for S to make a complete circuit about the earth.

Solution. The initial velocity vector and the center of the earth determine a plane in which the motion takes place, for all of the forces involved act in this plane. Using $F_d = ma_d$ applied in the direction of the radius vector and also at right angles to the radius vector, we get

$$m(\ddot\rho - \rho\dot\theta^2) = \frac{-mgR^2}{\rho^2} \tag{1}$$

$$m\frac{1}{\rho}\frac{d}{dt}(\rho^2\dot\theta) = 0 \tag{2}$$

The initial conditions are

$$v = v_0 \qquad \rho_0 = a \qquad \theta_0 = 0 \qquad \dot\rho_0 = v_0 \cos \alpha$$

$$\dot\theta_0 = \frac{v_0 \sin \alpha}{a} \qquad \text{when } t = 0 \tag{3}$$

From (2) and (3) we easily obtain

$$\rho^2\dot\theta = c = a^2\dot\theta_0 = \frac{a^2 v_0 \sin \alpha}{a} = av_0 \sin \alpha \tag{4}$$

Substituting c/ρ^2 for $\dot\theta$ from (4) in (1) and transforming slightly, we obtain

$$\ddot\rho = \frac{-gR^2}{\rho^2} + \frac{c^2}{\rho^3} \tag{5}$$

Replacing $\ddot\rho$ by $\dot\rho\, d\dot\rho/d\rho$ in (5), integrating the result, and using (3) to determine the constant of integration, we get

$$\frac{2\dot\rho^2}{2} = \frac{2gR^2}{\rho} - \frac{2c^2}{2\rho^2} + c_1 \tag{6}$$

$$c_1 = v_0^2 \cos^2 \alpha - \frac{2gR^2}{a} + \frac{c^2}{a^2} \tag{7}$$

Replacing c in (7) by its value $av_0 \sin \alpha$ from (4) and simplifying slightly, obtain

$$c_1 = v_0^2 - \frac{2gR^2}{a} \tag{8}$$

Solving (6) for $\dot{\rho}$ and completing a square, we get

$$\dot{\rho} = \pm \sqrt{-\left(\frac{c^2}{\rho^2} - \frac{2gR^2}{\rho} + \frac{g^2R^4}{c^2}\right) + \frac{g^2R^4}{c_2} + c_1} \tag{9}$$

Replacing $\dot{\rho}$ by $d\rho/dt$, dt by $\rho^2 d\theta/c$ from (4), and integrating the result, we obtain

$$\sin^{-1}\left(\frac{gR^2/c - c/\rho}{\sqrt{c_1 + g^2R^4/c^2}}\right) = \pm\theta + c_2 \tag{10}$$

Equate the sines of the members of (10), solve the result for ρ, transform the result, and get

$$\rho = \frac{c^2/(gR^2)}{1 - \sqrt{1 + c_1c^2/(g^2R^4)} \sin(\pm\theta + c_2)} \tag{11}$$

This represents a conic section for which eccentricity

$$e = \sqrt{1 + c_1c^2/(g^2R^4)} \tag{12}$$

Now a conic section represented by (11) is an ellipse, a parabola, or a hyperbola according as $e < 1$, $e = 1$, or $e > 1$. Hence if c_1, or from (8) $v_0^2 - 2gR^2/a$, satisfies

$$v_0^2 - \frac{2gR^2}{a} < 0 \qquad \text{(11) represents an ellipse} \tag{13}$$

$$v_0^2 - \frac{2gR^2}{a} = 0 \qquad \text{(11) represents a parabola} \tag{14}$$

$$v_0^2 - \frac{2gR^2}{a} > 0 \qquad \text{(11) represents a hyperbola} \tag{15}$$

It is left as exercise 4 below to show that if

$$v_0^2 = \frac{gR^2}{a} \qquad \text{and} \qquad \alpha = 90° \qquad \text{(11) represents a circle } \rho = a \tag{16}$$

If A and B represent the semiaxes of (11) in the case (13) and T the time for a revolution of S, then

$$\text{Area } \pi AB = \frac{1}{2}\int_0^{2\pi} \rho^2\, d\theta = \frac{1}{2}\int_0^T \rho^2\dot{\theta}\, dt = \frac{1}{2}\int_0^T c\, dt = \frac{1}{2}cT$$

or period

$$T = \frac{2\pi AB}{c} \tag{17}$$

Since the semimajor axis of (11) is $\frac{1}{2}(\rho_{\max} + \rho_{\min})$ we have

$$A = \frac{1}{2}\left[\frac{c^2}{gR^2(1 - e)} + \frac{c^2}{gR^2(1 + e)}\right] = \frac{c^2}{gR^2(1 - e^2)} \tag{18}$$

Also

$$B = A\sqrt{1 - e^2} \tag{19}$$

Substituting these values for A and B in (17), and using (12), (8), and (4), we derive*

$$T = \frac{2\pi gR^2}{[(2R^2g/a) - v_0^2]^{3/2}} \tag{20}$$

when $v_0^2 < 2R^2g/a$.

Exercises

1. Does the time T required by a satellite to circuit the earth change when α is changed but a and v_0 are not changed? Use equation (20).

2. If $v_0^2 = 2R^2g/a$, equation (20) indicates an infinite value for T. Does this agree with (14)? What is the significance of the imaginary answer given for T by (20) when $v_0^2 > 2gR^2/a$?

3. Why is $\sqrt{2gR^2/a}$ called the *escape velocity* for a satellite a miles from the center of the earth? Find the escape velocity for a satellite 4,960 mi from the center of the earth if $R = 3{,}960$ mi and $g = (32.2/5{,}280)$ mi/sec². Also find the escape velocity for the moon if $R = 1{,}080$ mi and $g = 32.2/[6(5{,}280)]$ mi/sec². Take $R = a = 1{,}080$ mi.

4. Using (4), (8), and (12) show that the eccentricity $e = 0$ only when $v_0^2 = gR^2/a$ and $\alpha = 90°$. Show in this case that (11) represents a circle of radius a.

5. If two satellites are moving about the earth in circular orbits, will the speed of the satellite nearer the earth be greater than the speed of the other? Why?

6. For elliptical orbits does the time T per revolution increase or decrease as v_0 increases while a remains fixed?

* See exercise 10 on page 267.

7. Solve the equations

$$\ddot{\rho} - \rho\theta^2 = -\frac{5}{9\rho^2} \qquad \frac{d}{dt}(\rho^2\dot{\theta}) = 0$$

subject to the conditions

$$\rho = 1 \text{ ft} \qquad \theta = 0 \qquad \dot{\rho} = 0 \qquad \dot{\theta} = 1 \text{ rad/sec}$$

when $t = 0$

Find the time T for the point (ρ ft, θ rad) to move through all its positions once, that is, its orbital period.

8. Solve the equations

$$\ddot{\rho} - \rho\dot{\theta}^2 = -\frac{gR^2}{\rho^2} \qquad \frac{d}{dt}(\rho^2\dot{\theta}) = 0$$

subject to the initial conditions

$$\rho = a \qquad \theta = 0 \qquad v = v_0 \qquad \dot{\rho} = 0 \qquad \dot{\theta} = v_0/a$$

all when $t = 0$

Show, in this case, that (20) applies without change.

★**9.** If a force of magnitude R/x^2 acts along the X axis always toward the origin on a 1-slug mass and if $v = v_0$ when $x = a$, find the complete period of the motion. First find a quarter of the period.

10. Derive (20) from (17), (18), (19), (12), (8), and (4).

95 Review problems

In the following problems ρ and θ are used for polar coordinates, and dots denote derivatives with respect to time.

1. Using the polar coordinates ρ and θ, find the equation of the curve for which the curvature equals $d\theta/ds$, where s denotes arc length. Use the formula for curvature $K = [\rho^2 + 2(\rho')^2 - \rho'']/[\rho^2 - (\rho')^2]^{3/2}$. Here $\rho' = d\rho/d\theta$ and $ds/d\theta = [\rho^2 + (\rho')^2]^{1/2}$.

★**2.** A particle moves at 10 ft/sec along a curve having a curvature equal to the reciprocal of the x-component of velocity. Find the equation of the curve if at time $t = 0$ it passes through (0,0) tangent

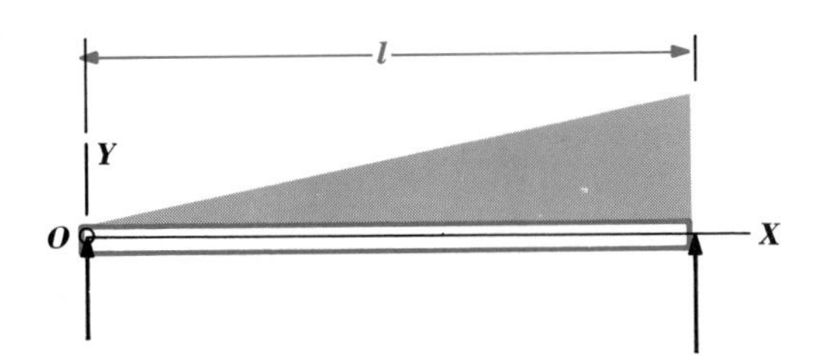

Figure 1

to the X axis with $\dot{x}$ positive. Use the formula for curvature $K = (\dot{x}\ddot{y} - \dot{y}\ddot{x})/(\dot{x}^2 + \dot{y}^2)^{3/2}$.

3. Find a formula for the equation of the elastic curve of a beam l ft long, simply supported at the ends, and carrying a load proportional to distance from the left end (see Fig. 1). Note that at x ft from the left $M = \frac{1}{6}wl^2x - \int_0^x (wy\,dy)(x - y)$.

4. Remembering that $v_0^2 = gR^2/a$ for a satellite in a circular orbit about the earth, assuming that the radius of the orbit is 4,260 mi and that of the earth 3,960 mi, show that the orbital speed is 4.74 mi/sec and the orbital period is 94.2 min. Take $g = (32.2/5{,}280)$ mi/sec^2.

5. Solve exercise 3 modified by the specification that the beam be fixed horizontally in a wall at: (*a*) the left end; (*b*) both ends.

6. A particle moves in a plane under the action of a force always directed toward a fixed point A in its plane; the velocity of the particle is $k\rho^n$, where ρ is its distance from A and the values of k* and n are constant. Find the equation of its path. Tell the nature of its path when: (*a*) $n = 0$; (*b*) $n = -2$; (*c*) $n = 1$; (*d*) $n = -3$.

7. Prove that the force in exercise 6 is $k^2nm\rho^{2n-1}$, where m is the mass of the particle.

8. If the velocity of an m-lb mass B at a distance a ($a > R$) from the center of the earth is v_0 ft/sec directly away from the center and if the attractive force of the earth is $-mgR^2/\rho^2$, show that the maximum value of ρ is $2gR^2/[(2gR^2/a) - v_0^2]$ provided that $v_0^2 < 2gR^2/a$. Use the principle that change in kinetic energy of B, $\frac{1}{2}mv^2 - \frac{1}{2}mv_0^2$, is equal to the work done, $\int_a^\rho F\,ds$, on B by the gravitational field of force. Also find ρ when the velocity of B is $\frac{1}{2}v_0$.

* The dimensions of k are $L^{1-n}T^{-1}$.

11

Existence Theorems and Applications

96 Foreword

Most of the theory used in the preceding chapters relates to the solutions of special types of differential equations in two unknowns. This chapter will provide a broad view of the field of differential equations by means of general existence theorems. It will indicate new points of view and methods. Also, it will give deeper insight into the general methods of the following chapters.

97 Replacement of differential equations by a system of the first order and the first degree

The solution of each set of equations already treated consisted of as many nondifferential relations as there were differential equations in the set. Also, the number of constants of integration was the same as the number of given equations. In general, it can be proved that *the general solution of a set of n first-order and first-degree differential equations in $n + 1$ variables consists of n nondifferential and linearly independent relations among these variables and n constants of integration.*

In this section, we shall indicate how the problem of finding the solution of n differential equations in $n + 1$ unknowns can be reduced to that of solving a set of differential equations of the first order and the first degree.

To get an idea of the method of procedure, consider the special system

$$\frac{d^3y}{dx^3} = x + \frac{d^2y}{dx^2}\frac{dz}{dx} \qquad \frac{d^2z}{dx^2} = zy + \frac{dy}{dx}\frac{dz}{dx} \tag{1}$$

Making the substitutions $dy/dx = y_1$, $dz/dx = z_1$, $dy_1/dx = y_2$ in (1), we obtain the equivalent system of the first order and the first degree:

$$\begin{aligned} &\frac{dy}{dx} = y_1 \qquad \frac{dz}{dx} = z_1 \qquad \frac{dy_1}{dx} = y_2 \\ &\frac{dy_2}{dx} = x + y_2z_1 \qquad \frac{dz_1}{dx} = zy + y_1z_1 \end{aligned} \tag{2}$$

Solving equations (2) and eliminating y_1, z_1, and y_2 from the resulting five equations, we obtain two equations in x, y, z, and the five constants of integration. These two equations constitute the general solution of equations (1).

Now, consider a set of n independent differential equations in n dependent variables and an independent variable. Substitute a new variable for each derivative of a dependent variable, up to the next to the highest-ordered one, and solve the resulting system of the first order for the derivatives contained in it. This set of equations, together with the substitution set, constitutes a system of differential equations of the first order and first degree which is equivalent to the original system.

Exercises

1. Replace, in accordance with the principle stated above, each of the following sets of differential equations by an equivalent set of the first order and the first degree:

(*a*) $\dfrac{d^2y}{dx^2} + x^2\dfrac{dy}{dx} + x^3y = 0$ (*b*) $\dfrac{d^2y}{dx^2} + P(x)\dfrac{dy}{dx} + Q(x)y = 0$

(*c*) $\dfrac{d^3y}{dt^3} = 3y + \dfrac{d^2y}{dt^2} + \dfrac{d^2x}{dt^2}, \dfrac{d^3y}{dt^3} = 3x - \dfrac{d^2y}{dt^2}.$

2. To solve $d^2y/dx^2 + y = 0$, derive, from the equivalent system $dy/dx = z$, $dz/dx = -y$, first $y^2 + z^2 = c_1^2$, and then, from $dy = \pm\sqrt{c_1^2 - y^2}\,dx$, get $y = c_1 \sin(x + c_2)$.

For each equation, write an equivalent system of the first order, and, using the method suggested in exercise 2, find the solution of the given equation:

3. $x^2 \dfrac{d^2y}{dx^2} - x\dfrac{dy}{dx} - x^2 - 2 = 0$ **4.** $\dfrac{d^2y}{dx^2} + x\dfrac{dy}{dx} + x = 0$

5. Write a system of equations of the first order and first degree which is equivalent to a linear nth-order differential equation in two variables. How many constants of integration would appear in the solution of this system?

98 Existence theorems*

Theorem I, stated below but not proved, includes as special cases the theorem of §6 and other theorems of this section; it gives sufficient but not necessary conditions for the existence of solutions of systems of differential equations; and it also indicates the nature of those solutions.

THEOREM I. *For a system of differential equations*

$$\frac{dy_i}{dx} = f_i(x, y_1, y_2, \ldots, y_n) \qquad i = 1, 2, \ldots, n \tag{1}$$

there exists a unique set of continuous solutions $y_1(x), y_2(x), \ldots, y_n(x)$ *of the given equations which take on the values* $y_1^0, y_2^0, \ldots, y_n^0$ *when* $x = x_0$, *provided that the functions*

$$f_1, f_2, \ldots, f_n \qquad \frac{\partial f_i}{\partial y_1}, \ldots, \frac{\partial f_i}{\partial y_n} \qquad i = 1, 2, \ldots, n \tag{2}$$

are continuous and single-valued† *in the regions defined by*

$$|x - x_0| \leqq a \qquad |y_1 - y_1^0| \leqq b_1, \ldots, |y_n - y_n^0| \leqq b_n \tag{3}$$

where the values of a and the b's are all greater than zero.

In any discussion it is highly important to write the set of functions corresponding to (2). The set of functions f_i, $i = 1, 2, \ldots, n$, are

* For a proof of Theorem I consult E. L. Ince, "Ordinary Differential Equations," pp. 62–83.

† A function $f(x_1, x_2, \ldots, x_n)$ is, by definition, single-valued for every set of values $x_1, x_2, \ldots, x_n$ in its domain. For emphasis we specify *single-valued.*

given or are easily obtained by using the plan of §97; then, in addition, write the partial derivative of each function f_i with respect to every variable in it except x. Next, determine the regions in which the set corresponding to (2) are continuous and single-valued. In a set of regions in which all the functions (2) are continuous and single-valued, Theorem I guarantees a continuous set of solutions $y_1(x)$, $y_2(x)$, . . . $y_n(x)$ which take on respective values y_1^0, y_2^0, . . . , y_n^0 when $x = x_0$ all centered in regions defined by (3).

Consider, for example,

$$\frac{d^2y}{dx^2} = \sqrt{x}\, y^{-1} + \frac{dx}{dy} \tag{4}$$

Let

$$\frac{dy}{dx} = y_1 \qquad \text{then } \frac{dy_1}{dx} = \sqrt{x}\, y^{-1} + (y_1)^{-1} \tag{1'}$$

These are the equations corresponding to (1). The functions corresponding to (2) are

$$y_1, \sqrt{x}\, y^{-1} + y_1^{-1}, 1, -\sqrt{x}\, y^{-2}, -y_1^{-2} \tag{2'}$$

If these functions (2′) are to be single-valued and continuous, we must have

$$x \geqq 0 \qquad y \neq 0 \qquad y_1 \neq 0 \tag{5}$$

Hence a unique continuous solution $y = y(x)$, $y_1 = y_1(x)$ of (1′), and therefore of (4), exists; and it takes on the values y^0, y_1^0 when $x = x_0$ if

$$|x - x_0| \leqq a \qquad |y - y^0| \leqq b_0 \qquad |y_1 - y_1^0| \leqq b_1 \tag{6}$$

provided the regions (6) all lie within the regions defined by (5)

In accordance with §97 the equation

$$\frac{d^ny}{dx^n} = f\left(x, y, \frac{dy}{dx}, \frac{d^2y}{dx^2}, \ldots, \frac{d^{n-1}y}{dx^{n-1}}\right) \tag{7}$$

is equivalent to the system

$$\begin{aligned} &\frac{dy}{dx} = y_1, \frac{dy_1}{dx} = y_2, \ldots, \frac{dy_{n-2}}{dx} = y_{n-1} \\ &\frac{dy_{n-1}}{dx} = f(x, y_1, y_2, \ldots, y_{n-1}) \end{aligned} \tag{8}$$

For the system (8) the functions of (2) are

$$y_2, y_3, \ldots, y_{n-1}, \frac{\partial f}{\partial y_i} \qquad i = 1, 2, \ldots, n-1 \tag{9}$$

Hence, we have the following theorem:

THEOREM II. *For a differential equation*

$$\frac{d^ny}{dx^n} = f\left(x,y, \frac{dy}{dx}, \frac{d^2y}{dx^2}, \ldots, \frac{d^{n-1}y}{dx^{n-1}}\right) \tag{10}$$

there exists a unique continuous solution $y(x)$ *having continuous derivatives* d^ky/dx^k, $k = 1, 2, \ldots, n-1$, such that

$$y = a_0, \frac{dy}{dx} = a_1, \frac{d^2y}{dx^2} = a_2, \ldots, \frac{d^{n-1}y}{dx^{n-1}} = a_{n-1} \tag{11}$$

when $x = x_0$, provided that the functions

$$f, \frac{\partial f}{\partial y}, \frac{\partial f}{\partial(d^ky/dx^k)} \qquad k = 1, 2, \ldots, n-1 \tag{12}$$

are continuous and single-valued in the regions defined by

$$|x - x_0| \leqq b, |y - a_0| \leqq b_0, \left|\frac{dy}{dx} - a_1\right| \leqq b_1, \ldots, \left|\frac{d^{n-1}y}{dx^{n-1}} - a_{n-1}\right| \leqq b_{n-1} \tag{13}$$

where the values b, b_i, $i = 0, 1, \ldots, n-1$, *are all greater than zero.*

Consider, for example, $x(x-4)\, dy/dx = 4$. Here there is only the one equation $dy/dx = 1/[x(x-4)]$. The functions (12) are $1/[x(x-4)]$, 0. This is continuous in every region not containing points for which $x = 1$ or $x = 4$; therefore, in every such region there is a unique solution through every point in it. The solution of $x(x-4)\, dy/dx = 4$ is $y = \ln |(x-4)/x| + c$, and y does not exist for points having abscissas 0 or 4.

Let f_1, f_2, and f_3 represent functions of x, and consider

$$\frac{d^2y}{dx^2} + f_1\frac{dy}{dx} + f_2y = f_3 \tag{14}$$

A system equivalent to (14) is

$$\frac{dy}{dx} = y_1 \qquad \frac{dy_1}{dx} = -f_1y_1 - f_2y + f_3 \tag{15}$$

and the expressions corresponding to (12) are

$$-f_1y_1 - f_2y + f_3, y_1, 1, -f_1, -f_2 \tag{16}$$

If the functions x, f_1, f_2, and f_3 are polynomials, functions (16) are single-valued and continuous everywhere and there is a solution of (15), and therefore of (14), for every choice of a point in the plane and a direction associated with it.

If the functions $f_1(x)$, $f_2(x)$, $f_3(x)$ are single-valued and continuous in a region $|x - a| \leqq b$, the functions in (16) are single-valued and continuous in x, y, and y_1, and a unique solution $y(x)$ of (14) exists such that $y_1(x) = y'(x)$, $y = \beta$, $y_1 = \gamma$ when $x = \alpha$ provided that $|\alpha - a| < b$, $|\beta| < m$, $|y_1| < n$, where m and n are any positive numbers.

Theorems I and II may not indicate solutions which exist. Thus $dy/dx - y/x = 0$ has the solution $y = cx$ through (0,0) although the condition of Theorem II that $\partial(y/x)/\partial y$ be continuous at $x = 0$ is not satisfied. Note that $y = cx$ is not unique, however, since c is arbitrary. Also, note that there is no solution through $(0,a)$, $a \neq 0$. Of course, there is a unique solution through every point (a,b), $a \neq 0$.

In dealing with a system of m linearly independent equations in $n + 1$ variables, it is possible to replace $n - m$ variables by arbitrary functions of the remaining $m + 1$ unknowns and then apply Theorem II to the result. A single total differential equation in three variables is a case in point. As in the case of total differential equations, a number of equations connecting the variables may often be found by solving exact equations or integrable equations derived from the given set.

Exercises

1. State Theorem II for $n = 1$, and observe that the result is the theorem of §6.

2. State Theorem II for $n = 2$.

3. Is a unique solution of $x\, d^2y/dx^2 - y = 0$ satisfying the condition $y = 2$ when $x = 0$ to be expected? Why?

4. For what values of x and y are solutions of

$$\sqrt{\sin x}\, \frac{d^2y}{dx^2} + (\sin y)y = 0$$

through (x,y) not guaranteed?

5. For what values of x are solutions of

$$f_1(x)\frac{d^2y}{dx^2} + f_2(x)\frac{dy}{dx} + f_3(x)y = f_4(x)$$

through (x,y) not guaranteed by Theorem II, if $f_1(x), \ldots, f_4(x)$ denote polynomials in x?

6. For $dy/dx = y^{4/3}$, the conditions of Theorem II with $n = 1$ are satisfied in the whole xy plane. What solution passes through (0,0)?

7. What condition of Theorem II with $n = 1$ is not satisfied by $dy/dx = 3y/x$ at $(0,0)$? $y = y_0(x/x_0)^3$ satisfies $dy/dx = 3y/x$ if $x_0 \neq 0$. Does $y = cx^3$ have at $(0,0)$ the slope given by $dy/dx = 3y/x$?

8. Discuss the existence of solutions of

$$P(x)\frac{d^2y}{dx^2} + Q(x)\frac{dy}{dx} + R(x)y = 0$$

by using Theorem II.

9. Does Theorem II show that the graph of the solution of

$$\frac{dy}{dx} = 2 \qquad \frac{dz}{dx} = x + y + z$$

consists of a unique curve through every point of space? Solve the equations, and find the solution for which $y = 5$ and $z = -10$ when $x = 0$.

10. Show that the solution of

$$x^2\frac{d^2y}{dx^2} - 2x\frac{dy}{dx} + 2y = 2x^3$$

could be obtained from the solution of

$$\frac{dy}{dx} = z \qquad \frac{dz}{dx} = \frac{2xz - 2y}{x^2} + 2x$$

In accordance with Theorem II, would you expect a unique solution of the pair of equations for which $y = 0$ and $z = 1$ when $x = 0$?

Show that the general solution of the pair of equations is $y = c_1x^2 + c_2x + x^3$, $z = 2c_1x + c_2 + 3x^2$ and that an infinite number of solutions $y = c_1x^2 + x + x^3$, $z = 2c_1x + 1 + 3x^2$ satisfy the conditions $y = 0$ and $z = 1$ when $x = 0$.

11. For $d^3y/dx^3 = 1/(d^2y/dx^2)$, write the corresponding expressions (12), and observe that $\partial[1/(d^2y/dx^2)]/\partial(d^2y/dx^2)$ is not continuous at points where $d^2y/dx^2 = 0$. What feature of the solution $y = \frac{1}{15}(2x + c_1)^{5/2} + c_2x + c_3$ of $d^3y/dx^3 = 1/(d^2y/dx^2)$ corresponds to this discontinuity?

12. Use Theorem II to find the values of a at which irregularities are to be expected in the solution of

$$x(x-1)\frac{d^2y}{dx^2} + (4x-2)\frac{dy}{dx} + 2y = 0$$

subject to the conditions $y = 0$ and $dy/dx = 1$ when $x = a$. Does the general solution $y = (c_1/x) + c_2/(x-1)$ bear out your answer?

99 Differential equations of the first order and first degree in the unknowns

By using §97, we are enabled to write a set of equations of form (1), §98, equivalent to a very general set of equations. For this reason and for future reference, we are considering equations of type (1), §98, not readily solvable by methods used earlier.

Consider the equations

$$\begin{aligned} P_1\,dx + Q_1\,dy + R_1\,dz &= 0 \\ P_2\,dx + Q_2\,dy + R_2\,dz &= 0 \end{aligned} \tag{1}$$

in which the P's, Q's, and R's represent functions of x, y, and z. A solution consists of two nondifferential equations in x, y, and z which satisfy both equations. *The main method of solution consists in* (1) *combining the given equations and others derived from them so as to obtain an equation in two unknowns, or an equation which is the result of equating to zero the total derivative of some expression involving the variables;* (2) *integrating the equations thus obtained and combining the results with the given differential equations to obtain other relations among the variables.*

Dividing equations (1) by dz and solving the results for dx/dz and dy/dz, we obtain

$$\frac{dx}{dz} = \frac{P}{R} \qquad \frac{dy}{dz} = \frac{Q}{R} \qquad \text{or} \qquad \frac{dx}{P} = \frac{dy}{Q} = \frac{dz}{R} \tag{2}$$

where

$$P\!:\!Q\!:\!R = \begin{vmatrix} Q_1 & R_1 \\ Q_2 & R_2 \end{vmatrix} : \begin{vmatrix} R_1 & P_1 \\ R_2 & P_2 \end{vmatrix} : \begin{vmatrix} P_1 & Q_1 \\ P_2 & Q_2 \end{vmatrix} \tag{3}$$

From (2), we may write

$$\frac{l\,dx}{lP} = \frac{m\,dy}{mQ} = \frac{n\,dz}{nR} \tag{4}$$

and then apply the theorem that in a continued proportion the sum of the antecedents is to the sum of the consequents as any antecedent is to its consequent, to obtain

$$\frac{dx}{P} = \frac{dy}{Q} = \frac{dz}{R} = \frac{l\,dx + m\,dy + n\,dz}{lP + mQ + nR} \tag{5}$$

where l, m, and n are functions of x, y, and z at our disposal. (1) *It may happen that, by suitably choosing l, m, and n, we can find an equation in only two variables or an equation readily integrable.* In either case, an integration gives us a required relation. (2) *Again we may be able to choose* l, m, *and* n *so that* $lP + mQ + nR = 0$ *and so that* $l\,dx + m\,dy + n\,dz = du$, *where* u *is some function of* x, y, *and* z. In this case, since $lP + mQ + nR = 0$ and since the last fraction in (5) is finite, it follows that $l\,dx + m\,dy + n\,dz = 0$; that is, $du = 0$, and $u = c$ *(constant) is a required relation.* The examples will illustrate these methods of procedure and also an elimination method.

Example 1. Solve

$$\frac{dx}{xz} = \frac{dy}{yz} = \frac{2\,dz}{x+y} \tag{a}$$

Solution. Integration of the equation of the first two ratios gives

$$y = c_1x \tag{b}$$

Writing equations (5) for (*a*) with $l = 1$, $m = 1$, $n = 0$, we obtain

$$\frac{2\,dz}{x+y} = \frac{dx+dy}{z(x+y)} \tag{c}$$

Multiplying through by $z(x + y)$ and integrating, we have

$$x + y = z^2 + c_2 \tag{d}$$

Equations (*b*) and (*d*) constitute the solution.

It is instructive to note that, after finding equation (*b*), we could have substituted c_1x from (*b*) for y in $dx/xz = 2\,dz/(x + y)$ and integrated the resulting equation in x and z to obtain

$$x + c_1x = z^2 + c \qquad \text{or} \qquad x + y = z^2 + c$$

Also, we could have noticed that

$$P + Q - 2zR = xz + yz - xz - yz = 0$$

and therefore that

$$l\,dx + m\,dy + n\,dz = dx + dy - 2z\,dz = 0$$

Hence

$$x + y - z^2 = c$$

Example 2. Solve

$$dx = \frac{dy}{-5x + 12y - 5z} = \frac{dz}{x + 2y + z} \tag{a}$$

Solution. Write

$$\frac{dy}{dx} = -5x + 12y - 5z \qquad \frac{dz}{dx} = x + 2y + z \tag{b}$$

Differentiate the first of (*b*) to obtain

$$\frac{d^2y}{dx^2} = -5 + 12\frac{dy}{dx} - 5\frac{dz}{dx} \tag{c}$$

in (*c*) replace dz/dx by its value from the second equation of (*b*), in the result replace z by its value from the first of (*b*), and simplify to obtain

$$\frac{d^2y}{dx^2} - 13\frac{dy}{dx} + 22y = -5 \tag{d}$$

Solve (*d*) for y, and substitute the result in the first of (*b*) to obtain, after slight simplification

$$\begin{aligned} y &= c_1e^{2x} + c_2e^{11x} - \tfrac{5}{22} \\ 5z &= 10c_1e^{2x} + c_2e^{11x} - 5x - \tfrac{30}{11} \end{aligned} \tag{e}$$

Equations (*e*) constitute the solution of (*a*).

Exercises

Solve:

1. $\dfrac{dx}{y} = \dfrac{dy}{x} = \dfrac{dz}{z}$
2. $\dfrac{dx}{ayz} = \dfrac{dy}{bzx} = \dfrac{dz}{cxy}$
3. $y\,dx + (y - 2x)\,dy + yz\,dz = 0,\ (x - y)\,dy - yz\,dz = 0.$
4. $z\,dx + (y - x)\,dy + z\,dz = 0,\ 2z\,dx - (2x + y)\,dy - z\,dz = 0.$

Using the method of example 2, solve:

5. $dy/dx = xz,\ x\,dz/dx = x^2 + y - z.$
6. $dy = (10x - y + 5z)\,dx,\ dz = (2x - y + z)\,dx.$

Solve by any method:

7. $\dfrac{dx}{y} = \dfrac{dy}{x} = \dfrac{2\,dz}{l - z^2}$
8. $\dfrac{dx}{x^2 - y^2 - z^2} = \dfrac{dy}{2xy} = \dfrac{dz}{2xz}$
9. $\dfrac{dx}{ny - mz} = \dfrac{dy}{lz - nx} = \dfrac{dz}{mx - ly}$
10. $\dfrac{dx}{y} = \dfrac{dy}{x} = \dfrac{dz}{w} = \dfrac{dw}{z}$

11. $(x + z)\,dx + (x - z)\,dy - (x + z)\,dz = 0$, $x(x + z)\,dx + y(z - x)\,dy - z(x + z)\,dz = 0$.

12. $\dfrac{dx}{x(y - z)} = \dfrac{dy}{y(z - x)} = \dfrac{dz}{z(x - y)}$

13. $\dfrac{dx}{y - xz} = \dfrac{dy}{x + yz} = \dfrac{dz}{x^2 + y^2}$

14. $\dfrac{dx}{x^2 - y^2 - yz} = \dfrac{dy}{x^2 - y^2 - xz} = \dfrac{dz}{z(x - y)}$

15. $\dfrac{dx}{w - z} = \dfrac{dy}{w - z} = \dfrac{dz}{x + y - 2w} = \dfrac{dw}{2z - x - y}$

Using the method of example 2, solve the following systems of equations:

16. $dx = \dfrac{dy}{10x - y + 5z} = \dfrac{dz}{2x - y + z}$

17. $dx = \dfrac{dy}{2e^x + y + z} = \dfrac{dz}{4e^x + y + z}$

18. $dx = \dfrac{dy}{6x^2 - y + 3z} = \dfrac{dz}{2x^2 + y + z}$

19. $dx = \dfrac{dy}{a \sin 2x + y + 2z} = \dfrac{dz}{-a \cos 2x - y - z}$

20. State sufficient conditions that equations (1) have a solution.

100 Total differential equations

The number of variables in a system of differential equations may exceed the number of equations by more than one. We shall consider here only the special case of one first-order first-degree differential equation in three variables. Under certain conditions, such an equation is integrable; for example, the solution of

$$2x\,dx + 2y\,dy + 2z\,dz = 0 \tag{1}$$

evidently is

$$x^2 + y^2 + z^2 = c \tag{2}$$

and, in general, the solution of

$$\frac{\partial f(x,y,z)}{\partial x}\,dx + \frac{\partial f(x,y,z)}{\partial y}\,dy + \frac{\partial f(x,y,z)}{\partial z}\,dz = 0 \tag{3}$$

is

$$f(x,y,z) = c \tag{4}$$

The object of this section is to arrive at the condition of integrability of an equation

$$P\,dx + Q\,dy + R\,dz = 0 \tag{5}$$

where P, Q, and R are continuous functions of x, y, and z, possessing continuous first partial derivatives with respect to x, y, and z.

If (5) is integrable, there exists a function $\mu(x,y,z)$ such that the expression

$$\mu P\,dx + \mu Q\,dy + \mu R\,dz \tag{6}$$

is exactly the derivative of some function, say $f(x,y,z)$. Hence, comparing the left member of (3) with (6) and using subscripts to indicate partial differentiation, we get

$$f_x = \mu P \qquad f_y = \mu Q \qquad f_z = \mu R \tag{7}$$

Since $f_{xy} = f_{yx}$, we have, from (7),

$$\mu P_y + P\mu_y = \mu Q_x + Q\mu_x \tag{8}$$

Similarly

$$\mu Q_z + Q\mu_z = \mu R_y + R\mu_y \tag{9}$$
$$\mu R_x + R\mu_x = \mu P_z + P\mu_z \tag{10}$$

Multiply equation (8) through by R, (9) by P, and (10) by Q, add the results member by member, simplify this result, and rearrange terms to obtain

$$P(Q_z - R_y) + Q(R_x - P_z) + R(P_y - Q_x) = 0 \tag{11}$$

This equation states a necessary condition that (5) be integrable; we shall prove that it is also a sufficient one.

Since the equation $P\,dx + Q\,dy = 0$ is always integrable if z is considered constant, there will be no loss in generality in assuming that $P\,dx + Q\,dy$ is an exact equation with respect to x and y. The solution of

$$P\,dx + Q\,dy = 0 \tag{12}$$

considering z as constant, may now be written in the form

$$f(x,y,z) + \varphi(z) = 0 \tag{13}$$

where $Q(z)$* represents an arbitrary function of z. Since $P\,dx + Q\,dy$ has been assumed to be an exact differential, we may write

$$P_y = Q_x \qquad P = f_x \qquad Q = f_y \tag{14}$$

* Since z is constant for the integration, $Q(z)$ functions as the constant of integration.

Taking account of (14), we may write $P\,dx + Q\,dy + R\,dz = 0$ in the form

$$f_x\,dx + f_y\,dy + f_z\,dz + (R - f_z)\,dz = 0 \tag{15}$$

or

$$df + (R - f_z)\,dz = 0 \tag{16}$$

Equation (16) can be integrated if there exists a relation independent of x and y between $R - f_z$ and f; that is, $R - f_z$ is a function of f and z, say $\psi(f,z)$. In this case

$$(R - f_z)_x = \psi_f f_x \qquad (R - f_z)_y = \psi_f f_y \tag{17}$$

Eliminating ψ_f from (17) and performing indicated partial differentiations on $R - f_z$, we get the following necessary and sufficient condition that $\psi(f,z)$ exist:

$$f_x(R_y - f_{zy}) - f_y(R_x - f_{zx}) = 0 \tag{18}$$

From (14),

$$f_x = P \qquad f_y = Q \qquad f_{zy} = (f_y)_z = Q_z \qquad f_{zx} = (f_x)_z = P_z$$

Substituting these results in (18), we get

$$P(R_y - Q_z) - Q(R_x - P_z) = 0 \tag{19}$$

Since, from (14), $P_y - Q_x = 0$, it appears that (19) is the same as (11) with the signs changed. Hence, if (11) holds, (16) can be expressed in terms of two variables f and z and solved. Since (16) and (15) are the same equation, this solution with f replaced by its value in terms of x, y, and z will be the integral of (5).

The proof just given suggests the following rule:

RULE. *To integrate a total differential equation*

$$P\,dx + Q\,dy + R\,dz = 0 \tag{20}$$

which satisfies the condition (11), *first integrate the equation*

$$P\,dx + Q\,dy = 0$$

treating z as constant, to obtain

$$f(x,y,z) = C \tag{21}$$

Find df, and change the given equation (20) *to the form* (16) *with R and f expressed in terms of f and z, solve this equation, and in the result replace f by its equal in terms of x, y, and z.*

If (20) is not integrable, we may replace one of the variables by an arbitrary function of the other two and attempt to solve the resulting equation in two unknowns. This is not generally practicable. Particular solution curves may be obtained, however, by expressing one variable in terms of the other two, directly or indirectly, by some relation, and solving it with the given equation. Example 2 illustrates the latter procedure; it finds sets of solution curves in parallel planes.

Example 1. Solve

$$yz^2\,dx - xz^2\,dy - (2xyz + x^2)\,dz = 0 \tag{a}$$

Solution. Substitution from (a) in (11) shows that (a) is integrable. A solution of

$$yz^2\,dx - xz^2\,dy = 0 \tag{b}$$

got by considering z as constant, is $f(x,y,z) = y/x = c$. Hence, we write

$$f = \frac{y}{x} \qquad x^2\,df = x\,dy - y\,dx \tag{c}$$

Substituting, from (c), $x^2\,df$ for $x\,dy - y\,dx$ and xf for y in (a), we get

$$z^2x^2\,df + (2x^2fz + x^2)\,dz = 0 \tag{d}$$

Dividing (d) through by x^2 and solving the resulting linear equation by a method of §25, we get

$$fz^2 = c - z \tag{e}$$

In this, replace f by y/x from (c), change slightly, and obtain

$$yz^2 = x(c - z)$$

Example 2. Prove that

$$dx + dy + y\,dz = 0 \tag{a}$$

is not integrable, and then solve it simultaneously with

$$x - y + z = m \tag{b}$$

Solution. Equation (11) is not satisfied by (a). From (b) we get

$$dx - dy + dz = 0 \tag{c}$$

Solving (a) and (c) simultaneously for dx/dz and dy/dz, we find

$$\frac{dx}{y+1} = \frac{dy}{y-1} = \frac{dz}{-2} \tag{d}$$

The solution of (d) is

$$x = y - z + c_1 \qquad y = 1 + ce^{-z/2} \tag{e}$$

Equation (b) and either of the equations (e), considered simultaneously, constitute a solution.

Exercises

Apply the condition (11), §100, of integrability, and find the solutions of the differential equations numbered 1 to 11:

1. $x\,dy - y\,dx - 2x^2z\,dz = 0$
2. $(2y - z)\,dx + 4\,dy - 2\,dz = 0$
3. $x\,dy + y\,dx + (2xy - z)\,dz = 0$
4. $z(y\,dx - x\,dy) - (2xy + 3x^2z^3)\,dz = 0$
5. $2x\,dx + 2y\,dy + (x^2 + y^2 + e^{-z})\,dz = 0$
6. $2xy^2z\,dx + (1 + 2x^2yz)\,dy + x^2y^2\,dz = 0$
7. $(2x - yz)\,dx - xz\,dy + (x^2 - xyz - xy)\,dz = 0$
8. $z\,dx + z\,dy + [2(x + y) + \sin z]\,dz = 0$
9. $(1 + z)(x\,dy + y\,dx) + (2xy - 4z)\,dz = 0$
10. $\sin z\,dx + \cos z\,dy + [(x + y)\cos z + (x - y)\sin z - e^{-z}\tan z]\,dz = 0$
11. $yz\,dx - (xz + x^2)\,dy - (xy + x^2)\,dz = 0$

12. Find the condition that $dz = M\,dx + N\,dy$ be integrable: (*a*) if M and N represent functions of x, y, and z; (*b*) if M and N are functions of x and y only.

13. Solve simultaneously

$$z\,dx + x\,dy + y\,dz = 0$$
$$ax + 2by - (a + 2b)z = c$$

101 Geometrical interpretation

It is proved in calculus that the normal to a surface $f(x,y,z) = 0$ at a point (x,y,z) on the surface has as direction numbers $\partial f/\partial x$, $\partial f/\partial y$, $\partial f/\partial z$ evaluated at (x,y,z). Hence, if $f(x,y,z) = 0$ is a solution of equation (20) of §100, equations (7), §100, show that a line through (x,y,z) having direction numbers P, Q, R meets $f(x,y,z) = 0$ in (x,y,z) at right angles. Now, the curves having tangents with direction numbers P, Q, R are defined by

$$\frac{dx}{P} = \frac{dy}{Q} = \frac{dz}{R} \tag{1}$$

Hence *the curves defined by* (1) *meet at right angles the surface defined by the solution of an integrable equation* (20), §100. Figure 1 indicates the relation.

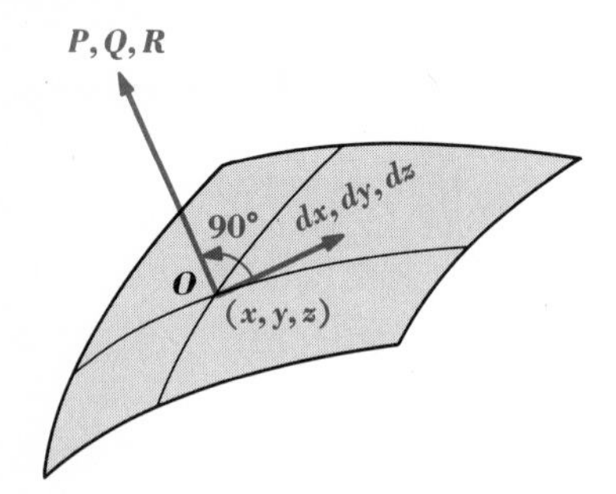

Figure 1

Example. Find the equation of the surfaces cutting at right angles the curves defined at points (x,y,z) by

$$\frac{dx}{2x} = \frac{dy}{2y} = \frac{dz}{z} \qquad (a)$$

Also, find the equations of the curves.

Solution. The equation of the required system of surfaces is

$$2x\,dx + 2y\,dy + z\,dz = 0 \qquad (b)$$

and, by inspection, we see that its solution is

$$2x^2 + 2y^2 + z^2 = c \qquad (c)$$

This represents a family of ellipsoids. The following solution of equations (a) is easily found:

$$x = c_1 y \qquad x = c_2 z^2 \qquad (d)$$

102 Fields of force in space

Integrable differential equations in three unknowns have important applications to fields of force in space. A force having magnitude r and direction making angles α, β, and γ with the X, Y, and Z axes, respectively, is denoted by the symbol $[X,Y,Z]$, where

$$X = r\cos\alpha \qquad Y = r\cos\beta \qquad Z = r\cos\gamma \qquad (1)$$

The sum of two forces $[X_1,Y_1,Z_1]$ and $[X_2,Y_2,Z_2]$ is defined by

$$[X_1,Y_1,Z_1] + [X_2,Y_2,Z_2] = [X_1 + X_2,\ Y_1 + Y_2,\ Z_1 + Z_2] \qquad (2)$$

Just as in §34, a field of force is defined by

$$[P(x,y,z),Q(x,y,z),R(x,y,z)]$$

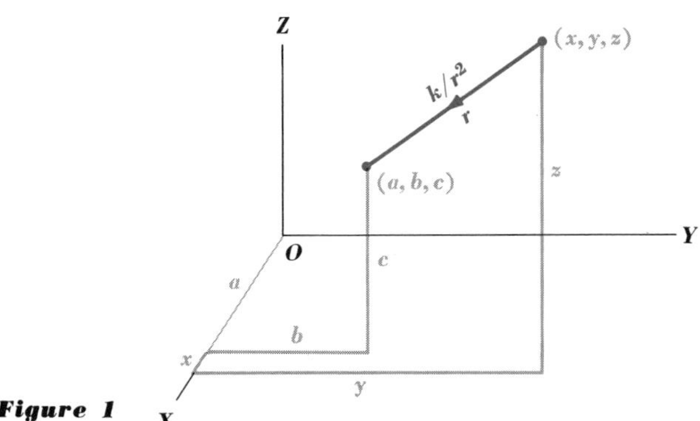

Figure 1

and the lines of force of the field are given by

$$\frac{dx}{P} = \frac{dy}{Q} = \frac{dz}{R} \tag{3}$$

Just as in §34 we assume that the work done on a body by a field of force $[X,Y,Z]$ is given by the line integral

$$W = \int F\,ds = \int (X\,dx + Y\,dy + Z\,dz) \tag{4}$$

evaluated along the path described by the moving particle. If

$$X\,dx + Y\,dy + Z\,dz = dQ(x,y,z) \tag{5}$$

then $Q(x,y,z)$ is called the potential of $[X,Y,Z]$ and field of force $[X,Y,Z]$ is called a conservative field of force. Note from (5) that if the potential function $Q(x,y,z)$ is known, then the corresponding field of force has components

$$X = \frac{\partial Q}{\partial x} \qquad Y = \frac{\partial Q}{\partial y} \qquad Z = \frac{\partial Q}{\partial Z} \tag{6}$$

The equipotential surfaces are represented by

$$Q(x,y,z) = c \tag{7}$$

for a conservative field of force.

Example. Find the equations of the lines of force and of the equipotential surfaces of a field of force in space due to a negative charge of electricity fixed at the point (a,b,c).

Solution. Figure 1 represents a negative charge of electricity fixed at point (a,b,c), point (x,y,z) at a distance r from (a,b,c), and the force at (x,y,z), due to the charge, of magnitude k/r^2 and directed from (x,y,z) toward (a,b,c). From Fig. 1, we get

$$r = \sqrt{(x-a)^2 + (y-b)^2 + (z-c)^2} \tag{a}$$

Here the work W done by the force is $\int -k/r^2\,dr = k/r + c$. Hence the potential function is

$$Q(x,y,z) = \frac{k}{r} + c \tag{b}$$

and the equipotential surfaces are given by

$$\frac{k}{r} = c \qquad \text{or} \qquad (x-a)^2 + (y-b)^2 + (z-c)^2 = \left(\frac{k}{c}\right)^2 = h^2 \tag{c}$$

The field of force, by (6), is represented by

$$[P,Q,R] = \left[-\frac{x-a}{r^2}, -\frac{y-b}{r^2}, -\frac{z-c}{r^2}\right] \tag{d}$$

Substituting P, Q, and R from (d) in (3), we get

$$\frac{dx}{x-a} = \frac{dy}{y-b} = \frac{dz}{z-c} \tag{e}$$

Solving (e) we obtain as equations of the lines of force

$$x - a = c_1(y-b) \qquad x - a = c_2(z-c) \tag{f}$$

These represent all straight lines through (a,b,c).

Exercises

Find the equations of the set of curves cutting at right angles the family of surfaces:

1. $x^2 + y^2 + z^2 = c^2$ **2.** $x^2 + 3y^2 - z^2 = c^2$
3. $2x^2 - 3y^2 - 4z^2 = c$ **4.** $xyz = c$

Find the equation of the family of surfaces orthogonal to the family of curves defined by:

5. $\dfrac{dx}{2x+y} = \dfrac{dy}{x+z} = \dfrac{dz}{y}$ **6.** $\dfrac{dx}{y} = \dfrac{dy}{-x} = \dfrac{dz}{2x^2z}$
7. $x\,dx = (3-z)\,dz = y\,dy$ **8.** $x^2 + 2y^2 + 3z^2 = a^2$, $z = b$.

(9–13) Find the fields of force and the equations of the lines of force and of the equipotential surfaces associated with the potential functions:

9. $Q = x^2 + y^2 + z^2$
10. $Q = k/r$, where $r = \sqrt{x^2 + y^2 + z^2}$.

11. $Q = a \ln x + b \ln y + c \ln z$ **12.** $Q = xyz$

13. $k/r_1 - k/r_2$; where $r_1 = \sqrt{(x+a)^2 + y^2 + z^2}$, $r_2 = \sqrt{(x-a)^2 + y^2 + z^2}$.

14. Find the equations of the lines of force and of the equipotential surfaces for the field of force $[x, y, 2x^2 + 2y^2 + 3z]$.

103 Review exercises

The material of this chapter consists essentially of existence theorems, §98, and total differential equations, §100.

1. Use §97 to write a system of differential equations of the first order equivalent to

$$(x-2)x\,D^3y - x\,D^2z + y + 2z = 0$$
$$(x-3)\,D^2y + (x-4)\,D^2z = 0$$

If (a,b,c) represents a point through which a solution curve of this system is to pass, for what values of a does Theorem I, §98, fail to guarantee a solution?

2. For the system

$$(x-2)(y+2)\frac{dy}{dx} = (3x-2z)y$$

$$(z^2 + 2x^2)\frac{dz}{dx} = y^{1/2} + z^{1/2}$$

state, for each of the following regions, whether Theorem I, §98, guarantees a solution curve through each of its interior points: (*a*) $|x-4| \leqq 1$, $|y-6| \leqq 1$, $|z| > 0.1$; (*b*) $|x| \leqq 1$, $|y+2| \leqq 1$, $|z| > 0$; (*c*) $|x| \leqq 1$, $|y+4| \leqq 1$, $|z| \leqq 1$; (*d*) $|x - 2{,}003| \leqq 2{,}000$, $|y - 2{,}003| \leqq 2{,}000$, $|z| \geqq 0.01$.

3. For which of the following equations does an integrating factor exist?

(*a*) $(0.4x - 0.2y)\,dx + (0.1y + z)\,dy + (10z + 2x)\,dz = 0$

(*b*) $(2x - 2y)\,dx + (2y - 2x)\,dy + (3x^2 - 6xy + 3y^2 - 10\cos z)\,dz = 0$

(*c*) $(3y + z^3)\,dx + xz\,dy + (xz^3 - 2xy)\,dz = 0$

(*d*) $2zy\,dx + 3xz\,dy - xy\,dz = 0$

4. Find the equations of the lines of force and the equation of the equipotential surfaces for a field of force having $Q(x,y,z) = k(x^2 + y^2 + z^2)^{-n}$ as potential function.

5. Solve $(zy + z^4)\,dx + xz\,dy + (xz^3 - 2xy)\,dz = 0$.

6. Solve $dy/dx = x + z$, $dz/dx = x + y$. Use the method of example 2, §99.

7. Solve $dx/(y - z) = dy/(z - x) = dz/(x - y)$.

8. Solve $dy/dx = 2x + y + z$, $dz/dx = 4x - y + z$.

Find the equations of the lines of force and of the equipotential surfaces for the fields of force:

9. $[x + y, x - y, z]$ **10.** $[3zx, zy, 3x^2 + 2y^2 - z^2]$

11. For which $Q(x,y,z) = k/r_1 + k/r_2$; where

$$r_1 = \sqrt{(x + a)^2 + y^2 + z^2} \qquad r_2 = \sqrt{(x - a)^2 + y^2 + z^2}$$

12. In exercise 11 with $k/r_1 + k/r_2$ replaced by $k/r_1^3 + k/r_2^3$.

12

Solution by Series

104 Introduction

The preceding chapters have been concerned mainly with solving special types of differential equations. This chapter is concerned with a general method called integration in series. This method may be applied to solve a large variety of differential equations, and it generally gives the complete solution.

It will be convenient to use the summation notation, which is expressed by

$$\sum_{n=p}^{q} Q(n) = Q(p) + Q(p+1) + Q(p+2) + \cdots + Q(q) \qquad (1)$$

where $q - p$ is a positive integer, and the series is endless if q is infinite. By expanding each member we can easily check that

$$\sum_{n=0}^{\infty} Q(n)a_n x^{n+p} = \sum_{n=p}^{\infty} Q(n-p)a_{n-p}x^n \qquad (2)$$

where p is an integer, positive or negative. For example, by using (2) we obtain

$$\sum_{n=0}^{\infty} (n^2 a_n x^{n-2} + n a_n x^{n+5}) = \sum_{n=-2}^{\infty} (n+2)^2 a_{n+2} x^n + \sum_{n=5}^{\infty} (n-5) a_{n-5} x^n \quad (3)$$

In dealing with functions defined by infinite series, the question of convergence is highly important. Hence an outstanding test should be mentioned.

Cauchy's ratio test states that a series of constants $\sum_{n=0}^{\infty} A_n$ is absolutely convergent if

$$\lim_{n\to\infty} \left| \frac{A_{n+1}}{A_n} \right| < 1 \quad (4)$$

and is divergent if

$$\lim_{n\to\infty} \left| \frac{A_{n+1}}{A_n} \right| > 1 \quad (5)$$

Another important feature is based on the following statement. If for all values of x in a finite interval

$$\sum_{n=0}^{\infty} a_n x^n = \sum_{n=0}^{\infty} b_n x^n \qquad \text{then } a_n = b_n \qquad n = 0, 1, \ldots, \infty \quad (6)$$

Thus if $\sum_{n=0}^{\infty} n c_n x^n = c_0 x + c_1 x^2 + c_2 x^3 + \cdots + c_n x^{n+1} + \cdots$, then $c_1 = c_0$, $2c_2 = c_1$, $3c_3 = c_2$, . . . , $nc_n = c_{n-1}$, . . . or $c_1 = c_0$, $c_2 = c_0/2!$, $c_3 = c_0/3!$, . . . , $c_n = c_0/n!$, ...*

Exercises

1. If, for every value of x, $\sum_{n=5}^{\infty} \frac{5x^n}{n!} = \sum_{m=0}^{\infty} a_m x^m$, find the values of the constants a_m.

* By definition, $0! = 1$, $1! = 1$, $n! = 1 \cdot 2 \cdot 3 \cdots n$.

2. Show that $\sum_{n=0}^{\infty} \frac{x^n 2^n}{n!} = 1 + 2x + 2x^2 + \sum_{n=0}^{\infty} \frac{x^{n+3}2^{n+3}}{(n+3)!}$.

3. Using (2), show that $\sum_{r=0}^{\infty} \frac{rx^r}{r!} = \sum_{r=1}^{\infty} \frac{x^r}{(r-1)!} = \sum_{s=0}^{\infty} \frac{x^{s+1}}{s!}$.

4. Using (2), show that $\sum_{r=6}^{\infty} \frac{(x+3)x^r}{r!} = \sum_{s=0}^{\infty} \frac{(x^{s+7} + 3x^{s+6})}{(s+6)!}$.

5. Using (4) and the alternating sign test, find the interval of convergence of

$$1 + \frac{1}{2}\frac{x}{3} + \frac{1}{3}\left(\frac{x}{3}\right)^2 + \frac{1}{4}\left(\frac{x}{3}\right)^3 + \cdots + \left(\frac{1}{n+1}\right)\left(\frac{x}{3}\right)^n + \cdots$$

Find the values of the a's:

6. $\sum_{n=0}^{\infty} (2n+1)a_{2n+3}x^{2n+1} = x + a_3x^3 + a_5x^5 + \cdots + a_{2n+1}x^{2n+1} + \cdots$

7. $\sum_{n=0}^{\infty} 2(n+1)^2 a_{2n+2}x^{2n} = 1 + a_2x^2 + a_4x^4 + \cdots + a_{2n}x^{2n} + \cdots$

8. $\sum_{n=0}^{\infty} (n+3)(n+2)a_{n+2}x^{n+1} = x + a_2x^2 + a_3x^3 + \cdots$

9. Use (4), (5), and inspection to show that $\sum_{n=0}^{\infty} (n+1)x^n$ converges if $-1 < x < 1$ and diverges if $|x| \geqq 1$.

10. Show that

$$\sum_{n=0}^{\infty} [n(n-1)a_n x^{n-2} - a_n x^{n+1}] = \sum_{n=-3}^{\infty} (n+3)(n+2)a_{n+3}x^{n+1} - \sum_{n=0}^{\infty} a_n x^{n+1}$$

11. Use (4) in determining the interval of absolute convergence of each series:

(*a*) $\sum_{n=0}^{\infty} 9^{-n}x^{2n}$ (*b*) $\sum_{n=0}^{\infty} \frac{4^{-n}(x-3)^{2n}}{n+1}$ (*c*) $\sum_{n=1}^{\infty} \frac{(-1)^{n+1}x^{2n-1}}{(2n-1)!}$

105 Representation of functions by infinite series*

The following theorems are of basic importance in the use of infinite series to represent functions.

THEOREM I. If a series of continuous functions is uniformly convergent in an interval, its sum is a continuous function.

An infinite power series with sum $f(x)$ has the form

$$f(x) = c_0 + c_1(x - a) + c_2(x - a)^2 + \cdots = \sum_{n=0}^{\infty} c_n(x - a)^n \tag{1}$$

Cauchy's ratio test with others enables us to find an interval of absolute convergence of a power series.

THEOREM II. If a power series is absolutely convergent in a first interval, then it is also uniformly convergent in any interval interior to the first.

Since $(x - a)^n$ is continuous, a power series which converges uniformly in an interval has a continuous sum by Theorem I. Using this fact and others we could deduce the following theorem.

THEOREM III. If a power series (1) is absolutely and uniformly convergent in an interval I, its sum $f(x)$ is a continuous function in I, and

$$\frac{d^k f(x)}{dx^k} = \sum_{n=0}^{\infty} \frac{d^k}{dx^k} [c_n(x - a)^n]$$

$$\int_a^x f(t)\, dt = \sum_{n=0}^{\infty} \int_a^x c_n(t - a)^n\, dt$$

where a and x are numbers in I. Also the sum, difference, and product of two functions will be represented by the algebraic sum, difference, or product of the corresponding series inside I. A like statement applies for quotients of series, but the interval of convergence depends upon the values of x for which the denominator vanishes.

These theorems may be used to show that a derived solution of a differential equation satisfies it.

* For proofs consult books on advanced calculus. For example, see R. Courant, "Differential and Integral Calculus," vol. I, pp. 391–404, 1937.

106 Solving differential equations by power series

If a linear equation of the type

$$P_0(x)D^ny + P_1(x)\,D^{n-1}y + \cdots + P_{n-1}(x)\,Dy + P_n(x)y = 0 \tag{1}$$

has the functions $P_0(x)$, $P_1(x)$, . . . , $P_n(x)$ continuous, single-valued, and bounded in an interval I, $b_1 \leqq x \leqq b_2$, and if $P_0(x_0) \neq 0$ where x_0 is in I, then it possesses n linearly independent solutions each of which may be represented in a Taylor's series convergent in a certain interval $x_0 - h < x < x_0 + h$. For simplicity we shall deal mainly with series in powers of x, but the same general plan may be used to obtain series solutions in successive integral powers of $x - x_0$. Modified series will be used in §§108 and 109 to obtain solutions of equations of type (1) when $P_0(0) = 0$.

Example 1. Solve the following differential equation for y as a power series in x:

$$\frac{d^2y}{dx^2} - xy = 0 \tag{a}$$

Solution. Assume that there exists a solution having the form

$$y = c_0 + c_1x + c_2x^2 + c_3x^3 + c_4x^4 + c_5x^5 + \cdots + c_nx^n + \cdots \tag{b}$$

or, using the summation notation

$$y = \sum_{n=0}^{\infty} c_nx^n \tag{c}$$

where the c's are constants to be determined. Substitute y from (b) in (a) and arrange the result in order of ascending powers of x to obtain

$$\begin{aligned} &2c_2 + (3 \cdot 2c_3 - c_0)x + (4 \cdot 3c_4 - c_1)x^2 + (5 \cdot 4c_5 - c_2)x^3 \\ &+ (6 \cdot 5c_6 - c_3)x^4 + \cdots + [n(n-1)c_n - c_{n-3}]x^{n-2} + \cdots = 0 \end{aligned} \tag{d}$$

Equation (d) is an identity. Hence, equating coefficients of powers of x to zero, we get

$$\begin{aligned} &2c_2 = 0,\ 3 \cdot 2c_3 - c_0 = 0,\ 4 \cdot 3c_4 - c_1 = 0,\ 5 \cdot 4c_5 - c_2 = 0, \\ &\qquad 6 \cdot 5c_6 - c_3 = 0,\ \ldots,\ n(n-1)c_n - c_{n-3} = 0^* \end{aligned} \tag{e}$$

* Formulas, such as $c_n = c_{n-3}/[n(n-1)]$, to be used with various integral values of n in finding equations involving the c's, are called recurrence formulas.

The solutions of equations (e) for the c's are

$$c_2 = 0,\ c_3 = \frac{c_0}{2 \cdot 3},\ c_4 = \frac{c_1}{3 \cdot 4},\ c_5 = 0,$$
$$c_6 = \frac{c_3}{5 \cdot 6} = \frac{c_0}{2 \cdot 3 \cdot 5 \cdot 6},\ \cdots,\ c_n = \frac{c_{n-3}}{n(n-1)} \qquad (f)$$

We may write the first six terms of the solution by replacing the values of c_2, c_3, c_4, and c_5 of (b) by their values from (f) to obtain

$$y = c_0 + c_1 x + \frac{c_0}{2 \cdot 3} x^3 + \frac{c_1}{3 \cdot 4} x^4 + \frac{c_0}{2 \cdot 3 \cdot 5 \cdot 6} x^6 + \frac{c_1}{3 \cdot 4 \cdot 6 \cdot 7} x^7 + \cdots \qquad (g)$$

This is the important part of the solution for values of x near zero.

However, a law for writing any number of terms is desired. By rearranging the values for the c's found in (f) and by using the last equation of (f) successively for different values of n, we find

$$\begin{array}{llll} c_0 = c_0 & c_1 = c_1 & c_2 = 0 & c_3 = \dfrac{c_0}{2 \cdot 3} \\ c_4 = \dfrac{c_1}{3 \cdot 4} & c_5 = 0 & c_6 = \dfrac{1 \cdot 4 c_0}{6!} & c_7 = \dfrac{2 \cdot 5 c_1}{7!} \\ c_8 = 0 & c_{3n} = \dfrac{1 \cdot 4 \cdot 7 \cdots (3n-2) c_0}{(3n)!} & & \\ c_{3n+1} = \dfrac{2 \cdot 5 \cdot 8 \cdots (3n-1) c_1}{(3n+1)!} & & c_{3n+2} = 0 & \end{array} \qquad (h)$$

Substituting the values of the c's from (h) in (b), we obtain

$$y = c_0 \left[1 + \frac{x^3}{3!} + \cdots + \frac{1 \cdot 4 \cdot 7 \cdots (3n-2)}{(3n)!} x^{3n} + \cdots \right] + c_1 \left[x + \frac{2x^4}{4!} + \cdots + \frac{2 \cdot 5 \cdot 8 \cdots (3n-1) x^{3n+1}}{(3n+1)!} + \cdots \right] \qquad (i)$$

or, using the summation notation

$$y = c_0 + c_1 x + c_0 \sum_{n=1}^{\infty} \frac{1 \cdot 4 \cdot 7 \cdots (3n-2)}{(3n)!} x^{3n} + c_1 \sum_{n=1}^{\infty} \frac{2 \cdot 5 \cdot 8 \cdots (3n-1) x^{3n+1}}{(3n+1)!} \qquad (j)$$

By applying Cauchy's ratio test we find that the two series of (i) are absolutely convergent for all values of x. Hence, in accordance with §105, each series represents, for all values of x, a function of x, their sum represents a function of x, and their derivatives may be found by using term-by-term differentiation. We can see by the method of derivation of (i), or by direct trial, that (i) satisfies (a). Also, we know by Theorem II, §98, that there is a unique solution of (a) satisfying the condition $y = c_0$, $dy/dx = c_1$ when $x = 0$, and since (i) satisfies these conditions, it is that unique solution. Like arguments could be used to validate all the other solutions of this chapter.

Alternative solution. A shorter solution results from using the summation notation. Substituting $y = \sum_{n=0}^{\infty} c_n x^n$ in (a), we get directly

$$\frac{d^2y}{dx^2} - xy = \sum_{n=0}^{\infty} [n(n-1)c_n x^{n-2} - c_n x^{n+1}]$$

$$= \sum_{n=-3}^{\infty} (n+3)(n+2)c_{n+3}x^{n+1} - \sum_{n=0}^{\infty} c_n x^{n+1} = 0 \qquad (k)$$

Substituting -3, -2, and -1 for n in the first sum of (k) we may represent the coefficients of powers of x in (k) by

$$0,\ 0,\ 2c_2 = 0,\ c_{n+3} = \frac{c_n}{(n+2)(n+3)} \qquad n = 0, \ldots, \infty \qquad (l)$$

Let c_0 be an arbitrary constant, and then use (l) with $n = 0, 3, 6, 9, \ldots$ to get

$$c_3 = \frac{c_0}{3!} \qquad c_6 = \frac{c_3}{5 \cdot 6} = \frac{c_0}{2 \cdot 3 \cdot 5 \cdot 6} = \frac{4c_0}{6!} \qquad (m)$$

Continuing in this manner we obtain

$$c_{3n} = c_0 \left[\frac{4 \cdot 7 \cdot 10 \cdots (3n-2)}{(3n)!}\right] \qquad (n)$$

Substituting the values of the c's in (c) we get the first solution of (j). Similarly using c_1 as any constant and then (l) successively with $n = 1, 4, 7, \ldots, 3k+1$ we obtain the second solution of (j). Also from the fact that $c_2 = 0$ from (l), we use (l) with $n = 2, 5, 8, \ldots$ to obtain $c_{3k+2} = 0$, $k = 0, 1, 2, \ldots$. The sum of the two solutions is a solution by §46.

Example 2. Find, by using a Maclaurin series, the solution of

$$\frac{d^2y}{dx^2} - x\frac{dy}{dx} - 2y = 0 \qquad (a)$$

for which $y' = 0$, $y = 1$ when $x = 0$.

Solution. Using the notation $y^{(k)} = (d^k y)/dx^k$, we obtain from (a),

$$
\begin{aligned}
y^{(2)} &= xy^{(1)} + 2y \\
y^{(3)} &= xy^{(2)} + 3y^{(1)} \\
y^{(4)} &= xy^{(3)} + 4y^{(2)} \\
&\cdots\cdots\cdots\cdots \\
y^{(n)} &= xy^{(n-1)} + ny^{(n-2)}
\end{aligned} \tag{b}
$$

In (b), substitute 0 for x, 1 for y, 0 for $y^{(1)}$, and solve the resulting equations for $y_0^{(2)}, y_0^{(3)}, \ldots, y_0^{(2n-1)}, y_0^{(2n)}$, to get

$$
\begin{aligned}
&y_0^{(2)} = 2,\ y_0^{(3)} = 0,\ y_0^{(4)} = 2\cdot 4,\ y_0^{(5)} = 0, \\
&\qquad y_0^{(6)} = 2\cdot 4\cdot 6,\ y_0^{(7)} = 0,\ y_0^{(8)} = 2\cdot 4\cdot 6\cdot 8, \\
&\qquad\qquad \ldots,\ y_0^{(2n-1)} = 0,\ y_0^{(2n)} = 2\cdot 4\cdot 6\cdots 2n
\end{aligned} \tag{c}
$$

Observing that $2\cdot 4\cdot 6\cdots 2n = 2^n n!$ and substituting the values (c) in Maclaurin's series,

$$
y = y_0 + y_0^{(1)}x + \frac{1}{2!}y_0^{(2)}x^2 + \frac{1}{3!}y_0^{(3)}x^3 + \cdots + \frac{1}{n!}y_0^{(n)}x^n + \cdots \tag{d}
$$

we get after slight simplification

$$
y = 1 + \frac{x^2}{1} + \frac{x^4}{1\cdot 3} + \frac{x^6}{1\cdot 3\cdot 5} + \cdots + \frac{x^{2n}}{1\cdot 3\cdot 5\cdots(2n-1)} + \cdots = \sum_{n=1}^{\infty} \frac{n!2^n}{(2n)!}x^{2n}
$$

Remark. Most of the problems in the following exercises can be solved by using Maclaurin's series or Taylor's series.

Exercises

Solve the differential equations 1 to 6 by using infinite series of the form (b) of example 1. D means d/dx.

1. $Dy - y = 0$
2. $Dy - 2xy = 0$
3. $D^2y - x^2y = 0$
4. $(x^2 + 1)\,D^2y + 6x\,Dy + 6y = 0$
5. $(x^2 - 1)\,D^2y - 6y = 0$
6. $x\,D^2y - Dy - x^3y = 0$

If, in searching for a recurrence formula, we obtain a relation such as $c_n(n)(n-3)(n-7) = 0$, the values of n are fixed and each one gives a term. From the illustrated relation we would write $y = c_0x^0 + c_3x^3 + c_7x^7$ as the required solution. Use recurrence formulas to solve exercises 7, 8, and 9:

7. $x^2\, D^2y - 12y = 0$ **8.** $x\, Dy - 9y = 0$

9. $x^2\, D^2y - 4x\, Dy + 6y = 0$

10. Find a solution of $(x^2 - 2x)\, D^2y + 6(x-1)\, Dy + 6y = 0$ in a series having the form $y = c_0 + c_1(x-1) + c_2(x-1)^2 + \cdots + c_n(x-1)^n + \cdots$.

11. Find a solution of $(x^2 + 2x)\, D^2y + 8(x+1)\, Dy + 12y = 0$ in a series having the form $y = c_0 + c_1(x+1) + c_2(x+1)^2 + \cdots + c_n(x+1)^n \cdots$.

Use series proceeding in powers of x to find the first five terms in the solutions of exercises 12 and 13. Avoid the method using summation symbols.

12. $(x-1)\, D^2y + y = 0$ **13.** $(x^2+1)\, D^2y + x\, Dy + xy = 0$

***14.** The theory of the oscillator in quantum mechanics uses those solutions of the equation

$$-\frac{d^2u}{dx^2} + x^2u = (2n+1)u \qquad n \text{ constant} \qquad (a)$$

that remain finite as x increases without limit. Find these solutions.

First show that $u = e^{-x^2/2}$ satisfies (a) when $n = 0$. Then let

$$u = ve^{-x^2/2}$$

in (a), and deduce the equation

$$\frac{d^2v}{dx^2} - 2x\frac{dv}{dx} + 2nv = 0 \qquad (b)$$

Next solve (b) to obtain the solutions v_1 and v_2 as infinite series. Now show that the solutions of (a)

$$u = v_1e^{-x^2/2} \qquad u = v_2e^{-x^2/2} \qquad (c)$$

satisfy the required condition when and only when n is zero or a positive integer.

107 Series solutions in descending powers of x

A series proceeding in descending powers of x has the form

$$a_0 + a_1x^{-1} + a_2x^{-2} + \cdots + a_nx^{-n} + \cdots = \sum_{n=0}^{\infty} a_nx^{-n} \tag{1}$$

Observe that

$$\sum_{n=0}^{\infty} f(n)a_nx^{-n-p} = \sum_{n=p}^{\infty} f(n-p)a_{n-p}x^{-n} \tag{2}$$

and this is true whether p is a positive or a negative integer. Generally when a linear differential equation has polynomial coefficients and a solution of the type (*b*), §106, converging in $|x| < r$, it also has a solution of the type (1) converging in $|x| > r$. Also a solution in the form (1) may exist in other cases. The procedure of finding in the form (1) a solution of a linear differential equation is similar to that used in §106.

Example. Find solutions in the form (1) of

$$(x^2 + 1)\, D^2y + 6x\, Dy + 6y = 0 \tag{a}$$

Solution. Substitute (1) for y in (*a*), and collect coefficients of like terms to obtain

$$\Sigma\{[-n(-n-1) - 6n + 6]a_nx^{-n} + n(n+1)a_nx^{-n-2}\} = 0 \tag{b}$$

Simplifying (*b*) and using (2) we get

$$\left[\sum_{n=0}^{\infty} (n-2)(n-3)a_n + \sum_{n=2}^{\infty} (n-2)(n-1)a_{n-2}\right] x^{-n} = 0 \tag{c}$$

In (*c*) substitute $n = 0, 1, 2,$ and 3, express the rest of the terms by an infinite sum, and obtain

$$6a_0 + 2a_1x^{-1} + 2a_1x^{-3} + \sum_{n=4}^{\infty} [(n-2)(n-3)a_n + (n-2)(n-1)a_{n-2}]x^{-n} = 0 \tag{d}$$

From (*d*) derive

$$a_0 = 0,\ a_1 = 0,\ a_n = -\frac{n-1}{n-3}a_{n-2} \qquad \text{when } n \geqq 4 \tag{e}$$

Using the recurrence formula in (*e*) we obtain

$$a_2 = a_2,\ a_4 = -3a_2,\ a_6 = 3 \cdot \tfrac{5}{3}a_2,\ \ldots,\ a_{2n} = (-1)^{n+1}(2n-1)a_2 \tag{f}$$

$$a_3 = a_3,\ a_5 = -2a_3,\ a_7 = 2 \cdot \tfrac{6}{4}a_3,\ \ldots,\ a_{2n+1} = (-1)^{n+1}na_3 \tag{g}$$

and the corresponding solutions are

$$y_1 = a_2[x^{-2} - 3x^{-4} + 5x^{-6} - 7x^{-8} + \cdots + (-1)^{n+1}(2n - 1)x^{2n} + \cdots] \quad (h)$$

$$y_2 = a_3[x^{-3} - 2x^{-5} + 3x^{-7} - 4x^{-9} + \cdots + (-1)^{n+1}nx^{2n+1} + \cdots] \quad (i)$$

These satisfy (a) and converge absolutely in $|x| > 1 \cdot y = y_1 + y_2$ is a general solution.

Exercises

1. Substitute (1) for y in $x^2Dy + y = 0$, derive the recurrence formula $c_{n+1} = [1/(n + 1)]c_n$, and obtain the solution $y = c_0[1 + x^{-1} + x^{-2}/2! + x^{-3}/3! + \cdots]$.

Derive one solution of each equation in descending powers of x.

2. $x^3 D^2y - y = 0$ **3.** $x^4 D^2y - y = 0$

4. Derive two solutions of $(x^2 - 1) D^2y + x Dy - y = 0$ in positive powers of x and one in descending powers of x.

5. Derive two solutions of $x^4 D^2y + 2x^3 Dy + y = 0$ in descending powers of x.

108 Solution involving a more general type of series

A function $f(x)$ is analytic at a value a of x if it can be expanded in an infinite series having the form

$$f(x) = a_0 + a_1(x - a) + a_2(x - a)^2 + \cdots + a_n(x - a)^n + \cdots \quad (1)$$

which converges absolutely in an interval $a - r < x < a + r$, where $r > 0$.

Series solutions in powers of $x - a$ of a differential equation

$$D^2y + p(x)\, Dy + q(x)y = 0 \quad (2)$$

exist if D means d/dx and $p(x)$ and $q(x)$ are analytic at $x = a$. They can be found by the method used in §106.

This section and the next will deal mainly with differential equations having the form

$$D^2y + \frac{p(x)}{x} Dy + \frac{q(x)}{x^2} y = 0 \tag{3}$$

where at least one of the functions $p(x)/x$, $q(x)/x^2$ is not analytic at $x = 0$. In this section, we shall derive solutions of equations of the type (3) by using series having the form

$$y = x^m(c_0 + c_1x + c_2x^2 + \cdots + c_nx^n + \cdots) \tag{4}$$

where m is a real number. The next section will deal with special cases of (3) in which series of the form (4) play a prominent role.

If y in (3), cleared of fractions, is replaced by x^m, the equation obtained by equating to zero the coefficient of the lowest (or highest, in case a series of descending powers is used) power of x in the result is called the indicial equation

Example. Solve $2x^2 D^2y + 3x\, Dy - (x^2 + 1)y = 0$. (a)

Solution. Replacing y by x^m in (a) we get after slight simplification

$$(2m^2 + m - 1)x^m - x^{m+2}$$

The indicial equation is

$$2m^2 + m - 1 = 0 \qquad \text{or} \qquad m = \tfrac{1}{2} \qquad m = -1 \tag{b}$$

When we begin the regular procedure, we shall meet the term $(2m^2 + m - 1)x^m$; and, to cause it to vanish, it will be necessary to set $2m^2 + m - 1 = 0$, find $m = \frac{1}{2}$ and -1, and take one of these roots as the value of m. A solution will be associated with each root.

Substituting y from (4) in (a) we get after some transformation

$$c_0[2m^2 + m - 1]x^m + c_1[2(m + 1)^2 + (m + 1) - 1]^{m+1} + \sum_{n=2}^{\infty} \{[2(n + m)^2 + (n + m) - 1]c_n - c_{n-2}\}x^{n+m} = 0 \tag{c}$$

The number m, being at our disposal, is chosen to make the first term of (c) vanish. Accordingly we use m from (b). To dispose of the second part of (c), take $c_1 = 0$. From the third part of (c), obtain

$$c_n = \frac{c_{n-2}}{(n + m + 1)[2(n + m) - 1]} \tag{d}$$

Substituting -1 from (b) for m in (c), obtain

$$c_n = \frac{c_{n-2}}{n(2n - 3)} \qquad n \geqq 2 \tag{e}$$

Now using $m = -1$, $c_0 = c_0$, $c_1 = 0$, and (e) with $n = 2, 4, 6, \ldots$

in (4) we obtain

$$y_1 = c_0 x^{-1}\left(1 + \frac{x^2}{2 \cdot 1} + \frac{x^4}{2 \cdot 4 \cdot 1 \cdot 5} + \frac{x^6}{2 \cdot 4 \cdot 6 \cdot 1 \cdot 5 \cdot 9} + \cdots\right) \qquad (f)$$

Similarly, substituting $\frac{1}{2}$ from (b) for m in (d), obtain

$$c_n = \frac{c_{n-2}}{(2n+3)n} \qquad n \geqq 2 \qquad (g)$$

Then, using in (4) $m = \frac{1}{2}$, $c_0 = b$, $c_1 = 0$, and (g) with $n = 2, 4, 6, \ldots$, we get

$$y_2 = bx^{1/2}\left(1 + \frac{x^2}{2 \cdot 7} + \frac{x^4}{2 \cdot 4 \cdot 7 \cdot 11} + \frac{x^6}{2 \cdot 4 \cdot 6 \cdot 7 \cdot 11 \cdot 15} + \cdots\right) \qquad (h)$$

The general solution is

$$y = y_1 + y_2 \qquad (i)$$

When the degree in m of the indicial equation is less than the order of the equation to be solved, a solution may be found by using a series of descending powers of x, having the form

$$y = x^m(c_0 + c_1x^{-1} + c_2x^{-2} + \cdots) = \sum_{n=0}^{\infty} c_n x^{m-n} \qquad (5)$$

When the roots of the indicial equation of (2) are equal, and often when the roots differ by an integer, the method used above gives only one solution. In this case the technique of the next section is used.

Exercises

Solve the following differential equations:

1. $2x\, D^2y + Dy - 2y = 0$
2. $(x^3 - x)\, D^2y + (8x^2 - 2)\, Dy + 12xy = 0$
3. $x\, D^2y + 3\, Dy - x^2y = 0$
4. $x^2\, D^2y + (x + 2x^2)\, Dy - 4y = 0$
5. $x^2\, D^2y - x^2\, Dy + (x - 2)y = 0$
6. $(x^3 - x)\, D^2y + (4x^2 - 2)\, Dy + 2xy = 0$

7. $(x - x^2)\, D^2y - (x + 1)\, Dy + y = 0$

8. $x^4\, D^2y + x\, Dy + y = 0$. *Hint:* Let $y = x^m(c_0 + c_1x^{-1} + c_2x^{-2} + \cdots)$.

9. $(x^4 - x^2)\, D^2y + 2x\, Dy - (2 + 2x^2)y = 0$. Find two solutions in positive powers of x and a solution in descending powers of x.

10. $x^4\, D^2y + x\, Dy - 2y = 0$. Find two solutions in descending powers of x.

11. $(x^4 - x^2)\, D^2y - (2x^3 - 3x)\, Dy + (2x^2 - 3)y = 0$

★12. $x^3\, D^3y + 6x^2\, D^2y + 6x\, Dy + a^3x^3y = 0$

★13. Find a particular solution of $x^2\, D^2y + x\, Dy - (1 + x^2)y = x^{3/2}$.

109 Indicial equation has roots differing by an integer

While the statements made below refer to second-order equations at $x = 0$, they may easily be extended to apply for linear equations of any order at $x = a$. When two roots of the indicial equation for an equation of the type (3), §108, are equal, the process of §108 fails to give the general solution at $x = 0$ and the same thing may be true when two roots differ by an integer. The following example will illustrate the procedure to be used in solving such equations:

Example. Solve

$$x^2D^2y + 2x\, Dy - xy = 0 \tag{a}$$

Solution. In the given equation substitute

$$y = x^m(c_0 + c_1x + c_2x^2 + \cdots + c_nx^n + \cdots) \tag{b}$$

Collect the coefficients of like terms, and simplify to obtain

$$c_0(m^2 + m) + \sum_{n=1}^{\infty} [c_n(n + m)(n + m + 1) - c_{n-1}]x^n = 0 \tag{c}$$

Consider m as any real number. Equating the coefficient of x^n to 0 in (c) and solving for c_n, we obtain

$$c_n = \frac{c_{n-1}}{(m + n)(m + n + 1)} \qquad n \geqq 1 \tag{d}$$

Solve (d) for the c's, take $c_0 = c(m + 1)$ and substitute the results in (*b*) to obtain

$$Y = cx^m\left[m + 1 + \frac{x\cancel{(m+1)}}{\cancel{(m+1)}(m+2)} + \frac{x^2}{(m+2)^2(m+3)} + \cdots\right]$$

$$= cx^m\left[m + 1 + \sum_{n=1}^{\infty} \frac{x^n}{(m+2)^2(m+3)^2 \cdots (m+n)^2(m+n+1)}\right] \quad (e)$$

Substituting Y for y in (a) and noting that the result is (c) with the sigma sum giving $-cx^m(m+1)$, we get

$$x^2 D^2Y + 2x\, DY - xY = cm(m+1)^2x^m \quad (f)$$

Taking $m = 0$ in (e) and (f) we see that

$$Y_{m=0} = y_1 = x^0\left[1 + \sum_{n=1}^{\infty} \frac{x^n}{n!(n+1)!}\right] \quad (g)$$

is a solution of (a). Also, since $\partial(D^ky)/\partial m = D^k(\partial Y/\partial m)$, we obtain from ($f$)

$$x^2 D^2\left(\frac{\partial Y}{\partial m}\right) + 2x\, D\left(\frac{\partial Y}{\partial m}\right) - x\frac{\partial Y}{\partial m} = c(m+1)(3m+1)x^m + cm(m+1)^2x^m \ln x \quad (h)$$

Since, when $m = -1$, the right member of (h) is zero, we see that

$$y_2 = \left(\frac{\partial Y}{\partial m}\right)_{m=-1} \quad (i)$$

is a solution of (a). The general solution of (a) then is

$$y = Ay_1 + B\left(\frac{\partial Y}{\partial m}\right)_{m=-1} \quad (j)$$

In finding $\partial Y/\partial m$, equation (2), got by logarithmic differentiation, is useful. If

$$f(m) = \frac{(m-a_1)(m-a_2) \cdots (m-a_n)}{(m-b_1)(m-b_2) \cdots (m-b_t)} \quad (1)$$

then

$$\frac{\partial f(m)}{\partial m} = f(m)\left(\sum_{k=1}^{n} \frac{1}{m-a_k} - \sum_{k=1}^{t} \frac{1}{m-b_k}\right) \quad (2)$$

From (e) we get

$$\frac{\partial Y}{\partial m} = Y \ln x + cx^m\left[1 - \frac{x}{(m+2)^2} + x^2\frac{-2/(m+2) - 1/(m+3)}{(m+2)^2(m+3)} + x^3\frac{-2/(m+2) - 2/(m+3) - 1/(m+4)}{(m+2)^2(m+3)^2(m+4)} + \cdots\right] \quad (k)$$

By finding a few terms from (g) and from (3) with $m = -1$, we can indicate the answer in the form

$$y = c_0\left(1 + \frac{x}{1!2!} + \frac{x^2}{2!3!} + \frac{x^3}{3!4!} + \cdots\right) + c_1x^{-1}\ln x\left(x + \frac{x^2}{1!2!} + \frac{x^3}{2!3!} + \cdots\right) + c_1x^{-1}\left[1 - x + \frac{x^2}{1!2!}\left(-\frac{2}{1} - \frac{1}{2}\right) + \cdots\right] \qquad (l)$$

Observing that the second parenthesized expression in (l) is x times the quantity in the first parentheses, and generalizing, we can write (l) in the form

$$y = (c_0 + c_1 \ln x)\sum_{n=0}^{\infty} \frac{x^n}{n!(n+1)!} + c_1x^{-1}\left[1 - x - \sum_{n=2}^{\infty} \frac{x^n}{(n-1)!n!}\left(\frac{1}{n} + \sum_{k=1}^{n-1} \frac{2}{k}\right)\right] \qquad (m)$$

Remark. To deal with equations of the type (3), §108, for which the method of §108 fails, use a general series of the type (b), write the general equation corresponding to (e) with the constant term $c(m - \beta)$, where β is the lesser root of the indicial equation, and then use the formula

$$y = A(Y)_{m=\alpha} + B\left(\frac{\partial Y}{\partial m}\right)_{m=\beta} \qquad (3)$$

A similar kind of procedure is effective for many differential equations of higher order. Series in decreasing powers of x are effective in some cases (see exercise 8 below).

Exercises

1. Using the regular procedure for $(x^2 - x)\,D^2y + x\,Dy - y = 0$ and using cm for c_0 obtain

$$Y = cx^m\left(m + \frac{m(m-1)}{m}x + \frac{m(m-1)}{m+1}x^2 + \cdots + \frac{m(m-1)}{m+n-1}x^n + \cdots\right) \qquad (a)$$

Show that the roots of the indicial equation are 0, 1. Get one solution by substituting 1 for m in (a). Use $y_2 = (\partial Y/\partial m)_{m=0}$ on the result to get a second solution.

2. For the equation $(x^3 + x^2)\, D^2y + x\, Dy - 2xy = 0$, obtain

$$Y = c_0x^m \left[1 - \frac{m-2}{m+1}x + \frac{(m-2)(m-1)}{(m+1)(m+2)}x^2 + \sum_{n=3}^{\infty} \frac{(-1)^n(m-2)(m-1)mx^n}{(m+n-2)(m+n-1)(m+n)}\right]$$

Show that the roots of the indicial equation are 0 and 0. Now, find the solution by using $y = Y_{m=0} + (\partial Y/\partial m)_{m=0}$.

Solve the following equations:

3. $x\, D^2y + Dy - xy = 0$ **4.** $x\, D^2y - y = 0$

5. $x\, D^2y + Dy + y = 0$ **6.** $x\, D^2y + Dy - x^2y = 0$

7. $x\, D^2y + 3\, Dy + xy = 0$

★8. $x^3\, D^2y - y = 0$. Use a series having the form $y = x^m(c_0 + c_1x^{-1} + c_2x^{-2} + \cdots)$.

110 The gamma function

The gamma function is widely used in the applications of differential equations involving infinite series. For $p > 0$, we define the gamma function $\Gamma(p)$ by

$$\Gamma(p) = \int_0^\infty x^{p-1}e^{-x}\, dx \tag{1}$$

Applying integration by parts to the right member, we get

$$\int_0^\infty x^{p-1}e^{-x}\, dx = [-x^{p-1}e^{-x}]_0^\infty + (p-1)\int_0^\infty x^{p-2}e^{-x}\, dx \tag{2}$$

The first term in the left member is zero if $p > 1$. Therefore, from (2) and (1) we get

$$\Gamma(p) = (p-1)\Gamma(p-1) \tag{3}$$

From (1), we get by direct integration

$$\Gamma(1) = 1 \tag{4}$$

Applying (3) repeatedly to $\Gamma(5)$, we obtain

$$\Gamma(5) = 4\Gamma(4) = 4 \cdot 3\Gamma(3) = 4 \cdot 3 \cdot 2\Gamma(2) = 4 \cdot 3 \cdot 2 \cdot 1 = 4!$$

Evidently the same procedure for p a positive integer gives

$$\Gamma(p+1) = p! \tag{5}$$

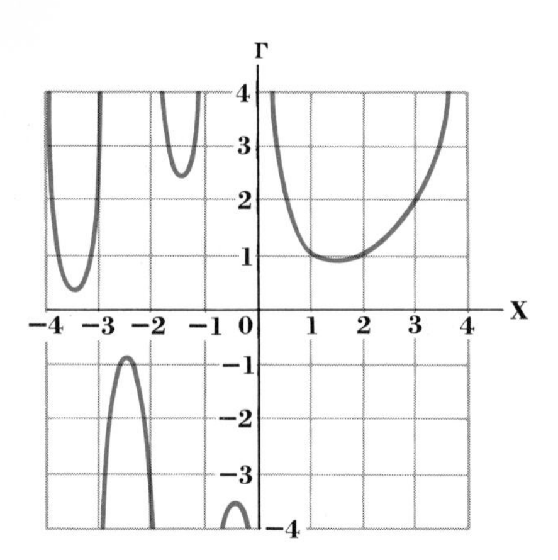

Figure 1

Certain values of $\Gamma(p)$, where $0 < p \leqq 1$, have been computed from (1) by means of infinite series.* A few values are shown in Table 1:

Table 1

p	0.1	0.2	0.3	0.4	0.5	0.6	0.7	0.8	0.9
$\Gamma(p)$	9.5	4.59	2.99	2.22	$\sqrt{\pi}$	1.49	1.30	1.16	1.07

The integral (1) does not define a value if $p \leqq 0$. However, we define the values of $\Gamma(p)$ for all real numbers as follows:

$\Gamma(p) = (p - 1)\Gamma(p - 1)$ *if* p *is neither zero nor a negative integer.*
$1/\Gamma(p) \to 0$ *as* p *approaches zero or a negative integer.*

$\Gamma(p)$ *is found from a table or computed directly from* (1) *if* $0 < p \leqq 1$.

Figure 1 shows the graph of $\Gamma(x)$.

To find $\Gamma(-\frac{3}{2})$, for example, we apply (3), written in the form $\Gamma(p) = p^{-1}\Gamma(p + 1)$ repeatedly, and Table 1 to get

$$\Gamma(-\tfrac{3}{2}) = (-\tfrac{3}{2})^{-1}\Gamma(-\tfrac{1}{2}) = (-\tfrac{3}{2})^{-1}(-\tfrac{1}{2})^{-1}\Gamma(\tfrac{1}{2}) = \tfrac{4}{3}\sqrt{\pi}$$

Exercises

Evaluate:

1. $\Gamma(4)$ **2.** $\Gamma(2.2)$ **3.** $\Gamma(-\frac{1}{2})$ **4.** $\Gamma(-1.3)$

Express in terms of t if t is not an integer:

5. $\Gamma(3 + t)$ **6.** $\Gamma(-3 + t)$ **7.** $\Gamma(-1 + t) \cdot \Gamma(1 + t)$

* A proof of the fact that $\Gamma(0.5) = \sqrt{\pi}$ is found in a footnote of §69.

Use (1), (3), and Table 1 to evaluate:

8. $\int_0^\infty t^{1.6}e^{-t}\,dt$ **9.** $\int_0^\infty t^{-0.5}e^{-t}\,dt$

10. $\int_0^\infty t^2e^{-t^2}\,dt$. *Hint:* Let $x = t^2$.

11. Evaluate $\int_0^\infty x^{9/2}e^{-x^3}\,dx$. *Hint:* Let $t = x^3$.

12. For what integral values of t will

$$(t + 100)(t + 99)(t + 98) \cdots (t + 56) = \frac{\Gamma(t + 101)}{\Gamma(t + 56)}$$

13. In what intervals is $\Gamma(t)$ negative?

14. By using reduction formulas, we can show that

$$\int_0^{\frac{1}{2}\pi} \sin^m \theta \cos^n \theta \, d\theta = \frac{\Gamma[\frac{1}{2}(m + 1)]\Gamma[\frac{1}{2}(n + 1)]}{2\Gamma[\frac{1}{2}(m + n + 2)]}$$

where m and n are integers with $m \geqq 0$ and $n \geqq 0$. Using this formula, (3), and Table 1 evaluate:

(*a*) $\int_0^{\frac{1}{2}\pi} \sin^2 \theta \cos^4 \theta \, d\theta$ (*b*) $\int_0^{\frac{1}{2}\pi} \sin^3 \theta \cos^5 \theta \, d\theta$

(*c*) $\int_0^{\frac{1}{2}\pi} \sin^{0.6} \theta \cos^{0.4} \theta \, d\theta$

111 Bessel's equation

Just as $d^2y/dx^2 + k^2y = 0$ defines the trigonometric functions and $x\,dy/dx = k$ the logarithmic functions, so also does

$$x^2 \frac{d^2y}{dx^2} + x\frac{dy}{dx} + (x^2 - k^2)y = 0 \tag{1}$$

define Bessel's functions. These have a wide field of applications; they are effective in solving the problems of flow of heat and of electricity in cylinders, the problems of the vibration of membranes, and many other problems. Volumes have been written on Bessel's functions and their applications, and extensive tables of their values have been computed. Some theory of these functions and of other outstanding functions will be developed in this chapter, and applications will be considered in later chapters.

The theory developed in §§108 and 109 may be used to solve (1). First get the indicial equation by substituting x^m for y in (1) and equating the coefficient of x^m to zero. This gives

$$m(m-1)+m-k^2=0 \qquad \text{or} \qquad m=\pm k \tag{2}$$

Next substitute $\sum_{r=0}^{\infty} c_r x^{m+r}$ for y in (1), equate the coefficient of x^{m+r} to zero, solve for c_r, and obtain

$$c_r = \frac{-c_{r-2}}{(m+r+k)(m+r-k)} \tag{3}$$

Using this with $m = k$, we obtain as a solution of (1)

$$\begin{aligned} y_1 &= c_0 x^k \left[1 - \frac{(x/2)^2}{1(1+k)} + \frac{(x/2)^4}{1\cdot 2(1+k)(2+k)} - \cdots \right. \\ &\qquad \left. + \frac{(-1)^r \Gamma(k+1)(x/2)^{2r}}{r!\Gamma(r+k+1)} + \cdots \right] \\ &= 2^k c_0 \sum_{r=0}^{\infty} \frac{(-1)^r \Gamma(k+1)(x/2)^{2r+k}}{r!\Gamma(r+k+1)} \end{aligned} \tag{4}$$

Similarly, if k is not an integer, we obtain from (4), with $m = -k$,

$$y_2 = 2^{-k} c_0 \sum_{r=0}^{\infty} \frac{(-1)^r \Gamma(1-k)(x/2)^{2r-k}}{r!\Gamma(r-k+1)} \tag{5}$$

In this case, the solution is

$$y = c_1 y_1 + c_2 y_2 \tag{6}$$

If k is an integer, (5) fails to give a solution and the method of §109 may be applied. From (3) with $c_0 = m + k$, we get

$$Y = x^m \left[m + k + \sum_{r=1}^{\infty} \frac{(-1)^r (m+k) x^{2r}}{\prod_{n=1}^{r} (m+2n+k)(m+2n-k)} \right] \tag{7}$$

where the sign Π indicates the product of the $2r$ factors obtained by replacing n in the denominator of (7) by the numbers $1, 2, \ldots, r$ in succession. Observe that, when $r \geqq k$, the factor $m + k$ cancels. Hence, taking the partial derivative of (7) with respect to m, replacing m by $-k$ in the result, multiplying by $(k-1)!$, and simplifying, we get

$$\begin{aligned} \left(\frac{\partial Y}{\partial m}\right)_{m=-k} &= x^{-k} \ln x \sum_{r=k}^{\infty} \frac{2(-1)^{r+k-1}(x/2)^{2r}}{r!(r-k)!} \\ &\quad + x^{-k} \sum_{r=0}^{k-1} \frac{(k-r-1)!(\frac{1}{2}x)^{2r}}{r!} \\ &\quad + x^{-k} \sum_{r=k}^{\infty} \left\{ \frac{2(-1)^{r+k}(\frac{1}{2}x)^{2r}}{r!(r-k)!} \sum_{n=1}^{r} \left[\frac{1}{2n} + \frac{1}{2(n-k)} \right] \right\} \end{aligned} \tag{8}$$

where the n in $n - k$ takes on all integral values from 1 to r except k, and $0! = 1$. The solution for the case when $k = 0$ cannot be obtained from (8) by replacing k by zero. This case will be considered in exercise 2. The general solution of (1) when k is any integer not zero is

$$y = c_1 y_1 + c_2 \left(\frac{\partial Y}{\partial m}\right)_{m=-k} \tag{9}$$

Exercises

1. Write the solution of (1) when (*a*) $k = \frac{1}{2}$; (*b*) $k = 3$.

2. Find the general solution of (1) when $k = 0$.

3. Note that the factor $m + k$ cancels from the coefficient of x^{2k} and all succeeding coefficients in (7). Let $m = -k$ in (7), consider k as a positive integer, and derive the solution

$$y_{(m=-k)} = x^{-k} \sum_{r=k}^{\infty} \frac{2(-1)^{r-k+1}(\frac{1}{2}x)^{2r}}{r!(k-1)!(n-k)!} = \sum_{r=0}^{\infty} \frac{(-1)^{r+1}(\frac{1}{2}x)^{2r+k}}{2^{k-1}(r+k)!(k-1)!r!}$$

Is this dependent upon (4)?

112 Bessel's functions

The particular form of (4), §111, obtained by setting c_0 equal to $1/[2^n\Gamma(n+1)]$ is denoted by J_n. Hence

$$J_n(x) = \sum_{r=0}^{\infty} \frac{(-1)^r(\frac{1}{2}x)^{2r+n}}{r!\Gamma(r+n+1)} \tag{1}$$

This is called Bessel's function of the first kind, and the function (8), §111, is called Bessel's function of the second kind. Series (1) is absolutely convergent for all values of x, except $x = 0$ when $n < 0$.

Denote by $J_{-n}(x)$ the result of replacing n by $-n$ in the right member of (1). Now, let $-n$ approach zero or a negative integer; then, by §110, $1/\Gamma(r - n + 1) \to 0$ when $r \leqq n - 1$, and we get

$$J_{-n}(x) = \sum_{r=0}^{\infty} \frac{(-1)^r(\frac{1}{2}x)^{2r-n}}{r!\Gamma(r-n+1)} = \sum_{r=n}^{\infty} \frac{(-1)^r(\frac{1}{2}x)^{2r-n}}{r!\Gamma(r-n+1)} \tag{2}$$

When n is a positive integer, replace r by $s + n$ in this and obtain

$$J_{-n}(x) = \sum_{s=0}^{\infty} \frac{(-1)^{s+n}(\frac{1}{2}x)^{2s+n}}{(s+n)!s!} = (-1)^n J_n(x)$$

or, when n is an integer

$$J_{-n}(x) = (-1)^n J_n(x) \tag{3}$$

$J_{1/2}(x)$ may be reduced to a simple and instructive form. Taking $n = \frac{1}{2}$ in (1), we get

$$J_{1/2}(x) = \sum_{r=0}^{\infty} \frac{(-1)^r(\frac{1}{2}x)^{2r+1/2}}{r!\Gamma(r+\frac{3}{2})} = \sum_{r=0}^{\infty} \frac{(-1)^r 2^{1/2}(\frac{1}{2}x)^{2r+1}}{x^{1/2}r!(r+\frac{1}{2})(r-\frac{1}{2}) \cdots \frac{1}{2}\Gamma(\frac{1}{2})} \tag{4*}$$

From Table 1, §110, $\Gamma(\frac{1}{2}) = \sqrt{\pi}$. Then the denominator of the last fraction of (4) takes the form

$$x^{1/2}2^{-(2r+1)}2 \cdot 4 \cdots (2r)[1 \cdot 3 \cdot 5 \cdots (2r+1)]\sqrt{\pi} = x^{1/2}2^{-(2r+1)}(2r+1)!\sqrt{\pi}$$

Using this value in (4) and recalling that $\sin x = x - x^3/3! + x^5/5! - \cdots$, we get

$$J_{1/2}(x) = \sqrt{\frac{2}{\pi x}} \sin x \tag{5}$$

Similarly, we can show that

$$J_{-1/2}(x) = \sqrt{\frac{2}{\pi x}} \cos x \tag{6}$$

From (5), we see that $n\pi$, where n represents integers not zero, are roots of $J_{1/2}(x)$. Also, from (6) it appears that $(2n+1)\frac{1}{2}\pi$, n an integer, represents the roots of $J_{-1/2}(x)$. The roots of $J_{1/2}(x)$ lie between corresponding roots of $J_0(x)$ and $J_1(x)$ as indicated by Table 2. The difference between successive roots of $J_0(x)$ approaches π from below, and

* The series (4) is easily proved absolutely convergent for all real values of x.

Table 2

Function	Roots				
$J_0(x)$	2.40	5.52	8.65	11.79	...
$J_{1/2}(x)$	3.14	6.28	9.42	12.57	...
$J_1(x)$	3.85	7.02	10.17	13.32	...

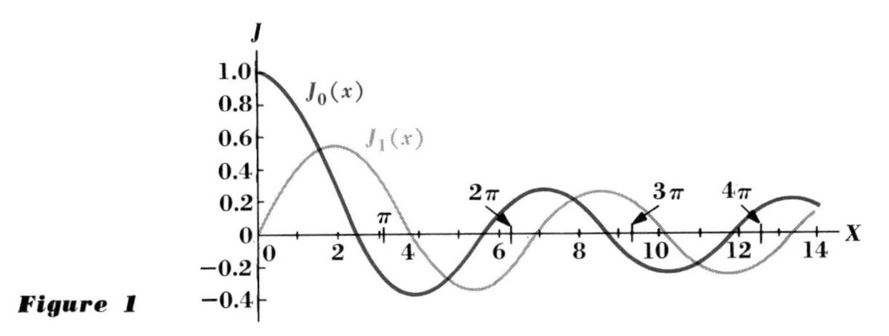

Figure 1

that for $J_1(x)$ approaches π from above. Figure 1 represents $J_0(x)$ and $J_1(x)$.

Some formulas, called recurrence formulas, are of basic importance. We have

$$
\begin{aligned}
J_n'(x) &= \sum_{r=0}^{\infty} \frac{(-1)^r \frac{1}{2}(2r+n)(\frac{1}{2}x)^{2r+n-1}}{r!\Gamma(r+n+1)} \\
&= \sum_{r=1}^{\infty} \frac{(-1)^r (\frac{1}{2}x)^{2r+n-1}}{(r-1)!\Gamma(r+n+1)} + \sum_{r=0}^{\infty} \frac{(\frac{1}{2}n)(-1)^r (\frac{1}{2}x)^{2r+n}(\frac{1}{2}x)^{-1}}{r!\Gamma(r+n+1)} \\
&= \sum_{s=0}^{\infty} \frac{(-1)^{s+1}(\frac{1}{2}x)^{2s+n+1}}{s!\Gamma(s+n+2)} + \frac{n}{x} J_n \qquad (7)
\end{aligned}
$$

$$\frac{dJ_n(x)}{dx} = J_n'(x) = -J_{n+1}(x) + \frac{n}{x} J_n(x) \qquad (8)$$

Also, from the first fraction of (7) we get

$$
\begin{aligned}
J_n' &= \sum_{r=0}^{\infty} \frac{(-1)^r \frac{1}{2}(2r+2n-n)(\frac{1}{2}x)^{2r+n-1}}{r!(r+n)\Gamma(r+n)} \\
&= \sum_{r=0}^{\infty} \frac{(-1)^r x^{2r+n-1}}{r!\Gamma(r+n)} - \tfrac{1}{2}n(\tfrac{1}{2}x)^{-1} \sum_{r=0}^{\infty} \frac{(-1)^r (\frac{1}{2}x)^{2r+n}}{r!\Gamma(r+n+1)} \\
&= J_{n-1} - \frac{n}{x} J_n
\end{aligned}
$$

or

$$J_n'(x) = J_{n-1}(x) - \frac{n}{x} J_n(x) \qquad (9)$$

Adding (8) and (9) member by member, we get

$$2J_n'(x) = J_{n-1}(x) - J_{n+1}(x) \qquad (10)$$

Eliminating J_n' between (8) and (9), we get

$$\frac{2n}{x} J_n(x) = J_{n-1}(x) + J_{n+1}(x) \qquad (11)$$

Taking $n = 0$ in (8), we get

$$J_0'(x) = -J_1(x) \tag{12}$$

Example. Develop recurrence formulas for integrals involving $J_n(ax)$.

Solution. Equation (10) with x replaced by ax is

$$2\,\frac{dJ_n(ax)}{d(ax)} = J_{n-1}(ax) - J_{n+1}(ax)$$

From this, by integration and slight transformation, we get

$$\int_0^x J_{n+1}(ax)\,dx = \int_0^x J_{n-1}(ax)\,dx - \frac{2}{a}\,[J_n(ax) - J_n(0)] \tag{a}$$

To get a recurrence formula applying to $\int_0^x x^k J_{n+1}(ax)\,dx$, $k \geqq 0$, proceed as follows: Applying integration by parts to $\int_0^x x^k J_n'(ax)\,dx$, we get

$$\int_0^x x^k J_n'(ax)\,dx = \frac{1}{a}\,x^k J_n(ax) - \frac{k}{a}\int_0^x x^{k-1} J_n(ax)\,dx \tag{b}$$

In (8) replace x by ax, multiply the result through by $x^k\,dx$, equate the integrals of the members with limits 0 to x, replace the left member of this second result by the right member of (*b*), solve this last equation for $\int_0^x x^k J_{n+1}(ax)\,dx$, and obtain

$$\int_0^x x^k J_{n+1}(ax)\,dx = \frac{n+k}{a}\int_0^x x^{k-1} J_n(ax)\,dx - \frac{x^k}{a}\,J_n(ax) \qquad k \geqq 1 \tag{c}$$

Equations (*b*) and (*c*) apply for $k \geqq 1$ and $n \geqq 0$, or n a negative integer.

The method of using equations (*b*) and (*c*) will be suggested in the following exercises.

Exercises

Using equation (1), show that for the range $0 < x < 1$:

1. $J_2(x) \cong^* \frac{1}{8}x^2(1 - \frac{1}{12}x^2)$, $|\text{error}| \leqq x^6/[2^6(2!)(4!)]$. *Hint:* $|\text{Error}| \leqq$ (absolute value of first neglected term).

2. $J_3(x) \cong (x^3/48)[1 - x^2/16]$; $|\text{error}| < x^7/[2^7(2!)(5!)]$.

3. $J_4(x) \cong (x^4/384)[1 - x^2/20]$; $|\text{error}| < x^8/[2^8(2!)(6!)]$.

* The symbol $\cong$ means *is approximately equal to.*

4. Find the values of $J_0(0)$, $J_k(0)$ when $k > 0$, and $J_{-2}(0.1)$ accurate to three decimal places.

5. Using (11), express $J_2(x)$, $J_3(x)$, and $J_4(x)$ in terms of $J_0(x)$ and $J_1(x)$. Show that the value of $J_n(x)$, n a positive integer, is determined by the values of $J_0(x)$ and $J_1(x)$.

6. Express $J_{5/2}(x)$ in terms of $J_{1/2}(x)$ and $J_{-1/2}(x)$. Express $J_{7/3}(x)$ and $J_{-7/3}(x)$ in terms of $J_{\pm 2/3}(x)$ and $J_{\pm 1/3}(x)$. Is the value of $J_k(x)$ when $x \neq 0$ determined by the values of $J_m(x)$, $|m| \leqq 1$? Explain.

7. Using (8) to (11) and the equations obtained by differentiating them, find in terms of $J_0(x)$ and $J_1(x)$: (*a*) $J_0''(x)$; (*b*) $J_1''(x)$; (*c*) $J_2'''(x)$. Are all values of $(d^s/dx^s)[J_k(x)]$ when $x \neq 0$ expressible in terms of the values of $J_m(x)$, $|m| \leqq 1$?

8. Using (10), (11), and the fact that $J_0(1) \cong 0.765$ and $J_1(1) = 0.440$, find the approximate value of: (*a*) $J_1'(1)$; (*b*) $J_2'(1)$; (*c*) $J_2''(1)$.

9. Derive equation (6).

10. Obtain, from (12), $\int_0^x J_1(ax)\,dx = [1 - J_0(ax)]/a$. Now, take $n = 0$ in equation (*a*) of the example, note that $J_{-1}(x) = -J_1(x)$, and get the same result.

Using equation (*a*) of the example, derive:

11. $$\int_0^x J_2(ax)\,dx = \int_0^x J_0(ax)\,dx - \frac{2}{a}J_1(ax)$$

12. $$\int_0^x J_3(ax)\,dx = \frac{1 - J_0(ax) - 2J_2(ax)}{a}$$

Using equations (3), (12), and (*c*) of the example for various values of n and k, derive

$$\int_0^x xJ_{n+1}(ax)\,dx = \frac{n+1}{a}\int_0^x J_n(ax)\,dx - \frac{x}{a}J_n(ax)$$

and obtain from this equation:

13. $$\int_0^x xJ_0(ax)\,dx = -\frac{x}{a}J_{-1}(ax) = \frac{x}{a}J_1(ax)$$

14. $$\int_0^x xJ_1(ax)\,dx = \frac{\int_0^x J_0(ax)\,dx - xJ_0(ax)}{a}$$

15. $$\int_0^x xJ_2(ax)\,dx = \frac{2 - 2J_0(ax) - axJ_1(ax)}{a^2}$$

16. $\int_0^x x^2 J_{-1}(ax)\,dx = -\frac{x^2}{a} J_{-2}(ax)$ or

$$\int_0^x x^2 J_1(ax)\,dx = \frac{x^2}{a} J_2(ax)$$

17. $\int_0^x x^2 J_0(ax)\,dx = \dfrac{-\int_0^x J_0(ax)\,dx + xJ_0(ax) + ax^2 J_1(ax)}{a}$

18. $\int_0^x x^k J_{k-1}(ax)\,dx = \int_0^x (-1)^{k+1} x^k J_{1-k}(ax)\,dx = (x^k/a)J_k(ax)$, k a positive integer.

★19. Using equations (8) to (11), prove that $J_n(x)$ satisfies $x^2(d^2J_n/dx^2) + x(dJ_n/dx) + (x^2 - n^2)J_n = 0$.

Use equation (a) of the example to derive:

20. $\int_0^x J_4(ax)\,dx = \int_0^x J_0(ax)\,dx - \dfrac{2[J_1(ax) + J_3(ax)]}{a}$

21. $\int_0^x J_5(ax)\,dx = \dfrac{1 - J_0(ax) - 2J_2(ax) - 2J_4(ax)}{a}$

Using (5), (6), and (8) to (11), show that:

22. $J_{-3/2}(x) = -\sqrt{\frac{2}{\pi x}}\,(x^{-1}\cos x + \sin x)$

23. $J_{-7/2}(x) = \sqrt{\frac{2}{\pi x}}\,[(1 - 15x^{-2})\sin x + (6x^{-1} - 15x^{-3})\cos x]$

★24. $J''_{-3/2}(x) = \sqrt{\frac{2}{\pi x}}\,[(1 - \frac{15}{4}x^{-2})\sin x + (2x^{-1} - \frac{15}{4}x^{-3})\cos x]$

★25. The *modified Bessel equation* is

$$x^2\frac{d^2y}{dx^2} + x\frac{dy}{dx} - (x^2 + n^2)y = 0$$

Show that, when n is not an integer, the solution of this equation is

$$y = c_1 I_n(x) + c_2 I_{-n}(x)$$

$$= c_1 \sum_{r=0}^{\infty} \frac{(\frac{1}{2}x)^{2r+n}}{r!\Gamma(r+n+1)} + c_2 \sum_{n=0}^{\infty} \frac{(\frac{1}{2}x)^{2r-n}}{r!\Gamma(r-n+1)}$$

113 Expansion of functions in terms of Bessel's functions

A set of functions $f(n,x)$ are called orthogonal functions for the range a to b if

$$\int_a^b f(m,x)f(n,x)\,dx = 0 \qquad m \neq n \tag{1}$$

For example, the functions sin nx, *n an integer*, are orthogonal for the range 0 to 2π since

$$\int_0^{2\pi} \sin mx \sin nx \, dx = 0 \qquad n \neq m$$

We shall show that the functions $\sqrt{x}\, J_k(\alpha_n x)$, α_n a root of $J_k(x)$, are orthogonal for the range 0 to 1.

Denoting derivatives by primes and replacing x first by ax and then by bx in (1), §111, we get after slight simplification

$$x^2y'' + xy' + (a^2x^2 - k^2)y = 0 \tag{2}$$
$$x^2y'' + xy' + (b^2x^2 - k^2)y = 0 \tag{3}$$

Let

$$u = J_k(ax) \qquad v = J_k(bx) \tag{4}$$

Since $y = u(x)$ satisfies (2) and $y = v(x)$ satisfies (3), we have

$$x^2u'' + xu' + (a^2x^2 - k^2)u = 0 \tag{5}$$
$$x^2v'' + xv' + (b^2x^2 - k^2)v = 0 \tag{6}$$

Multiplying (5) through by v and (6) by u and subtracting the second result from the first, we obtain

$$x^2(u''v - v''u) + x(u'v - uv') + x^2(a^2 - b^2)uv = 0 \tag{7}$$

Note that

$$\frac{d}{dx}[x(u'v - uv')] = x(u''v - uv'') + u'v - uv' \tag{8}$$

Divide (7) through by x, and integrate the result to obtain

$$(b^2 - a^2)\int_0^x xuv\, dx = x[u'v - uv']_0^x \tag{9}$$

In (9), substitute for u and v their values from (4), with k replaced by n and x by at and get

$$(b^2 - a^2)\int_0^x tJ_n(at)J_n(bt)\, dt = x[aJ_n'(ax)J_n(bx) - bJ_n(ax)J_n'(bx)] \tag{10}$$

In this, let a and b be distinct roots of $J_n(x) = 0$ so that $J_n(a) = 0$, $J_n(b) = 0$, and take $x = 1$. Then the right member of (10) becomes

$$aJ_n'(a)J_n(b) - bJ_n(a)J_n'(b) = 0$$

Therefore, *if a and b are roots of $J_n(x) = 0$ and $b \neq a$,*

$$\int_0^1 tJ_n(at)J_n(bt)\, dt = 0 \tag{11}$$

Accordingly, $\sqrt{x}\, J_n(\alpha_m x)$, where $J_n(\alpha_m) = 0$, $m = 0, 1, 2, \ldots, \infty$, represents an orthogonal set of functions for the range 0 to 1.

If $a = b$, (11) does not hold. Consider a as fixed, and let b approach a as a limit. Equating the derivatives of the members of (10) with respect to b and setting $b = a$, $x = 1$, and $J_n(a) = 0$ in the result, use (9) and (11), §112, and get after slight simplification

$$\int_0^1 t[J_n(at)]^2\, dt = \tfrac{1}{2}J_n'^2(a) = \tfrac{1}{2}J_{n+1}^2(a) \qquad \text{if } J_n(a) = 0 \tag{12}$$

Replacing n by zero in (12) and using the fact that $J_0'(a) = -J_1(a)$ from (12), §112, we get

$$\int_0^1 t[J_0(at)]^2\, dt = \tfrac{1}{2}J_1^2(a) \qquad \text{if } J_0(a) = 0 \tag{13}$$

Also, in (11) replace n by 0 to obtain

$$\int_0^1 tJ_0(at)J_0(bt)\, dt = 0 \qquad \text{if } J_0(a) = 0 \qquad J_0(b) = 0 \qquad a \neq b \tag{14}$$

Equations (13) and (14) are used to expand functions in a series of Bessel's functions. Assume that for a given function $f(x)$ and the interval $0 \leqq x \leqq 1$

$$f(x) = a_1J_0(\alpha_1 x) + a_2J_0(\alpha_2 x) + \cdots + a_nJ_0(\alpha_n x) + \cdots \tag{15}$$

where α_m, $m = 0, 1, 2, \ldots,$ are roots of $J_0(x)$ and $\alpha_m < \alpha_{m+1}$. Also, assume that integration term by term of the series (15) multiplied by $xJ_0(nx)$ is valid for the interval $0 \leqq x \leqq 1$. Then, by using (13) and (14), we get

$$\int_0^1 xf(x)J_0(\alpha_n x)\, dx = \int_0^1 \sum_{m=0}^{\infty} a_m xJ_0(\alpha_n x)J_0(\alpha_m x)\, dx = \tfrac{1}{2}a_nJ_1^2(\alpha_n)$$

or

$$a_n = \frac{2}{J_1^2(\alpha_n)} \int_0^1 xf(x)J_0(\alpha_n x)\, dx \tag{16}$$

To expand $f(x) = 1$ in a series of the type (15), replace $f(x)$ by 1 in (16) to obtain

$$\begin{aligned} a_n &= \frac{2}{J_1^2(\alpha_n)} \int_0^1 xJ_0(\alpha_n x)\, dx = \frac{2}{J_1^2(\alpha_n)} \left[\frac{x}{\alpha_n} J_1(\alpha_n x)\right]_0^1 \\ &= \frac{2}{\alpha_n J_1(\alpha_n)} \end{aligned} \tag{17}^*$$

* $\int_0^x xJ_0(ax)\, dx = (x/a)J_1(ax)$. See exercise 13, §112.

Using this result in (15) with $n = 0, 1, 2, \ldots$, we get

$$1 = \sum_{n=0}^{\infty} \frac{2J_0(\alpha_n x)}{\alpha_n J_1(\alpha_n)} \qquad 0 \leqq x < 1 \tag{18}$$

Evidently (18) does not hold when $x = 1$ for $J_0(\alpha_n) = 0$. It holds when $x = 0$; therefore, $\sum_{n=0}^{\infty} 1/[\alpha_n J_1(\alpha_n)] = \frac{1}{2}$.

Exercises

1. Expand x^2 in a series of the form (15).
2. Expand x^4 in a series of the form (15).
3. Expand x in a series of the form $\sum_{r=1}^{\infty} c_r J_1(\alpha_r x)$.
4. Using (11), (12), and $\int_0^x x^{k+1} J_k(ax)\,dx = (x^{k+1}/a) J_{k+1}(ax)$, show that

$$x^k = \sum_{r=1}^{\infty} \frac{2}{\alpha_r J_{k+1}(\alpha_r)} J_k(\alpha_r x)$$

where $J_k(\alpha_r) = 0$, k is zero or a positive integer, and $0 < x < 1$.

114 Legendre's functions

Legendre's functions are adapted to spherical coordinates. They are highly important in their applications to problems of physics, quantum mechanics, and engineering generally.

Legendre's equation has the form

$$(1 - x^2)\,D^2y - 2x\,Dy + k(k+1)y = 0 \tag{1}$$

We seek solutions in descending powers of x having the form

$$y = c_0 x^m + c_1 x^{m-1} + \cdots + c_r x^{m-r} + \cdots \tag{2}$$

Setting $y = x^m$ in (1) and equating to zero the coefficient of x^m in the result, we get

$$-m^2 - m + k(k+1) = 0 \qquad \text{or} \qquad m = k, -k-1 \tag{3}$$

Proceeding in the usual manner, we get

$$c_n = \frac{(m - n + 2)(m - n + 1)}{(m - n - k)(m - n + k + 1)} c_{n-2} \tag{4}$$

Now, using (4), first with $m = k$ and then with $m = -k - 1$, and (2), we get

$$y_k = x^k \left[1 - \frac{k(k-1)}{2(2k-1)} x^{-2} + \frac{k(k-1)(k-2)(k-3)}{2 \cdot 4(2k-1)(2k-3)} x^{-4} + \cdots \right] \tag{5}$$

$$y_{-k-1} = x^{-k-1} \left[1 + \frac{(k+1)(k+2)}{2(2k+3)} x^{-2} + \frac{(k+1)(k+2)(k+3)(k+4)}{2 \cdot 4(2k+3)(2k+5)} x^{-4} + \cdots \right] \tag{6}$$

The general solution is

$$y = c_1 y_k + c_2 y_{-k-1}$$

and this converges for all values of x satisfying $|x| \geqq 1$. Evidently, if k is zero or a positive integer, the series (5) terminates. The polynomial solutions thus obtained, when multiplied by $(2k)!/[2^k(k!)^2]$, are called Legendre polynomials and are denoted by $P_k(x)$. Hence

$$P_k(x) = \frac{(2k)!}{2^k(k!)^2} y_k \tag{7}$$

where k is zero or a positive integer. The coefficient $(2k)!/[2^k(k!)^2]$ was chosen so that

$$P_k(1) = 1 \tag{8}$$

From (7) and (5), we can derive

$$P_k(x) = \sum_{r=0}^{L} \frac{(-1)^r(2k - 2r)!}{2^k r!(k - 2r)!(k - r)!} x^{k-2r} \tag{9}$$

where $L = (k - 1)/2$ if k is odd and $L = k/2$ if k is even. From (5) or (9), we easily obtain

$$\begin{aligned} &P_0(x) = 1 \qquad P_1(x) = x \qquad P_2(x) = \tfrac{1}{2}(3x^2 - 1) \\ &P_3(x) = \tfrac{1}{2}(5x^3 - 3x) \qquad P_4(x) = \tfrac{1}{8}(35x^4 - 30x^2 + 3) \\ &P_5(x) = \tfrac{1}{8}(63x^5 - 70x^3 + 15x) \end{aligned} \tag{10}$$

Another important property, permitting the expression of any function in Legendre polynomials, will now be considered.

The following proof shows that the Legendre polynomials constitute an orthogonal set in the interval $-1 \leqq x \leqq 1$ by demonstrating that

$$\int_{-1}^{1} P_r(x)P_k(x)\,dx = 0 \qquad r \neq k \tag{11}$$

Equation (1) may be written

$$\frac{d}{dx}[(1 - x^2)y'] + k(k + 1)y = 0$$

In this, replace y by the solution $P_k(x)$, multiply through by $P_r(x)$, and integrate to obtain

$$\int_{-1}^{1} P_r(x)\frac{d}{dx}[(1 - x^2)P_k'(x)]\,dx + k(k + 1)\int_{-1}^{1} P_k(x)P_r(x)\,dx = 0$$

Apply integration by parts to the first integral to get

$$[(1 - x^2)P_k'(x)P_r(x)]_{-1}^{1} - \int_{-1}^{1}(1 - x^2)P_r'P_k'\,dx + k(k + 1)\int_{-1}^{1} P_kP_r\,dx = 0 \quad (12)$$

The first term of this vanishes. Now write (12) with r and k interchanged to obtain

$$-\int_{-1}^{1}(1 - x^2)P_k'P_r'\,dx + r(r + 1)\int_{-1}^{1} P_rP_k\,dx = 0$$

Subtract this from (12), member by member, to get

$$(k - r)(k + r + 1)\int_{-1}^{1} P_rP_k\,dx = 0$$

Since $r \geqq 0$, $k \geqq 0$, and $k \neq r$, we see that $(k - r)(k + r + 1) \neq 0$. Therefore the integral must vanish and (11) holds true.

It can be proved that*

$$\varphi(s,x) = (1 + s^2 - 2xs)^{-1/2} = \sum_{k=0}^{\infty} P_ks^k \quad (13)$$

The student may check this by expanding φ in powers of s in a Maclaurin series and equating the coefficients of like powers of s to obtain

$$P_0(x) = \varphi(0,x) = 1 \qquad P_1(x) = \frac{\partial\varphi(0,x)}{\partial s} = x$$

$$P_2(x) = \frac{1}{2!}\frac{\partial^2\varphi(0,x)}{\partial s^2} = \frac{1}{2}(3x^2 - 1) \qquad \cdots\cdot$$

The relation (13) may be used to prove that

$$\int_{-1}^{1} P_k^2(x)\,dx = \frac{2}{2k + 1} \quad (14)$$

Equating the squares of the sides of (13), integrating these squares,

* See Louis A. Pipes, "Applied Mathematics for Engineers and Physicists," 2d ed., pp. 367–368, McGraw-Hill Book Company, New York, 1958.

using limits -1 to $+1$, and taking account of (11), we get

$$\int_{-1}^{1} \frac{dx}{1 + s^2 - 2sx} = \sum_{k=0}^{\infty} s^{2k} \int_{-1}^{1} P_k^2(x)\, dx \tag{15}$$

From the left member, we get

$$\begin{aligned}\int_{-1}^{1} \frac{dx}{1 + s^2 - 2sx} &= \left[-\frac{1}{2s} \ln (1 + s^2 - 2sx) \right]_{-1}^{1} \\ &= \frac{1}{s} [\ln (1 + s) - \ln (1 - s)] \\ &= \frac{1}{s} \sum_{r=1}^{\infty} (-1)^{r+1} \frac{s^r}{r} + \frac{1}{s} \sum_{r=1}^{\infty} \frac{s^r}{r} = \sum_{k=0}^{\infty} \frac{2s^{2k}}{2k + 1}\end{aligned}$$

Using this result in (15) and equating the coefficients of s^{2k}, we get (14). Equations (11) and (14) are used to expand functions $f(x)$* in terms of Legendre functions. Multiplying

$$f(x) = a_0 P_0(x) + a_1 P_1(x) + \cdots + a_n P_n(x) + \cdots \tag{16}$$

by $P_n(x)$, equating the integrals of the resulting members with limits -1 to $+1$, and taking account of (14) and (11), we get

$$a_n = \frac{2n + 1}{2} \int_{-1}^{1} f(x) P_n(x)\, dx \tag{17}$$

Using these values in (16), we get

$$f(x) = \sum_{n=0}^{\infty} \left[\frac{2n + 1}{2} \int_{-1}^{1} f(x) P_n(x)\, dx \right] P_n(x) \tag{18}$$

Example. Show that $\int_{-1}^{1} P_n(x)\, dx$ is 2 when $n = 0$ and is 0 when $n = 1, 2, \ldots$. Show that $\int_{-1}^{1} x P_n(x)\, dx$ is 0 when $n = 0, 2, 3, \ldots$ and is $\frac{2}{3}$ when $n = 1$.

Solution. Observing that $P_0(x) = 1$, $P_1(x) = x$, we have

$$\int_{-1}^{1} 1 \cdot P_0(x)\, dx = \int_{-1}^{1} P_0^2(x)\, dx = \frac{2}{2 \cdot 0 + 1} = 2 \qquad \text{by (14)}$$

$$\int_{-1}^{1} P_n(x)\, dx = \int_{-1}^{1} P_0(x) P_n(x)\, dx = 0 \qquad n = 1, 2, \ldots \qquad \text{by (11)}$$

$$\int_{-1}^{1} x P_n(x) = \int_{-1}^{1} P_1(x) P_n(x)\, dx = 0 \qquad \text{when } n = 0, 2, 3, \ldots \qquad \text{by (11)}$$

$$\int_{-1}^{1} x P_1(x)\, dx = \int_{-1}^{1} P_1^2(x)\, dx = \frac{2}{2 \cdot 1 + 1} = \frac{2}{3} \qquad \text{by (14)}$$

* Sufficient conditions for the expansion are that $f(x)$ and $f'(x)$ are sectionally continuous in $-1 \leqq x \leqq 1$.

Exercises

Assume that n represents a nonnegative integer in these exercises.

1. Check by direct operation that: (*a*) $P_0(1) = P_3(1) = P_5(1) = 1$; (*b*) $\int_{-1}^{1} P_1(x)P_2(x)\,dx = 0$, $\int_{-1}^{1} P_2^2(x)\,dx = 2/(2 \cdot 2 + 1) = \frac{2}{5}$.

2. Find $P_6(x)$ by using (5) and (7).

3. Find $P_7(x)$ by using (9).

4. Using (9), find the degree of $P_k(x)$ in x.

5. Express x^5 in terms of P_1, P_3, and P_5.

6. Using (10), show that $1 = P_0$, $x = P_1$, $x^2 = \frac{1}{3}(2P_2 + 1)$, $x^3 = \frac{1}{5}(2P_3 + 3P_1)$, $x^4 = \frac{1}{35}(8P_4 + 20P_2 + 7)$. Show that the expression for x^n in terms of Legendre polynomials involves P_n but not P_k with $k > n$.

7. Using (11), (14), and exercise 6, find the values of: (*a*) $\int_{-1}^{1} x^2P_n(x)\,dx$; (*b*) $\int_{-1}^{1} x^3P_n(x)\,dx$; (*c*) $\int_{-1}^{1} x^4P_n(x)\,dx$.

8. Using (18), expand in terms of $P_k(x)$, $k = 0, 1, 2, \ldots$: (*a*) x^2; (*b*) x^3; (*c*) x^4; (*d*) $ax^2 + bx + c$.

9. Which of the functions $P_0(x)$, $P_1(x)$, . . . will not appear in the expansion of $x^n + 2x^{n-1} + 3x^{n-2} + \cdots + nx + n + 1$ in terms of Legendre functions?

10. Show that $\int_{-1}^{1} x^kP_{k+r}\,dx = 0$ if $r > 0$.

13

Numerical Solutions of Differential Equations

115 Introduction

By a numerical solution of a differential equation in x and y, we mean a table of values of y opposite corresponding values of x for a particular solution of the equation. The methods considered are step-by-step procedures which use at each stage the results previously obtained. They require much computation, but they can be applied when other methods fail, and the results are furnished in great detail. Moreover, the computation can be carried out by computing machines. Numerical methods are used for diverse purposes. A suggestion of their usefulness is given by the fact that at proving grounds giant computers and numerical methods of solving differential equations are used to plot the paths of projectiles fired from big guns.

116 Methods of successive approximations

The following two examples illustrate step-by-step procedures of approximating solutions of differential equations. They are very instructive but are not generally practical.

Example 1. Approximate the solution of

$$\frac{dy}{dx} = 0.3x + 0.45 \tag{a}$$

passing through the point $(-1,-1)$.

Solution. A process of approximating a solution of $dy/dx = f(x,y)$ consists in starting at a point P_0, computing the slope at P_0,

$$\text{Slope at } P_0(x_0,y_0) = f(x_0y_0)$$

drawing a short line P_0P_1 having slope $f(x_0,y_0)$, computing the slope $f(x_1,y_1)$ at P_1, drawing a short line P_1P_2 having slope $f(x_1,y_1)$, and continuing the process any desirable number of times. Figure 1 represents an approximate solution of equation (a) through $P_0(-1,-1)$. The slope at $(-1,-1)$ is $0.3(-1) + 0.45 = 0.15$. Through P_0, draw a straight line P_0P_1 with slope 0.15 so that x changes $\frac{1}{2}$. By (a), the slope at P_1 is $0.3(-\frac{1}{2}) + 0.45 = 0.30$. Draw P_1P_2 with slope 0.3 so that the change in x is $\frac{1}{2}$. The slope at P_2 is $0.3(0) + 0.45 = 0.45$; through P_2, draw P_2P_3 with slope 0.45, and so on. The exact solution passes through (3,2), whereas P_8 is (3,1.7). More accuracy would be obtained by using shorter line segments; theoretically, any desirable accuracy is attainable by the use of large scales and small line segments.

Example 2. Find approximately the solution of $(d^2y/dx^2) - 2x(dy/dx) = 2x$ which satisfies the initial conditions $y = 1$, $dy/dx = 0$ when $x = 0$.

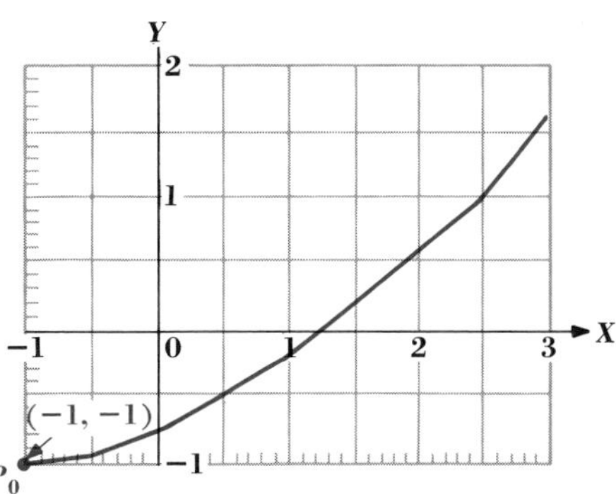

Figure 1

Solution. Writing $D = d/dx$, $z = dy/dx$, $Dz = d^2y/dx^2$ in the statement of the problem, we have the system

$$Dy = z \qquad Dz = 2x + 2xz \tag{a}$$

and the initial conditions

$$y = 1 \qquad z = 0 \qquad \text{when } x = 0 \tag{b}$$

The process to be performed repeatedly consists in substituting the approximations for y and z in terms of x in the right members of (a), integrating the results, and determining the constants of integration by using (b).

Substituting the first approximation $y = 1$, $z = 0$ in the right members of (a), we get

$$Dy = 0 \qquad Dz = 2x \tag{c}$$

Integrating these equations and determining the constants of integration by using (b), we get

$$y = 1 \qquad z = x^2 \tag{d}$$

as the second approximation.* Substituting these values in the right members of (a), integrating, and determining the constants, we get the third approximation

$$y = 1 + \frac{x^3}{3} \qquad z = x^2 + \frac{x^4}{2} \tag{e}$$

The fourth approximation is

$$y = 1 + \frac{x^3}{3} + \frac{x^5}{5 \cdot 2!} \qquad z = x^2 + \frac{x^4}{2} + \frac{x^6}{3!} \tag{f}$$

and the $(n + 2)$nd $(n > 1)$ approximation for y is found to be

$$y = 1 + \frac{x^3}{3} + \frac{x^5}{5 \cdot 2!} + \cdots + \frac{x^{2n+1}}{(2n + 1)n!} \tag{g}$$

Here again, the limit approached by the nth approximation, as n becomes infinite, is the series solution which satisfies the given initial conditions.

In general, the method applied to the system (a) of example 2 may

* The first approximation (b) agrees with the required integral curve at $(0,1)$ in slope. Successive approximations have the values of higher and higher ordered derivatives at $(0,1)$ in common with the required solution.

be applied to approximate a solution of any system of n first-order equations in $n + 1$ unknowns, provided that the conditions of Theorem I of §98 are satisfied for the system.*

Exercises

1. Use the method of example 1 to approximate the solution of $y' = -0.2(x + 2) - 0.2$ which passes through $(-2,2)$. Extend the approximation from $(-2,2)$ to the point for which $x = 3$. Take one unit as the projection of each line segment on the X axis. The point reached on line $x = 3$ should be $(3,-1.0)$. The exact solution passes through $(3,-1.5)$.

2. Solve exercise 1 by using as slope of each line segment the average (one-half the sum) of the values of y' at its ends instead of the slope at its initial point. The terminal point of your fifth line segment should be $(3,-1.5)$.

3. Using the method of example 2, find a five-term approximation to the solution of

$$\frac{d^2y}{dx^2} + 3x\frac{dy}{dx} - 6y = 0$$

for which $dy/dx = 1$ and $y = 1$ when $x = 0$.

4. To approximate the solution of $dy/dx = xy$ which satisfies $y = 1$ when $x = 0$, proceed as follows: Take $y = 1$ as the first trial approximate solution, substitute 1 for y in the right member of $dy/dx = xy$, solve the resulting equation, determine the constant of integration so that $y = 1$ when $x = 0$, and obtain

$$y_2 = 1 + \tfrac{1}{2}x^2$$

Now, replace y in the right member of $dy/dx = xy$ by $1 + \frac{1}{2}x^2$, integrate the result, determine the constant of integration, and obtain

$$y_3 = 1 + \frac{x^2}{2} + \frac{x^4}{2^2(2!)}$$

Repeat this process three more times, and then obtain by inspection the term of the solution containing x^{2n}.

5. Use the method of exercise 4 to derive the solution of $dy/dx = xy + 1$ through $(0,1)$.

* See E. L. Ince, "Ordinary Differential Equations," pp. 63–74.

117 Newton's interpolation formula

A widely used method of numerical approximation of solutions* of differential equations is based on one of Newton's interpolation formulas. This section relates to this formula.

A pair of values x and y will be referred to as the point (x,y). In Table 1, each point (0,0), (1,2), . . . , (5,250) satisfies $y = 2x^3$, and each number in any column except the first two equals the adjacent number on its left minus the one immediately above the latter. Check the table.

A general table, called a horizontal difference table, suggested by Table 1, follows. For Table 2, we assume that the x values are equally spaced so that

$$x_1 - x_0 = x_2 - x_1 = x_3 - x_2 = \cdots = x_n - x_{n-1} = h \tag{1}$$

* The development used in this text follows, in a general way, the plan set forth by James B. Scarborough in "Numerical Mathematical Analysis," 4th ed., The Johns Hopkins Press, Baltimore, 1958.

Table 1

x	y	Δ_1	Δ_2	Δ_3	Δ_4	Δ_5
0	0					
1	2	2				
2	16	14	12			
3	54	38	24	12		
4	128	74	36	12	0	
5	250	122	48	12	0	0

Table 2

x	y	Δ_1	Δ_2	Δ_3	Δ_4	Δ_n
x_0	y_0					
x_1	y_1	$\Delta_1 y_1$				
x_2	y_2	$\Delta_1 y_2$	$\Delta_2 y_2$			
x_3	y_3	$\Delta_1 y_3$	$\Delta_2 y_3$	$\Delta_3 y_3$		
x_4	y_4	$\Delta_1 y_4$	$\Delta_2 y_4$	$\Delta_3 y_4$	$\Delta_4 y_4$	
.	.	.	.	.	.	
.	.	.	.	.	.	
.	.	.	.	.	.	
x_n	y_n	$\Delta_1 y_n$	$\Delta_2 y_n$	$\Delta_3 y_n$	$\Delta_4 y_n$	. . . $\Delta_n y_n$

and that *each element in any column except the first two is equal to the adjacent element on its left minus the element above the adjacent element.* Thus

$$\Delta_1 y_1 = y_1 - y_0,\ \Delta_1 y_2 = y_2 - y_1,\ \cdots,\ \Delta_1 y_k = y_k - y_{k-1} \tag{2}$$

$$\begin{aligned} \Delta_2 y_2 &= \Delta_1 y_2 - \Delta_1 y_1 = (y_2 - y_1) - (y_1 - y_0) = y_2 - 2y_1 + y_0 \\ &\cdots \\ \Delta_2 y_k &= y_k - 2y_{k-1} + y_{k-2} \end{aligned} \tag{3}$$

$$\begin{aligned} \Delta_3 y_3 &= \Delta_2 y_3 - \Delta_2 y_2 = (y_3 - 2y_2 + y_1) - (y_2 - 2y_1 + y_0) \\ &= y_3 - 3y_2 + 3y_1 - y_0 \\ &\cdots \\ \Delta_3 y_k &= y_k - 3y_{k-1} + 3y_{k-2} - y_{k-3} \end{aligned} \tag{4}$$

and so on. We now search for a polynomial $\varphi(x)$ such that (x_0,y_0), (x_1y_1), . . . , (x_n,y_n) satisfies $y = \varphi(x)$. Take $\varphi(x)$ in the form

$$\begin{aligned} y = \varphi(x) &= a_0 + a_1(x - x_n) + a_2(x - x_n)(x - x_{n-1}) + \cdots \\ &+ a_n(x - x_n)(x - x_{n-1})(x - x_{n-2}) \cdots (x - x_1) \end{aligned} \tag{5}$$

Substituting in (5) the points (x_n,y_n), (x_{n-1},y_{n-1}), (x_{n-2},y_{n-2}), (x_{n-3},y_{n-3}), and taking account of (1), we obtain

$$\begin{aligned} y_n &= a_0 \\ y_{n-1} &= a_0 + a_1(x_{n-1} - x_n) = a_0 - a_1 h \\ y_{n-2} &= a_0 + a_1(-2h) + a_2(-2h)(-h) \\ y_{n-3} &= a_0 + a_1(-3h) + a_2(-3h)(-2h) \\ &\qquad + a_3(-3h)(-2h)(-h) \end{aligned} \tag{6}$$

The solution of equations (6) for a_0, a_1, a_2, and a_3 is

$$\begin{aligned} a_0 &= y_n,\ a_1 = \frac{y_n - y_{n-1}}{h},\ a_2 = \frac{y_n - 2y_{n-1} + y_{n-2}}{2!h^2}, \\ a_3 &= \frac{y_n - 3y_{n-1} + 3y_{n-2} - y_{n-3}}{3!h^3},\ \cdots \end{aligned} \tag{7}$$

Comparing these values with (2), (3), and (4), and assuming, as can be proved, that the suggested law holds generally, we get

$$a_0 = y_n,\ a_1 = \frac{\Delta_1 y_n}{h},\ a_2 = \frac{\Delta_2 y_n}{2!h^2},\ a_3 = \frac{\Delta_3 y_n}{3!h_3},\ \cdots,\ a_n = \frac{\Delta_n y_n}{n!h^n}$$

Substituting these values in (5), we obtain Newton's interpolation formula

$$\begin{aligned} y = \varphi(x) &= y_n + \frac{\Delta_1 y_n}{h}(x - x_n) + \frac{\Delta_2 y_n}{2!h^2}(x - x_n)(x - x_{n-1}) \\ &+ \frac{\Delta_3 y_n}{3!h^3}(x - x_n)(x - x_{n-1})(x - x_{n-2}) + \cdots \\ &+ \frac{\Delta_n y_n}{n!h^n}(x - x_n)(x - x_{n-1}) \cdots (x - x_1) \end{aligned} \tag{8}$$

Observe that y_n and the Δ's are those of the last line of Table 2. But any line is the last one of a table with the ones below deleted, and therefore (8) applies to any line of Table 2. For example, applying (8) to the line in Table 1 beginning with 5, we use $h = 1$, $y_n = 250$, $x_n = 5$, $x_{n-1} = 4$, $x_{n-2} = 3$, $x_{n-3} = 2$, $x_{n-4} = 1$ and obtain

$$y = 250 + 122(x - 5) + \tfrac{48}{2}(x - 5)(x - 4) + \tfrac{12}{6}(x - 5)(x - 4)(x - 3) \quad (9)$$

a cubic satisfied by all points (x,y) of the table. In fact, it simplifies to $y = 2x^3$. Also, the result of substituting 2.5 for x in (9) is $y = 31.25$, and we say that 31.25 is the value of $\varphi(2.5)$ interpolated between $y = 16$ and $y = 54$. Also, (8), for the line beginning with 2, considered as a last line, is

$$y = 16 + 14(x - 2) + \tfrac{12}{2}(x - 2)(x - 1) \quad (10)$$

an equation satisfied by (2,16), (1,2), and (0,0), but not by (3,54). When $x = 1.3$ in (9), $y = 4.394$ and 4.394 is a value interpolated between $y = 2$ and $y = 16$.

Observe that *if all the kth differences* $\Delta_k y_s$ *are equal, all* Δ*'s of higher order than k would be zero and the polynomial* (8) *would be of the kth order.* Thus, in Table 1, all third differences are 12, and the corresponding polynomial obtained from (8) is a cubic. The converse is also true. In a table typified by Table 2 but derived from a polynomial of the kth degree, the differences of the kth order would be equal. Table 1 illustrates this fact.

Exercises

1. Apply (8) to the line of Table 1 beginning with 3, and simplify to get $y = 2x^3$. Do the same for the line beginning with 4.

2. From $y = 2x^2 + 4x$, find the values of y when x has the values 2, 4, 6, 8, and 10; make a horizontal difference table like Table 1 for the results; and compute differences. Apply (8) for the line beginning with 10 and also for the line beginning with 6. Find a value of y from this latter equation for $x = 5$ and one for $x = 4.5$.

3. Make a horizontal difference table for the points (0,2), (1,3), (2,18), (3,83), (4,258), (5,627), and (6,1,298). The fourth differences should all be 24. Apply (8) for the line beginning with 4, and show that (6,1,298) satisfies it. Apply (8) to the line beginning with 3, and show that (2,18), (1,3), and (0,2), satisfy it but that (4,258) does not. Explain.

118 Interpolation

If a function $y = f(x)$ makes the association indicated by the points (x_0,y_0), $(x_0 + h, y_1)$, $(x_0 + 2h, y_2)$, . . . , $(x_0 + nh, y_n)$, then equation (8), §117, called an interpolation formula of $f(x)$, approximates $f(x)$. Finding values of y by applying (8), §117, is called interpolation. If $f(x)$ is continuous and finite in the interval of x considered, the errors of interpolation vary with h and approach zero as h approaches zero. If the kth differences are nearly constant, then equation (8), §117, with $n = k$ will generally give approximations as accurate as the values of y, $i = 1, 2, \ldots, n$, used. Thus, if third differences are nearly constant, a cubic polynomial will be employed.

A form of (8), §117, obtained by means of the substitution

$$x - x_n = uh \tag{1}$$

simplifies computation. From (1), we get

$$x - x_k = x - x_n + (x_n - x_k) = [u + (n - k)]h \tag{2}$$

Making the substitution (2) in (8), §117, we get

$$\begin{aligned} y = \varphi(x) = \varphi(x_n + uh) &= y_n + \frac{\Delta_1 y_n}{1!} u + \frac{\Delta_2 y_n}{2!} u(u+1) \\ &+ \frac{\Delta_3 y_n}{3!} u(u+1)(u+2) + \cdots \\ &+ \frac{\Delta_n y_n}{n!} u(u+1)(u+2) \cdots (u+n-1) \end{aligned} \tag{3}$$

Interpolating to find y when $x = x_n - \frac{1}{2}h$, that is, when $u = -\frac{1}{2}$, we get, from (3),

$$\begin{aligned} y\left(\text{at } u = -\frac{1}{2}\right) &= y_n + \Delta_1 y_n\left(-\frac{1}{2}\right) + \frac{\Delta_2 y_n}{2!}\left(-\frac{1}{2}\right)\left(\frac{1}{2}\right) \\ &+ \frac{\Delta_3 y_n}{3!}\left(-\frac{1}{2}\right)\left(\frac{1}{2}\right)\left(\frac{3}{2}\right) \\ &+ \frac{\Delta_4 y_n}{4!}\left(-\frac{1}{2}\right)\left(\frac{1}{2}\right)\left(\frac{3}{2}\right)\left(\frac{5}{2}\right) + \cdots \end{aligned} \tag{4}$$

An illustration will indicate the interpolative power of equation (3). Table 3 is a horizontal difference table for $y = \log_{10} x$. In general, the accuracy of an interpolation in a table like Table 3 will not exceed that of the data, in this case four decimal places. However, the terms coming from (3) or (4) should be computed to five decimal places and their sum rounded to four decimal places. Applying (4) to the last

Table 3

x	$y = \log_{10} x$	Δ_1	Δ_2	Δ_3	Δ_4	Δ_5
4.0	0.6021					
4.2	0.6232	0.0211				
4.4	0.6435	0.0203	−0.0008			
4.6	0.6628	0.0193	−0.1010	−0.0002		
4.8	0.6812	0.0184	−0.0009	+0.0001	0.0003	
5.0	0.6990	0.0178	−0.0006	+0.0003	0.0002	−0.0001

line of Table 3, we get

$$\begin{aligned}
&\log_{10}\left[5 + \left(-\frac{1}{2}\right)(0.2)\right] \\
&= 0.6990 + 0.0178\left(-\frac{1}{2}\right) - \frac{0.0006}{2!}\left(-\frac{1}{2}\right)\left(\frac{1}{2}\right) \\
&\qquad + \frac{0.0003}{3!}\left(-\frac{1}{2}\right)\left(\frac{1}{2}\right)\left(\frac{3}{2}\right) + \frac{0.0002}{4!}\left(-\frac{1}{2}\right)\left(\frac{1}{2}\right)\left(\frac{3}{2}\right)\left(\frac{5}{2}\right) \\
&\qquad - \frac{0.0001}{5!}\left(-\frac{1}{2}\right)\left(\frac{1}{2}\right)\left(\frac{3}{2}\right)\left(\frac{5}{2}\right)\left(\frac{7}{2}\right) \\
&= 0.6990 - 0.0089 + 0.00008 - 0.00002 - 0 + 0 \\
&= 0.6902 \qquad \text{nearly}
\end{aligned}$$

a result correct to four decimal places. Similarly, from (3) and Table 3 with $u = -\frac{3}{2}$ we get

$$\log_{10}[5 - \tfrac{3}{2}(0.2)] = \log_{10} 4.7 = 0.6721$$

Exercises

1. Make a horizontal difference table for $y = \sin x$ from the data $(0,0)$, $(\frac{1}{6}\pi,0.50)$, $(\frac{2}{6}\pi,0.87)$, and $(\frac{3}{6}\pi,1)$. Using (3), write an interpolation formula, and compute from it $\sin(\frac{5}{12}\pi)$ and $\sin(\frac{3}{12}\pi)$ by using in order $u = -\frac{1}{2}$ and $u = -\frac{3}{2}$. The respective results should be 0.97 and 0.71.

2. Look up the logarithms to the base 10 of 3, 3.5, 4, 4.5, and 5 accurate to three figures, make a corresponding difference table, write the corresponding equation (3), and from this find in order logarithms to the base 10 of 4.75, 4.25, and 3.73.

3. If on a certain day the temperature T at time $t = 0$ (noon) was 80°, at 3 P.M. it was 90°, at 6 P.M. it was 80°, and at 9 P.M. it was 50°, find, by means of (2) and (3), an expression approximating T in terms of t.

4. Show that the sum of $1 + 2 + 3 + \cdots + n$ is given by $s = 6 + 3u + \frac{1}{2}u(u + 1)$ where $u = n - 3$, $n = 1, 2, \ldots, n$. *Hint:* $y_1 = 1$, $y_2 = 1 + 2$, $y_3 = 1 + 2 + 3$, etc.

5. Show that the sum of $1^2 + 2^2 + 3^2 + \cdots + n^2$ is $30 + 16u + \frac{7}{2}u(u + 1) + \frac{1}{3}u(u + 1)(u + 2)(u + 3)$ where $u = n - 4$, $n = 1, 2, \ldots, n$. Read the hint to exercise 4.

119 Formulas for approximate integration*

The general plan of numerical integration of a function $y = f(x)$ consists in forming the approximating polynomial (3), §118, for $f(x)$ and integrating this polynomial. We shall derive some basic formulas by integrating the first five terms of polynomial (3), §118, with y replaced by y', over various intervals of length h. From (1), §118, $u = (x - x_n)/h$; therefore, $dx = h\,du$; when $x = x_n$, $u = 0$, and when $x = x_n + h$, $u = 1$. Hence, if $y' = \varphi(x)$ and if Y_n^{n+1} denotes $\int_{x_n}^{x_n+h} y'\,dx$, we get

$$Y_n^{n+1} = \int_{x_n}^{x_n+h} y'\,dx = h\int_0^1 y'\,du = h\int_0^1 \left[y'_n + \Delta_1 y'_n u + \frac{\Delta_2 y'_n}{2!}u(u+1) + \frac{\Delta_3 y'_n}{3!}u(u+1)(u+2) + \frac{\Delta_4 y'_n}{4!}u(u+1)(u+2)(u+3)\right] du \tag{1}$$

or, evaluating the integral

$$Y_n^{n+1} = h(y'_n + \tfrac{1}{2}\Delta_1 y'_n + \tfrac{5}{12}\Delta_2 y'_n + \tfrac{3}{8}\Delta_3 y'_n + \tfrac{251}{720}\Delta_4 y'_n) \tag{2}$$

Similarly, denote $\int_{x_{n-1}}^{x_n} y'\,dx = h\int_{-1}^{0} y'\,du$ by Y_{n-1}^n, evaluate the integral of (1) with limits 0 to 1 replaced by limits -1 to 0, and obtain

$$Y_{n-1}^n = h(y'_n - \tfrac{1}{2}\Delta_1 y'_n - \tfrac{1}{12}\Delta_2 y'_n - \tfrac{1}{24}\Delta_3 y'_n - \tfrac{19}{720}\Delta_4 y'_n) \tag{3}$$

Also, using (1) with limits 0 to 1 replaced by respective limits -2 to 0, -2 to -1, -3 to -2, and -4 to -3, obtain the following equations:

$$Y_{n-2}^n = 2h(y'_n - \Delta_1 y'_n + \tfrac{1}{6}\Delta_2 y'_n - \tfrac{1}{180}\Delta_4 y'_n) \tag{4}$$

$$Y_{n-2}^{n-1} = h(y'_n - \tfrac{3}{2}\Delta_1 y'_n + \tfrac{5}{12}\Delta_2 y'_n + \tfrac{1}{24}\Delta_3 y'_n + \tfrac{11}{720}\Delta_4 y'_n) \tag{5}$$

* For a more complete discussion of the derivation and use of equations (1) to (7), consult James B. Scarborough, "Numerical Mathematical Analysis."

$$Y_{n-3}^{n-2} = h(y'_n - \tfrac{5}{2}\Delta_1 y'_n + \tfrac{23}{12}\Delta_2 y'_n - \tfrac{3}{8}\Delta_3 y'_n - \tfrac{19}{720}\Delta_4 y'_n) \tag{6}$$
$$Y_{n-4}^{n-3} = h(y'_n - \tfrac{7}{2}\Delta_1 y'_n + \tfrac{53}{12}\Delta_2 y'_n - \tfrac{55}{24}\Delta_3 y'_n + \tfrac{251}{720}\Delta_4 y'_n) \tag{7}$$

The two formulas in color are very important and are used regularly. The others are used only to check starting lines of a solution or a doubtful figure. The solution of a differential equation in the next section will illustrate methods of using them.

120 Illustration and discussion of formulas (2) to (7), §119

Consider a numerical solution accurate to four decimal places of

$$\frac{dy}{dx} = y' = 2x + y \tag{a}$$

from (0,2) to a point having abscissa 1.*

There are many ways of finding the first few lines of a solution. For example, we can often use an approximate formula got by deriving several terms of a Taylor's series or by the method of exercise 4, §116. The method used here applies repeatedly the formula

$$y_{x_0+h} \cong y_{x_0} + \tfrac{1}{2}(y'_{x_0} + y'_{x_0+h})h \tag{1}$$

where a letter with a subscript p indicates the value of the letter when $x = p$ and $\cong$ means *equals approximately*. When $x = 0$, $y = 2$, and, from (*a*), $y'_0 = 2(0) + 2 = 2$. We now desire $y_{0.05}$. Estimate that $y_{0.05} = 2$, use (*a*) to get $y'_{0.05} = 2(0.05) + 2 = 2.1$, and then use (1) to get

$$y_{0.05} = 2 + \tfrac{1}{2}(2 + 2.1)(0.05) = 2.1025 \tag{b}$$

To get an improved approximation for $y_{0.05}$, take $y_{0.05} = 2.1025$, use (*a*) to get $y' = 2(0.05) + 2.1025 = 2.2025$, and then (1) to get

$$y_{0.05} \cong 2 + \tfrac{1}{2}(2 + 2.2025)(0.05) \cong 2.1051 \tag{c}$$

Now, use (1) again with $y_{0.05} = 2.1051$, $y'_{0.05} \cong 2.2051$ from (*a*), and get the same value 2.1051 as $y_{0.05}$ accurate to four decimal places. Therefore, we accept 2.1051 as the value of $y_{0.05}$.

To evaluate $y_{0.1}$, use (*a*) and (1) repeatedly, as before, with $x_0 = 0.05$ to obtain in succession

$$y_{0.1} \cong 2.1051 + \tfrac{1}{2}(2.2051 + 2.4051)(0.05) = 2.2204$$
$$y_{0.1} \cong 2.1051 + \tfrac{1}{2}(2.2051 + 2.4204)(0.05) = 2.2207 \tag{d}$$

Further trials yield the same result for $y_{0.1}$.

* Equation (*a*) is easily solved by the method of §25. A simple problem was chosen to illustrate the process of numerical solution and avoid undue complication.

Table 4

Line	x	y	y'	$\Delta_1 y'$	$\Delta_2 y'$	$\Delta_3 y'$	$\Delta_4 y'$
1	−0.1	1.8193	1.6193				
2	−0.05	1.9049	1.8049	0.1856			
3	0	2.0000	2.0000	0.1951	0.0095		
4	0.05	2.1051	2.2051	0.2051	0.0100	0.0005	
5	0.1	2.2207	2.4207	0.2156	0.0105	0.0005	0.0000
6	0.15	2.3474	2.6474	0.2267	0.0111	0.0006	0.0001
7	0.20	2.4857	2.8857	0.2383	0.0116	0.0005	−0.0001
8	0.25	2.6362	3.1362	0.2505	0.0122	0.0006	0.0001
9	0.30	2.7995	3.3995	0.2633	0.0128	0.0006	0.0000

Again, using (a) and (1) repeatedly with $h = -0.05$, $y_{-0.05} = 2$, we obtain in succession

$$\begin{aligned} y_{-0.05} &\cong 2 + \tfrac{1}{2}(2 + 1.9)(-0.05) = 1.9025 \\ y_{-0.05} &\cong 2 + \tfrac{1}{2}(2 + 1.8025)(-0.05) = 1.9049 \end{aligned} \qquad (e)$$

Further attempts to approximate $y_{-0.05}$ yield the same result. Similarly, using (a) and (1) twice with $x_0 = -0.05$, $y_{-0.05} = 1.9049$, $h = -0.05$, we get

$$y_{-0.1} \cong 1.8193 \qquad (f)$$

Using the values of the y's from (c), (d), (e), and (f) and the corresponding values for the y''s from (a), and computing differences, we get the first five lines of Table 4. To check these we use equations (7), (6), (5), (4), and (3) of §119 in that order. Observe that Y_m^n *is the increment which added to* y_m *(*y *in the* m*th line) gives* y_n.

Applying (7), §119, to line 5 of Table 4, we get

$$\begin{aligned} Y_1^2 &= 0.05[2.4207 - \tfrac{7}{2}(0.2156) + \tfrac{53}{12}(0.0105) - \tfrac{55}{24}(0.0005) + 0] \\ &= 0.08556^* \end{aligned}$$

Adding 0.0856 to 1.8193 = y_1, we get 1.9049 = y_2, thus checking y_1. Applying (6), §119, with $n = 5$ to line 5 of Table 4, we get

$$\begin{aligned} Y_2^3 &= 0.05[2.4207 - \tfrac{5}{2}(0.2156) + \tfrac{23}{12}(0.0105) - \tfrac{3}{8}(0.0005) + 0] \\ &= 0.09508 \end{aligned}$$

Adding 0.0951 to 1.9049 (= y_2), we get 2.0000, thus checking y_2. Similarly, using (5), §119, on line 5, we get $Y_3^4 = 0.10510$, and this added to 2 (= y_3) gives 2.1051 = y_4. Also, using (3), §119, on line 5, we get $Y_4^5 = 0.1156$, which added to 2.1051 (= y_4) gives 2.2207 = y_5.

* We compute the numbers inside the brackets to five decimal places and round off final results to four decimal places.

This completes the check of the first five lines of Table 4. The checks are important if one doubts the accuracy of his initial table.

To get a new line, use equations (2), §119, and (*a*) and then use (3) or (4), §119, to check the new value of y. Desiring line 6 for Table 4, we use (2), §119, with $n = 5$ on line 5 to obtain

$$\begin{aligned} Y_5^6 &= 0.05[2.4207 + \tfrac{1}{2}(0.2156) + \tfrac{5}{12}(0.0105) + \tfrac{3}{8}(0.0005) + 0] \\ &= 0.12665 \end{aligned}$$

Adding 0.1267 to 2.2207 ($= y_5$), we get $2.3474 = y_6$. To complete the sixth line, use (*a*) to get $y_6' = 2(0.15) + 2.3474 = 2.6474$, and then complete the differences for this line. Now, applying (3), §119, to line 6, we get

$$\begin{aligned} Y_5^6 &= 0.05[2.6474 - \tfrac{1}{2}(0.2267) - \tfrac{1}{12}(0.0111) - \tfrac{1}{24}(0.0006) + 0] \\ &= 0.12665 \end{aligned}$$

and, adding 0.1267 to 2.2207, obtain $2.6474 = y_6$. As another check, use (4) on line 6 to obtain

$$Y_4^6 = 0.1[2.6474 - 0.2267 + \tfrac{1}{6}(0.0111) + 0] = 0.24225$$

and adding 0.2423 to $2.1051 = y_4$, obtain $2.3474 = y_6$. Similarly, using (2), (3), and (4), §119, we may compute and check lines 7, 8, and 9.

When the change for a new line due to third differences $\tfrac{3}{8}h\,\Delta_3 y_n'$ *is very slight, the interval h on x may be doubled.* Hence, we shall double the interval by writing the x's, y's, and y''s of the odd numbered lines of Table 4 and computing the differences. This gives the first five lines of Table 5. These lines may be checked by using equations (7), (6), (5), and (3) of §119. The others may be found and checked by using (2), (3), and (4), §119, just as we derived lines 6 to 9 of Table 4.

Table 5

Line	x	y	y'	$\Delta_1 y'$	$\Delta_2 y'$	$\Delta_3 y'$	$\Delta_4 y'$
1	−0.1	1.8193	1.6193				
2	0	2.0000	2.0000	0.3807			
3	0.1	2.2207	2.4207	0.4207	0.0400		
4	0.2	2.4856	2.8856	0.4649	0.0442	0.0042	
5	0.3	2.7994	3.3994	0.5138	0.0489	0.0047	0.0005
6	0.4	3.1673	3.9673	0.5679	0.0541	0.0052	0.0005
7	0.5	3.5949	4.5949	0.6276	0.0597	0.0056	0.0004
8	0.6	4.0885	5.2885	0.6936	0.0660	0.0063	0.0007
9	0.7	4.6550	6.0550	0.7665	0.0729	0.0069	0.0006
10	0.8	5.3022	6.9022	0.8472	0.0807	0.0078	0.0009
11	0.9	6.0385	7.8385	0.9363	0.0891	0.0084	0.0006
12	1.0	6.8732	8.8732	1.0347	0.0984	0.0093	0.0009

Observe that equations (7), (6), (5), and (4) of §119 are used for checking the first five lines of a solution and that (2) and (3), §119, are used for finding additional lines of the tabular solution. Equation (4), §119, is simple to compute and may be used as an extra check. Of course, any of the equations (1) to (7), §119, may be used to check any part of a solution.

Exercises

1. Apply equations (7), (6), (5), and (3) of §119 with $n = 5$ to line 5 of Table 5 to check the first five values of y.

2. Use (2), §119, applied to line 5 of Table 5 to obtain y_6 for the sixth line, then compute y'_6 by (a), and find the differences in line 6. Check y in line 6 by applying (3), §119, to line 6. Also, check y_6 in line 6 by applying (4), §119, to line 6, finding Y_4^6, and adding it to y_4.

3. Compute y_7 of Table 5 by using (2), §119; then compute y'_7 by using (7), §119, and the differences of line 7. Then check y_7 by using (3), §119, and also by using (2), §119.

4. Compute a thirteenth line for Table 5.

5. To begin the numerical solution of

$$\frac{dy}{dx} = y' = x + 0.1y \tag{a}$$

through (0,1), take $h = 0.1$, use (1), §120, as was done to find the first five lines of Table 4, and compute y_0, $y_{0.1}$, $y_{0.2}$, $y_{0.3}$, and $y_{0.4}$ accurate to four decimal places.

6. Using the values obtained from exercise 5, compute $y'_{0.1}$, $y'_{0.2}$, $y'_{0.3}$, and $y'_{0.4}$, and then make a five-line horizontal difference table for approximating the solution of $dy/dx = x + 0.1y$ through (0,1). Check your table by using (7), (6), (5), and (3) of §119 applied to the fifth line.

7. Compute and check four more lines by using (2), (3), and (4), §119, and the result of exercise 6.

8. From the table of exercises 6 and 7, make a five-line horizontal difference table for doubling the interval and extend the table by three additional lines.

***9.** Find by numerical methods the solution of $dy/dx = x + y$ for which $y = 1$ when $x = 0$. First use intervals of $h = 0.05$ for x. After finding nine lines of the tabular solution, double the interval and compute three lines in addition to the five lines copied from the first part of the problem.

★**10.** Find the solution of

$$\frac{dy}{dx} = 0.1x + \log_{10} y$$

if $y = 2$ when $x = 0$. First, find the values of y and y' for which x is 0.0, 0.1, 0.2, 0.3, and 0.4. Then compute differences, and, using equations (2) and (3), §119, compute four more lines of the table. Now, double the interval, and compute four new lines by using (2) and (3), §119.

★**11.** Use Maclaurin's series to show that the solution of $dy/dx = 0.1x + \log_{10} y$ through (0,2) is $y \cong 2 + 0.30103x + 0.08268x^2 + 0.004343x^3 + \cdots$. Use this series to check your result in exercise 10 for the values 0.1, 0.2, 0.3, and 0.4 of x.

121 Halving the interval h of x

When the value of the fourth difference in (2), §119, $\frac{251}{720}h\,\Delta_4 y'_n$ is more than $\frac{1}{2}$ in the last figure of the computed values, the interval h should be shortened. If $h = 0.1$ and the data are accurate to four decimal places, the difference $\frac{251}{720}h\,\Delta_4 y'_n$ should be less than 0.00005. Hence if

$$\tfrac{251}{720}h(0.1)\,\Delta_4 y'_n > 0.00005 \qquad \text{or} \qquad \Delta_4 y'_n > 0.00014$$

the interval h is too large, and the general plan is to halve the interval, that is, to cut the interval from h to $\frac{1}{2}h$. An illustration will indicate the method of doing this.

Three lines of the solution of (*a*), §120, follow.

Now we apply the right member of interpolation formula (3), §118, with y replaced by y', y_n by y'_n, and u by $-\frac{1}{2}$, to the third line of Table 6. From this, we get as the value of y' at $x = 1.95$

$$y'_{1.95} = 27.5562 - \frac{1}{2}(2.8126) + \left(-\frac{1}{2}\right)\left(\frac{1}{2}\right)\frac{0.2676}{2!}$$
$$+ \left(-\frac{1}{2}\right)\left(\frac{1}{2}\right)\left(\frac{3}{2}\right)\frac{0.0254}{3!} + \left(-\frac{1}{2}\right)\left(\frac{1}{2}\right)\left(\frac{3}{2}\right)\left(\frac{5}{2}\right)\frac{0.0023}{4!}$$
$$= 26.11477$$

Table 6

x	y	y'	$\Delta_1 y'$	$\Delta_2 y'$	$\Delta_3 y'$	$\Delta_4 y'$
1.8	18.5986	22.1986	2.3028	0.2191	0.0207	0.0016
1.9	20.9436	24.7436	2.5450	0.2422	0.0231	0.0024
2.0	23.5562	27.5562	2.8126	0.2676	0.0254	0.0023

Table 7

Line	x	y	y'	$\Delta_1 y'$	$\Delta_2 y'$	$\Delta_3 y'$	$\Delta_4 y'$
1	1.80	18.5986	22.1986				
2	1.85	19.7393	23.4393	1.2407			
3	1.90	20.9436	24.7436	1.3043	0.0636		
4	1.95	22.2148	26.1148	1.3712	0.0669	0.0033	
5	2.00	23.5562	27.5562	1.4414	0.0702	0.0033	0.0000
6	2.05	24.9716	29.0716	1.5154	0.0740	0.0038	0.0005
7	2.10	26.4647	30.6647	1.5931	0.0777	0.0037	−0.0001

Similarly, we apply to the third line of Table 6 equation (3), §118, with y replaced by y', y_n by y'_n, and u by $-\frac{3}{2}$ to obtain $y'_{1.85} = 23.4393$. From (a), §120, we compute $y_{1.95} = y'_{1.95} - 2x = 26.11477 - 3.9 = 22.21477$, $y_{1.85} = 23.43929 - 3.7 = 19.73929$. Making a horizontal difference table from the values of x, y, y' just found and those in Table 6 and computing two new lines, we get Table 7.

Simpson's rule furnishes an excellent check. It may be written for our purposes

$$\int_a^{a+2kh} y'\,dx = \frac{h}{3}(y'_a + 4y'_{a+1} + 2y'_{a+2} + 4y'_{a+3} + \cdots + 4y'_{a+2k-1} + y'_{a+2k}) = y_{a+2kh} - y_a \quad (1)$$

For example, taking $a = 1.8$ and $k = 2$, we get from (1) and Table 7

$$\frac{0.05}{3}[22.1986 + 4(23.4393) + 2(24.7436) + 4(26.1148) + 27.5562] = 4.9576$$

and

$$y_{2.00} - y_{1.80} = 23.5562 - 18.5986 = 4.9576$$

Exercises

1. Using (1) with $a = 2.00$ and $k = 1$, make a check involving the last three lines of Table 7.

2. Using the last three lines of Table 5 and the right member of (4), §118, halve the interval; that is, construct a horizontal difference table for the five lines of the solution of $dy/dx = 2x + y$ through (0,2) associated with the values 0.8, 0.85, 0.9, 0.95, and 1.0 of x. Check your table by means of (1) and also by means of equations (4) to (7), §119.

Table 8

x	y	y'	$\Delta_1 y'$	$\Delta_2 y'$	$\Delta_3 y'$	$\Delta_4 y'$
1.0	1.8394	2.1606	0.4072	−0.0903	0.0197	−0.0049
1.2	2.3060	2.4940	0.3334	−0.0738	0.0165	−0.0032
1.4	2.8330	2.7670	0.2730	−0.0604	0.0134	−0.0031

3. Three lines in the solution of $dy/dx = 4x - y$ through (0,1) are given in Table 8. Apply (1) with $a = 1.0$ and $k = 1$ as a check. Use (4), §118, to find y' when $x = 1.1$ and when $x = 1.3$. Now make a five-line horizontal difference table based on the interval 0.1, and check the table. Use (2), (3), and (4), §119, to compute and check two more lines of this table. Then check all seven lines by means of (1).

122 Numerical solution of a system of simultaneous equations

The general plan of solving numerically two equations having the form

$$\frac{dy}{dx} = f_1(x,y,z) \qquad \frac{dz}{dx} = f_2(x,y,z)$$

consists in computing two difference tables, one for y and one for z, by means of equations (1) to (7), §119, the results from each being used in the other as they become available. Consider the solution of the equation

$$x\frac{d^2y}{dx^2} = y \tag{a)*$$

for which $x = 1$, $y = 1$, and $y' = 0$, or the equivalent set (§97),

$$y' = \frac{dy}{dx} = z \qquad z' = \frac{dz}{dx} = \frac{y}{x} \tag{b}$$

* The solution of this equation by §109 involves a complicated procedure and result, and the constants of integration involve difficulties. The solution by numerical methods is straightforward.

In beginning the solution, note that $y'_{x_0} + y'_{x_0+h} = z_{x_0} + z_{x_0+h}$ and $z'_{x_0} + z'_{x_0+h} = (y/x)_{x_0} + (y/x)_{x_0+h}$, and use the formulas

$$y_{x_0+h} \cong y_{x_0} + \tfrac{1}{2}(z_{x_0} + z_{x_0+h})h \tag{1}$$

$$z_{x_0+h} \cong z_{x_0} + \frac{1}{2}\left[\left(\frac{y}{x}\right)_{x_0} + \left(\frac{y}{x}\right)_{x_0+h}\right]h \tag{2}$$

where, as in equation (1), §120, a letter with a subscript p indicates the value of the letter when $x = p$. For the first line of the solution, we have

$$x = 1 \qquad y_1 = 1 \qquad y'_1 = z_1 = 0 \qquad x_1 = 1 \qquad \begin{aligned} z_1 &= 0 \\ z'_1 &= 1 \end{aligned} \tag{c}$$

Taking $h = 0.05$, $y_{1.05} = 1$, $x_1 = 1$, $z_1 = 0$, $x_{1.05} = 1.05$ in (2), we get $z_{1.05} \cong 0 + \frac{1}{2}[1 + (1/1.05)]0.05 = 0.04881$. Using $z = 0.04881$ in (1), we get $y_{1.05} \cong 1 + \frac{1}{2}(0 + 0.04881)(0.05) = 1.00122$.* Using $x = 1$, $y = 1.00122$, $h = 0.05$, $z = 0$ in (2), we get as a new estimate $z_{1.05} = 0 + \frac{1}{2}(1 + 1.00122/1.05)\ (0.05) = 0.0488$. Since this new estimate of z is the same as the old one, accurate to four decimal places, we accept it and the corresponding value of y. The new line then is

$$x = 0.05 \qquad y = 1.0012 \qquad z = 0.0488 \tag{d}$$

Now, assume that z changes by the same amount 0.0488 while x changes 0.05, and take $z_{1.1} = 0.0976$, $z_1 = 0.0488$, and $y_1 = 1.0012$ in (1) to obtain $y = 1.0049$. Then use (2) with $z_{1.05} = 0.0488$, $(y/x)_{1.05} = 1.0012/1.05$, and $(y/x)_{1.10} = 1.0049/1.1$ to obtain $z_{1.1} = 0.09552$. Another use of (1) with $z_{1.1} = 0.09552$ gives $y_{1.1} = 1.0048$. Further trials for $y_{1.1}$ and $z_{1.1}$ give the same value. Hence we accept $y_{1.1} = 1.0048$ and $z_{1.1} = 0.0955$. The three lines thus computed with differences are given in Table 9. To get two more lines, we continue to use

* In the computation, we use five decimal places and then round off results to four places.

Table 9

Line	x	y	z	$\Delta_1 z$	$\Delta_2 z$
1	1	1	0		
2	1.05	1.0012	0.0488	0.0488	
3	1.10	1.0048	0.0955	0.0467	0.0021

Table 10

Line	x	$z = y'$	$z' = y/x$	$\Delta_1 z'$	$\Delta_2 z'$	$\Delta_3 z'$	$\Delta_4 z'$
1	1	0.0000	1.0000				
2	1.05	0.0488	0.9535	−0.0465			
3	1.10	0.0955	0.9135	−0.0400	0.0065		
4	1.15	0.1403	0.8789	−0.0346	0.0054	−0.0011	
5	1.20	0.1835	0.8490	−0.0299	0.0047	−0.0007	0.0004
6	1.25	0.2253	0.8232	−0.0258	0.0041	−0.0006	0.0001
7	1.30	0.2659	0.8010	−0.0222	0.0036	−0.0005	0.0001
8	1.35	0.3054	0.7819	−0.0191	0.0031	−0.0005	0.0000
9	1.40	0.3441	0.7656	−0.0163	0.0028	−0.0003	0.0002

Table 11

Line	x	y	$y' = z$	$\Delta_1 y'$	$\Delta_2 y'$	$\Delta_3 y'$	$\Delta_4 y'$
1	1.00	1	0.0000				
2	1.05	1.0012	0.0488	0.0488			
3	1.10	1.0048	0.0955	0.0467	−0.0021		
4	1.15	1.0107	0.1403	0.0448	−0.0019	0.0002	
5	1.20	1.0188	0.1835	0.0432	−0.0016	0.0003	0.0001
6	1.25	1.0290	0.2253	0.0418	−0.0014	0.0002	−0.0001
7	1.30	1.0413	0.2659	0.0406	−0.0012	0.0002	0.0000
8	1.35	1.0556	0.3054	0.0395	−0.0011	0.0001	−0.0001
9	1.40	1.0718	0.3441	0.0387	−0.0008	0.0003	0.0002

(1) and (2). However, if we assume that, for line 4, $\Delta_2 z = 0.0021$, we get $\Delta_1 z = 0.0467 - 0.0021 = 0.0446$ and $z_{1.15} = 0.0446 + 0.0955 = 0.1401$. Using this value for $z_{1.15}$ in (1) and proceeding as before, we find $y_{1.15} = 1.0107$ and then $z_{1.15} = 0.1403$. Now add the fourth line to Table 9, compute differences, and proceeding as before, compute the fifth line. For each value of z we find a value for dz/dx by using (*b*). Tables 10 and 11 exhibit the five lines thus computed with corresponding differences; applying equations (7), (6), (5), and (3), §119, we check the first five lines of these tables.

Now apply (2), §119, to the fifth line of Tables 10 and 11 to obtain $y_{1.25} = 1.0290$ and $z_{1.25} = 0.2253$, enter $z_{1.25} = y'_{1.25} = 0.2253$ in Table 11, compute $z'_{1.25} = 1.0290/1.25 = 0.8232$ and enter it in Table 10, compute differences for the sixth lines, and check by using (3), §119, on these sixth lines. Proceeding in the same way, compute and check lines 7, 8, and 9 of Tables 10 and 11.

Exercises

1. Compute lines 7, 8, and 9 of Tables 10 and 11.

2. Complete Table 9 by computing and adding lines 4 and 5 to it.

3. Check the values of $y_{1.4}$ and $z_{1.4}$ by applying Simpson's rule, equation (1), §121, to the second and third columns of Tables 10 and 11.

4. Double the interval from Tables 10 and 11, and compute and check lines for $y_{1.5}$, $z_{1.5}$, $y_{1.6}$, and $z_{1.6}$.

5. To find an approximate solution of

$$y' = \frac{dy}{dx} = z \qquad z' = \frac{dz}{dx} = y - xz \tag{a}$$

satisfying $y = 1$ and $z = 2$ when $x = 0$, use the formulas

$$y_{i+1} = 1 + \int_0^x z_i\, dx \qquad z_{i+1} = 2 + \int_0^x (y_i - xz_i)\, dx \tag{b}$$

where the subscripts i refer to ith approximations. For example,

$$y_1 = 1 + \int_0^x 2\, dx = 1 + 2x$$

$$z_1 = 2 + \int_0^x [1 + 2x - x(2)]\, dx = 2 + x \tag{c}$$

To get y_2 and z_2, use (*b*) and (*c*) with $i = 1$. Similarly, find y_3 and z_3.

6. From the formulas derived in exercise 5, compute the values of

Table 12

x	-0.2	-0.1	0	0.1	0.2
y	0.6199	0.8050	1	1.2050	1.4199
z	1.8013	1.9002	2	2.0998	2.1987

Table 12. From this table, compute values for y' and z', make corresponding difference tables similar to Tables 10 and 11, and then using (2) and (3), §119, compute and check

$$y_{0.3},\ z_{0.3},\ y_{0.4},\ z_{0.4},\ y_{0.5},\ z_{0.5},\ y_{0.6},\ z_{0.6}$$

123 The Runge-Kutta method

The method set forth in this section was devised by C. Runge about 1894 and extended by W. Kutta a few years later. It may well be used in obtaining the first few values required for a solution. It may be used for an extended solution also, but this use involves much computation.

Let x_0 and y_0 be initial values of the variables x and y for a solution of the equation

$$\frac{dy}{dx} = f(x,y) \tag{1}$$

A change h is made in x, and the corresponding change Δy in y is found by computing in order k_1, k_2, k_3, k_4, and Δy by means of the following formulas:

$$\begin{aligned} k_1 &= f(x_0,y_0)h \\ k_2 &= f(x_0 + \tfrac{1}{2}h, y_0 + \tfrac{1}{2}k_1)h \\ k_3 &= f(x_0 + \tfrac{1}{2}h, y_0 + \tfrac{1}{2}k_2)h \\ k_4 &= f(x_0 + h, y_0 + k_3)h \\ \Delta y &= \tfrac{1}{6}(k_1 + 2k_2 + 2k_3 + k_4) \end{aligned} \tag{2}$$

This gives the values $x_1 = x_0 + h$ and $y_1 = y_0 + \Delta y$. To find a third pair of values, use the same equations (2) with x_1 and y_1 taking the place of x_0 and y_0, respectively. When x_2 and y_2 are found, use (2) with x_0 and y_0 replaced by x_2 and y_2, etc.

Observe that the change Δy in y for a change h in x is an average of six values of $(dy/dx)h$. The first, k_1, is $(dy/dx)h$ evaluated at the beginning of the interval; the last, k_4, is $(dy/dx)h$ evaluated approximately at the end of the interval; and two other values, each doubled, are computed at midpoints of the segment from (x_0,y_0) to approximated end points. If f is a function of x only, the value Δy of (2) becomes

$$y = \tfrac{1}{6}h[f(x_0) + 4f(x_0 + \tfrac{1}{2}h) + f(x_0 + h)]$$

the value obtained by applying Simpson's rule to $y' = f(x)$ for the interval from x_0 to $x_0 + h$.

The error inherent in the Runge-Kutta method is roughly that of h^5.* The following example will illustrate the procedure in using the method.

Example. Find three pairs of values of x and y for the solution of $dy/dx = -x + y$ by means of equations (2) if the initial values of the variables are $x = 0$, $y = 2$.

* See F. A. Willers, "Numerisches Integration," pp. 91–92. Also consult H. Levy and E. A. Baggott, "Numerical Solutions," Dover Publications Inc., New York.

Solution. The first value pair is $x = 0$, $y = 2$. Take $h = 0.1$, $x_0 = 0$, and $y_0 = 2$ in (2), keep five decimal places during the computation, but round off the value of Δy to four decimal places. This gives

$$k_1 = (-0 + 2)(0.1) = 0.2$$
$$k_2 = \left(-0 - \frac{0.1}{2} + 2 + \frac{0.2}{2}\right)(0.1) = 0.205$$
$$k_3 = \left(-0 - \frac{0.1}{2} + 2 + \frac{0.205}{2}\right)(0.1) = 0.20525$$
$$k_4 = (-0 - 0.1 + 2 + 0.20525)(0.1) = 0.21053$$
$$\Delta y = \tfrac{1}{6}[0.2 + 2(0.205) + 2(0.20525) + 0.21053] = 0.2052$$

Hence

$$x_1 = 0.1 \qquad y_1 = 2.2052$$

Substituting these values of x_1 and y_1 for x_0 and y_0 and 0.1 for h in (2), we obtain $k_1 = 0.21052$, $k_2 = 0.21605$, $k_3 = 0.21632$, $k_4 = 0.22215$, and $\Delta y = \frac{1}{6}(k_1 + 2k_2 + 2k_3 + k_4) = 0.21624$. Hence, the third pair of values of x and y is

$$x_2 = 0.2 \qquad y_2 = 2.4214$$

The next pair of values could be computed by using x_2 and y_2 for x_0 and y_0 in (2).

Exercises

1. Check the values for x_2 and y_2 in the solution just given. Then compute $x_3 = 0.3$ and $y_3 = 2.6499$ by using (2) with $x_0 = x_2$, $y_0 = y_2$, and $h = 0.1$. Finally, compute y when $x = 0.4$.

2. For the equation $dy/dx = -x + y$ of the example, find y'_0, $y'_{0.1}$, $y'_{0.2}$, $y'_{0.3}$, and $y'_{0.4}$, make the corresponding horizontal difference table of x, y, and y', check this table by using (7), (6), (5), and (3) of §119, and then use (2) and (3), §119, to compute two new lines.

3. Find three value pairs of x and y by using (2) for the solution of $dy/dx = xy + 1$, in which $y = 1$ when $x = 0$. Use $h = 0.1$.

4. The Runge-Kutta equations for solving two simultaneous equations of the type

$$\frac{dy}{dx} = f(x,y,z) \qquad \frac{dz}{dx} = F(x,y,z)$$

are

$$
\begin{aligned}
k_1 &= f(x_0,y_0,z_0)\,\Delta x \\
l_1 &= F(x_0,y_0,z_0)\,\Delta x \\
k_2 &= f(x_0 + \tfrac{1}{2}\,\Delta x,\ y_0 + \tfrac{1}{2}k_1,\ z_0 + \tfrac{1}{2}l_1) \\
l_2 &= F(x_0 + \tfrac{1}{2}\,\Delta x,\ y_0 + \tfrac{1}{2}k_1,\ z_0 + \tfrac{1}{2}l_1) \\
k_3 &= f(x_0 + \tfrac{1}{2}\,\Delta x,\ y_0 + \tfrac{1}{2}k_2,\ z_0 + \tfrac{1}{2}l_2) \\
l_3 &= F(x_0 + \tfrac{1}{2}\,\Delta x,\ y_0 + \tfrac{1}{2}k_2,\ z_0 + \tfrac{1}{2}l_2) \\
k_4 &= f(x_0 + \Delta x,\ y_0 + k_3,\ z_0 + l_3) \\
l_4 &= F(x_0 + \Delta x,\ y_0 + k_3,\ z_0 + l_3) \\
\Delta y &= \tfrac{1}{6}(k_1 + 2k_2 + 2k_3 + k_4) \\
\Delta z &= \tfrac{1}{6}(l_1 + 2l_2 + 2l_3 + l_4)
\end{aligned}
$$

In using these equations, first find k_1 and l_1, then k_2 and l_2, then k_3 and l_3, then k_4 and l_4, and finally Δy and Δz. Use these equations to find the first five lines of the solution of

$$\frac{dy}{dx} = z - x \qquad\qquad \frac{dz}{dx} = y + x$$

satisfying $y = 1$ and $z = 1$ when $x = 0$. Use $h = 0.05$. Then use (2) and (3), §119, to compute two more lines.

14

Partial Differential Equations

124 Introduction

Partial differential equations are likely to enter into any serious investigation involving more than two variables. We have already met them in the discussions of potential theory, envelopes of curves, exact differential equations, and existence theorems. They play a particularly important role in most sciences dealing with wave motions, such as, for example, heat, light, electricity, magnetism, radio, radar, television, and weather. This chapter will deal mainly with the theory of partial differential equations, and the next with their applications.

125 Solution of a partial differential equation

A *solution*, or *integral*, of a partial differential equation is a nondifferential relation among the variables which satisfies the equation. Its general form involves one or more arbitrary functions.* The following examples will illustrate these facts.

Example 1. Prove that

$$z = ax + a^2y^2 + b \qquad (a)$$

in a solution of

$$\frac{\partial z}{\partial y} = 2y\left(\frac{\partial z}{\partial x}\right)^2 \qquad (b)$$

Solution. Differentiating (a) partially with respect to x and with respect to y, we obtain

$$\frac{\partial z}{\partial x} = a \qquad \frac{\partial z}{\partial y} = 2a^2y \qquad (c)$$

Substitution from (c) in (b) gives the identity

$$2a^2y = 2ya^2$$

Example 2. Prove that

$$y = \varphi(ct - x) + \psi(ct + x) \qquad (a)$$

where φ and ψ are arbitrary functions,† is a solution of

$$\frac{\partial^2 y}{dt^2} = c^2 \frac{\partial^2 y}{\partial x^2} \qquad (b)$$

Solution. It will be convenient to use the notation

$$\frac{d\varphi}{d(ct - x)} = \varphi' \qquad \frac{d^2\varphi}{[d(ct - x)]^2} = \varphi''$$
$$\frac{d\psi}{d(ct + x)} = \psi' \qquad \frac{d^2\psi}{[d(ct + x)]^2} = \psi''$$

Partial differentiation of (a) gives

$$\frac{\partial y}{\partial t} = c\varphi' + c\psi' \qquad \frac{\partial y}{\partial x} = -\varphi' + \psi'$$
$$\frac{\partial^2 y}{\partial t^2} = c^2\varphi'' + c^2\psi'' \qquad \frac{\partial^2 y}{\partial x^2} = \varphi'' + \psi'' \qquad (c)$$

* Throughout the chapters on partial differential equations, we shall assume that the arbitrary functions and their derivatives involved are continuous and single-valued in all regions under consideration.

† Equation (a) in example 2 represents a motion compounded of two wave motions having equal wavelengths and periods but opposite directions. Because of this fact, equation (b) of example 2 plays a very important role in mathematical physics.

Substituting from (*c*) in (*b*), we obtain the identity

$$c^2\varphi'' + c^2\psi'' = c^2\varphi'' + c^2\psi''$$

The following observations pertain to the wave equation (*a*): When a smooth wave moves on a lake, the individual drops of water merely move up and down, and thus each surface drop is at troughs and crests in succession. Any individual crest appears to move across the lake. Now consider a motion in the xy plane if the distance y of a point from the X axis is defined by $y = a \sin (ct + x)$, where t represents time. Fix attention on the point (x_0,y). Since $y = a \sin (ct + x_0)$, the point (x_0,y) merely moves up and down parallel to the Y axis with period $2\pi/c$. At time t_0 all points $(x, a \sin [ct_0 + x])$ lie on the fixed curve $y = a \sin (ct_0 + x)$ of wavelength $\lambda = 2\pi/1$. Consider the crest moving at $y = a$. Then $x + ct = \frac{1}{2}\pi$ say, and we get $dx/dt = -c$ as the velocity of the crest parallel to the X axis. This indicates why wave equation (*a*) represents two superimposed waves moving in opposite directions with speed c.

Throughout the chapters on partial differential equations we shall use the notation

$$p = \frac{\partial z}{\partial x} \qquad q = \frac{\partial z}{\partial y} \qquad r = \frac{\partial^2 z}{\partial x^2} \qquad s = \frac{\partial^2 z}{\partial x\, \partial y} \qquad t = \frac{\partial^2 z}{\partial y^2} \tag{1}$$

Example 3. Find the differential equation having as solution

$$\varphi(x^2 - z^2, x^3 - y^3) = 0 \tag{a}$$

where φ represents an arbitrary function.

Solution. Consider z as a function of x and y defined by (*a*), and let

$$x^2 - z^2 = u \qquad x^3 - y^3 = v$$

Differentiating (*a*) partially with respect to x and then partially with respect to y, we get

$$\frac{\partial\varphi}{\partial u}(2x - 2zp) + \frac{\partial\varphi}{\partial v}(3x^2) = 0 \tag{b}$$

$$\frac{\partial\varphi}{\partial u}(-2zq) + \frac{\partial\varphi}{\partial v}(-3y^2) = 0 \tag{c}$$

Eliminating $\partial q/\partial u$ and $\partial q/\partial v$ from (*b*) and (*c*) we get

$$z(x^2q + y^2p) = xy^2$$

Exercises

In the following exercises, φ and ψ represent arbitrary functions of the indicated variables.

Verify the fact that each equation on the left is a solution of the partial differential equation written opposite it:

1. $az = a^2x + y + b,\qquad pq = 1.$
2. $2z = 2axe^y + a^2e^{2y} + b,\qquad q = xp + p^2.$
3. $6az = x^3y + x\varphi(y) + \psi(y),\qquad ar = xy.$

Using the method of example 3, derive the differential equation having a solution:

4. $z = \varphi(x + y)$
5. $z = x\varphi(x + y)$
6. $y = z\varphi(x + y)$
7. $y = \varphi(x) + \psi(z)$
8. $y = \varphi(x + y) + \psi(z)$
9. $z = \varphi(x + 2y) + \psi(x - y)$
10. $\varphi(x^2 + y^2, y^2 - z^2) = 0$
11. $\varphi(x^2 - z^2, y^2 - z^2) = 0$

12. Prove that (*a*) $z = \varphi(y/x) + \psi(xy)$ is a solution of $x^2r - y^2t = qy - px$; (*b*) $x = \varphi(z) + \psi(y)$ is a solution of $ps - qr = 0$.

13. Show that $q = 0$ for all cylindrical surfaces having their elements parallel to the Y axis

14. Show that, for all surfaces of revolution about the Z axis, $yp = xq$ and, for all surfaces of revolution about the X axis, $y + zq = 0$.

15. For a wave motion defined by $y = a \cos(\pi t + x)$, x and y in feet and t in seconds, find the wave length, the period, and the velocity of the wave.

126 Equations easily integrable

Some partial differential equations may be solved by inspection. Thus, the equation

$$\frac{\partial z}{\partial x} = x^2 + y \tag{1}$$

evidently has as a solution

$$z = \tfrac{1}{3}x^3 + xy + \varphi(y) \tag{2}$$

where φ represents an arbitrary function. Since only differentiation with respect to x was indicated in (1), we got (2) from it by integrating with respect to x while treating y as a constant. Observe that $\varphi(y)$ plays the role of a constant since y was considered constant for the integration.

Again consider the equation

$$ys + p = 4xy \tag{3}$$

where s and p have the meanings defined in (1) of §125. Writing this in the form

$$y\frac{\partial p}{\partial y} + p = 4xy$$

we see that it is an ordinary linear equation of the first degree in p and y if x is considered constant. Hence, its solution is

$$py = 2xy^2 + \varphi(x) \tag{4}$$

In (4), replace p by its equal $\partial z/\partial x$, and integrate again, considering y as a constant, to obtain

$$z = \int \left[2xy + \frac{1}{y}\varphi(x)\right] dx \qquad y \text{ constant}$$

or

$$z = x^2y + \frac{1}{y}\varphi_1(x) + \psi(y)$$

Here, $\varphi_1(x)[= \int\varphi(x)\,dx]$ and $\psi(y)$ are arbitrary functions.

As a case requiring the solution of a second-order equation, consider

$$r - y^2z = xy^2 \tag{5}$$

Writing this in the form

$$\frac{\partial^2 z}{\partial x^2} - y^2z = xy^2$$

integrating it, considering y as constant, and using functions of y as constants of integration, we get

$$z = \varphi_1(y)e^{xy} + \varphi_2(y)e^{-xy} - x$$

These illustrations indicate how a large class of partial differential equations may be integrated by using the methods of solving ordinary differential equations. It is worthy of note that *the constant of integration consists of an arbitrary function of the variable considered constant during the integration.*

Exercises

Solve:

1. $\dfrac{\partial z}{\partial x} = 3x^2 + y^2$
2. $y\dfrac{\partial z}{\partial y} + z = x^2$
3. $y\dfrac{\partial p}{\partial y} + p = 2x$
4. $r = f(x,y)$
5. $ys = x + ay$
6. $t - q = e^x + e^y$
7. $p + r = xy$
8. $xr + p = xy$

(9–14) Using the plan of solution suggested for equation (5), solve the following equations:

9. $r - 4z = 8y^2$
10. $t + 4z = 8x^2$
11. $r - 2p - 3z = y^2(2 + 3x)$
12. $t + 2q - 3z = x^2(2 - 3y)$
13. $2r - 3yp - 5y^2z = y^2(3 + 5x)$
14. $t - 2xq + 5x^2z = 5x^3$

Solve:

15. $r + p^2 = y^2$
16. $zp + z^2 = xy^2$
17. $pr + p^2 = a$
18. $\dfrac{\partial r}{\partial y} + \dfrac{\partial p}{\partial x} = 12$

127 Equations having the form $Pp + Qq = R$

In this section and elsewhere, we shall use the notation

$$\frac{\partial u}{\partial x} = u_x \qquad \frac{\partial u}{\partial y} = u_y \qquad \frac{\partial^2 u}{\partial x\,\partial y} = u_{xy} \qquad \text{etc.} \tag{1}$$

The content of this section relates to the following theorem.

THEOREM. If $u(x,y,z) = a$, $v(x,y,z) = b$ *are two independent integrals of the ordinary differential equations*

$$\frac{dx}{P(x,y,z)} = \frac{dy}{Q(x,y,z)} = \frac{dz}{R(x,y,z)} \tag{2}*$$

then

$$\varphi(u,v) = 0 \qquad \text{or} \qquad u = \psi(v) \tag{3}$$

* The theory of §99 applies to solve equations of the type (2).

is a general solution of

$$Pp + Qq = R \tag{4}$$

provided that $\partial\varphi/\partial z \neq 0$.

Assume that a solution of equation (2) consists of two independent equations

$$u(x,y,z) = a \qquad v(x,y,z) = b \tag{5}$$

derived from (2) by the methods of §99 or other methods.

The theorem will be proved by showing that $\varphi(u,v) = 0$ satisfies

$$P\varphi_x + Q\varphi_y + R\varphi_z = 0 \tag{6}$$

and then that any solution of (6) is also a solution of (4).

From (3) we get

$$\begin{aligned} \varphi_x(u,v) &= \varphi_u u_x + \varphi_v v_x \qquad \varphi_y = \varphi_u u_y + \varphi_v v_y \\ \varphi_z &= \varphi_u u_z + \varphi_v v_z \end{aligned} \tag{7}$$

Substituting the values of φ_x, φ_y, and φ_z from (7) in the left member of (6) and transforming the result, we get

$$\begin{aligned} &P\varphi_x + Q\varphi_y + R\varphi_z \\ &\quad = P(\varphi_u u_x + \varphi_v v_x) + Q(\varphi_u u_y + \varphi_v v_y) + R(\varphi_u u_z + \varphi_v v_z) \\ &\quad = \varphi_u(Pu_x + Qu_y + Ru_z) + \varphi_v(Pv_x + Qv_y + Rv_z) \end{aligned} \tag{8}$$

Equating the ratios of (2) to λ we get

$$dx = P\lambda \qquad dy = Q\lambda \qquad dz = R\lambda \tag{9}$$

Also from (5) we get

$$u_x\,dx + u_y\,dy + u_z\,dz = 0 \qquad v_x\,dx + v_y\,dy + v_z\,dz = 0 \tag{10}$$

Substituting the values of dx, dy, and dz from (9) in (10), we get

$$\lambda(Pu_x + Qu_y + Ru_z) = 0 \qquad \lambda(Pv_x + Qv_y + Rv_z) = 0 \tag{11}$$

Now $\lambda \neq 0$ because dx, dy, and dz are not all zero, and we get from (11)

$$Pu_x + Qu_y + Ru_z = 0 \qquad Pv_x + Qv_y + Rv_z = 0 \tag{12}$$

Using these values in (8) we get (6).

If $\varphi(x,y,z) = c$ is any solution of (6), it is also a solution of (4). To prove this, note that

$$\varphi_x + \varphi_z p = 0 \qquad \varphi_y + \varphi_z q = 0 \tag{13}$$

replace φ_x and φ_y in (6) by their values from (13), divide the result through by $-\varphi_z$,* and obtain (4). This completes the proof of the theorem.

If $\varphi(u,v)$ is solved for u in terms of v, the result has the form

$$u = \psi(v) \tag{14}$$

and this is often a more convenient form than $\varphi(u,v) = 0$.

Since $\partial u/\partial x$, $\partial u/\partial y$, $\partial u/\partial z$ are direction numbers of a normal to surface $u(x,y,z) = c$ at (x,y,z), equation (6) shows that *a line with direction numbers p, q, -1 is normal to a surface represented by an integral $\varphi(x,y,z) = c$ of* (4), and (2) shows that *the surface is tangent at (x,y,z) to a line passing through (x,y,z) and having direction numbers P, Q, R.* Figure 1 is suggestive.

By the same line of reasoning we could show that *if*

$$u_i(x_1,x_2, \ldots ,x_n,z) = c_i \qquad i = 1, 2, \ldots , n \tag{15}$$

are independent solutions of the differential equations

$$\frac{dx_1}{P_1} = \frac{dx_2}{P_2} = \cdots = \frac{dx_n}{P_n} = \frac{dz}{R} \tag{16}$$

then

$$\varphi(u_1,u_2, \ldots ,u_n) = 0,$$

φ representing an arbitrary function, is a solution of

$$P_1 z_{x_1} + P_2 z_{x_2} + \cdots + P_n z_{x_n} = R \tag{17}$$

Example 1. Solve $(y + z)p + (x + z)q = x + y$.

* $\varphi_z \neq 0$, for any function that contains both u and v must contain z; and therefore φ_z cannot vanish.

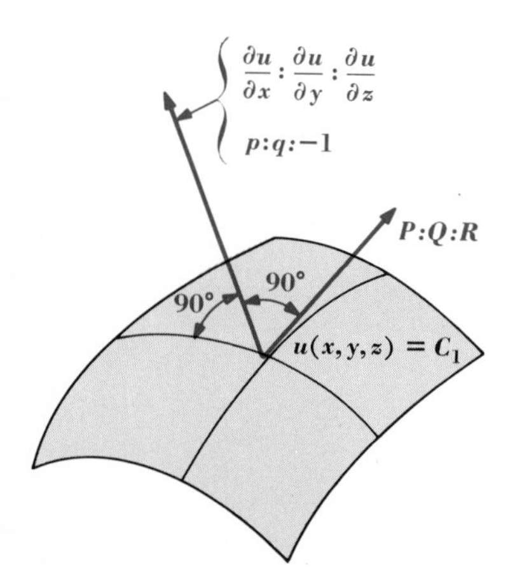

Figure 1

Solution. The first step is to find two integrals of

$$\frac{dx}{y+z} = \frac{dy}{x+z} = \frac{dz}{x+y} \tag{a}$$

By the theory of proportion, we find from (*a*)

$$\frac{dx + dy + dz}{2x + 2y + 2z} = \frac{dx - dy}{-(x - y)} = \frac{dx - dz}{-(x - z)} \tag{b}$$

Two integrals of (*b*) evidently are

$$\ln (x + y + z) + 2 \ln (x - y) = C_1$$

or

$$(x + y + z)(x - y)^2 = c_1 \tag{c}$$

and

$$\ln (x - y) - \ln (x - z) = C_2 \qquad \text{or} \qquad x - y = c_2(x - z) \tag{d}$$

Hence, the solution required is

$$\varphi\left[(x + y + z)(x - y)^2, \frac{x - y}{x - z}\right] = 0$$

or it may be written

$$x - y = (x - z)\varphi[(x + y + z)(x - y)^2]$$

Example 2. Solve

$$2xz_{xx} - yz_{xy} + 2x + 2z_x = 0 \tag{a}$$

Solution. In (*a*), let us set $z_x = p$ to obtain

$$2xp_x - yp_y = -2p - 2x \tag{b}$$

Solving (*b*) for p by the method of this section, we get

$$2px + x^2 = \varphi(xy^2) \qquad \text{or} \qquad p = \frac{\partial z}{\partial x} = -\tfrac{1}{2}x + \tfrac{1}{2}x^{-1}\varphi(xy^2) \tag{c}$$

Integrating the second equation, considering y as constant, we get

$$z = -\frac{x^2}{4} + \int \frac{1}{2xy^2}\,\varphi(xy^2)y^2\,dx = -\frac{x^2}{4} + \psi_1(xy^2) + \psi_2(y)$$

Here, $\psi_1(xy^2)$, being defined by $\psi_1'(xy^2) = [1/(2xy^2)]\varphi(xy^2)$, is an arbitrary function.

Exercises

Solve the following differential equations:

1. $p + q = z$
2. $xp + yq = 2z$
3. $xp + zq = y$
4. $ap + bq = c$
5. $xp + zq + y = 0$
6. $xyq - x^2p = y^2$
7. $x^2p + y^2q = axy$
8. $yzp + xzq + 2xy = 0$

9. If $R = 0$ in (4), show that $z = c$ is one solution.

10. $y\,\partial^2 z/(\partial x\,\partial y) + \partial^2 z/\partial x^2 = 4x$. *Hint:* First integrate with respect to x, treating y as constant. Observe that $\int_a^x \psi_1(ye^{-x})\,dx = \int_a^x \psi'(ye^{-x})(-ye^{-x}\,dx) = \psi(ye^{-x}) + \varphi(y)$.

11. Determine the equations of the surfaces that cut orthogonally the system of ellipsoids $\frac{1}{5}x^2 + \frac{1}{3}y^2 + z^2 = c^2$. *Hint:* $P = \frac{2}{5}x$, $Q = \frac{2}{3}y$, $R = 2z$.

Read the solution of example 2, and then solve the following equations:

12. $xr - ys = 2y$
13. $xr + ys = 0$
14. $xs - yt = 2x$
15. $xs - yt + q = 12xy$
16. $xs - 2x^2t = q$
17. $xs - 2yt = 2q + x$
18. $r + ys = p$
19. $xr + s + p = 0$

20. The condition that $u(x,y)$ be an integrating factor of

$$M(x,y)\,dx + N(x,y)\,dy = 0$$

is $\partial(uM)/\partial y - \partial(uN)/\partial x = 0$. Show that this is equivalent to

$$Nu_x - Mu_y = u(M_y - N_x) \qquad (a)$$

Using (*a*), find an integrating factor of

(*a*) $(y + y^2)\,dx - (x + y^2 + 2xy)\,dy = 0$
(*b*) $(\ln x - 2xy)\,dx + (2xy - 2x^2)\,dy = 0$

21. Use equation (*a*) of exercise 20 to find an integrating factor of $M\,dx + N\,dy = 0$ if $(\partial N/\partial x) - (\partial M/\partial y)$ is equal to

(*a*) $Mf(y)$ (*b*) $Nf(x)$ (*c*) $\dfrac{k(xM - yN)}{xy}$

(*d*) $\dfrac{axM - byN}{xy}$

22. Solve $xy^2p - y^3q + axz = 0$.

23. Solve $(x + y)(p - q) = z$.

24. Solve $xp + yq = 0$. Show that $z = c_1$ is one solution and $y/x = c_2$ is another. Check by direct substitution that $z = y^2/x^2$ and also $z = x/y$ are solutions.

25. Solve $x(1 + y)q = z$.

26. Solve $x\, \partial^2 z/\partial x^2 + y\, \partial^2 z/(\partial x\, \partial y) + \partial z/\partial x = 2x + y$.

27. Find the equation of all surfaces whose normals intersect the line $x = y$, $z = 0$. *Hint:* $p/(x - a) = q/(y - a) = -1/z$. Eliminate a.

128 Finding particular solutions satisfying given conditions

The solution of a differential equation

$$P(x,y,z)p + Q(x,y,z)q = R(x,y,z) \tag{1}$$

involves an arbitrary function and therefore has great generality. The solution of $xp + yq = z$ is

$$u = \varphi(v) \qquad \text{where } u = \frac{z}{x} \qquad v = \frac{y}{x} \tag{2}$$

This represents any surface containing all lines from (0,0,0) to points on any space curve. When various conditions are imposed, the form of $\varphi(v)$ in (2) must be determined to get the corresponding solution.

Example 1. Find the solution of $xp + yq = z$ which represents a surface through the curve

$$x = t \qquad y = 2t^2 + t \qquad z = 1 + t^3 \tag{a}$$

Solution. Here the general solution, from (2), is

$$z = x\varphi\left(\frac{y}{x}\right) \tag{b}$$

Substituting x, y, and z from (a) into (b), we get

$$1 + t^3 = t\varphi\left(\frac{2t^2 + t}{t}\right) = t\varphi(2t + 1) \tag{c}$$

In (c) let $w = 2t + 1$, or $t = \frac{1}{2}(w - 1)$, and obtain

$$1 + \tfrac{1}{8}(w - 1)^3 = \tfrac{1}{2}(w - 1)\varphi(w)$$

or

$$\varphi(w) = 2(w - 1)^{-1} + \tfrac{1}{4}(w - 1)^2 \tag{d}$$

Using the form of $\varphi(w)$ from (d) with $w = y/x$, we get

$$z = x\left[2\left(\frac{y}{x} - 1\right)^{-1} + \frac{1}{4}\left(\frac{y}{x} - 1\right)^{2}\right] \qquad (e)$$

or

$$4xz = \frac{8x^3}{y - x} + (y - x)^2 \qquad (f)$$

Example 2. Find a solution of $xp - yq = z$ such that $p = 3q$ when $x = 2$.

Solution. The general solution of $xp - yq = z$ may be written

$$z = x\varphi(xy) \qquad (a)$$

Here

$$p = \varphi + xy\varphi' \qquad \text{where } \varphi'(u) = \frac{d\varphi(u)}{du} \qquad (b)$$

and

$$q = x^2\varphi' \qquad (c)$$

Expressing the condition that $p = 3q$ when $x = 2$, by using (b) and (c) we get

$$\varphi(2y) + 2y\varphi'(2y) = 12\varphi'(2y) \qquad (d)$$

In (d) replace $2y$ by w, solve for $\varphi(w)$, and obtain

$$\varphi(w) + (w - 12)\frac{d\varphi(w)}{dw} = 0 \qquad (e)$$

$$\varphi(w)(w - 12) = c \qquad \text{or} \qquad \varphi(w) = \frac{c}{(w - 12)} \qquad (f)$$

Substituting $c/(xy - 12)$ from (f) for $\varphi(xy)$ in (a), we get

$$z = \frac{cx}{xy - 12} \qquad (g)$$

Exercises

Find the solution of each equation subject to the indicated conditions:

1. $xp + yq = z$; satisfied by $x = t$, $y = t$, $z = t^3 + 5$.
2. $xp - yq = z$; satisfied by $x = 2t + 2$, $y = t^{-1}$, $z = t^2$.
3. $xp + yq = y^2$; if $p = q$ when $y = 2$.
4. $p + q = z$; if it contains the curve $x + y = 1$, $x^2 - y^2 = z$.

Hint: The curve in parametric form is given by $x = t$, $y = 1 - t$, $z = 2t - 1$.

5. $p + q = z$; if it contains the curve $x^2 + z^2 = a^2$, $y = 0$.

6. $p + q = z$; if it contains the curve $z = \sin x$, $y = 0$.

7. Find the equation of all surfaces having tangent planes with x-intercept 1. Also, find the equation of the particular surface passing through the circle $x = 0$, $y^2 + z^2 = 25$.

8. Find the equation of a surface for which

$$xyq - x^2p = y^2$$

and satisfying the condition $p = q$ when $x = 1$.

129 Separation of variables

In this section, we shall denote differentiation by subscripts. The process considered will be applied only to linear partial differential equations, that is, equations linear in a dependent variable and its partial derivatives with respect to independent variables. Thus,

$$A(x,y)z_{xx} + B(x,y)z_{xy} + C(x,y)z_{yy} = E(x,y)$$

is an equation in x, y, and z linear in z. Just as in §46, we can show that, if $z_i = f_i(x,y)$, $i = 1, 2, \ldots, n$, are solutions of an equation linear in z and having no term free of z, then $z = c_1f_1 + c_2f_2 + \cdots + c_nf_n$ is also a solution.

A process often used in the investigations of physical science is called separation of variables. The following example illustrates the process and the type of solutions obtained by using it:

Example. Solve

$$z_{xx} + 4z_{yy} = 0 \qquad (a)$$

Solution. Assume a solution having the form

$$z = XY \qquad (b)$$

where X is a function of x only and Y a function of y only. Denoting derivatives by dots, we use such expressions as

$$\frac{dX}{dx} = \dot{X} \qquad \frac{d^2X}{dx^2} = \ddot{X} \qquad \frac{dY}{dy} = \dot{Y} \qquad \text{etc.} \qquad (c)$$

Substituting z from (b) in (a), we obtain

$$\ddot{X}Y + 4\ddot{Y}X = 0 \tag{d}$$

From this we easily obtain

$$\frac{\ddot{X}}{X} = -\frac{4\ddot{Y}}{Y} \tag{e}$$

Here $\ddot{X}/X$, being a function of x, does not change when y alone changes and $-4\ddot{Y}/Y$ does not change when x alone changes. Hence these two ratios will be equal when and only when both are equal to the same constant k, where k may be a suitable constant, that is, when

$$\ddot{X} = kX \qquad \text{and} \qquad -4\ddot{Y} = kY \tag{f}$$

The solutions of (f) for $k > 0$ are

$$X = a_1e^{\sqrt{k}x} + a_2e^{-\sqrt{k}x} \qquad Y = a_3 \sin\left(\tfrac{1}{2}\sqrt{k}\,y\right) + a_4 \cos\left(\tfrac{1}{2}\sqrt{k}\,y\right) \qquad k > 0 \tag{g}$$

Now (b) and (g) satisfy (a) for k any value. Hence, in accordance with the principle of adding solutions of linear equations, we may write as a solution of (a)

$$z = \sum^{k} X_kY_k = \sum^{k} \{e^{\sqrt{k}x}[a_k \sin\left(\tfrac{1}{2}\sqrt{k}\,y\right) + b_k \cos\left(\tfrac{1}{2}\sqrt{k}\,y\right)] + e^{-\sqrt{k}x}[c_k \sin\left(\tfrac{1}{2}\sqrt{k}\,y\right) + d_k \cos\left(\tfrac{1}{2}\sqrt{k}\,y\right)]\} \tag{h}$$

where k may take in succession the values of any finite set of positive numbers. If k takes the values k_1, k_2, k_3, . . . , then equation (h), involving an infinite series, is a valid solution within its region of absolute convergence.

If k is negative, the sine-cosine part of the solution would go with X and the exponential part with Y. When $k = 0$, we have

$$X = c_1x + c_2 \qquad Y = c_3y + c_4$$

Thus, it appears that the solution is made up of three types. For any application, appropriate types are used.

Exercises

Use the method of separation of variables to find the various types of solutions of exercises 1 to 4:

1. $z_{xx} + z_{yy} = 0$ **2.** $z_{xx} - z_y = 0$

3. $z_x + z_y = 0$ **4.** $z_x + z_y - 3z = 0$

5. Solve $z_{xx} + z_{yy} + z = 0$, assuming that $\ddot{X}/X = k - 1$, where $k > 1$. What values of k will involve other types of solutions?

6. If, in solving $z_{xx} + z_{yy} + z_x + 2z_y = 0$, we use $\ddot{X}/X + \dot{X}/X = k$, state the ranges of k associated with oscillation of: (*a*) the x-factor; (*b*) the y-factor; (*c*) neither factor.

★7. For Laplace's equation,*

$$(\partial^2 u/\partial x^2) + (\partial^2 u/\partial y^2) + (\partial^2 u/\partial z^2) = 0$$

assume that $u = XYZ$, and obtain $\ddot{X}/X + \ddot{Y}/Y + \ddot{Z}/Z = 0$. This will be satisfied if

$$\frac{\ddot{X}}{X} = l \qquad \frac{\ddot{Y}}{Y} = m \qquad \frac{\ddot{Z}}{Z} = n \qquad l + m + n = 0$$

Find a solution, assuming that $l > 0$, $m > 0$, and $n = -l - m$. How many different types of solutions, each associated with a set of values for l, m, and n, exist?

★8. Laplace's equation for two dimensions and polar coordinates is

$$\frac{\partial^2 u}{\partial \rho^2} + \frac{1}{\rho}\frac{\partial u}{\partial \rho} + \frac{1}{\rho^2}\frac{\partial^2 u}{\partial \theta^2} = 0$$

Show that one solution has the form (*h*), of the illustrative example, with x replaced by θ and y by $\ln \rho$.

9. To solve $z_{xx} + z_{xy} + z_{yy} = 0$, let $z = XY$, and obtain

$$\frac{\ddot{X}}{X} + \frac{\dot{X}}{X}\frac{\dot{Y}}{Y} + \frac{\ddot{Y}}{Y} = 0$$

Let $\dot{X}/X = k$, and solve for X; show that $\ddot{X}/X = k^2$; then solve for Y, and write a solution of the given equation.

10. Laplace's equation for cylindrical coordinates in space is

$$\frac{\partial^2 u}{\partial \rho^2} + \frac{1}{\rho}\frac{\partial u}{\partial \rho} + \frac{1}{\rho^2}\frac{\partial^2 u}{\partial \theta^2} + \frac{\partial^2 u}{\partial z^2} = 0$$

Show that a solution of this has the form $u = R(\rho)\Theta(\theta)Z(z)$, where

$$\frac{\ddot{Z}}{Z} = k_1 \qquad \frac{\ddot{\Theta}}{\Theta} = k_2 \qquad \rho^2\ddot{R} + \rho\dot{R} + (k_2 + k_1\rho^2)R = 0$$

* This equation, called Laplace's equation, is of basic importance because it must hold for a flow of substance, such as heat, air, or water, if the quantity entering a region is equal to the quantity leaving it. Exercise 1 relates to Laplace's equation for two dimensions.

130 Hyperbolic, parabolic, elliptic equations

Partial differential equations of the second order are of great importance, as we shall see in the next chapter.

Consider a partial differential equation having the form

$$az_{xx} + bz_{xy} + cz_{yy} = f(x,y,z,z_x,z_y) \tag{1}$$

where a, b, and c represent continuous functions of x and y having all derivatives involved continuous and f represents a polynomial function in the indicated variables. Equation (1) is called

Hyperbolic if $b^2 - 4ac > 0$
Parabolic if $b^2 - 4ac = 0$
Elliptic if $b^2 - 4ac < 0$

We shall derive simple forms, called canonical forms, of (1) for these cases and solve some particular examples.

Let us transform (1) by means of the substitution

$$u = u(x,y) \qquad v = v(x,y) \tag{2}*$$

Thinking of z as a function of u and v, we have, in accordance with the laws of transformation from calculus,

$$z_x = z_u u_x + z_v v_x \qquad z_y = z_u u_y + z_v v_y \tag{3}$$

Note that $z_{xx} = (z_x)_u u_x + (z_x)_v v_x$ and therefore, from (3), that

$$z_{xx} = (z_{uu}u_x + z_{uv}v_x)u_x + z_u u_{xx} + (z_{vu}u_x + z_{vv}v_x)v_x + z_v v_{xx} \tag{4}$$

Note that $z_{uv} = z_{vu}$, change (4) slightly, then derive expressions for z_{xy} and z_{yy} by the process applied for z_{xx}, and obtain

$$z_{xx} = z_{uu}u_x^2 + 2z_{uv}u_x v_x + z_{vv}v_x^2 + z_u u_{xx} + z_v v_{xx} \tag{5}$$
$$z_{xy} = z_{uu}u_x u_y + z_{uv}(u_x v_y + u_y v_x) + z_{vv}v_x v_y + z_u u_{xy} + z_v v_{xy} \tag{6}$$
$$z_{yy} = z_{uu}u_y^2 + 2z_{uv}u_y v_y + z_{vv}v_y^2 + z_u u_{yy} + z_v v_{yy} \tag{7}$$

Substituting the values of z_{xx}, z_{xy}, z_{yy}, z_x, and z_y from (3), (5), (6) and (7) in (1), we get

$$\alpha z_{uu} + \beta z_{uv} + \gamma z_{vv} + \delta z_u + \epsilon z_v = F(u,v,z,z_u,z_v) \tag{8}$$

where

$$\alpha = au_x^2 + bu_x u_y + cu_y^2 \qquad \gamma = av_x^2 + bv_x v_y + cv_y^2 \tag{9}$$
$$\beta = 2au_x v_x + b(u_x v_y + u_y v_x) + 2cu_y v_y \tag{10}$$
$$\delta = au_{xx} + bu_{xy} + cu_{yy} \qquad \epsilon = av_{xx} + bv_{xy} + cv_{yy} \tag{11}$$

* We assume that $u_x v_y - v_x u_y \neq 0$ so that (2) may be solved for x and y in terms of u and v.

and F is the result of transforming f by (2). In making a transformation, compute α, γ, β, δ, ϵ, and F by using (9), (10), and (11) and substituting the values thus obtained in (8).

Hyperbolic Case. To get canonical forms we generally find the roots λ and μ of

$$ar^2 + br + c = 0 \tag{12}$$

and take

$$\frac{u_x}{u_y} = \lambda \qquad \frac{v_x}{v_y} = \mu \tag{13}$$

Since $r_1 = \lambda$ and $r_2 = \mu$ satisfy (12), we see that u_x/u_y and v_x/v_y from (13) satisfy $\alpha = 0$, $\gamma = 0$ of (9), and accordingly (8) is reduced to the canonical form

$$\beta z_{uv} = \varphi(u,v,z,z_u,z_v) \tag{14}$$

where $\varphi = F - \delta z_u - \epsilon z_v$. To get definite expressions for u and v, think of the curves, called characteristics, $u = c_1$, $v = c_2$, and consider that for these

$$-\frac{u_x}{u_y} = \frac{dy}{dx} \qquad \frac{-v_x}{v_y} = \frac{dy}{dx}$$

or, *if* λ *and* μ *are the roots of* (12), *the solutions of*

$$\frac{dy}{dx} + \lambda = 0 \qquad \frac{dy}{dx} + \mu = 0 \tag{15}$$

may be taken as u *and* v.

Consider, for example

$$z_{xx} + xz_{xy} - 6x^2z_{yy} = x^{-1}z_x \tag{a}$$

Here $a = 1$, $b = x$, $c = -6x^2$, and corresponding to (12) and its roots λ and μ are

$$r^2 + xr - 6x^2 = 0 \qquad r_1 = 2x \qquad r_2 = -3x$$

Substituting these in (15) and integrating the results, we get $y + x^2 = c_1$, $y - \frac{3}{2}x^2 = c_2$, and

$$u = y + x^2 \qquad v = y - \tfrac{3}{2}x^2 \tag{b}$$

Note first that the right member of (*a*) is

$$x^{-1}z_x = x^{-1}(z_uu_x + z_vv_x) = x^{-1}(2xz_u - 3xz_v) = 2z_u - 3z_v \tag{c}$$

Substituting from (*b*) in (9), (10), and (11), we get

$$\alpha = \gamma = 0 \qquad \beta = -25x^2 \qquad \delta = 2z_u \qquad \epsilon = -3z_v \tag{d}$$

Substituting these values in (8), we get

$$-25x^2 z_{uv} + 2z_u - 3z_v = 2z_u - 3z_v \qquad \text{or} \qquad z_{uv} = 0$$

From $z_{uv} = 0$, we easily get

$$z = \varphi(u) + \psi(v) = \varphi(y + x^2) + \psi(y - \tfrac{3}{2}x^2)$$

Parabolic Case. For the parabolic case, we have $b^2 - 4ac = 0$. Hence, $ar^2 + br + c = 0$ has equal roots λ and λ. Just as in the hyperbolic case, we take

$$\frac{dy}{dx} + \lambda = \frac{dy}{dx} - \frac{b}{2a} = 0 \tag{e}$$

and get as its solution $u = c_1$. When this value of u is used in (9) and (10), we see at once that $\alpha = 0$ and it happens that $\beta = 0$ also,* independently of the choice of v. Hence, *in the parabolic case* (8) *is reduced by the substitution derived from* (*e*) *to the canonical form*

$$\gamma z_{vv} = \varphi(u,v,z,z_u,z_v) \tag{16†}$$

Consider, for example, the equation

$$z_{xx} - 4xz_{xy} + 4x^2 z_{yy} = 0 \tag{f}$$

Here the root of $r^2 - 4xr + 4x^2 = 0$ is $r = 2x$. Hence, we write

$$\frac{dy}{dx} + 2x = 0 \qquad y + x^2 = c \tag{g}$$

Accordingly, we make the substitution

$$u = y + x^2 \qquad v = x \tag{h}$$

where $v = x$, a simple substitution, was used, since v may be taken arbitrarily. Using (*h*) together with (3), (8), (9), (10), and (11), we get, from (*f*),

$$z_{vv} + 2z_u = 0 \tag{i}$$

This may be solved by separation of the variables to obtain

$$\begin{aligned} z &= \sum_k e^{ku/2}(c_{1k} \sin \sqrt{k}\, v + c_{2k} \cos \sqrt{k}\, v) \\ &= \sum_k e^{k(y+x^2)/2}(c_{3k} \sin \sqrt{k}\, x + c_{4k} \cos \sqrt{k}\, x) \end{aligned} \tag{j}$$

* The proof that $\beta = 0$ is left for exercise 11.

† A substitution $v = v(x,y)$ based on $dy/dx + \lambda = 0$ would have led to $\alpha z_{uu} = \varphi(u,v,z,z_u,z_v)$.

Elliptic Case. Since for this case $b^2 - 4ac < 0$, the roots of $ar^2 + br + c = 0$ will have the form $\lambda \pm i\mu$. Hence, the corresponding substitution based on

$$\frac{u_x}{u_y} = \lambda + i\mu \qquad \frac{v_x}{v_y} = \lambda - i\mu \tag{17}$$

is like the hyperbolic case, but the solution involves imaginary functions. Accordingly, we derive functions u and v based on

$$\frac{dy}{dx} + \lambda + i\mu = 0 \qquad \frac{dy}{dx} + \lambda - i\mu = 0 \tag{18}$$

and then make the substitution

$$u_1 = \tfrac{1}{2}(u + v) \qquad v_1 = -\tfrac{1}{2}i(u - v) \tag{19}$$

This leads to the canonical form (see exercise 12)

$$z_{uu} + z_{vv} = \varphi(u,v,z,z_u,z_v)$$

Consider the equation

$$z_{xx} + y^2 z_{yy} + y z_y = 0 \tag{k}$$

The roots of $r^2 + y^2 = 0$ are $\pm iy$. Hence, we derive from

$$\frac{dy}{dx} + iy = 0 \qquad \frac{dy}{dx} - iy = 0 \tag{l}$$

the function

$$u = ye^{ix} = y(\cos x + i \sin x) \qquad v = y(\cos x - i \sin x) \tag{m}$$

We now make the substitution

$$u_1 = \tfrac{1}{2}(u + v) = y \cos x \qquad v_1 = -\tfrac{1}{2}i(u - v) = y \sin x \tag{n}$$

Proceeding in the usual manner to use the substitution (n) in (k) by means of (3), (9), (10), and (11), we get

$$y^2(z_{u_1u_1} + z_{v_1v_1}) = 0 \tag{o}$$

Solving this by the method of separation of the variables and replacing u_1 and v_1 by their values from (n), we get

$$z = \sum^{k} [c_{1k} \sin (\sqrt{k}\, y \cos x) + c_{2k} \cos (\sqrt{k}\, y \cos x)](c_{3k}e^{\sqrt{k}y \sin x} + c_{4k}e^{-\sqrt{k}y \sin x})$$

Exercises

Use substitutions to solve the following equations:

1. $z_{xx} + z_{xy} - 2z_{yy} = 0$. *Hint:* The roots of $r^2 + r - 2 = 0$ are $r = 1, -2$; from $dy/dx + 1 = 0$ and $dy/dx - 2 = 0$, obtain $u = y + x$ and $v = y - 2x$. Now use (3), (8), (9), (10), and (11).

2. $z_{xx} - (m + n)z_{xy} + mnz_{yy} = 0$, where m and n are constants.

3. $z_{xx} + 2mz_{xy} + m^2z_{yy} = 0$. *Hint:* From $dy/dx + (-m) = 0$, obtain $u = y - mx$. Make the substitution $u = y - mx$, $v = x$.

4. $z_{xx} - 2z_{xy} + 5z_{yy} = 0$. *Hint:* Using (17), (18), and (19), deduce the substitution $u_1 = x + y$, $v_1 = 2x$.

5. $z_{xx} - 2mz_{xy} + (m^2 + n^2)z_{yy} = 0$

6. $x^2z_{xx} + xyz_{xy} - 2y^2z_{yy} = \frac{4}{3}(yz_y - xz_x)$

7. $z_{xx} - 4xz_{xy} + 4x^2z_{yy} = 2z_y$

8. $z_{xx} - 2yz_{xy} + 5y^2z_{yy} + 5yz_y = 0$

9. $\rho^2 \dfrac{\partial^2 z}{\partial \rho^2} + \dfrac{\partial^2 z}{\partial \theta^2} + 2\rho \dfrac{\partial z}{\partial \rho} = 0$

10. $x^2y^2z_{xx} + 2xyz_{xy} + z_{yy} = -x(1 + y^2)z_x$. *Hint:* Take $v = y$.

11. Show in the parabolic case that the equation $ar^2 + br + c$ has $-b/2a$ as a double root. Also, show that the substitution $u = u(x,y)$, $v = v(x,y)$, for which $u_x/u_y = -b/2a$, has $\beta = 0$ as a result provided that u and v are independent functions.

★12. Show in the elliptic case that the substitution (19) in (1) results in $\alpha = \gamma$ and $\beta = 0$. *Hint:* $u_{1x} = \frac{1}{2}(u_x + v_x), \ldots, v_{1y} = -\frac{1}{2}i(u_y - v_y)$. Also, remember that $au_x^2 + bu_xu_y + cu_y^2 = av_x^2 + bv_xv_y + cv_y^2 = 0$.

13. How could the canonical form $z_{uu} = \psi(u,v,z,z_u,z_v)$ be obtained for the parabolic case?

15

Applications of Partial Differential Equations

131 Fourier series

Orthogonal functions have already been considered in §§111 to 114. Perhaps the most important set is the one based on $\sin nx$ and $\cos nx$, where n is an integer. A series having the form

$$a_0 + a_1 \cos x + a_2 \cos 2x + \cdots + a_n \cos nx + \cdots + b_1 \sin x + b_2 \sin 2x + \cdots + b_n \sin nx + \cdots \tag{1}$$

is called a trigonometric series. First let us assume that a certain function $f(x)$ can be expanded in a series of the form (1),

$$f(x) = a_0 + \sum_{n=1}^{\infty} (a_n \cos nx + b_n \sin nx) \tag{2}$$

which is uniformly convergent in the interval $-\pi \leqq x \leqq \pi$, and then attempt to find expressions for the coefficients a_n and b_n.

First, we find by direct evaluation of integrals

$$\int_{-\pi}^{\pi} \sin mx\, dx = \int_{-\pi}^{\pi} \cos mx\, dx = \int_{-\pi}^{\pi} \sin mx \cos nx\, dx = 0 \tag{3}$$

$$\int_{-\pi}^{\pi} \sin mx \sin nx\, dx = \int_{-\pi}^{\pi} \cos mx \cos nx\, dx = 0 \qquad m \neq n$$

$$\int_{-\pi}^{\pi} \sin^2 mx\, dx = \int_{-\pi}^{\pi} \cos^2 mx\, dx = \pi \tag{4}$$

where m and n represent integers.*

Since the right member of (2) multiplied by $\cos nx$ is uniformly convergent, term-by-term integration may be applied to integrate its right member. Multiplying (2) through by $\cos nx\, dx$, equating the definite integrals of its members from $-\pi$ to π, and taking account of (3) and (4), we get

$$\int_{-\pi}^{\pi} f(x) \cos nx\, dx = a_n \int_{-\pi}^{\pi} \cos^2 nx\, dx = \pi a_n$$

$$a_n = \frac{1}{\pi} \int_{-\pi}^{\pi} f(x) \cos nx\, dx \tag{5}$$

Similarly, multiplying (2) through by $\sin nx\, dx$ and proceeding as before, we obtain

$$b_n = \frac{1}{\pi} \int_{-\pi}^{\pi} f(x) \sin nx\, dx \tag{6}$$

Also, in like manner, we get

$$a_0 = \frac{1}{2\pi} \int_{-\pi}^{\pi} f(x)\, dx \tag{7}$$

The series (2) with coefficients defined by (5), (6), and (7) is called a Fourier expansion of $f(x)$, the coefficients are called Fourier coefficients, and the series is called a Fourier series The following theorem has been proved.†

THEOREM. If $f(x)$ is single-valued and finite in the interval $-\pi < x < \pi$ *and has only a finite number of discontinuities and of maxima and minima in this interval, then the Fourier series resulting from* (2) by substituting in it the values of a_n, b_n, and a_0 *from* (5),

* Throughout this chapter, m and n represent integers unless the context indicates otherwise.

† Expansion in Fourier series is treated in books on advanced calculus. Consult R. Courant, "Differential and Integral Calculus", vol. I, pp. 437–456; W. E. Byerly, "Fourier's Series and Spherical Harmonics," Ginn and Company, Boston, 1893; H. S. Carslaw, "Introduction to the Theory of Fourier's Series and Integrals," Macmillan & Co., Ltd, London, 1930.

(6), *and* (7),

$$a_0 = \frac{1}{2\pi}\int_{-\pi}^{\pi} f(x)\,dx \qquad a_m = \frac{1}{\pi}\int_{-\pi}^{\pi} f(x)\cos mx\,dx$$
$$b_m = \frac{1}{\pi}\int_{-\pi}^{\pi} f(x)\sin mx\,dx \qquad m = 1, 2, \ldots \tag{8}$$

is equal to $f(x)$ *for all values of* x *in the interval* $-\pi < x < \pi$ *except at points of discontinuity.* At a point of discontinuity where $x = a$, the value of the series is

$$\tfrac{1}{2}\lim_{\epsilon \to 0}\,[f(a - \epsilon) + f(a + \epsilon)] \qquad \epsilon > 0 \tag{9}$$

When $x = -\pi$ and when $x = \pi$, the value of the series for $f(x)$ is

$$\tfrac{1}{2}[f(-\pi) + f(\pi)] \tag{10}$$

Both $\sin mx$ and $\cos mx$ have the period 2π, since $\sin m(x + 2k\pi) = \sin mx$ and $\cos m(x + 2k\pi) = \cos mx$, where m and k are integers. Hence, *the values assumed by the series* (2) *in the interval* $-\pi < x < \pi$ *are assumed by it in any other interval* $(2k - 1)\pi < x < (2k + 1)\pi$. *Otherwise stated,* (2) *together with* (8), (9), *and* (10) *defines a function* $f(x + k2\pi)$, k *an integer.* Figure 1 indicates this situation by showing the graph of a function from $-\pi$ to π and repetitions of it.

The Fourier series representing $\int_{-\pi}^{x} f(x)\,dx$, $-\pi < x < \pi$, *may be obtained by integrating, term by term, the Fourier series for* $f(x)$, *but only under certain conditions* will the Fourier series for* $df(x)/dx$ *be obtained by differentiating the Fourier series for* $f(x)$, *term by term.*

Example. Expand x in a Fourier series. Find from the result by integration a Fourier series for x^2.

Solution. Using (7), (5), and (6) with $f(x) = x$, we get

$$a_0 = \frac{1}{2\pi}\int_{-\pi}^{\pi} x\,dx = 0$$
$$a_n = \frac{1}{\pi}\int_{-\pi}^{\pi} x\cos nx\,dx = \frac{1}{\pi}\left[\frac{x\sin nx}{n} + \frac{\cos nx}{n^2}\right]_{-\pi}^{\pi} = 0$$
$$b_n = \frac{1}{\pi}\int_{-\pi}^{\pi} x\sin nx\,dx = \frac{1}{\pi}\left[-\frac{x\cos nx}{n} + \frac{\sin nx}{n^2}\right]_{-\pi}^{\pi} = \frac{-2\cos n\pi}{n}$$

* If $f(x + 2\pi n) = f(x)$ and if $f'(x)$ is continuous and single-valued and has only a finite number of maxima and minima in $-\pi < x < \pi$, then $f'(x)$ is represented for all values of x by the term-by-term derivative of the Fourier series for $f(x)$.

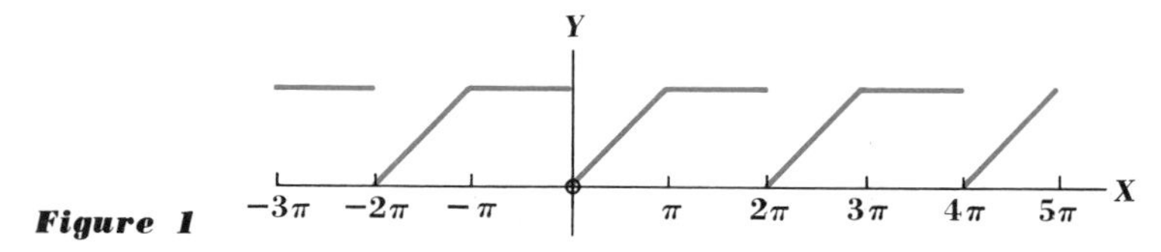

Figure 1

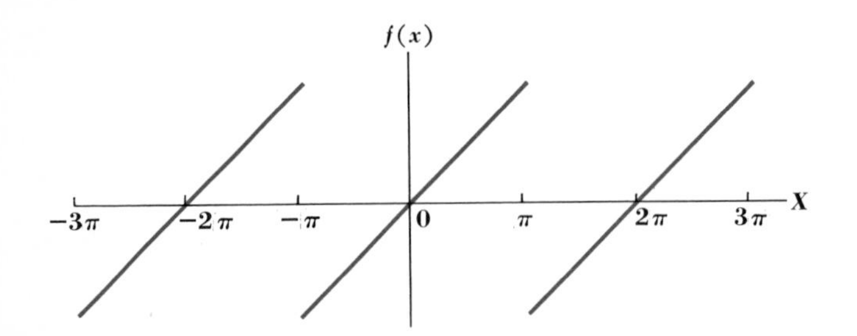

Figure 2

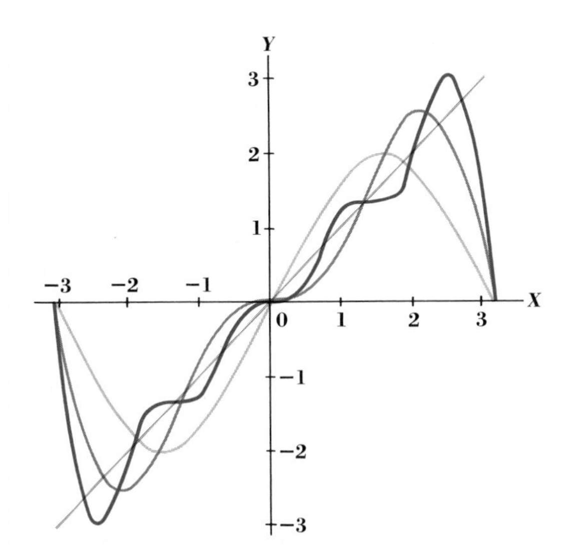

Figure 3

Substituting in (2) these values for a_0, a_n, and b_n and x for $f(x)$, we get

$$x = \sum_{n=1}^{\infty} \frac{-2\cos n\pi}{n} \sin nx = 2\sin x - \tfrac{2}{2}\sin 2x + \tfrac{2}{3}\sin 3x - \tfrac{2}{4}\sin 4x + \cdots \tag{11}$$

Actually, (11) represents a function $\varphi(x)$ such that $\varphi(x + m2\pi) = f(x)$ and $\varphi(x) = x$ in $-\pi < x < \pi$, $f(-\pi) = 0$, $f(\pi) = 0$. Figure 2 shows the graph of $f(x)$ for x in $-3\pi < x < 3\pi$. In Fig. 3, the straight line represents the graph of $y = x$, the lightest blue curve represents the first approximation $y = 2\sin x$, the medium blue curve represents the approximation

$$y = 2\sin x - \sin 2x$$

and the dark blue curve represents the four-term approximation

$$y = 2\sin x - \sin 2x + \tfrac{2}{3}\sin 3x - \tfrac{1}{2}\sin 4x$$

From (11), we get by integration

$$\int_{-\pi}^{x} x\,dx = \tfrac{1}{2}x^2 - \tfrac{1}{2}\pi^2 = a_0 + \sum_{n=1}^{\infty} \frac{2\cos n\pi}{n^2}\cos nx \tag{12}$$

By using (7) with $f(x) = \frac{1}{2}(x^2 - \pi^2)$, we get $a_0 = -\frac{1}{3}\pi^2$. Using this value of a_0 in (12) and transforming slightly, we obtain

$$x^2 = \tfrac{1}{3}\pi^2 - 4\left(\frac{1}{1^2}\cos x - \frac{1}{2^2}\cos 2x + \frac{1}{3^2}\cos 3x - \cdots\right) \tag{13}$$ *

It is interesting to substitute 0 for x in (13), transform the result, and obtain

$$\tfrac{1}{12}\pi^2 = \frac{1}{1^2} - \frac{1}{2^2} + \frac{1}{3^2} - \frac{1}{4^2} + \cdots$$

and to substitute π for x in (13) and then derive

$$\tfrac{1}{6}\pi^2 = \frac{1}{1^2} + \frac{1}{2^2} + \frac{1}{3^2} + \cdots$$

Exercises

1. Substitute $\frac{1}{2}\pi$ for x in (11), and show that $\frac{1}{4}\pi = 1 - \frac{1}{3} + \frac{1}{5} - \frac{1}{7} + \cdots$.

2. Draw the part of the complete graph situated in the interval $-3\pi \leqq x \leqq 3\pi$ for the right member of (13).

3. Expand in a Fourier series the function $f(x)$ for which

$$f(x + m2\pi) = f(x) \qquad f(x) = -1 \text{ in } -\pi < x < 0$$

$f(x) = 1$ in $0 < x < \pi$, and $f(x) = 0$ when x is -1, 0, or 1. *Hint:* By (7), $a_0 = (1/2\pi)\int_{-\pi}^{0}(-1)\,dx + (1/2\pi)\int_0^{\pi} dx = 0$ and $b_n = (1/\pi)\int_{-\pi}^{0} -\sin nx\,dx + (1/\pi)\int_0^{\pi}\sin nx\,dx = (2/n\pi)(1 - \cos n\pi)$.

4. Expand in a Fourier series the function $f(x)$ for which $f(x) = 0$ in the interval $-\pi < x < 0$, $f(x) = 1$ in the interval $0 < x < \pi$, and $f(x) = \frac{1}{2}$ when $x = \pi$.

5. Expand $f(x)$ in a Fourier series if $f(x) = \pi + x$ in $-\pi \leqq x \leqq 0$ and $f(x) = \pi - x$ in $0 \leqq x \leqq \pi$. Using your result and the fact that $f(x) = \pi$ when $x = 0$ deduce that $1/1^2 + 1/3^2 + 1/5^2 + \cdots = \frac{1}{8}\pi^2$.

* Actually, (13) represents a function $\psi(x)$ such that $\psi(x + m2\pi) = \psi(x)$ and $\psi(x) = x^2$ in the interval $-\pi \leqq x \leqq \pi$.

6. Find a Fourier series for $\sin \frac{1}{2}x$ by using (2), (5), (6), and (7). Does it represent $\sin \frac{1}{2}x$ for all values of x? Draw a figure representing the Fourier series in $-3\pi < x < 3\pi$.

7. If $f(x)$ is such that $f(x) = f(-x)$, that is, if $f(x)$ is an even function, show that the corresponding Fourier series defined by (2), (5), (6), and (7) contains no sine terms. State the nature of the Fourier series for a function $\varphi(x)$ if $\varphi(x) = -\varphi(-x)$, that is, for an odd function. Check your answer by means of series (11) and (13).

8. Equate $\int_{-\pi}^{x} x^2\,dx$ to the definite integral of the right member of (13) with limits $-\pi$ to x, and in the result replace the first term $\frac{1}{3}\pi^2 x$ in the right member by $\frac{1}{3}\pi^2$ times the series for x from (11), and then write the Fourier series for x^3.

132 Cosine series. Sine series

A function considered on an interval $-a \leqq x \leqq a$ is even if $f(x) = f(-x)$ and odd if $f(x) = -f(-x)$. From these definitions we can easily verify that a function $\varphi(x)$ is:

$$\text{even} \qquad \text{if } \varphi(x) = \begin{cases} f(x) & 0 \leqq x \leqq a \\ f(-x) & -a \leqq x \leqq 0 \end{cases} \tag{1}$$

$$\text{odd} \qquad \text{if } \varphi(x) = \begin{cases} f(x) & 0 \leqq x \leqq a \\ -f(-x) & -a \leqq x \leqq 0 \end{cases} \tag{2}$$

If $f(x)$ is an even function

$$\int_{-\pi}^{\pi} f(x)\,dx = \int_{-\pi}^{0} f(x)\,dx + \int_{0}^{\pi} f(x)\,dx = 2\int_{0}^{\pi} f(x)\,dx \qquad f(x) \text{ even}$$

for the elements $f(x_i)\,\Delta x_i$ and $f(-x_i)\,\Delta x_i$ in one kind of sum having the first integral as a limit are equal. Similarly, if $f(x)$ is an odd function

$$\int_{-\pi}^{\pi} f(x)\,dx = \int_{-\pi}^{0} f(x)\,dx + \int_{0}^{\pi} f(x)\,dx = 0 \qquad f(x) \text{ odd}$$

for a consideration of elements $f(x_i)\,\Delta x_i$ and $f(-x_i)\,\Delta x_i$ shows that they cancel. Taking account of the equations above and observing that $\cos mx$ is even and $\sin mx$ odd, we see that equations (8), §131, may be written

$$a_0 = \frac{1}{\pi}\int_{0}^{\pi} f(x)\,dx \qquad a_m = \frac{2}{\pi}\int_{0}^{\pi} f(x)\cos mx\,dx \qquad b_m = 0 \tag{3}$$

$$a_0 = 0 \qquad a_m = 0 \qquad b_m = \frac{2}{\pi}\int_{0}^{\pi} f(x)\sin mx\,dx \tag{4}$$

Since the expansions based on (3) and (4) are special cases of (8), §131, they exist under the conditions mentioned in the theorem of §131.

Observe that (3) may be used to expand an even function and (4) to expand an odd function. If (3) or (4) is applied to expand any function $f(x)$, the expansion of $\varphi(x)$ (defined by (1) or (2), respectively) will be obtained. In all cases the expansions based on (3) and (4) will apply for the interval $0 \leqq x \leqq \pi$.

It is well to keep in mind that the product of two even or of two odd functions is even, and the product of an even and an odd function is odd. Also note that the graph of an even function is symmetric to the Y axis and that of an odd function is symmetric to the origin.

Note that, if $x = Lz/\pi$, $x = -L$ when $z = -\pi$ and L when $z = \pi$. Hence, a *Fourier expansion representing $f(x)$ on the interval $-L < x < L$* can be obtained by replacing x in $f(x)$ by Lz/π,*

$$x = \frac{Lz}{\pi} \tag{5}$$

expanding $f(Lz/\pi)$ by means of (8), §131, *and replacing z by $\pi x/L$ in the result.* The same procedure may be employed with (3) and (4) to obtain an expansion relating to the interval $0 < x < L$.

Example 1. Expand $f(x) = 1$ in a sine series by using (2), §131, and (4).

Solution: Using (4), we get

$$a_0 = 0 \qquad a_m = 0 \qquad b_m = \frac{2}{\pi}\int_0^\pi \sin mx\,dx = \frac{2(1 - \cos m\pi)}{\pi m}$$

Substituting these values in (2), §131, we obtain

$$1 = \frac{4}{\pi}\left(\sin x + \frac{\sin 3x}{3} + \frac{\sin 5x}{5} + \cdots\right) \tag{a}$$

Example 2. By using (3) and (5), expand x in a cosine series over the range $0 < x < c$. Define completely the function represented by the expansion.

Solution. Let $x = cz/\pi$, and apply (3) to obtain

$$a_0 = \frac{1}{\pi}\int_0^\pi \frac{c}{\pi} z\,dz = \frac{c}{2}$$

$$a_m = \frac{2}{\pi}\int_0^\pi \frac{cz}{\pi}\cos mz\,dz = \frac{2c}{\pi^2 m^2}(\cos m\pi - 1)$$

* Here $f(x)$ on the range $-L < x < L$ is subject to the limitations of $f(x)$ on $-\pi < x < \pi$ specified in §131. Also, the expansions represent a function $F(x)$ for which $F(x) = f(x)$ on $0 < x < \pi$ and $F(x + m2\pi) = F(x)$.

Substitute these values in (3) and obtain

$$\frac{cz}{\pi} = \frac{c}{2} + \frac{2c}{\pi^2}\left(\frac{-2\cos z}{1^2} - \frac{2\cos 3z}{3^2} - \frac{2\cos 5z}{5^2} - \cdots\right) \qquad (a)$$

or, replacing z by $\pi x/c$ and simplifying slightly,

$$x = \frac{c}{2} - \frac{4c}{\pi^2}\left(\frac{1}{1^2}\cos\frac{\pi x}{c} + \frac{1}{3^2}\cos\frac{3\pi x}{c} + \cdots\right) \qquad (b)$$

Denoting by $f(x)$ the right member of (*b*), we get $f(x) = x$ on $0 \leqq x \leqq c$, $f(x) = -x$ on $-c \leqq x \leqq 0$, and $f(x + m2c) = f(x)$.

Exercises

1. (*a*) Replace x by Lz/π, expand Lz/π by (4), and then replace z by $\pi x/L$ in the result. (*b*) Using integration on the expansion of (*a*) and also the fact that $a_0 = (1/\pi)\int_0^\pi (L^2z^2/\pi^2)\,dz = \frac{1}{3}L^2$, obtain an expansion of x^2 in a series of cosines. (*c*) Define in terms of x^2 the function $F(x)$ represented by the expansion from (*b*). (*d*) Use the result from (*b*) to deduce that $\pi^2/6 = 1/1^2 + 1/2^2 + 1/3^2 + \cdots$.

2. Using (4), expand x^2 in a series of sines. If $F(x)$ represents the expansion obtained, define $F(x)$ in terms of x^2 for all values of x.

3. Show that for the interval $0 < x < L$

$$mx(L - x) = \frac{8L^2m}{\pi^3}\left(\frac{1}{1^3}\sin\frac{\pi x}{L} + \frac{1}{3^3}\sin\frac{3\pi x}{L} + \frac{1}{5^3}\sin\frac{5\pi x}{L} + \cdots\right)$$

If $F(x)$ represents the expansion, define $F(x)$ in terms of $mx(L - x)$ for all values of x.

4. Find a Fourier series representing $mx(L^2 - x^2)$ on the range $-L < x < L$.

★5. Expand $f(x) = |\pi + \frac{1}{2}x|$ in a Fourier series representing $|\pi + \frac{1}{2}x|$ on the interval $-4\pi < x < 4\pi$. *Hint:* $|\pi + \frac{1}{2}x| = -\pi - \frac{1}{2}x$ if $x < -2\pi$, and $|\pi + \frac{1}{2}x| = \pi + \frac{1}{2}x$ if $x > -2\pi$ (see Fig. 1). Use $x = 4z$ and (2), (5), (6), and (7), §131.

6. Show that $F(x) = \alpha(x) + \alpha(-x)$ is an even function and that $G(x) = \alpha(x) - \alpha(-x)$ is an odd function.

7. Expand $\cos\frac{1}{2}x$ in a Fourier series. Use your answer to show that $\frac{1}{4}\pi - \frac{1}{2} = 1/(1\cdot 3) - 1/(3\cdot 5) + 1/(5\cdot 7) - 1/(7\cdot 9) + \cdots$

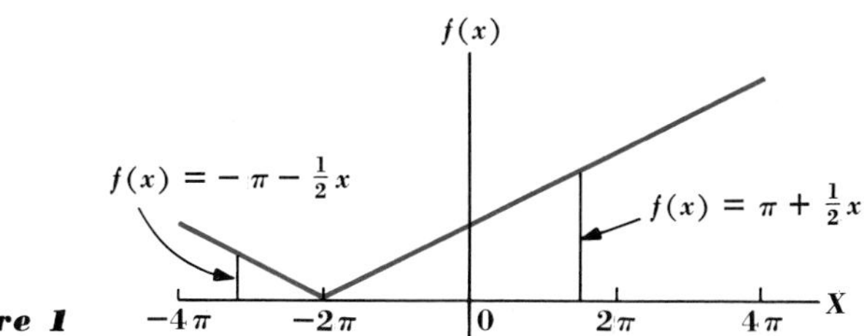

Figure 1

133 Vibrations of a string

Sums, differences, and constant multiples of the following formulas may be used to construct desired ones:

$$1 = \frac{4}{\pi}\left(\sin\frac{\pi x}{L} + \frac{1}{3}\sin\frac{3\pi x}{L} + \frac{1}{5}\sin\frac{5\pi x}{L} + \cdots\right) \tag{A}$$

$$x = \frac{2L}{\pi}\left(\frac{1}{1}\sin\frac{\pi x}{L} - \frac{1}{2}\sin\frac{2\pi x}{L} + \frac{1}{3}\sin\frac{3\pi x}{L} - \frac{1}{4}\sin\frac{4\pi x}{L}\cdots\right) \tag{B}$$

$$x^2 = \frac{2L^2}{\pi^3}\left[\left(\frac{\pi^2}{1} - \frac{4}{1^3}\right)\sin\frac{\pi x}{L} - \frac{\pi^2}{2}\sin\frac{2\pi x}{L} + \left(\frac{\pi^2}{3} - \frac{4}{3^3}\right)\sin\frac{3\pi x}{L} - \frac{\pi^2}{4}\sin\frac{4\pi x}{L} + \cdots\right] \tag{C}$$

Figure 1 represents a string L units long fastened at A and B. Assume that the vibrations are so small that the tension T in the string may be considered constant, that the weight of the string is small in comparison with T,* that the length of the string may be considered as a constant L for each of its positions, and that each point in the string moves parallel to the Y axis. Consider the motion of a small piece PQ (see Fig. 1) of the string Δx units long. Two forces of magnitude T act at its ends, one inclined θ and the other $\theta + \Delta\theta$ to the X axis. Since θ is small, $\sin\theta = \tan\theta = \partial y/\partial x$, approximately. Therefore, applying Newton's law parallel to the Y axis, we get

$$T[\sin(\theta + \Delta\theta) - \sin\theta] = T\left[\frac{\partial y(x + \Delta x, t)}{\partial x} - \frac{\partial y(x,t)}{\partial x}\right] = \frac{\rho\,\Delta x}{g}\frac{\partial^2 y}{\partial t^2} \tag{1}$$

* Assume in this section that all strings considered are taut.

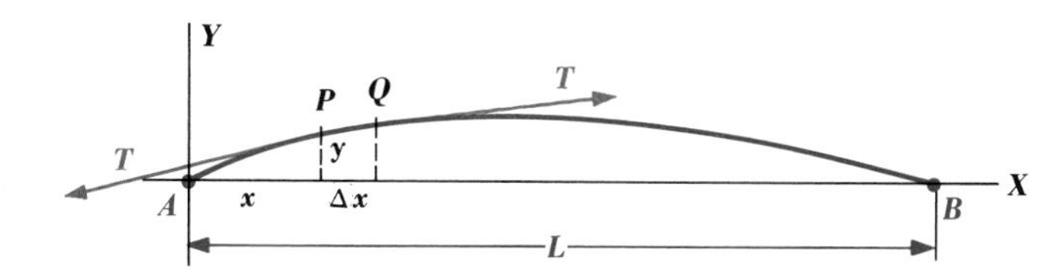

Figure 1

where ρ is the weight per unit length of the string. Dividing (1) through by $\rho\,\Delta x/g$ and equating the limits of its members, we have

$$a^2\frac{\partial^2 y(x,t)}{\partial x^2} = \frac{\partial^2 y}{\partial t^2} \qquad a^2 = \frac{Tg}{\rho} \tag{2}$$

The following example relates to the string of Fig. 1.

Example. A string is stretched along the X axis, to which it is attached at $x = 0$ and at $x = L$. Find y in terms of x and t, assuming that $y = mx(L - x)$ when $t = 0$.

Solution. Using the method of separation of variables, §129, substitute

$$y = X(x)T(t) \tag{a}$$

in (2), and divide the result by XT to obtain

$$\frac{a^2\, d^2X/dx^2}{X} = \frac{d^2T/dt^2}{T} \tag{b}$$

Equating each member of (b) to $-\omega^2$ and solving the resulting equations, obtain

$$X = c_1 \sin\frac{\omega x}{a} + c_2\cos\frac{\omega x}{a} \tag{c}$$

$$T = c_3 \sin\omega t + c_4 \cos\omega t \tag{d}$$

It now appears from (a), (c), and (d) that solutions of (2) for y can consist of sums of terms having the forms

$$\begin{matrix} A\cos\dfrac{\omega}{a}x\cos\omega t & B\cos\dfrac{\omega}{a}x\sin\omega t \\ C\sin\dfrac{\omega}{a}x\cos\omega t & F\sin\dfrac{\omega}{a}x\sin\omega t \end{matrix} \tag{e}$$

Here y for a point on the string with abscissa x depends on x and t, and we denote it by $y(x,t)$. When $x = 0$, $y = 0$ and when $x = L$, $y = 0$, at all times. Hence we write the initial conditions in the form

$$y(0,t) = 0 \qquad y(L,t) = 0 \qquad y(x,0) = mx(L - x) \tag{f}$$

The first two conditions of (f) will be satisfied, if we take

$$\omega = \frac{n\pi a}{L} \qquad n \text{ an integer} \tag{g}$$

and restrict ourselves to terms having the form of those in the second line of (e). From exercise 3, §132, we have

$$mx(L - x) = \frac{8L^2m}{\pi^3}\left(\sin\frac{\pi x}{L} + \frac{1}{3^3}\sin\frac{3\pi x}{L} + \frac{1}{5^3}\sin\frac{5\pi x}{L} + \cdots\right) \tag{h}$$

Hence, conditions (2) and (f) are satisfied by

$$y = \frac{8L^2m}{\pi^3}\left(\frac{1}{1^3}\cos\frac{a\pi t}{L}\sin\frac{\pi x}{L} + \frac{1}{3^3}\cos\frac{3\pi at}{L}\sin\frac{3\pi x}{L} + \cdots\right) \qquad (i)$$

Exercises

1. If the string of the illustrative example is 3 ft long and weighs $\frac{1}{30}$ lb and if $T = 10$ lb and $m = 0.01$, find the equation of the moving string. Find the time frequency of the first harmonic, that is, the first term.

2. Show that $y(x,t)$ equated to the first term in (i) satisfies (2), and that $y(0,t) = 0$ and $y(L,t) = 0$ in this case. Also find $y(x,0)$, $y_t(0,0)$, and $y_t(L,0)$ in this case.

3. Replace *first term* in exercise 2 by kth *term*, and solve the resulting problem.

4. Show that $y(x,t) = A\{\sin[\pi(x + at)/L] + \sin[\pi(x - at)/L]\}$ satisfies (2) and $y(0,t) = 0$, $y(L,t) = 0$, $y_t(0,t) = 0$, and $y_t(L,t) = 0$. Find $y_t(x,0)$ in this case.

5. In the illustrative example, replace $y = mx(L - x)$ by $y = mx(L^2 - x^2)$, and solve the resulting problem. Use the answer to exercise 4, §132. If t is time in seconds, how often will the string return to the position it had at time $t = 0$?

6. Show that $y(x,t) = A\{\cos[\pi(x + at)/L] - \cos[\pi(x - at)/L]\}$ satisfies (2), $y(0,t) = 0$, $y(L,t) = 0$, $y_t(0,t) = 0$, and $y_t(L,t) = 0$. Also, find $y(x,0)$.

7. If the string of Fig. 1 is at rest when $t = 0$ on the lines $y = mx$ on $0 \leqq x \leqq \frac{1}{2}L$ and $y = m(L - x)$ on $\frac{1}{2}L \leqq x \leqq L$, express $y(x,t)$ in terms of x and t. At what time after $t = 0$ will the string be in the position it had at time $t = 0$? *Hint:* Expand $f(x) = y(x,0)$ in a series of sines.

134 Method involving Laplace transforms

Laplace transforms furnish powerful methods* of solving problems involving partial differential equations. Here a suggestion of method will be given by an example.

* An excellent presentation of the methods is given in Ruel V. Churchill, "Operational Mathematics," pp. 108–251.

Example. One end of an infinitely long string lying at rest along the X axis is fixed to the origin. It is acted upon by gravity parallel to the Y axis. Find $y(x,t)$ for any point (x,y) on it.

Solution. By the line of argument used in §133 we arrive at the equation.

$$a^2y_{xx} - y_{tt} = g \tag{1}$$

and the initial conditions

$$\begin{aligned} &y(x,0) = 0 \qquad y_t(x,0) = 0 \qquad y(0,t) = 0 \\ &\lim_{x\to\infty} y_x(x,t) = 0 \end{aligned} \tag{2}$$

If $Y(p,x)$ is the transform of $y(x,t)$ with respect to t, we have

$$T\left\{\frac{a^2\,\partial^2 y(x,t)}{\partial x^2}\right\} = \int_0^\infty e^{-pt}a^2\frac{\partial^2 y}{\partial x^2}\,dt = a^2\frac{\partial^2}{\partial x^2}\int_0^\infty e^{-pt}y\,dt^* = a^2\frac{\partial^2 Y}{\partial x^2} \tag{3}$$

Also

$$T\{y_{tt}\} = p^2Y - py(x,0) - y_t(x,0) \tag{4}$$

in accord with (5), §62. Equating the transforms of the members of (1) while taking account of (3), (4), and (2) we get

$$a^2Y_{xx} - p^2Y_{tt} = \frac{g}{p} \tag{5}$$

$$Y(0,p) = 0 \qquad \lim_{x\to\infty} Y_x(x,p) = 0 \tag{6}$$ †

The solution of (5) is

$$Y = c_1e^{px/a} + c_2e^{-px/a} - \frac{g}{p^3} \tag{7}$$

where c_1 and c_2 are considered as functions of p. Since, by (1) and (2), §60, $p > 0$, we see that $c_1 = 0$, by the second condition of (6). Therefore from (7) and (2), §68,

$$y(x,t) = \varphi\left(t - \frac{x}{a}\right) - \tfrac{1}{2}gt^2 \tag{8}$$

* The condition that the symbols $\partial^2/\partial x^2$ and $\int$ may be interchanged is that y_x and y_{xx} be continuous in a region $\alpha \leqq x \leqq \beta$, and $a \leqq y \leqq b$.

† Note that $Y(0,p) = \int_0^\infty e^{-pt}y(0,t)\,dt = 0$,

$$\lim_{x\to\infty} Y_x(x,p) = \lim_{x\to\infty}\int_0^\infty e^{-pt}y_x(x,t)\,dt = 0$$

where $\varphi(t - x/a) = 0$ if $x > at$. From (8) and the condition that $y(0,t) = 0$, we get

$$0 = \varphi(t) - \tfrac{1}{2}gt^2 \qquad \text{or} \qquad \varphi(t) = \tfrac{1}{2}gt^2 \tag{9}$$

From (8) and (9), we obtain

$$y(x,t) = \tfrac{1}{2}g\left(t - \frac{x}{a}\right)^2 - \tfrac{1}{2}gt^2 \qquad \text{if } x < at \tag{10}$$

Accordingly we have

$$\begin{aligned} y(x,t) &= -\tfrac{1}{2}gt^2 && x > at \\ y(x,t) &= \tfrac{1}{2}g\left(t - \frac{x}{a}\right)^2 - \tfrac{1}{2}gt^2 && x < at \end{aligned} \tag{11}$$

Observe that at points where $x > at$, the string is falling freely, but at points where $x < at$ the points lie on a curve.

Exercises

1. Solve $a^2y_{xx} = y_{tt}$, with $x \geqq 0$ and $t \geqq 0$ and subject to the conditions $y(x,0) = 0$, $y_t(x,0) = 1$, $y(0,t) = 0$, and $\lim\limits_{x\to\infty} y(x,t) = 0$.

2. Solve $y_{xx} - 2y_{tx} - 3y_{tt} = 0$ subject to the conditions $y(x,0) = 0$, $y_t(x,0) = 1$, $\lim\limits_{x\to\infty} y(x,t) = 0$, and $y(0,t) = te^{-t}$. *Hint:* $T\{y_{tx}\} = p\,\partial Y/\partial x$.

3. Solve $y_x + 3y_t = 0$, $y(x,0) = 0$, $y(0,t) = t^3$.

135 Vibrations of a rod

Figure 1 represents a straight, elastic, homogeneous rod of density ρ, modulus of elasticity E, length L, and cross-sectional area A. It is fixed at Q and R, but the particles of the rod between Q and R move along the line QR. It is assumed that the pressure on any cross section is uniformly distributed and that all particles on any cross section have the same velocity. Let M at a distance x from Q, and N, at a distance $x + \Delta x$ from Q, be the positions of two particles on the rod

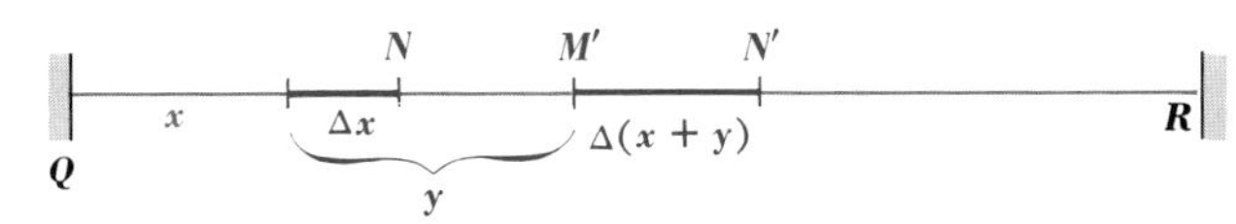

Figure 1

when it is unstretched; and let M' and N', at distances y and $y + \Delta y$ from Q, respectively, be the positions of the same particles when the rod is in motion. If $P(x,t)$ represents the rightward force on the cross section at M', we have by Hooke's law

$$E = \frac{PL}{A\mathcal{E}} \qquad \text{or} \qquad \mathcal{E} = \frac{PL}{AE} \tag{1}$$

where $\mathcal{E}$ is the amount that a rod of length L and cross-sectional area A is stretched by force P. Applying this equation to the piece $M'N'$ in Fig. 1, we obtain

$$y(x + \Delta x,\, t) - y(x,t) = \frac{P(x_1,t)}{AE}\,\Delta x \tag{2}$$

where x_1 satisfies $x < x_1 < x + \Delta x$. Dividing by Δx and equating the limits of the two members as $\Delta x \to 0$, we get

$$\frac{\partial y(x,t)}{\partial x} = \frac{P(x,t)}{AE} \tag{3}$$

Now, apply Newton's law of motion to the part $M'N'$ of the rod. This gives

$$P(x + \Delta x,\, t) - P(x,t) = \frac{A\rho\,\Delta x}{g}\,\frac{\partial^2 y_1}{\partial t^2} \tag{4}$$

where y_1 satisfies $y < y_1 < y + \Delta y$. Now, divide by Δx, and equate limits of the members of (4) to obtain

$$\frac{\partial P}{\partial x} = \frac{A\rho}{g}\,\frac{\partial^2 y}{\partial t^2} \tag{5}$$

Eliminate P between (3) and (5) to obtain

$$a^2\,\frac{\partial^2 y(x,t)}{\partial x^2} = \frac{\partial^2 y(x,t)}{\partial t^2} \qquad a^2 = \frac{Eg}{\rho} \tag{6}$$

Note that this equation has the same form as (2) in §133, and therefore involves the same mathematics used in §133. Of course the interpretation is different.

Exercises

1. Find $y(x,t)$ for the rod of Fig. 1 if (6) holds and if $y(0,t) = 0$, $y(L,t) = 0$, and $y(x,0) = x(L - x)$. Find the time between two successive positions when all the beam is instantaneously at rest.

2. For a steel rod $\rho = 490$ lb/ft³, $E = 4.3 \times 10^9$ lb/ft², and $g = 32$ ft/sec². Using the initial conditions $y(0,t) = 0$, $y(3,t) = 0$, $y(x,0) = 0.0001x(3 - x)$, and $(\partial y/\partial t)_{t=0} = 0$, find the corresponding solution of (6), and give the frequency of the harmonic represented by the first term.

136 Flow of heat

Let $\theta(x,y,z,t)$ represent the temperature at any point in space at time t, and assume that *the heat flows in the direction of decreasing temperature* and that *the rate* (*in calories per second*) *across any infinitesimal square is proportional to the area of the square and to* $\partial\theta/\partial s$, *where s is measured normal to the square.* Also, assume that *the quantity of heat in a small body is proportional to its mass and to its temperature* θ.

To get the partial differential equation of heat flow, express in mathematical symbols the relation that *rate at which heat enters the small block of Fig.* 1 *minus the rate at which it leaves is equal to the rate of increase of heat in the block.* The rate at which heat leaves through face AB is $k[\partial\theta(x,y_1,z_1,t)/\partial x]\,\Delta y\,\Delta z$, where k (in calories per centimeter per degree per second) is a constant and point (x,y_1,z_1) is a certain point in face AB. Similarly, the rate at which heat enters through face CD is approximately $k[\partial\theta(x + \Delta x, y_1, z_1,t)/\partial x]\,\Delta y\,\Delta z$. Hence, the rate at which heat enters through the faces of the block perpendicular to the X axis is

$$k\left[\frac{\partial\theta(x + \Delta x, y_1, z_1, t)}{\partial x} - \frac{\partial\theta(x,y_1,z_1,t)}{\partial x}\right]\Delta y\,\Delta z \tag{1}$$

Similarly, the rate at which heat enters the block through faces perpendicular to the Y axis is

$$k\left[\frac{\partial\theta(x_2, y + \Delta y, z_2, t)}{\partial y} - \frac{\partial\theta(x_2,y,z_2,t)}{\partial y}\right]\Delta x\,\Delta z \tag{2}$$

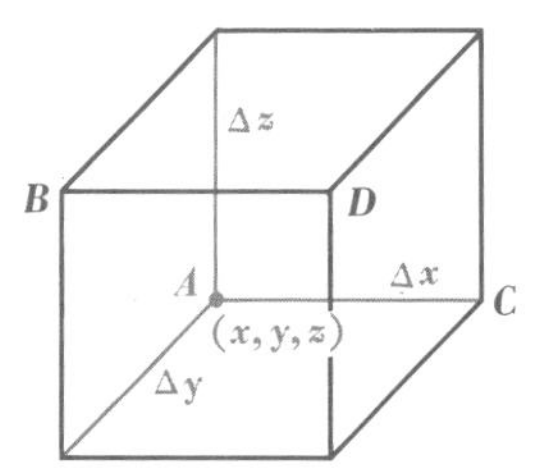

Figure 1

and the rate for the faces perpendicular to the Z axis is

$$k\left[\frac{\partial\theta(x_3, y_3, z + \Delta z, t)}{\partial z} - \frac{\partial\theta(x_3,y_3,z,t)}{\partial z}\right]\Delta y\,\Delta x \tag{3}$$

The rate of increase of heat in the block is

$$c\rho\,\Delta x\,\Delta y\,\Delta z\,\frac{\partial\theta(x_4,y_4,z_4,t)}{\partial t} \tag{4}$$

where ρ (in grams per cubic centimeter) is the density, c (in calories per gram per degree) is the specific heat, and (x_4,y_4,z_4) is a certain point in the block.

The limit of (1) divided by $\Delta x\,\Delta y\,\Delta z$ as $\Delta x \to 0$ is

$$\lim_{\Delta x\to 0} k\left[\frac{\partial\theta(x + \Delta x, y_1, z_1)/\partial x - \partial\theta(x,y_1,z_1)/\partial x}{\Delta x}\right] = k\,\frac{\partial^2\theta(x,y_1,z_1)}{\partial x^2}$$

and similar statements apply to expressions (2) and (3). Also, as Δx, Δy, and Δz approach zero, all the points (x,y_1,z_1), (x_2,y,z_2), (x_3,y_3,z), and (x_4,y_4,z_4) approach point (x,y,z). Equating the sum of the expressions (1) to (3) to the expression (4), dividing the result through by $\Delta x\,\Delta y\,\Delta z$, and equating the limits of the two members as Δx, Δy, and Δz approach zero, we obtain

$$k\left(\frac{\partial^2\theta}{\partial x^2} + \frac{\partial^2\theta}{\partial y^2} + \frac{\partial^2\theta}{\partial z^2}\right) = c\rho\,\frac{\partial\theta}{\partial t} \tag{5}$$

To get the equation of heat flow in a plate with insulated surfaces, omit $\partial^2\theta/\partial z^2$ from (5) to obtain

$$k\left(\frac{\partial^2\theta}{\partial x^2} + \frac{\partial^2\theta}{\partial y^2}\right) = c\rho\,\frac{\partial\theta}{\partial t} \tag{6}$$

and to get the equation for the flow in an insulated rod, omit $\partial^2\theta/\partial y^2$ from (6) to get

$$k\,\frac{\partial^2\theta}{\partial x^2} = c\rho\,\frac{\partial\theta}{\partial t} \tag{7}$$

After heat has flowed until the temperature at any point is constant, the steady state is reached. *To obtain the equations for flow of heat in the steady state, replace* $\partial\theta/\partial t$ *by zero in* (5) *to* (7). Thus, for steady-state flow we get from (5)

$$\frac{\partial^2\theta}{\partial x^2} + \frac{\partial^2\theta}{\partial y^2} + \frac{\partial^2\theta}{\partial z^2} = 0 \tag{8}$$

Example. Fourier's problem is to find the temperature θ at any point (x,y) of a thin plate (see Fig. 2), π units wide and infinitely long,

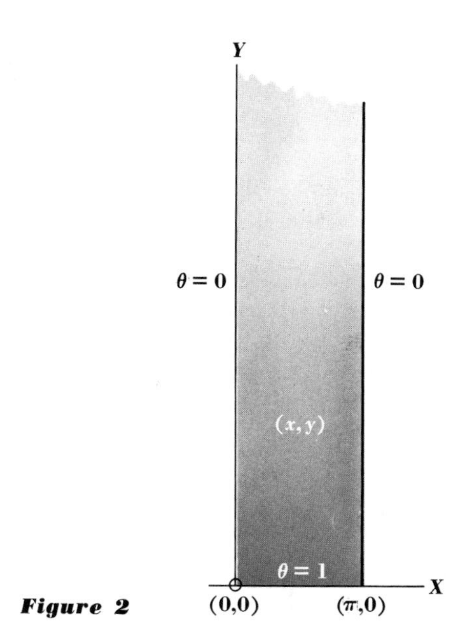

Figure 2

assuming: (1) the steady state so that

$$\frac{\partial^2\theta}{\partial x^2} + \frac{\partial^2\theta}{\partial y^2} = 0 \tag{a}$$

from (6) with $\partial\theta/\partial t = 0$; (2) perfectly insulated surfaces; (3) the short edge constantly at temperature unity; (4) the long edges at temperature zero.

Solution. Taking the Y axis along an infinite edge and the X axis along the short edge, we have the boundary conditions: (α) temperature $\theta = 0$ when $x = 0$; (β) $\theta = 0$ when $x = \pi$; (γ) $\theta = 0$ when $y = \infty$; (δ) $\theta = 1$ when $y = 0$.

To solve (a), substitute $\theta = X(x)Y(y)$ in it, and divide by XY to obtain

$$\frac{d^2X/dx^2}{X} + \frac{d^2Y/dy^2}{Y} = 0$$

Equating the first term to $-\omega^2$ and the second to ω^2, solving the resulting equations for X and Y, and forming $\theta = XY$, obtain

$$\theta = (c_1e^{\omega y} + c_2e^{-\omega y})(c_3 \sin \omega x + c_4 \cos \omega x) \tag{b}$$

Hence, any one of the terms

$$\begin{matrix} Ae^{\omega y} \sin \omega x & Be^{\omega y} \cos \omega x & Ce^{-\omega y} \sin \omega x \\ Ge^{-\omega y} \cos \omega x & & \end{matrix} \tag{c}$$

is a solution of (a), and any sum of such terms is a solution. It remains to choose such a sum that the initial conditions will be satisfied by it.

Conditions (α) and (β) will be satisfied by

$$\theta = \sum_{\omega=1}^{\infty} (c_{1\omega}e^{\omega y} + c_{2\omega}e^{-\omega y}) \sin \omega x \tag{d}$$

and condition (γ) will also be satisfied by (d) if $c_{1\omega} = 0$. The expansion of unity in a Fourier series is

$$1 = \frac{4}{\pi}\left(\sin x + \frac{1}{3}\sin 3x + \frac{1}{5}\sin 5x + \cdots\right) \tag{e}$$

Hence, conditions (α), (β), (γ), and (δ) are satisfied by

$$\theta = \frac{4}{\pi}\left(\frac{1}{1}e^{-y}\sin x + \frac{1}{3}e^{-3y}\sin 3x + \frac{1}{5}e^{-5y}\sin 5x + \cdots\right)$$

Exercises

1. Solve the problems obtained from the illustrative example by replacing condition (3) by: (a) the short edge has temperature $\theta(x,0) = A \sin 3x$; (b) $\theta(x,0) = Ax$.

2. In the example, replace π units wide by L units wide, and solve the resulting problem.

3. In the example, replace π units wide by L units wide and condition (3) by: the temperature at point $(x,0)$ is mx when $0 \leqq x \leqq \frac{1}{2}L$ and is $m(L - x)$ when $\frac{1}{2}L \leqq x \leqq L$. Solve the resulting problem.

137 One-dimensional heat flow

The temperature θ in an insulated rod through which heat is flowing parallel to the axis of the rod satisfies equation (7) in §136, namely

$$a^2 \frac{\partial^2\theta(x,t)}{\partial x^2} = \frac{\partial\theta(x,t)}{\partial t} \qquad a^2 = \frac{k}{c\rho} \tag{1}$$

To find solutions of (1), substitute in it

$$\theta = X(x)T(t) \tag{2}$$

and divide by XT to obtain

$$\frac{a^2\, d^2X/dx^2}{X} = \frac{dT/dt}{T}$$

Equate each member to $-a^2\omega^2$, solve the resulting equations for X and

T, substitute the solutions in (2), and conclude that solutions of (1) may consist of sums of terms having the forms

$$Ae^{-a^2\omega^2 t}\sin\omega x \qquad Be^{-a^2\omega^2 t}\cos\omega x \tag{3}$$

The following example will illustrate a method of solving simple problems relating to the flow of heat:

Example. A rod L cm long with insulated lateral surface is initially at temperature 20°C throughout. If one end is kept at 10°C and the other at 100°C, find the temperature θ as a function of time t and distance x from the end at 10°C.

Solution. The boundary conditions are

$$\theta(0,t) = 10 \qquad \theta(L,t) = 100 \qquad \theta(x,0) = 20 \tag{a}$$

A sum $\varphi(x,t)$ of terms having the first form of (3) will satisfy the condition $\varphi(0,t) = 0$, $\varphi(n\pi/\omega,t) = 0$, n an integer. Also, $\theta = A + Bx$ satisfies (1). Now, let the required solution be

$$\theta(x,t) = A + Bx + \varphi(x,t) \tag{b}$$

where

$$\varphi(x,t) = \sum_{n=1}^{\infty} A_n e^{-a^2n^2\pi^2t/L^2}\sin\frac{n\pi}{L}x \tag{c}$$

Using the conditions (a) with (b) and (c), we obtain

$$\begin{aligned} 10 &= A + \varphi(0,t) = A \qquad 100 = A + BL + \varphi(L,t) = A + BL \\ 20 &= A + Bx + \varphi(x,0) \end{aligned} \tag{d}$$

Solve (d) for A, B, and $\varphi(x,0)$ to get

$$A = 10 \qquad B = \frac{90}{L} \qquad \varphi(x,0) = 10 - \frac{90}{L}x \tag{e}$$

Next, expand $\varphi(x,0)$ in a Fourier series for the interval $0 < x < L$, to obtain

$$\varphi(x,0) = 10 - \frac{90x}{L} = -\frac{4}{\pi}\left(\frac{35}{1}\sin\frac{\pi x}{L} - \frac{45}{2}\sin\frac{2\pi x}{L} + \frac{35}{3}\sin\frac{3\pi x}{L} - \frac{45}{4}\sin\frac{4\pi x}{L} + \cdots\right) \tag{f}$$

To form $\varphi(x,t)$, write in front of the first, second, . . . terms in (f) the respective results of setting $n = 1, 2, \ldots$ in $e^{-n^2a^2\pi^2t/L^2}$. Then, $\theta(x,t)$ from (b) is given by

$$\theta(x,t) = 10 + \frac{90x}{L} - \frac{4}{\pi}\left(\frac{35}{1}e^{-a^2\pi^2t/L^2}\sin\frac{\pi x}{L} - \frac{45}{2}e^{-4a^2\pi^2t/L^2}\sin\frac{2\pi x}{L} + \cdots\right) \tag{g}$$

Exercises

By a *rod* in the following problems, we shall mean a straight rod with insulated lateral surface and with ends A and B. The letter θ will refer to temperature at a point in the rod, and x will refer to the distance of a point in the rod from end A.

1. Observe that in the steady state of heat flow for a rod $\partial\theta/\partial t = 0$ and θ does not contain t. Hence, $d^2\theta/dx^2 = 0$, and $\theta = Ax + B$. End A of a rod 120 cm long is kept at 56° and the other at 200°; find the temperature x cm from A if the steady state prevails. If end A of a rod L ft long is kept at P° and end B at Q°, find the temperature x cm from end A.

2. At time $t = 0$, a rod 100 cm long has temperature $4x + 20$ at x cm from end A. If the temperature at A is suddenly changed to and kept at 56° and that at B is changed to and kept at 200°, find θ at x cm from A at time t.

3. The temperature at end A of a rod 100 cm long is 20°, and that at end B is 200°, and the flow is in the steady state. At time $t = 0$, the temperatures at A and B are suddenly changed to and kept at 60° and 160°, respectively; find θ in terms of x and t at time t.

4. Assume that the rod of exercise 3 is cast iron, and find the temperature at $x = 50$ cm, $t = 1{,}000$ sec. For cast iron, $k = 0.17$ cal/cm deg sec, $c = 0.113$ cal/g deg, $\rho = 7.20$ g/cm^3.

5. At time $t = 0$, a rod AB 40 cm long has temperature $20x$ for the range $0 < x < 20$ cm and temperature $800 - 20x$ for the range $20 < x < 40$ cm, where x is measured from A. Also, at time $t = 0$, the ends A and B are changed to and kept at 800° and 0°, respectively. Find temperature θ in terms of x and t.

138* Applications to nuclear fission

Atoms consist of positively charged nuclei and negatively charged electrons, or particles of electricity. The nucleus contains protons and neutrons. A proton carries a positive charge of electricity which may bind a charge of electrons of equal magnitude. The neutrons

* There are many excellent books treating nuclear-reactor theory. Consult, for example: Samuel Glasstone, "Elements of Nuclear Reactor Theory," and M. A. Schultz, "Control of Nuclear Reactors and Power Plants," 2d ed., McGraw-Hill Book Company, 1961.

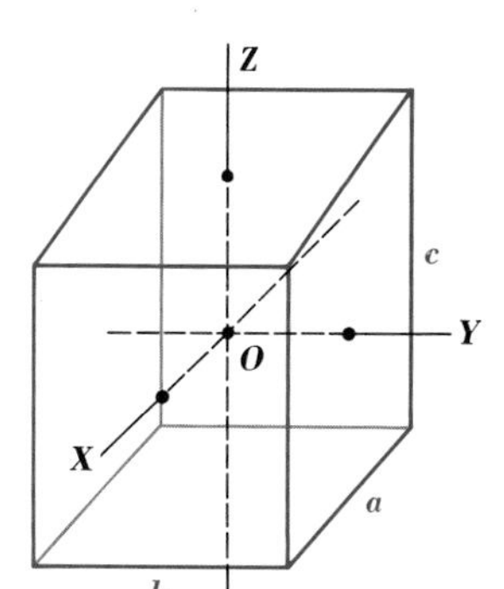

Figure 1

carry no charge. Energy, called nuclear energy, may be released by a fragmentation of the particles of a nucleus.

Consider some uranium in a reactor and an initial source of neutrons from a radioactive decay process. When a neutron of given energy is absorbed by a uranium nucleus, the nucleus is likely to split into two or more fragments with release of energy. This process is called fission. Two or more neutrons are released from the fragmentation, and these may function to create more fissions in other nuclei under proper conditions. If the average life of a free neutron is l, we may speak of the generations of neutrons and denote by k the ratio of the number of neutrons in any generation to the number of neutrons in the immediately preceding generation. If $k = 1$, the situation is called critical. If $k > 1$, a chain reaction can take place, resulting under suitable conditions in the explosion of an atomic bomb; if $k < 1$ the chain reaction will die down.

The following equation is basic in nuclear theory:

$$\frac{\partial^2\psi(x,y,z)}{\partial x^2} + \frac{\partial^2\psi(x,y,z)}{\partial y^2} + \frac{\partial^2\psi(x,y,z)}{\partial z^2} + B^2\psi(x,y,z) = 0 \qquad (1)$$

where B^2 is a positive constant called buckling and $\psi(x,y,z)$, called the neutron flux, is the sum of the distances traveled per second per cubic centimeter at (x,y,z) by bombarding neutrons. For a certain value of B^2, called material buckling, the arrangement of material is critical. Our problem will be to deduce relations among dimensions and minimum volumes of reactors filled with materials subject to (1).

First, consider a reactor in the shape of a rectangular box having dimensions a, b, and c as indicated in Fig. 1. Take the origin of coordinates at the center of the box and the coordinate axes parallel to its edges as indicated. Since there is no flux outside the reactor, we have as boundary conditions

$$\psi = 0 \qquad \text{at} \qquad x = \pm\tfrac{1}{2}a \qquad y = \pm\tfrac{1}{2}b \qquad z = \pm\tfrac{1}{2}c \qquad (2)$$

Also, assume symmetry of $\psi(x,y,z)$ with respect to the coordinate planes. Applying the method of separation of the variables, §129, to solve (1), we let

$$\psi(x,y,z) = X(x) \cdot Y(y) \cdot Z(z) \tag{3}$$

Substitute this value of ψ in (1), divide through by XYZ, and obtain

$$\frac{d^2X/dx^2}{X} + \frac{d^2Y/dy^2}{Y} + \frac{d^2Z/dz^2}{Z} + B^2 = 0 \tag{4}$$

Equating the first fraction to $-\alpha^2$* and solving the resulting equation, we get

$$X = c_1 \cos \alpha x + c_2 \sin \alpha x$$

The condition $\psi = 0$ when $x = \pm\frac{1}{2}a$ demands, because of (3) and symmetry, that $X = 0$ when $x = \pm\frac{1}{2}a$; therefore, we must take $c_2 = 0$ and $\alpha = (2n + 1)\frac{1}{2}(\pi/a)$. Hence, choosing $\frac{1}{2}\pi/a$ for α, the least positive value,† we have

$$X = c_1 \cos \frac{\pi}{a} x$$

Applying the same process to each of the fractions, we get

$$X = c_1 \cos \frac{\pi}{a} x \qquad Y = d_1 \cos \frac{\pi}{b} y \qquad Z = e_1 \cos \frac{\pi}{c} z \tag{5}$$

and, from (3),

$$\psi(x,y,z) = A \cos \frac{\pi}{a} x \cos \frac{\pi}{b} y \cos \frac{\pi}{c} z \tag{6}$$

The constant A depends on the power output of the reactor. Substituting from (6) in (4), we get

$$B^2 = \frac{\pi^2}{a^2} + \frac{\pi^2}{b^2} + \frac{\pi^2}{c^2} \tag{7}$$

For a given value of B^2, the volume of the reactor will be minimum if it

* The negative value $-\alpha^2$ is used because the resulting trigonometric solutions lend themselves to satisfying (2). Of course (4) requires negative quantities to neutralize $+B^2$.

† A little reflection indicates that a more complicated arrangement of material in the reactor would be represented by a value for α like $3\pi/a$ or $5\pi/a$, since this would mean a number of planes parallel to the yz plane on which ψ would vanish.

has the shape of a cube. In this case, we see, from (7), that

$$a = b = c = \frac{\pi\sqrt{3}}{B} \qquad \text{Vol } v = \frac{\pi^3 3\sqrt{3}}{B^3} = \frac{161}{B^3} \tag{8}$$

The arrangement will be critical if B^2 has a certain value called the *material buckling*.

Exercises

1. For a spherical reactor we use spherical coordinates (see Fig. 2) so that

$$x = r \sin\theta \cos\varphi \qquad y = r \cos\theta \cos\varphi \qquad z = r\cos\theta$$

In these coordinates, (1) takes the form

$$\frac{\partial^2\psi}{\partial r^2} + \frac{2}{r}\frac{\partial\psi}{\partial r} + \frac{1}{r^2\sin\theta}\frac{\partial}{\partial\theta}\left(\sin\theta\frac{\partial\psi}{\partial\theta}\right) + \frac{1}{r^2\sin^2\theta}\frac{\partial^2\psi}{\partial\varphi^2} + B^2\psi = 0 \tag{9}$$

where $\psi(r,\theta,\varphi)$ is the neutron flux. Assuming uniform and symmetrical distribution of material, so that ψ depends only upon r, delete all terms from (5) involving θ and φ to get

$$\frac{\partial^2\psi}{\partial r^2} + \frac{2}{r}\frac{\partial\psi}{\partial r} + B^2\psi = 0 \tag{10}$$

Show that the solution of this is

$$\psi = \frac{c_1}{r}\sin Br + \frac{c_2}{r}\cos Br$$

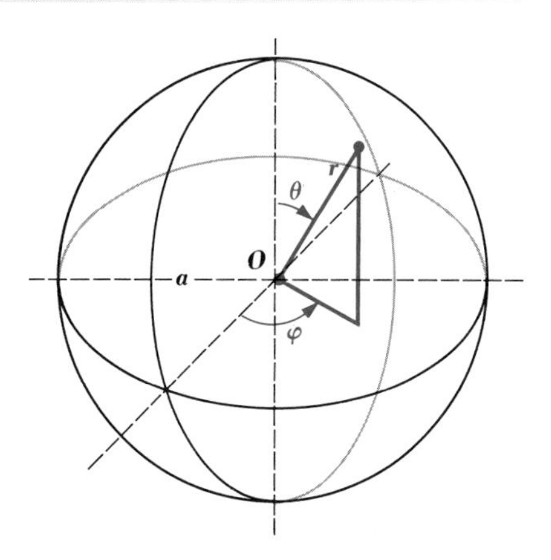

Figure 2

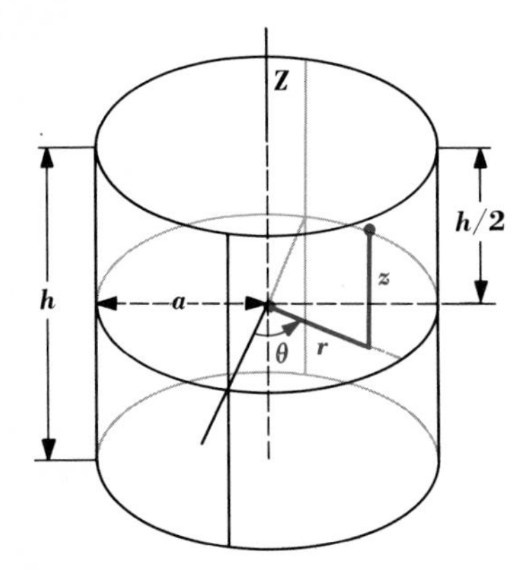

Figure 3

Assume as initial conditions

$$\psi(a) = 0 \qquad \psi(r) \text{ is bounded}$$

and deduce that

$$\psi(r) = \frac{A}{r}\sin\frac{\pi r}{a} \qquad (\text{Vol})_{\min} = \frac{130}{B^3} \tag{11}$$

If B^2 is properly determined, (11) applies when the arrangement of material is critical.

*2. For a cylindrical reactor, we use cylindrical coordinates (see Fig. 3) so that

$$x = r\cos\theta \qquad y = r\sin\theta \qquad z = z$$

In cylindrical coordinates, (1) takes the form

$$\frac{\partial^2\psi}{\partial r^2} + \frac{1}{r}\frac{\partial\psi}{\partial r} + \frac{1}{r^2}\frac{\partial^2\psi}{\partial\theta^2} + \frac{\partial^2\psi}{\partial z^2} + B^2\psi = 0 \tag{12}$$

Assuming that ψ depends only on r and z, delete from (12) the term involving θ, and obtain

$$\frac{\partial^2\psi}{\partial r^2} + \frac{1}{r}\frac{\partial\psi}{\partial r} + \frac{\partial^2\psi}{\partial z^2} + B^2\psi = 0 \tag{13}$$

Solve this by *separation of the variables*, using

$$\psi = R(r)Z(z)$$

and use the initial conditions

ψ is bounded
$\psi = 0 \qquad \text{when } z = \pm\tfrac{1}{2}h \qquad \text{or} \qquad r = a$

to obtain

$$Z = c_1 \cos \frac{\pi}{h} z \qquad R = c_2 J_0\left(\frac{2.405r}{a}\right)$$

where 2.405 is the smallest positive root of $J_0(x)$. Now, deduce that

$$\psi(r,z) = A \cos \frac{\pi}{h} z J_0\left(\frac{2.405r}{a}\right) \tag{14}$$

$$B^2 = \left(\frac{\pi}{h}\right)^2 + \left(\frac{2.405}{a}\right)^2 \tag{15}$$

and then that, for given B^2 and minimum volume $V_{\min}$ of the cylindrical reactor

$$2B^2a^2 = 3(2.405)^2 \qquad B^2h^2 = 3\pi^2 \qquad V_{\min} = \frac{148.2}{B^3} \tag{16}$$

Hint: $u = AJ_0(nr)$ is a solution of $d^2u/dr^2 + (1/r)\,du/dr + n^2u = 0$.

139 Vibrations of a membrane

Think of a right-circular cylindrical drum having a vertical axis and having as a top a tightly stretched, thin, homogeneous membrane. The membrane is depressed symmetrically to a vertical line through its center. Assume that the depression at all times is so slight that we may think of the particles of the membrane as moving vertically and of the tension T g/cm across a line in the surface as being constant and directed at right angles to the line. The problem is to find the equation of the motion of the membrane.

Let (r,θ,y) be coordinates of points in the membrane, where r and θ are polar coordinates of a system in the drum head with pole O at its center and y designates displacement from the drum head. Observe that y is independent of θ from symmetry and that for any curve cut out of the membrane (see Fig. 1) by a vertical plane through the

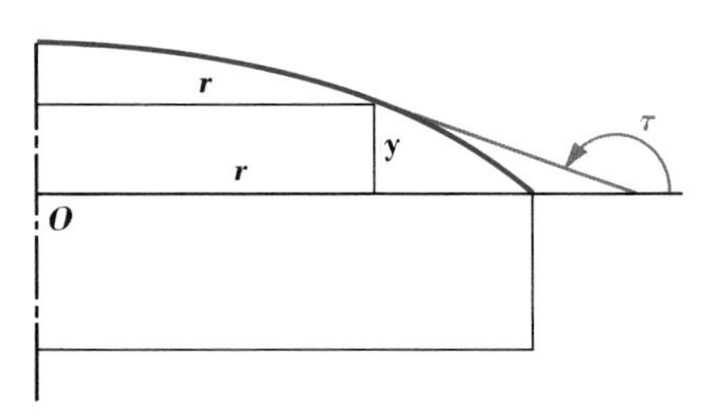

Figure 1

center O and having a tangent of inclination τ, we may use the approximate equations

$$\sin \tau = \tan \tau = \frac{\partial y}{\partial r} = \frac{T_v}{T} \tag{1}$$

where T_v is the vertical component of radial tension at a point. The vertical component of force on a small element $r\,\Delta\theta$ of a circle on the moving membrane y units above the drum head, from (1), is $T(\partial y/\partial r)r\,\Delta\theta$, and on the complete circle it is

$$F_v = 2\pi T r \frac{\partial y}{\partial r}$$

The vertical force on the zone of radial width Δr is then

$$\frac{\partial F_v}{\partial r}\Delta r = 2\pi T\,\Delta r\left(r_1 \frac{\partial^2 y}{\partial r^2} + \frac{\partial y}{\partial r}\right) \tag{2}$$

where r_1 is between r and $r + \Delta r$. The mass times the acceleration of the zone is $\rho 2\pi r_2\,\Delta r\,\partial^2 y/\partial t^2$, where ρ is the mass per square unit of the membrane and r_2 is between r and $r + \Delta r$. Now, using Newton's law of motion for the zone, equate the limits of the right member of (2) divided by Δr and $2\pi r_2\,\Delta r(\partial^2 y/\partial t^2)/\Delta r$, and obtain

$$2\pi T\left(r \frac{\partial^2 y}{\partial r^2} + \frac{\partial y}{\partial r}\right) = 2\pi\rho r \frac{\partial^2 y}{\partial t^2}$$

Dividing this through by $2\pi r T$, we get

$$\frac{\partial^2 y}{\partial r^2} + \frac{1}{r}\frac{\partial y}{\partial r} = a^2 \frac{\partial^2 y}{\partial t^2} \tag{3}$$

where $a^2 = \rho/T$. In (3), substitute $y = R(r)Z(t)$ for y, divide by RZ, and obtain

$$\frac{1}{R}\left(\frac{d^2R}{dr^2} + \frac{1}{r}\cdot\frac{dR}{dr}\right) = a^2 \frac{d^2Z}{dt^2}\cdot\frac{1}{Z} \tag{4}$$

In this, take $(d^2Z/dt^2)/Z = -\omega^2$, and obtain

$$Z = c_1 \sin \omega t + c_2 \cos \omega t \tag{5}$$

The other equation obtained from (4) is

$$\frac{d^2R}{dr^2} + \frac{1}{r}\frac{dR}{dr} + a^2\omega^2 R = 0 \tag{6}$$

In this, substitute $w/(a\omega)$ for r, and obtain after a slight simplification

$$\frac{d^2R}{dw^2} + \frac{1}{w}\frac{dR}{dw} + R = 0 \qquad w = a\omega r \qquad a^2 = \frac{\rho}{T} \tag{7}$$

This is Bessel's equation with $n = 0$. Its basic solutions (see exercise 2, §111) are

$$J_0(w) = \sum_{n=0}^{\infty} \frac{(\frac{1}{2}w)^{2n}(-1)^n}{(r!)^2}$$

$$Y_0(w) = \ln w \sum_{n=0}^{\infty} \frac{(\frac{1}{2}w)^{2n}(-1)^n}{(n!)^2} + \sum_{n=1}^{\infty} \frac{(\frac{1}{2}x)^{2n}(-1)^{n+1}}{(n!)^2} \sum_{m=1}^{n} \frac{1}{m}$$

and

$$R = c_1 J_0(w) + c_2 Y_0(w) \qquad w = a\omega r \qquad a = \sqrt{\frac{\rho}{T}} \tag{8}$$

Exercises

1. Why must c_2 in (8) be taken as zero generally? Might c_2 be different from zero if the membrane had the shape of a plane area bounded by two concentric circles?

2. If $\bar{R}$ is the radius of the circular membrane and $c_2 = 0$ in (8), why must $J_0(a\bar{R}) = 0$? Three roots of $J_0(w)$ are $w_1 = 2.40$, $w_2 = 5.52$, and $w_3 = 8.65$. State three corresponding values that $\bar{R}$ may have.

3. If $\bar{R} = 5.52/a$, Fig. 2 indicates the shape of a cross section of the moving membrane at time $t = 0$. The complete membrane would be generated by revolving curve $APBC$ about the Y axis. Considering (5), state what happens as t varies from 0 to $2\pi/\omega$: (*a*) to point A; (*b*) to point P; (*c*) to points B;* (*d*) to points C.

* Points such as B on the complete surface lie at rest on a circle called the nodal circle.

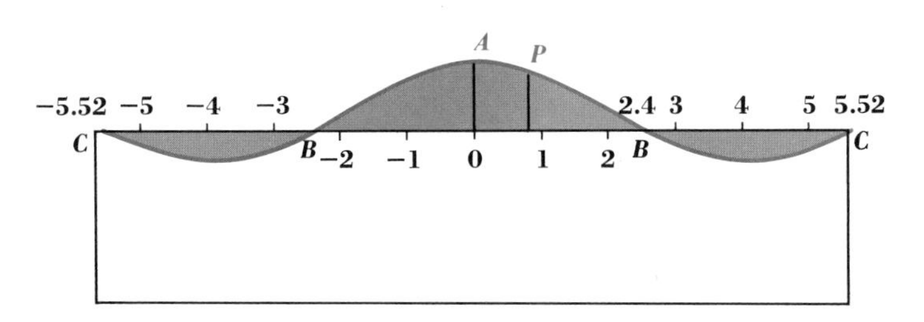

Figure 2

4. Draw a figure analogous to Fig. 2 for the situation when: (*a*) $\bar{R} = 2.40/a$; (*b*) $\bar{R} = 8.65/a$.

★**5.** If the vertical cross section through the center of the membrane at time $t = 0$ has the equation $y = m(L - x)$, where L is the radius of the drum, show that

$$y(r,t) = \sum_{i=1}^{\infty} \left[\frac{-mL}{\alpha_i^2} \int_0^1 J_0(\alpha_i z)\, dz \right] J_0\left(\frac{\alpha_i r}{L}\right) \cos \frac{\alpha_i}{aL} \frac{t}{\frac{1}{2}J_1^2(\alpha_i)}$$

$$J(\alpha_i) = 0$$

First expand $m(L - Lz)$ in an infinite series by the method of §113 and in the result replace z by r/L.

140 Telephone, telegraph, and radio equations

Figure 1 represents a long line carrying electricity. The current goes out through AB and returns through the ground from C to D. Let L (in henrys per mile) be the inductance of the line AB, let R (in ohms per mile) be its resistance, let C (in farads per mile) be its capacitance to ground, and let G (in mhos per mile) be the leakage of current or conductance to ground. Let $e(x,t)$ and $i(x,t)$ represent the voltage and current at a point in the line AB x mi from A, and derive relations between e and i by considering the flow of electricity in a small portion PQ of the cable having length Δx. The drop Δe in potential along PQ will be approximately

$$\Delta e = -R\, \Delta x i(x_1,t) - L\, \Delta x \frac{\partial i(x_1,t)}{\partial t} \tag{1}$$

where $x < x_1 < x + \Delta x$. Dividing this equation through by Δx and equating the limits approached by its members as Δx approaches zero,

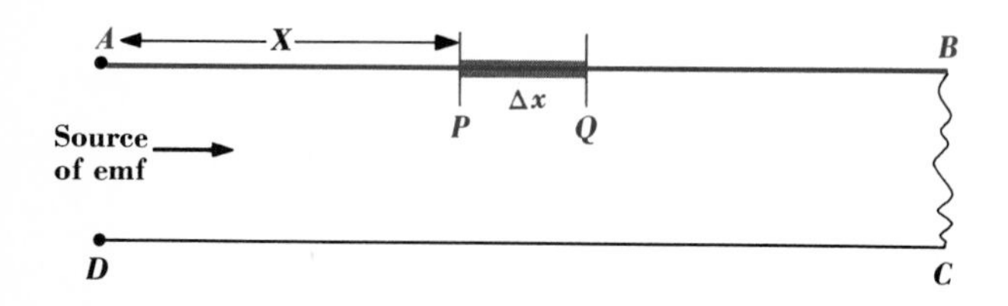

Figure 1

we get

$$\frac{\partial e}{\partial x} = -Ri(x,t) - L\frac{\partial i(x,t)}{\partial t} \tag{2}$$

Here, changes in e due to leakage and capacitance to ground are of second order and therefore have nothing to do with equation (2).

Similarly, the drop in current along PQ is approximately

$$\Delta i = -G\,\Delta x e(x_1,t) - C\,\Delta x\,\frac{\partial e(x_1,t)}{\partial t}$$

Dividing this through by Δx and equating the limits approached by its members as Δx approaches zero, we obtain

$$\frac{\partial i}{\partial x} = -Ge - C\frac{\partial e}{\partial t} \tag{3}$$

Equations (2) and (3) are basic equations. Three important sets will be derived from them.

Eliminate i between (2) and (3) by equating the partial derivatives with respect to x of the two members of (2), replacing $\partial i/\partial x$ in the result by its value from (3) and $\partial^2 i/(\partial x\,\partial t)$ by its value obtained from (3) by partial differentiation with respect to t, and simplifying; the result is

$$\frac{\partial^2 e}{\partial x^2} = RGe + (RC + LG)\frac{\partial e}{\partial t} + LC\frac{\partial^2 e}{\partial t^2} \tag{4}$$

Similarly, eliminate e between (2) and (3) to get

$$\frac{\partial^2 i}{\partial x^2} = RGi + (RC + LG)\frac{\partial i}{\partial t} + LC\frac{\partial^2 i}{\partial t^2} \tag{5}$$

The equations (2) to (5) are known as the telephone equations.

In many applications to telegraph signaling, G and L are negligible. Replacing G and L by 0 in equations (2) to (5), we obtain the telegraph equations which follow:

$$\begin{aligned} \frac{\partial e}{\partial x} &= -Ri \\ \frac{\partial i}{\partial x} &= -C\frac{\partial e}{\partial t} \\ \frac{\partial^2 e}{\partial x^2} &= RC\frac{\partial e}{\partial t} \\ \frac{\partial^2 i}{\partial x^2} &= RC\frac{\partial i}{\partial t} \end{aligned} \tag{6}$$

For high frequencies, we may place $G = R = 0$ in (2) to (5) to obtain the radio equations which follow:

$$\begin{aligned} \frac{\partial e}{\partial x} &= -L\frac{\partial i}{\partial t} \\ \frac{\partial i}{\partial x} &= -C\frac{\partial e}{\partial t} \\ \frac{\partial^2 e}{\partial x^2} &= LC\frac{\partial^2 e}{\partial t^2} \\ \frac{\partial^2 i}{\partial x^2} &= LC\frac{\partial^2 i}{\partial t^2} \end{aligned} \tag{7}$$

Exercises

1. Substitute $e = X(x)T(t)$ in the third of the radio equations (7), and, by the usual procedure, deduce that it is satisfied by the expressions for e

$$\begin{matrix} \cos(\omega\sqrt{LC}\,x)\cos\omega t & \cos(\omega\sqrt{LC}\,x)\sin\omega t \\ \sin(\omega\sqrt{LC}\,x)\sin\omega t & \sin(\omega\sqrt{LC}\,x)\cos\omega t \end{matrix} \tag{8}$$

If i and e are to satisfy the four radio equations (7) and $e = A\sin(\omega\sqrt{LC}\,x)\cos\omega t$, show that i must have the form $i = -A\sqrt{C/L}\cos(\omega\sqrt{LC}\,x)\sin\omega t + B$, where B is a constant.

2. If $i = A\cos(\omega\sqrt{LC}\,x)\sin\omega t$, find e so that i and e satisfy the radio equations (7).

3. Substitute $e = X(x)T(t)$ in the third equation of (6), and, by the regular procedure, deduce that some solutions of (6) have the form

$$A\epsilon^{-(\omega^2/RC)t}\cos\omega x \qquad B\epsilon^{-(\omega^2/RC)t}\sin\omega x \tag{9}$$*

4. (a) If $e = A\epsilon^{-(\omega^2/RC)t}\cos\omega x$, find a corresponding function $i(x,t)$ such that e and i satisfy (6). (b) If $i = A\epsilon^{-(\omega^2/RC)t}\sin\omega x$, find $e(x,t)$ such that i and e will satisfy (6).

5. In a steady-state condition for which i and e are functions of x only, solve (6). Since i and e depend on x only, $\partial i/\partial t$ and $\partial e/\partial t$ are zero.

6. In Fig. 1, take L mi as the length of AB, and solve the corresponding telegraph equations (6). Use as initial conditions $e(0,t) = 0$, $e(L,t) = 0$, and $e(x,0) = 2 + 3x/L$. Equations (A), (B), and (C) of §133 may be used to save time.

* In this section, $\epsilon = 2.7183$ approximately.

7. If $u = x + t/\sqrt{LC}$ and $v = x - t/\sqrt{LC}$, show that $i = \varphi(u) + \psi(v)$ satisfies the fourth equation of (7). Then, derive $e = \sqrt{L/C}\,[\psi(v) - \varphi(u)] + H$, where H is constant, from equations (7).

8. A line is called distortionless if $LG = RC$, or $G/C = R/L$. Make the substitution

$$e = E(x,t)\epsilon^{-Gt/C} \qquad i = I(x,t)\epsilon^{-Gt/C}$$

in (2) to (5) to obtain equations having the form (7) of the radio equations for a distortionless line. Also, check directly that (4) is satisfied by

$$e = A\epsilon^{-Gt/C} \sin (\omega \sqrt{LC}\, x) \cos \omega t$$

when $G/C = R/L$, and find the corresponding $i(x,t)$ to satisfy (2) and (3).

141 Fluid motion

Because of the importance of fluid motion, as exemplified by the flow of air over airplane wings and the flow of water near ships, and because the solution of fluid-motion problems involves partial differential equations, a brief introduction to the subject will be given.

Consider the motion of a homogeneous fluid with continuous structure and no viscosity.* In this case all forces exerted by the fluid on a surface will be normal to the surface. For simplicity, think of a fluid moving between two parallel planes, and assume that any particle remains in a plane parallel to the bounding planes and that the motions in all such planes are the same. The flow will then be two-dimensional.

The motion of the fluid will be due to pressure in the fluid and a force proportional to the mass like the pull of gravity. Thus, the pressure p in the fluid will be a function of x, y, and t, and the force per unit mass will have components $X(x,y,t)$ and $Y(x,y,t)$ parallel to the coordinate axes. Also, let $u(x,y,t)$ and $v(x,y,t)$ be the x- and y-components of the velocity at time t.

From calculus, we have for the x- and y-components of acceleration

$$\begin{aligned} a_x &= \frac{du}{dt} = \frac{\partial u}{\partial x} u + \frac{\partial u}{\partial y} v + \frac{\partial u}{\partial t} \\ a_y &= \frac{dv}{dt} = \frac{\partial v}{\partial x} u + \frac{\partial v}{\partial y} v + \frac{\partial v}{\partial t} \end{aligned} \tag{1}$$

* All fluids are viscous, but many, water for example, are only slightly viscous.

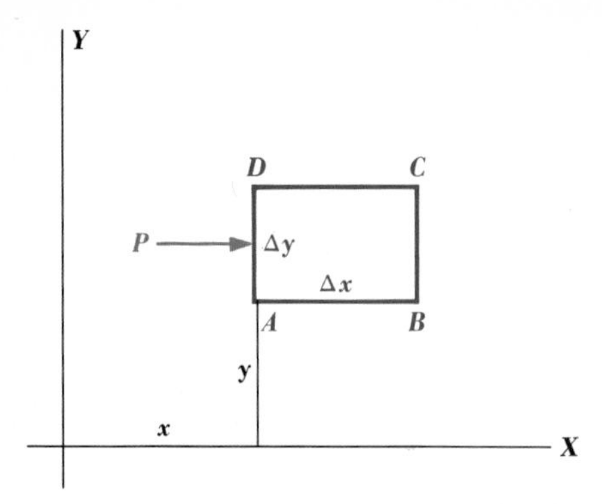

Figure 1

Now apply Newton's law of motion to the element of fluid represented by $ABCD$ in Fig. 1. The forces at time t on the faces represented by AD and BC may be expressed as an average pressure multiplied by the area $h\,\Delta y$, where h is the distance between the bounding planes. Accordingly, we write as the total force due to pressure on the surfaces represented by AD and BC

$$h\,\Delta y\, p(x,y_1,t) - h\,\Delta y\, p(x + \Delta x,\, y_1,\, t) \tag{2}$$

where $y < y_1 < y + \Delta y$. Also, the force proportional to mass acting on the element in the x-direction may be expressed by

$$\frac{h}{g}\,\Delta y\,\Delta x\,\bar{\rho}X(x_2,y_2,t) \tag{3}$$

where $\bar{\rho}$ is the average density of the fluid in the element and (x_2,y_2) is a point properly chosen in the element. Hence, we have

$$h\,\Delta y\, p(x,y_1,t) - h\,\Delta y\, p(x + \Delta x,\, y_1,\, t) + \frac{h}{g}\,\Delta y\,\Delta x\,\bar{\rho}X(x_2,y_2,t) = \frac{\bar{\rho}h\,\Delta y\,\Delta x}{g}\frac{du(x_3,y_3,t)}{dt} \tag{4}$$

where (x_3,y_3) is a properly chosen point in the element. Dividing through by $\Delta x\,\Delta y$, equating the limits of the two members as Δx and Δy approach zero, and simplifying slightly, we get

$$\frac{du}{dt} = X - \frac{g}{\rho}\frac{\partial p}{\partial x} \tag{5}$$

Applying Newton's law parallel to the Y axis, we obtain in a like manner

$$\frac{dv}{dt} = Y - \frac{g}{\rho}\frac{\partial p}{\partial y} \tag{6}$$

Finally, express the condition that the rate of change of the amount of fluid in the element is the rate at which fluid enters minus the rate at which it leaves. The rate of change of the quantity is

$$\frac{\partial}{\partial t}[h\,\Delta x\,\Delta y\,\rho(x_1,y_1,t)] = h\,\Delta x\,\Delta y\,\frac{\partial\bar{\rho}}{\partial t} \tag{7}$$

The rate of entering minus the rate of leaving is approximately

$$h\,\Delta y[\overline{\rho u}(x,y_1,t) - \overline{\rho u}(x+\Delta x,\,y_1,\,t)] + h\,\Delta x[\overline{\rho v}(x_1,y,t) - \overline{\rho v}(x_1,\,y+\Delta y,\,t)] \tag{8}$$

Equating the limit of (8) divided by $\Delta x\,\Delta y$ as Δx and Δy approach zero to the limit of (7) divided by $\Delta x\,\Delta y$, we get

$$-\frac{\partial(\rho u)}{\partial x} - \frac{\partial(\rho v)}{\partial y} = \frac{\partial\rho}{\partial t} \tag{9}$$

Equations (5), (6), and (9) are the differential equations of fluid flow for the special case considered.

To understand the simplest case, the idea of *rotation* will be required. Figure 2 shows positions $A'B'C'D'$, after Δt units of time, of four fluid particles originally at corners A, B, C, and D of a rectangle having side AB parallel to the X axis and side AD parallel to the Y axis. Disregarding infinitesimals of higher order than the first, we get for the angular velocity ω_x of $A'B'$

$$\omega_x = \frac{(\partial v/\partial x)\,\Delta x\,\Delta t}{\Delta t\,\Delta x} = \frac{\partial v}{\partial x} \tag{10}$$

and for the angular velocity ω_y of $A'D'$

$$\omega_y = \frac{-(\partial u/\partial y)\,\Delta y\,\Delta t}{\Delta t\,\Delta y} = \frac{-\partial u}{\partial y} \tag{11}$$

One-half the sum of the velocities ω_x and ω_y is called the rotation of the

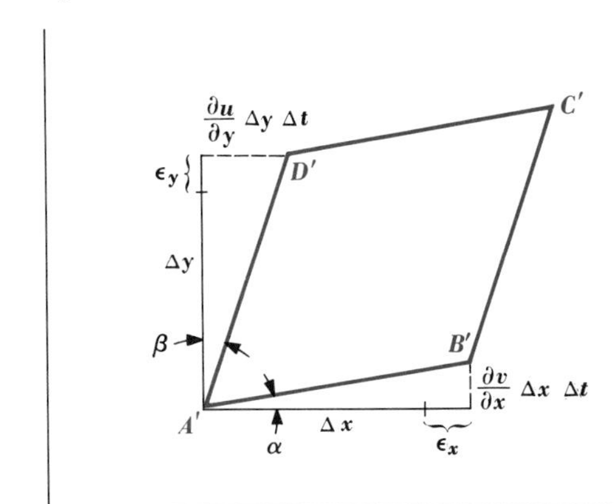

Figure 2

fluid, that is,

$$\text{Rotation} = \frac{1}{2}\left(\frac{\partial v}{\partial x} - \frac{\partial u}{\partial y}\right) \tag{12}$$

Now, consider the motion of a fluid for which the rotation is zero, that is

$$\frac{\partial v}{\partial x} = \frac{\partial u}{\partial y} \tag{13}$$

the fluid is incompressible, that is,

$$\rho = \text{constant} \tag{14}$$

the fluid is in the steady state, that is, u and v are functions of x and y only; and the force is conservative, that is, there exists a force function $U(x,y)$ such that

$$\frac{\partial U}{\partial x} = X \qquad \frac{\partial U}{\partial y} = Y \tag{15}$$

From (1), (13), and the fact that the steady state exists,

$$\frac{du}{dt} = \frac{\partial u}{\partial x}u + \frac{\partial u}{\partial y}v = \frac{\partial u}{\partial x}u + \frac{\partial v}{\partial x}v \tag{16}$$

Hence, taking account of (13) to (16), we may write (5) in the form

$$u\frac{\partial u}{\partial x} + v\frac{\partial v}{\partial x} = \frac{\partial U}{\partial x} - \frac{g}{\rho}\frac{\partial p}{\partial x} \tag{17}$$

The integral of (17) is

$$\frac{u^2 + v^2}{2} = U - \frac{g}{\rho}p + C \tag{18}$$

Exercises

1. From (13), deduce that in steady-state irrotational fluid motion there is a *velocity potential* $\varphi(x,y)$ such that

$$\frac{\partial \varphi}{\partial x} = u \qquad \frac{\partial \varphi}{\partial y} = v$$

The curves $\varphi(x,y) = c$ are called *curves of constant velocity potential.*

2. A streamline of a fluid in motion is a curve at each of whose points the direction of the velocity is the same as that of the curve.

Hence, along streamlines, $dy/dx = v/u$, or

$$v\,dx - u\,dy = 0$$

Show that, for steady-state irrotational motion of an incompressible fluid, this equation is exact because of (9), which, for the case in question, simplifies to $\partial u/\partial x + \partial v/\partial y = 0$. Hence, show that there are streamlines represented by

$$\psi(x,y) = C \qquad \frac{\partial \psi}{\partial x} = v \qquad \frac{\partial \psi}{\partial y} = -u$$

Also, show that the streamlines $\psi(x,y) = C$ are the orthogonal trajectories of the velocity potential curves considered in exercise 1.

3. If $f(z) = v(x,y) + iu(x,y)$, where u and v are real functions of x and y, is analytic, then

$$\frac{\partial v}{\partial x} = \frac{\partial u}{\partial y} \qquad \frac{\partial u}{\partial x} = -\frac{\partial v}{\partial y}$$

These are equivalent to (13) and (9) for steady-state irrotational motion of an incompressible fluid.

(*a*) Using $f(z) = (x + iy)^2 = x^2 - y^2 + 2ixy$, take $v = x^2 - y^2$ and $u = 2xy$, and find the corresponding equations of the streamlines and curves of constant velocity potential. Also, assuming that $X = 0$ and $Y = -g$, find p in terms of x and y. (*b*) Carry out the same process for $f(z) = 1/z$, assuming $X = 0$ and $Y = -g$.

Answers

§2, page 3

1. 1, 1. **2.** 1, 2. **3.** 1, 6. **4.** 2, 1. **5.** 2, 1. **6.** 2, 2. **7.** 3, 2. **8.** 2, 2.

§4, page 9

1. $x^2 + y^2 = c$. **2.** $x + y = c, x + y = 1$. **3.** $xy = c, xy = 2$. **4.** $(x - 1)(y - 2) = c$. **5.** $1 + y^2 = c(1 + 2x^2)$. **6.** $16S - 15 = Re^{-16t}$. **7.** $5y = 3(y + 2)e^{2x}$. No. **8.** $y^2 = -\frac{4}{3}(1 + x)(1 - x)^{-1}e^{1-2/x}$. **9.** $r\theta^2 = 1$. **10.** $2y + 1 = 2e^{2/x}$. **11.** $i = ce^{-Rt/L}$. **12.** $3e^x + e^{-3y} = 4$. **13.** $\sin^{-1} x - \sin^{-1} y = \frac{1}{3}\pi$. **14.** $y^4(2 + x) = -6 + 3x$. (*a*) $y = 0$ yes, others no. (*b*) $y = 0$ no, others yes. **15.** $xy = 2(x - 1)e^{2/x}$. No, no, yes.

§5, page 11

1. (*a*) 0. (*b*) -2. (*c*) y_0. **2.** $\frac{1}{2}x = m$. **5.** (*a*) $y = x^2 + x$. (*b*) $y = x^2 - 2x$. (*c*) No. Yes. **6.** (*c*) $x = 1/m$. (*e*) 10 units upward. **7.** (*a*) $x = c_1$ where slope $m = f(c_1)$. (*b*) $y = c_2$ where $m = f(c_2)$. (*c*) $y = x + c$.

§6, page 14

1. $\partial f/\partial y = 1/x$ does not exist when $x = 0$. Yes. **2.** All points in the xy plane except those on lines $x = 1$ and $y = 0$. Same. **3.** All points for which $-2 < y < 2$ except those on $y = 0$. All points for which $-2 \leqq y \leqq 2$ except those on $y = 0$. **4.** All points in xy plane except those for which $y = 0$. All points in the plane. **5.** (*a*) $\sqrt{25 - y^2} - \sqrt{25 - x^2} = c$. (*b*) All points satisfying both $|y| < 5$ and $|x| < 5$ except $(a,0)$ where $|a| < 5$. (*c*) Same as (*b*) except $|y| \leqq 5$ replaces $|y| < 5$. (*d*) Points having $y = \pm 5$, but $|x| < 5$. **6.** (x,y) where $x > 0$, $y > 0$. **7.** $y > x$, $y - x = 1$.

§7, page 17

1. $y' = 3x^2$. **2.** $xy' = 2y$. **3.** $(x + 2)y' = y$. **4.** $(x^2 + x)y' = (2x + 1)y$. **5.** $y' = y$. **6.** $y'' - y = 0$. **7.** $y''' = 0$. **8.** $y'' + 4y = 0$. **9.** $xy' = y$. **10.** $yy' + x = 0$. **11.** $yy'' - (y')^2 = 0$. **12.** $y'' + 4y = 3 \sin x$. **13.** $y'' + y = 0$. **14.** $x^2y'^2 - 2xyy' + (1 + x^2)y^2 = x^4$. **15.** $y'' = 0$. **16.** $y'(y^2 - x^2 + 2x) + 2y(x - 1) = 0$. **17.** $(y - x)^2 [1 + (y')^2] = (1 + y')^2$. **18.** $xy'^2 - yy' + 1 = 0$.

§8, page 19

1. $y + 9x = c_1$, $y = 2x + c_2$. **2.** $y = x^3 + c_1$, $y = x^4 + c_2$. **3.** $4y^2 + 3x^2 = c_1$, $4y^2 - 3x^2 = c_2$. **4.** $xy = c_1$, $x^2 + y^2 = c_2$. **5.** $x^2y = c_1$, $y = c_2x$. **6.** $y = c_1 - \ln x$, $2y = 3x^2 + c_2$. **7.** $2y + x^2 = c_1$, $y = 2 \ln x + c_2$. **8.** $y - x = c_1$, $y - x^2 = c_2$. **9.** $y^{1/2} = x^{1/2} + c_1$, $y^{1/2} = -x^{1/2} + c_2$. **10.** $y = 2x^{1/2} + c_1$, $y = 2x^{1/2} - \frac{1}{2}x^2 + c_2$, $x \geqq 0$. **11.** $y^{1/2} = x^{1/2} + 1$, $y^{1/2} + \ln |\sqrt{y} - 1| = \sqrt{x} + 1$, $x \geqq 0$, $y \geqq 0$, but not $(0,0)$.

§9, page 19

1. 1, 3. **2.** 1, 3. **3.** 2, 4. **7.** $y = xy' + y'^3$. **8.** $y' = 5y$. **9.** $y''' = 0$. **10.** $(x + 2y)y'' = (1 + y'^2)(y' - 2)$. **11.** (*a*) $y > 0, x > 1$. (*b*) $y \geqq 0, x > 1$. **12.** (*a*) $y \neq 0, x > -2$. (*b*) $x > -2$. **13.** (*a*) $|y| > |x|$. (*b*) $|y| > |x|$. **14.** $y = c(x + a)^{1-a}$. **15.** $x^2(y + 2) = cy$. **16.** $\ln y + \tan^{-1} y = x - \frac{1}{2}\pi$. **17.** $2y(x - 1) = x$. **18.** $xyy' + 25 - y^2 = 0$. **19.** $y'' = 0$. **20.** $(y' + 1)(y - xy') = 2y'$. **21.** $(2y^2 - 2x^2 + 2xy) + y'(y^2 - 4xy - x^2) = 0$. **22.** $y = xy' - y'^2$. **23.** $x^2(1 + y^2) = 17$. **24.** $27 \cot y = (1 + e^x)^3$. **25.** $y(\ln y - 1) + \ln (\ln x) = c$. **26.** $y + 3x = c_1$, $y - x = c_2$. **27.** $xy = c_1$, $x = c_2y$. **28.** $6y = 4x^3 + 3x^2 + c_1$, $6y = 4x^3 - 3x^2 + c_2$.

§10, page 23

2. (*a*) $y = (x - 1)^2$. (*b*) $(x - 1)^2 + (y + 1)^2 = 29$. (*c*) $(x - 1)(y - 1) = 6$. **4.** (*a*) $y = ce^x$. (*b*) $y = ce^{6x}$. (*c*) $y = 4 + ce^x$. **5.** $x^2 + y^2 = c$. **6.** $2y^2 + x^2 = c$. **7.** $xy = c$. **8.** $y^2 = \pm x^2 + c$. **9.** $y = ce^{x/k}$. **11.** $y = ce^{x/k}$. **12.** $y^3 = 6kx + c$. **13.** $x^2 - y^2 = c$. **14.** $y = cx^2$. **15.** $x = cy^2$. **16.** $3x^2 + y^2 = c$. **17.** $y^3 = cx^2$. **18.** $xy^{2a} = c$. **19.** $y^3 = cx^5$, $y^3x^5 = c$.

§11, page 27

1. $xy = c$. **2.** $x^2 + 2y^2 = c$. **3.** $3x^2 + 2y^2 = c$. **4.** $2y^2 + 2y - \ln|2y+1| + 8x = c$. **5.** $8y + 19x = c$. **6.** $8y - 17\ln(8y+1) - 32x = c$. **7.** $16y + 17\ln|(y-2)/(y+2)| + 4x = c$. **9.** $2y^2 + mx^2 = c$. **10.** $3y^2 = c - 2x$. **11.** $x + c = \int[(2+4y^3)/(8-y^3)]\,dy$. **13.** φ and φ_1, and ψ and ψ_1 are mutually orthogonal pairs.

§12, page 30

1. $\rho = ae^{\theta\cot\alpha}$. **2.** (*a*) $OT = \rho\sin\psi$, $TP = \rho\cos\psi$. (*b*) $\rho = h\sec(\theta + c)$. (*c*) $\rho = c\sin\theta$. (*d*) $\rho = ce^{2\theta}$. **3.** (*a*) $\rho = c\sin(\theta + \frac{2}{3}\pi)$. (*b*) $\rho = c\sin(\theta + \alpha + \frac{1}{6}\pi)$. **4.** (*a*) $\rho = c(\cos\frac{1}{2}\theta)^2$. (*b*) $\rho = c[\sin(\frac{1}{2}\theta + \alpha)]^2$. **5.** (*a*) $\rho(1 - \sin\theta) = c$. (*b*) $\rho[\sin(\frac{1}{2}\theta + \frac{1}{4}\pi - \alpha)]^2 = c$. **6.** $\rho = ee^{-\theta^2/2}$, $0 < \theta \leqq 2\pi$. (*b*) $\sec^2\alpha\ln(\tan\alpha + \theta) - \theta\tan\alpha = \ln(\rho c)$ **7.** (*a*) $\rho\sin(\theta + \frac{1}{3}\pi) = c$. (*b*) $\rho\sin(\frac{1}{6}\pi + \alpha - \theta) = c$. **8.** (*a*) $\rho^{n^2} = c\cos n\theta$. (*b*) $\ln\rho = \int[(n - \tan\alpha\tan n\theta)/(n\tan\alpha + \tan n\theta)]\,d\theta$. **9.** (*a*) $\rho\sin^3\theta = c(1-\cos\theta)^2$. (*b*) $\ln\rho = \int\{[\sin(\theta-\alpha) + 2\sin\alpha]/[\cos(\theta-\alpha) - 2\cos\alpha]\}\,d\theta + c$. **10.** $\rho\csc\psi$, $\rho\sec\psi$, $\rho\cot\psi$, $\rho\tan\psi$. **11.** $\rho = h\theta + c$. **12.** $\rho = \sqrt{h}\sin(\theta + c)$. **13.** $\rho = ce^{\pm\theta}$. **14.** $\varphi - \tan\frac{1}{2}\varphi = \frac{1}{2}\theta + c$, $\varphi + \cot\frac{1}{2}\varphi = c - \frac{1}{2}\theta$, $\rho = \frac{1}{2}\sin\varphi$.

§13, page 33

1. $x^2 + y^2 = 13$. **2.** $2x^3 + y^2 = 25$. **3.** $x + 4 = 3\ln|y|$.

§15, page 35

1. $T = 100\,(0.9)^{t/20}$, 22.4 min, 62.2°C. **2.** $T = 20 + 80(\frac{7}{8})^{t/20}$. 23.0 min, 63.9°C. **4.** $q = (\frac{1}{2})^{t/(4.5\times10^9)}$, $q = (\frac{1}{2})^{t/(8.8\times10^9)}$, $q = (\frac{1}{2})^{10^5 t}$. **5.** 1.91 hr. **6.** $i = 30(\frac{11}{30})^{100t}$. **7.** $\frac{16}{81}I_0$, $\frac{256}{6561}I_0$. **8.** $p = 14.7(10.1/14.7)^{h/10000}$, 8.37 lb/in². **9.** At the end of 3 days, 16 gal. **10.** 11.55 days, 23.10 days.

§16, page 39

3. $v = 20/(1 + 6t)$, 0 ft/sec; $v = 20e^{-0.3s}$; 0. **4.** (*a*) $7.51(1 - e^{-0.2t})$. (*b*) 6.5 ft/sec. (*c*) 7.5 ft/sec. **5.** 6.02 mi/sec, 6.95 mi/sec. **7.** $t = (2S^{3/2} - 2R^{3/2})/(3R\sqrt{2g})$, 2659 sec = 44.32 min. **8.** $\sqrt{2gR} = 1.47$ mi/sec, $\sqrt{\frac{3}{2}gR} = 1.27$ mi/sec. **9.** $\sqrt{gR}$ or 25900 ft/sec; $\frac{1}{2}\pi\sqrt{R/g}$ or 1266 sec. **10.** (*a*) $-H < V < H$. (*b*) $v > H$. (*c*) $v < -H$. **11.** $v = 35(1 - e^{-t/5})$, $s = 35(t + 5e^{-t/5}) - 175$, 30.3 ft/sec, 198.7 ft. **12.** $v = 3t^2$, $s = t^3$, $s = (1/\sqrt{27})v^{3/2}$. **14.** $208\frac{1}{3}$ hr, $\frac{5}{4}$ slugs.

§17, page 45

1. 1.37×10^8 cal. **2.** 1.66×10^6 cal. **3.** (*a*) 1 min 44 sec. (*b*) 1 min 48 sec. **4.** 26 min 28 sec. **5.** (*a*) 190 lb. (*b*) 200 lb. (*c*) 23.10 min. **6.** 75 lb. **7.** $\frac{80}{81}$. **8.** 54.4 lb. **9.** $r = -\frac{1}{600}t + \frac{1}{20}$. **10.** about 19.5 min. **11.** 1.18 gal.

§18, page 47

1. $x + y - xy = 7$. **2.** $y = 2e^{-x} - x + c$. $y = \ln(1 + e^{-x}) + c$. **3.** $(1 - 3y)^{-2} = ce^{3(x+y)}$, $3y^2 = 2x + c$. **4.** $\rho = ce^{-\theta^2/2}$. $\rho = ce^{-\theta}(1 + \theta)^2$. **5.** $\frac{1}{32}$. **6.** $p = cv^k$. **7.** $\rho = c \sin \theta$, $\rho = c \csc \theta$. **8.** $\rho = a \sec(\theta + c)$. **9.** $\rho = c/(1 + \cos \theta)$. **10.** $\rho \cos \theta = c, \rho \sec \theta = c$. **11.** (c) 858 sec. **13.** 9.33×10^5 cal, 7.21×10^4 cal, 1.17×10^5 cal. **14.** $\rho = ce^{\theta/(2k)} \csc \theta$. **15.** $v = 173(1 - e^{-gt/173})$; 151 ft/sec. **16.** 0.124 per cent. **17.** 7 min 59 sec. **18.** 112 sec. **19.** 43 min 50 sec. **21.** $\frac{1}{2}kwl$.

§19, page 52

1. $2x - \ln(2x - 2y + 1) = c$. **2.** $\ln(2x + 3y - 3) + y = c$. **3.** $(x + y)^2 - 4y = c$. **4.** $x + y + 6 \ln(6 - 2x - y) = c$. **5.** $(x - 2y)^2 + 10(x - 2y) + 2y = c$. **6.** $x = \tan^{-1}(x + \frac{1}{2}y) + c$. **7.** $x(x^3 + 4y^3) = 5$. **8.** $y^{(m-n+1)}/(m - n + 1) = \int_a^{xy} [z^m/\varphi(z)]\, dz + c$, $m - n + 1 \neq 0$. **9.** $y^2(x^2 - 3y^2) = 1$. **10.** $(x^2 + y^2)(10 - 9x) = 5x$. **11.** $(st - 2)^3 t = cs$. **12.** $x(x^2 + 3y^2) = 14$. **13.** $\rho^3\theta^3(c - e^\theta) = 1$. **14.** $(2y + cx)(x + y)^2 + x = 0$. **15.** $x^2y^2 = c(x^2 + y^2)$. **16.** $y^{-(m+n+1)}/-(m + n + 1) = \int_a^{y/x} [z^{-m-2}/\varphi(z)]\, dz$. **17.** $2x^{(n-2m+1)/2}/(n - 2m + 1) = \int_a^{xy^2} [z^{(n-1)/2}/\varphi(z)]\, dz + c$. **18.** $x + \int_a^{2x+3y} \psi(z + m)\, dz/[3\varphi(z + c) - 2\psi(z + m)] = c_1$.

§20, page 55

2. $x^2 - 3xy - y^2 = c$. **3.** $3x^2 + 4xy = c$. **4.** $xy^2 + y^3 = c$. **5.** $2x^3 - 7xy^2 = c$. **6.** $3\theta^2 + 4\rho\theta - 4\rho^2 = c$. **7.** $y^3 = 3x^3 \ln x$. **8.** $\rho = c \sin \theta$, $\rho = c \cos(\alpha + \theta)$. **9.** $2x + y \ln x = 3y$. **10.** $2y^2 \ln(y^3/x^2) + 2xy + x^2 = cy^2$. **11.** $y + \sqrt{y^2 + x^2} = cx^2$. **12.** $\ln x + \cos(y/x) = c$. **14.** (a) $y^2 - x^2 = c$, $(y^2 - 2x^2) \tan \alpha + 2xy = c$. (b) $x^2 - 4y^2 = c$, $\ln(4y^2 - 3xy \tan \alpha + x^2) + (10/\sqrt{16 - 9 \tan^2 \alpha}) \tan \alpha \tan^{-1}[(8y - 3x \tan \alpha)/(x\sqrt{16 - 9 \tan^2 \alpha})] = c$. **16.** $y^2 \ln y + 2x^2 = cy^2$. **18.** $9x(y - 1)^2 + 3x^2(y - 1) + 5(y - 1)^3 = c$.

§21, page 58

1. $(x + y - 2)^3 = c(x - y + 2)$. **2.** $(y - 2x - 3)^4 = c(x + 1)^3$. **3.** $(y + 2x - 4)^2 = c(x + y - 1)$. **4.** $5x - 10y + \ln(10x + 5y - 2) = c$. **5.** $x + 6y + \ln(2x - 3y) = c$. **6.** $4x^2 + 6xy - 7y^2 - 14x + 8y = c$. **7.** $\ln[(y - x)^2 + (x - 1)^2] + 2 \tan^{-1}[(y - x)/(x - 1)] = c$. **8.** $y^2 - 4(x - 2)y - (x - 2)^2 = c$. **9.** $x = x' + (b\gamma - c\beta)/(a\beta - b\alpha)$, $y = y' + (\alpha c - a\gamma)/(a\beta - b\alpha)$.

§22, page 59

1. $2x\, dx + 2y\, dy$. **2.** $(x\, dy - y\, dx)/x^2$. **3.** $x^{m-1}y^{n-1}(my\, dx + nx\, dy)$. **8.** $x^2y + 2x^2 + 3x + y^2 - 5y + c$. **9.** $x^2 - xy + 2x + \ln y + c$. **10.** $x^4 + 2x^4y^3 + 2y^5 + c$. **11.** $yx^{-1} - 4x \ln x + 4x + c$. **12.** $xy - \cos x - 2 \sin y + c$. **13.** $xy^{-2} + yx^{-1} - 2y + c$. **14.** $-\frac{1}{2}x^{-2}y^{-2} + y^2 + c$.

§23, page 63

1. $2x^2 - 2xy + 5x + y^2 = c$. **2.** $x^3 + \frac{3}{2}x^2y^2 - y^3 + y^2 = c$. **3.** $a^2x - x^2y - xy^2 - \frac{1}{3}y^3 = c$. **4.** $ax^2 + bxy + ey^2 + gx + hy = c$. **5.** $x = cy$. **6.** $y = cx$. **7.** $y^2 - x^3 = cx$. **8.** $y \ln (2x - 2) + \ln y = c$. **9.** $\cos y \ln (5x + 15) + \ln y = c$. **10.** $\rho^2(\sec 2\theta + 2) = c$. **11.** $\rho \sin 2\theta - \rho^2 \cos 2\theta = c$. **12.** $\ln x + 2x \ln 5y = c$. **13.** $3ye^{2x} = x^3 + c$. **14.** $ye^{x^2} = x^3 + c$. **15.** $4x + x^4y = cy$. **16.** $x + \sqrt{x^2 + y^2} = c$. **17.** $x^2y^2(x^2 - y^2) = c$. **18.** $y^3(1 + \cos 2x) = c$. **19.** $y^2 \ln [5x/(x + 3)] - 3 \cos y = c$.

§24, page 67

1. $xy = x^3 + c$. **2.** $(xy)^{-2} = 2x^{-1} + c$. **3.** $3 \text{ lin } (y/x) = y^3 + c$. **4.** $\ln [(x + 2y)/(x - y)] = 3x + c$. **5.** $y^2 + 3 + x + cy = 0$. **6.** $\tan^{-1} (y/x) = \frac{1}{4}y^4 + c$. **7.** $y^3(x^2 - 1) = c$. **8.** $5y = x^2(y^5 + c)$. **9.** $x^2y(y^2 + c) + 2 = 0$. **10.** $y^2 + x^4 = cx^6$. **11.** $6 \tan^{-1} (y/x) + (x^2 + y^2)^{-3} = c$. **14.** $x^3y^3 + 2x^2y = c$. **15.** $x^2y^3 - 2xy^2 = c$. **16.** (*a*) $ye^{x^2} = \frac{1}{2}x^2 + c$. (*b*) $x^3y = e^x + c$. **17.** $y = x^2 + 3 + cx$. **18.** $3xy = (x^2 + y^2)^{3/2} + c$. **19.** $\tan^{-1} (x/y) = \frac{1}{4}(x^2 + y^2)^2 + c$. **20.** $\tan^{-1} (3y/2x) = 6 \sqrt{4x^2 + 9y^2} + c$. **21.** $x^3(y^4 - x^2) = c$. **22.** $2y^4 + x = cx^3$. **23.** $3 \ln (xy^2) = x^3 + c$. **24.** $\ln |(ay - x)/(2ay - x)|^2 = a(x^2 + y^2) + c$. **25.** $\sqrt{6} \tan^{-1} [\sqrt{3/2}\,(y/x)] = (2x^2 + 3y^2)^3 + c$. **26.** $\ln (x/y) + xy/(x - y) = c$. **27.** $x^{1/2}(7y^4 + x^3) = c$. **29.** (*a*) $x = ce^{x/y}$. (*b*) $y^3 + 3x^3 \ln cx = 0$.

§25, page 70

1. (1). **2.** (5). **3.** Neither. **4.** $y = \frac{1}{5}x^4 - \frac{3}{2}x + cx^{-1}$. **5.** $15x^2y = 3x^5 + 10x^3 + c$. **6.** $x = \frac{2}{5}y^2 + cy^{-3}$. **7.** $x = 3y^2 \ln y + 1$. **8.** $y = x^3 + 2x^2 - x^{-1}$. **9.** $4x^3y + 2x \cos 2x = c + \sin 2x$. **10.** $y = ax + 2a\sqrt{1 + x^2}$. **11.** $\rho = \sin \theta - 1 + ce^{-\sin \theta}$. **12.** $3y = f(x) + c[f(x)]^{-2}$. **13.** $y = x^2 \ln x + 2x^2 - x$. **14.** $2x = y^4(y^2 + 7)$. **15.** $x = \sin^2 y(c - \cot y)$. **16.** $2s = e^{3t}(t^3 + ct)$. **17.** $4x + 2y = 1 - 5e^{-2y}$. **18.** $2xy = 5 + ce^{-x^2}$. **19.** $5xy = (y^5 + c)e^y$. **20.** $y\sqrt{1 + x^2} = \ln x + c$. **21.** $2x[f(y)]^3 = [f(y)]^2 + c$. **22.** $f(x) = x^3 + 3x^{-1} \ln x + cx^{-1}$. **23.** $x = 1 + ce^{-\varphi(y)}$.

§26, page 72

1. $xy(c - \frac{3}{2}x^2) = 1$. **2.** $1 = ye^x(c - x^2)$. **3.** $x^{-1}y^{-2} = c - 2y$. **4.** $xy^{-2} + x^5 = c$. **5.** $x^{-2}e^{2y^2} = c - 4y^3$. **6.** $x^{-3}y^{-3} + x^2 = c$. **8.** $y^{1/2} = (x - 2)^2 + c(x - 2)^{-1/2}$. **9.** $xe^{y^3}(c - y^2) = 2$. **10.** $3x^5 + 5(x^2y)^{-3} = c_1$. **11.** $y = x^{1/2}(c + y^{-1/2})$. **12.** $e^{2x/3} = y^2(c - e^{2x})$. **13.** $y^{-1}e^{-x} = 13 - 12e^x$. **14.** $y^3(x + 1) = \frac{8}{3}[(x + 1)^3 - 1]$. **15.** $\ln [\ln (ye^{\cos x})] = x + c$. **16.** $6y = 3x^4 - 3x^2 + 4x^2 \ln (\frac{3}{2}\, yx^{-2} + 1)$. **17.** $(3e^{-2x} + c)y^3e^{3x} = 2$. **18.** $8 \ln [x/(x + 3e^{y^2})] + 3e^{-4y^2} = c$.

§27, page 75

1. $x = t^2 + c_1$, $y = c_1t + c_2$. **2.** $x = c_1e^{-t} + c_2$, $y = (t + c_1)e^{-t}$. **3.** $x = 1000t + c_1$, $y = 500t - 8t^2 + c_2$. **4.** $\rho = \frac{1}{2}e^t + c_1e^{-t}$, $\theta = \frac{1}{2}e^t - c_1e^{-t} + c_2$. **5.** $x = 6t^2 + 12t$, $y = -2t^3 + 12t - 10$. **6.** $x = t + 2t^{-1}$, $y = 2 \ln t - t - 2t^{-1} + 5$. **7.** $x^2 = t^2(2 \ln t + c_1)$, $2y = t^3$

$(2 \ln t + c_1 - 1) + c_2 t$. **8.** $x = 4e^t(t - 1) + 2e^{2t} + 2$, $y = 4e^t - e^{2t}(2t - 1) - 3$. **9.** $\rho = c_1 t$, $\theta = (1 + c_1)t \ln t + c_2 t$. **10.** $x = c_1 \cos(at + c_2)$, $y = c_1 \sin(at + c_2)$. **11.** 160000 ft.

§28, page 77

1. $y = (c_2 - x)e^{-x} + c_1$. **2.** $y = x^3 + \frac{3}{2}x^2 + c_1 x + c_2$. **3.** $x^2 y = -\ln x + c_1 x^2 + c_2$. **4.** $y = 2x + c_1 + c_2 e^{-x}$. **5.** $xy = 12x - 24 + 12e^{1-x}$. **6.** $s = 2t^3 + 100t$. **7.** $xy = x^3 + x^2 - 1$. **8.** $2x = 9(1 - e^{-y^2})$. **9.** $y\varphi(x) = \psi(x) + c_1 x + c_2$. **10.** $y = (x^2 + x)e^{-x^2}$. **11.** $y = \cos x - 7 \sin x$. **12.** $y = e^x(1 - \cot x)$.

§29, page 78

1. $xy \ln(cx/y) = x + y$. **2.** $\ln x + \tan^{-1}[(y + x)/2x] = c$. **3.** $3y(x^2 + 3) + 5x^3 = c$. **4.** $4x - 2y = \ln[c(x + y - 1)/(x + y + 1)]$. **5.** $x^3 + x^2 y + \sin y = 1$. **6.** $x + y = \ln(cx\,y)$. **7.** $8y^{-1} = 1 - 2x^2 + ce^{-2x^2}$. **8.** $x = (t + 3c_1)e^{3t}$, $9y = (12t + 36c_1 - 1)e^{3t} + c_2$. **9.** $2y = 3x^2 - x + c_1$, $2t = x^3 + c_1 x + c_2$. **10.** $(cx^2 + x + 1)\,xy^2 = 1$. **11.** $5x^{-3}y^{-3} = c - 3y^5$. **12.** $5xy^2 = 18(y^2 - x^2)$. **13.** $xy = 3(x^2 - 1)^3 - 21(x^2 - 1)$. **14.** $25y^3 + 8(1 + 2y)^2(1 - y)e^{3x^2/2} = 0$. **15.** $x^4 = Ry^4 e^{1/y}$. **16.** $y = x^4 - x + c_1 + c_2 x^3$. **17.** $x = c_1 + c_2 e^y + y^3 e^y$. **18.** $\ln|xy| + y \ln|(x - 1)/(x + 1)| = c$. **19.** $(x^2 - y)^4 = 3\sqrt{x^2 - y^2} + c$. **20.** $x^2 + y^2 + c = \ln(3x^2 + 4y^2 - 2)$. **21.** $y^2 + x(\ln x - 4) = 0$. **22.** $y \sec^3 x = 2 \tan x + c$. **23.** $y \ln|\text{cx}| = \pm 1$. **24.** $x^{-1} = \frac{1}{2} + ce^{y^2}$. **25.** $y \ln 3x + x + y^2 = c$. **26.** $\ln(x^3 y^5) = \frac{1}{3}y^3 + c$. **27.** $(xy)^{-2} + 2 \ln(2x^2 + y^2 - 3) = c$. **28.** $y = cx$, $y^2 - 2x^2 = c$. **29.** $xy + 1 = 4.946\, e^{-x/y}$. **30.** $5y^2 = 8 \cos x + 4 \sin x - 4e^{\pi - 2x}$. **31.** $x^3 y = 8xy - 16$. **32.** $x + y + c = 4 \ln(2x + 3y + 7)$. **33.** $6x^2 y + 6xy^2 + a^2 y + b^2 x = c$. **34.** $(x + y + 5)^2 = 16x + 20$. **35.** $(2x + y - 5)^2 = (3x + 2y - 2)^2 + c$. **36.** $4y = \ln[c(4x - 3y - 3)/(4x - 3y + 1)]$. **37.** $y(\sec x + \tan x) = x + c$. **38.** $xy(x^2 + y^2 - 1) + 1 = 0$. **39.** $y = x + x \ln[(x + y)/4]$. **40.** $6\rho\theta - 2\rho^3 + 3 \sin^2 \theta = c$. **41.** $(y - x^2 - xy)(x + y)^3 = c(y + 2x^2 + 2xy)$. **42.** $xy(x^2 - xy + y^2) = c$. **43.** $x^2\sqrt{y} + \ln(cx) = 0$. **44.** $x^{3/2}(x^2 + y^2) + y^{3/2} = c$. **45.** $x[c - (y - 2x)^2] = 2y - 10x \tan^{-1}(y/x)$. **46.** $x - y = t - c_1$, $x \ln x - y = c_2 x$. **47.** $xy = t^2 + c_1$, $x + y = \ln(2t^2 + c_1) + c_2$. **48.** $y^2 = 2x + c_1$, $y\sqrt{y^2 + 1} + \ln(y + \sqrt{y^2 + 1}) \pm 10t = c_2$. **49.** $xyt = c_1$, $xy + yt + tx = c_2$.

§30, page 81

1. $4x^3 + 12xy^2 - 3y^4 = 204$. **2.** $x + c = 2\sqrt{2x + 3y} - 4 \ln(2 + \sqrt{2x + 3y})$. **3.** $x + ky^2 = cy$. **4.** $(x^2 + y^2)^{1-n/2} = (2 - n)(kx + c)$. **5.** $x^2 + y^2 = cx$. **6.** $18\frac{1}{3}$ lb. **7.** $x^2 + y^2 = cx$. **8.** $t = 10.91 \ln[(4 - x)/(4 - 2x)]$, 2. **9.** $t = 19.14\, \{\ln[(12 - 4x)/(12 - 3x)] + 1/(3 - x) - \frac{1}{3}\}$, 3. **10.** $t = 22.4 \ln[\frac{15}{16}(4 - x)^2/(x^2 - 8x - 15)]$, 3. **12.** $y^2 = 2cx + c^2$. **13.** The part above the X axis or the part below the X axis of any straight line. **14.** $2nk\, \rho^{n-2} = (n - 2)\theta$, $0 \leqq \theta \leqq 2\pi$. **15.** A parabola. **16.** $a\rho^{-1} = \sin \theta + ce^{\theta \cot \alpha}$. **17.** $c\rho^2 = a - \rho \sin \theta$. **18.** $\sqrt{l^2 - y^2} + l \ln[(l - \sqrt{l^2 - y^2})/y] = c \pm x$. **19.** $x^2 + y^2 = c(y + x \tan \alpha)$.

§31, page 86

1. $x = (t + 1)^2$, $y = 2(t + 1)$. **4.** $x = 50{,}000(1 - e^{-0.032t})$, $y = 81{,}250(1 - e^{-0.032t}) - 1{,}000t$. **5.** $6\frac{2}{3}$ lb. **7.** Since $M/(M - kt)$ must be real, $M - kt > 0$, or $t < M/k$. The rocket charge will be dissipated completely when $t = M/k$. $c \ln 10$. **8.** $0.670cM/k$. **9.** $v = c \ln[M/(M - kt)]$

$-gt$, $c \ln 10 - 0.9Mg/k$, $x = (c/k)(M - kt) \ln [(M - kt)/M] + ct - \frac{1}{2}gt^2$, $(cM/k)0.670 - \frac{1}{2}(0.81)gM^2/k^2$. **10.** $x^2y = -1$, $x = (2 - t)^{-1}$, $y = -(2 - t)^2$. **11.** $x = \frac{1}{2}t + \frac{25}{2}t^{-1}$, $y = \frac{1}{2}t - \frac{25}{2}t^{-1}$. **12.** $x = y = Re^{4t}$. **14.** 19.38 lb.

§32, page 90

1. $i = (E/R)(1 - e^{-Rt/L})$. **2.** (*a*) 0.316 volt. (*b*) 3.16 volts. **3.** (*a*) 0.282 henry. (*b*) 1.90 henrys. **4.** 24.5 volts. **5.** 63.3 volts. **6.** $q_\infty = 0$, $q_{t=RC} = 0.368\, q_0$, $q_{t=2RC} = 0.135\, q_0$, $t_{q=0.01q_0} = 4.61RC$. **7.** $0.368I_0$, $0.135I_0$, 0. **8.** (a) $i = (E/R)[1 - e^{-(R/L)t}]$, starting from 0, rapidly approaches E/R. (b) $i = [E/(R^2 + L^2\omega^2)](R \sin \omega t - \omega L \cos \omega t + \omega L e^{-Rt/L})$, starting from 0, rapidly approaches $[E/(R^2 + L^2\omega^2)](R \sin \omega t - \omega L \cos \omega t)$. **12.** $R = 15.9$ ohms. **13.** $q = q_0 \cos (t/\sqrt{LC})$, $i = (-q_0/\sqrt{LC}) \sin (t/\sqrt{LC})$.

§33, page 92

1. 1.05 lb/in^2. **2.** $(p_0^{2/7} - \frac{2}{7}k^{-5/7}h)^{3.5}$, $\frac{7}{2}p_0^{2/7}k^{5/7}$. **3.** 98000 ft.

§34, page 98

1. $\frac{1}{2}k(b^2 - a^2)$ ft-lb. **2.** (*a*) 20 ft-lb. (*b*) 4 ft-lb. **3.** (*a*) -120 ft-lb. (*b*) 5.5 ft-lb. **4.** $x^2 + xy^3$, -434 ft-lb. **5.** $x^2 + 2y^2 = c$, $y = cx^2$. **6.** $x^2 + y^2 = c$, $[2x,2y]$, $y = cx$. **7.** $x^2 + y^2 = cxy$, $[(x^2 - y^2)/(x^2y), (y^2 - x^2)/xy^2]$, $x^2 - y^2 = c$. **8.** $x^2y + y^3 = c$; $[2xy, x^2 + 3y^2]$, $y^2 = cx^3 - x^2$. **10.** (*a*) $\frac{1}{2}(x^2 + y^2)$. (*b*) $x^3 - x^2y + y^2$. (*c*) $x^2 + x \sin y$. **12.** (*b*). **13.** $[kx/\sqrt{(x^2 + y^2)}, ky/\sqrt{(x^2 + y^2)}]$; $k \ln \sqrt{x^2 + y^2} = c_1$ or $x^2 + y^2 = c$; $y = cx$. **14.** 0. **15.** $x^3 + 6xy^2 = c$, $2y^2 - 3x^2 = cy^{1/2}$. **17.** $[\partial/\partial x(k_1/r_1 + k_2/r_2 + k_3/r_3), \partial/\partial y\, (k_1/r_1 + k_2/r_2 + k_3/r_3)]$.

§35, page 102

1. $ax^3 + bx^2y - exy^2 - fy^3 = am^3$. **2.** $2\rho = \theta + \theta^3$. **3.** $x^2 + y^2 = cy$. **4.** $\rho = c \sin (\theta + \alpha + \delta)$ when $\tan \delta = \frac{1}{6}$. **5.** 40 lb. **6.** $s = 4 - 4 \cos 4t$. **7.** $q = 0.02(1 - e^{-500t})$, $q \to \frac{1}{50}$ coul as $i \to 0$. **8.** 52700 ft. **9.** $[3x^2 - 6xy, -3x^2]$, $(x + 2y)^2(x - y) = c$, $x^3 - 3x^2y = c$, 20. **10.** $[-5(x^2 + y^2)^{-7/2}x, -5(x^2 + y^2)^{-7/2}y]$ $y = cx$, $x^2 + y^2 = c$, $k(5^{-5/2} - 1)$. **11.** $[\frac{1}{4}\pi \cos \frac{1}{4}\pi x \cos \frac{1}{4}\pi y, -\frac{1}{4}\pi \sin \frac{1}{4}\pi x \sin \frac{1}{4}\pi y]$, $\sin (\frac{1}{4}\pi y) = c \cos (\frac{1}{4}\pi x)$, $\sin (\frac{1}{4}\pi x) \cos (\frac{1}{4}\pi y) = c$, $\sqrt{2}$. **13.** $\rho = 9/(5 + 4 \cos \theta)$. **14.** $s \pm \sin t$. **15.** $2 \ln (2Li + Rq) + Rq/(2Li + Rq) = c$. **16.** $38\frac{3}{4}$ lb, $28\frac{1}{8}$ lb. **17.** $r(1 + \mu^2) = \rho R[(1 - \mu^2) \sin \theta - 2\mu \cos \theta] + ce^{\mu\theta}$. **18.** (*c*) $2\sqrt{2}\,\mu = (1 - 2\mu - \mu^2)e^{3\pi\mu/4}$. **19.** $\ln (x^3y + x^3\sqrt{x^2 + y^2}) - 10y = c$, $\sqrt{x^2 + y^2} = 5x^2 + 4y + c$. **20.** $\sqrt{x^2 + y^2} = c + x$, $\sqrt{x^2 + y^2} = c - x$. **21.** $x^3y - y^3x = c$, $x^4 - 6x^2y^2 + y^4 = c$. **22.** $q = 10^{-3}[13 \sin 380t - 9.7(\cos 380t - e^{-500t})]$, $i = dq/dt$.

§37, page 107

1. $(y - x + c)(2y - x^2 + c)(y - x^3 + c) = 0$. **2.** $(9y + c)^2 - x^6 = 0$. **3.** $(3y + c)^2 = 25x^3$. **4.** $(\ln y + c)^2 = 4x^2$. **5.** $(y^2x - c)(yx^2 - c) = 0$. **6.** $y^2 = (x + c)^3$, $y = 0$. **7.** $y^2 = 2cx + c^2$. **8.** $(y - c)(2y - 3x^2 + c)(2y + x^2 + c) = 0$. **9.** Each family of the new system will be the orthogonal trajectories of a family of the old system. **10.** $4xy = (x + c)^2$. **11.**

$(y - c)(x + y - c)(x^2 + xy + y^2 - c) = 0$. **12.** $(y + c)^2 = \{\sqrt{k^2 - x^2} + k \ln |(\sqrt{k^2 - x^2} - k)/x|\}^2$, $|x| \leqq k$, $k > 0$. **13.** $[y + x + c + \frac{2}{3} \ln |3x - 1|][y + x + c - 2 \ln |x - 1|] = 0$.

§38, page 111

1. $xy + 1 = 0$. **2.** $(x - y)^2 = 2$. **3.** $y^2 = 4x^2$. **4.** $y^2 = x^2$. **5.** $x^2 + y^2 + x = 0$. **6.** $x^2 = 4y^3$. **7.** $x^2y^2 = 1$. **8.** $x^2y = 1$. **12.** $y = 3$, $x = c$.

§39, page 115

2. None. **3.** $y = \pm 2$; $y = 0$ is a tac-locus. **4.** $x = \pm 2$; $x = 0$ is a tac-locus. **5.** None. **6.** None. **7.** $y^2 = x^2$. **8.** $y = -x^2$. **9.** $y = 0$. **10.** $64y = -x^8$. **11.** $x = 0$.

§40, page 118

1. $2y = c^2 + 2cx - x^2$, $y = -x^2$. **2.** $2cy = c^2x^2 + 3$, $y^2 = 3x^2$. **3.** $y = cx - x^2 + c^2$, $4y + 5x^2 = 0$. **4.** $xy = c^2x + c$, $4x^2y + 1 = 0$. **5.** $3x = 2p + cp^{-1/2}$, $3y = p^2 - cp^{1/2}$. **6.** $x = 2p + cp^{-2}$, $y = p^2 + 2cp^{-1}$. **7.** $y = cx + c^3$, $27y^2 = -4x^3$. **8.** $y = cx + c^2 - \frac{1}{2}x^2$, $y = -\frac{3}{4}x^2$. **9.** $y = c \ln x + c^2$, $4y = -(\ln x)^2$. **10.** $y = c \ln x + c^2 - \frac{1}{2}x^2$, $4y = -(\ln x)^2 - 2x^2$. **11.** $x = -2p + cpe^{-p^2/2}$, $y = (p^2 + 1)(-2 + ce^{-p^2/2}) + p^2$.

§41, page 119

1. $cx = e^y + 4c^3$, $x = 3e^{2y/3}$. **2.** $x = y + \ln(1 + ce^{-y})$. **3.** $4cx = 2y^2 + c^3$, $64x^3 = 27y^4$. **4.** $(x + c)^2 = 2cy - c^2$, $x^2 + 2xy - y^2 = 0$. **5.** $2cx = 2 \sin y + c^3$, $2x = 3(\sin y)^{2/3}$. **6.** $x = c - 2p - 2 \ln(p - 1)$, $y = c - p^2 - 2p - 2 \ln(p - 1)$. **7.** $x = p^{-2} + 2cp$, $y = 2p^{-1} + cp^2$. **8.** $x = (3c)^{-1}y^3 + \phi(c)$. **9.** $x = (1/c)e^y + \phi(c)$. **10.** $3x = 4e^{3y/4}$.

§42, page 122

1. (*a*) $(y - cx)^2 = \sin(y - cx) + c^2$. (*b*) $e^{y-cx} = (y - cx)^2 + c^3$. **2.** $27y^2 = -4x^3$. **3.** $x^2 + 4y = 4$. **4.** $xy = -1$. **5.** $(y - \frac{1}{2})^3 = \frac{27}{16}x^2$. **6.** $8y^3 = 27x^2$. **7.** $e^y = c(x - 3c)$. **8.** $y^2 = cx + \frac{1}{2}c^3$. **9.** $y = px - \frac{1}{4}p^2$. **10.** $y = px + \frac{1}{4}p^{-1}$. **11.** $y = px \pm 5\sqrt{p^2 + 1}$. **12.** $y = c \ln x + \varphi(c)$. **13.** $y = cx - \frac{1}{2}x^2 + f(c)$. **14.** $y = cx - 2x^2 + f(c)$. **15.** $y = m(cx)^{1/m} + \varphi(c)$. **16.** $2x^2y = -1$.

§43, page 124

1. $(y - ce^{3x})(y - 2x^2 + c) = 0$. **2.** $(y - ce^x)(y - x^2 + c)(y + x + c) = 0$. **3.** $y = cx + c^2$, $x^2 + 4y = 0$. **4.** $y = cx - c^3$, $27y^2 = 4x^3$. **5.** $y = cx - 3x^2 + \frac{1}{2}c^2$, $2y + 7x^2 = 0$. **6.** $y^2 = 2cx + c^3$, $27y^4 = -32x^3$. **7.** $y = c \ln x + \frac{1}{4}c^2$, $y = -(\ln x)^2$. **8.** $cx = e^y + \frac{1}{3}c^3$, $4x^3 = 9e^{2y}$. **9.** $y^2 + (c - x)^2 = c$, $x + \frac{1}{4} = y^2$. **10.** $y(y - 3)^2 = (x - c)^2$, $y = 0$. $y = 1$ is a tac-locus. **11.** $512y^3 + 27x^4 = 0$. **12.** $8(y - 1)^3 = 27x^2$. **13.** $y = 3c^{1/3}x^{1/3} + f(c)$. **14.** $y = c \ln(\sin x) + \varphi(c)$. **15.** (*a*) $4p(y - px) = 1$, $27y^2 = 2(2x - 1)^3$. (*b*) $2p(y - px) = 2p^2 + 1$, $27y^2 = 8x^3$. **17.** $y = px \pm \sqrt{B^2 + A^2p^2}$.

§44, page 130

1. $2x$. **2.** $8x^2 + 16x$. **3.** $-2x^3 + 3x^2$. **4.** $12x^2$. **5.** $27e^{3x}$. **6.** $2e^{5x}$. **7.** $-2a^2 \sin ax$. **8.** 0. **9.** 0. **10.** $y = c_1 + c_2e^{2x}$. **11.** $y = c_1e^{2x} + c_2e^{-2x}$. **12.** $y = c_1 + c_2e^{2x} - \frac{3}{2}x^2 - \frac{3}{2}x$. **13.** $y = c_1 + c_2x + c_3e^{-3x} + \frac{1}{2}e^{-x}$. **14.** $y = c_1 + cx + x^4$. **15.** $y = c_1e^x + c_2e^{2x} + c_3e^{3x}$. **16.** $y = 2 + 3e^{-3x}$. **17.** $y = c_1e^{3x} + c_2e^{-4x} - 1$. **18.** $y = c_1 + c_2e^{3x} - \frac{1}{2}e^x$. **19.** $y = c_1e^{2x} + c_2e^{-2x} + \frac{1}{5}e^{3x}$. **20.** $y = (c_1 + \frac{1}{3}x)e^x + c_2e^{-2x}$. **21.** $y = (c_1 + c_2x)e^{2x} + x^2e^{2x}$. **22.** $y = 3e^{-2x} - 2e^{2x}$. **23.** $y = e^x(x^2 - 3x + 3)$. **24.** $2e^{-x}$. **25.** 0. **26.** $2e^{-x}$. **27.** $2e^{-2x}$. **28.** 0. **29.** 0.

§45, page 132

3. Independent. **4.** Dependent. **5.** Independent. **6.** Dependent. **7.** Dependent. **8.** Yes. The relation among the 20 functions of A is a relation among the elements of B since zeros can be used as coefficients of the 20 additional functions of B. **9.** Yes. $5(e^x) - 1(5e^x) + 0f(x) \equiv 0$. **10.** Yes, calling the last element $f_{n+1}(x)$, we have $f_1(x) + 2f_2(x) + \cdots + nf_n(x) + (-1)f_{n+1} \equiv 0$. **11.** (*a*) $f(x) \not\equiv 0$. (*b*) $f(x) = 0$. **12.** $f(x) = c$. **13.** (*a*) $f(x)$ is not a constant. (*b*) $f(x) = ae^{bx}$, a and b nonzero constants. **14.** (*a*) Yes. (*b*) Yes. (*c*) No.

§47, page 135

1. $y = c_1e^x + c_2e^{2x}$. **2.** $y = c_1e^x + c_2e^{-5x}$. **3.** $y = c_1e^{-3x} + c_2e^{-x}$. **4.** $y = e^{5x/2}(c_1e^{\sqrt{13}x/2} + c_2e^{-\sqrt{13}x/2})$. **5.** $y = c_1 + c_2e^{\sqrt{5}x/2} + c_3e^{-\sqrt{5}x/2}$. **6.** $y = c_1e^x + c_2e^{-x} + c_3$. **7.** $y = c_1e^{kx} + c_2e^{-kx}$. **8.** $y = c_1e^x + c_2e^{-x} + c_3e^{3x}$. **9.** $y = c_1e^x + c_2e^{2x} + c_3e^{-3x}$. **10.** $y = c_1 + c_2e^{3x} + e^{-x}(c_3e^{\sqrt{2}x} + c_4e^{-\sqrt{2}x})$.

§49, page 137

1. $y = e^{3x}(c_1 + c_2x)$. **2.** $y = e^{-2x}(c_1 + c_2x)$. **3.** $y = c_1 + c_2x + c_3e^x$. **4.** $y = c_1 + c_2x + c_3x^2 + c_4e^{2x} + c_5e^{-2x}$. **5.** $y = c_0 + e^x(c_1 + c_2x)$. **6.** $y = c_1 + c_2x + c_3x^2$. **7.** $y = (5 - 14x)e^x$. **8.** $y = e^{-x}$. **9.** $y = (c_1 + c_2x)e^x + c_3e^{-x}$. **10.** $y = e^x(c_1 + c_2x + c_3x^2)$. **11.** $y = c_1 + c_2x + e^{2x}(c_3 + c_4x) + e^{-2x}(c_5 + c_6x)$. **12.** $y = c_1 + c_2x + e^{2x}(c_3 + c_4x) + c_5e^{-4x}$. **13.** $y = x + 2e^x$. **14.** $y = 2x + 4e^{-x}$. **15.** $y = 1 + 2xe^{2x}$. **16.** $y = e^{-x} + (2x - 1)e^x$.

§50, page 140

1. $y = e^x(c_1 \sin x + c_2 \cos x)$. **2.** $y = c_0 + e^{2x}(c_1 \sin x + c_2 \cos x)$. **3.** $y = c_1e^{2x} + e^{-x}(c_2 \sin 3x + c_3 \cos 3x)$. **4.** $y = c_1 \sin 2x + c_2 \cos 2x$. **5.** $y = (c_1 + c_2x) \sin 2x + (c_3 + c_4x) \cos 2x$. **6.** $y = c_1 + c_2x + c_3 \sin x + c_4 \cos x$. **7.** $y = \sin x - \cos x$. **8.** $y = e^{-x} \sin x$. **9.** $y = c_1 + c_2x + (c_3 + c_4x) \sin \sqrt{3}\, x + (c_5 + c_6x) \cos \sqrt{3}\, x$. **10.** $y = c_0e^{-ax} + e^{ax/2}[c_1 \sin (\frac{1}{2}\sqrt{3}\, ax) + c_2 \cos (\frac{1}{2}\sqrt{3}\, ax]$. **11.** $y = 2 \sin 3x$. **12.** $y = 1 + e^x \sin x$. **13.** $y = 4e^x \sin x - 2e^{-2x}$. **14.** $y = 4 \cos 2(x - 1)$.

§51, page 142

1. $y = c_1e^{2x} + c_2e^{-2x} - 3$. **2.** $y = c_1e^{-3x} + c_2e^x + 2e^{4x}$. **3.** $y = c_1e^x + c_2e^{-2x} + 3x$. **4.** $y = c_1e^{2x} + c_2e^{-x} - 3e^x$. **5.** $y = c_1 \sin x + c_2 \cos x + 3 + 3e^x$. **6.** $y = e^{-2x}(c_1 + c_2x) + \frac{1}{4}(x -$

$1 + 2e^{2x})$. **7.** $y = c_1 + c_2e^{-x} - 1.2 \sin 2x - 0.6 \cos 2x$. **8.** $y = c_1 + c_2x + c_3e^x + \cos x - \sin x$. **9.** $y = c_1 \sin x + c_2 \cos x + e^x(2 \sin x - 4 \cos x)$. **10.** $y = e^{3x} - 1 - \cos 3x$. **11.** $y = c_1e^x + c_2e^{-x} - 2x^2 - 4$. **12.** $y = 2x + 3e^{-3x}$. **13.** $y = 1 + 2e^{-x} + \frac{1}{2}e^x$. **14.** $y = xe^x - \frac{3}{4}e^x - 2$. **15.** $y = e^{2x} + e^x + 2e^{-x}$.

§52, page 144

1. $y_p = Axe^{2x}$. **2.** $y = x(A + Bx)e^{-2x}$. **3.** $y_p = x[Ax \sin 2x + Bx \cos 2x + C \sin 2x + D \cos 2x]$. **4.** $y_p = Ax^2$. **5.** $y_p = x^2(Ax^2 + Bx + c)$. **6.** $y_p = x^3Ae^{2x} + Bx^2 + Cx$. **7.** $y = c_1 + c_2e^{-x} + 2x^2 - 4x$. **8.** $y = c_1e^x + c_2e^{-x} + \frac{5}{2}xe^x$. **9.** $y = c_1 + c_2e^{-x} - xe^{-x}$. **10.** $y = c_1 \sin x + c_2 \cos x - \frac{1}{2}x \cos x$. **11.** $y = (c_1 + c_2x + c_3x^2 + 2x^3)e^{2x}$. **12.** $y = e^x (c_1 \sin x + c_2 \cos x - \frac{1}{2}x \cos x)$. **13.** $y_p = 3 + \frac{1}{2}x^2e^x$. **14.** $y_p = \frac{1}{4}x \sin 2x$. **15.** $y_p = 2x^2 - 2x - \frac{1}{2}xe^{-2x}$. **16.** $y_p = x - x \cos 2x$. **17.** $y_p = 2x + \frac{1}{34}(15 \sin 2x + 60 \cos 2x)$.

§53, page 146

1. $y = (c_1 + c_2x + c_3x^2 + \frac{1}{6}x^3)e^{2x}$. **2.** $y = (c_1 + c_2x + \frac{1}{6}x^3)e^{-x}$. **3.** $y = (c_1 + c_2x + c_3x^2 + 2x^3)e^{-x}$. **4.** $y = (c_1 + c_2x - \sin x)e^{2x}$. **5.** $y = \frac{1}{5}e^{-2x} (2 \cos x - \sin x) + c_1e^{-2x} + c_2$. **6.** $y = (c_1e^x + c_2 + c_3x - 6x^2)e^{-x}$. **10.** $y = (c_1 \sin 2x + c_2 \cos 2x + \frac{1}{3} \sin x)e^{-2x}$. **11.** $y = (c_1 + c_2x + c_3x^2 - 2x^3)e^{-x}$. **12.** $y = [24(1 - \cos x) + x^2(x^2 - 12)]e^x$.

§54, page 151

1. $\frac{1}{4}x^4$. **2.** $2e^{2x}$. . **3.** $\frac{1}{6}x^3e^x$. **4.** $-\frac{3}{5} \cos 3x$. **5.** $x^4 - 4x^3 + 12x^2 - 24x + 24$. **6.** $-1/1{,}728 \cos 4x$. **7.** $\frac{1}{3}x^3 - \frac{1}{3}x^2 + \frac{2}{9}x - \frac{2}{27}$. **8.** $\frac{1}{2} \sin x\, e^{2x}$. **9.** $y_p = 8x^3e^{3x}$. **10.** $y_p = e^{-x} \ln (2x + 3)$. **11.** $y_p = (8x^2 - 4x + 1)e^{3x}$. **12.** $y_p = \frac{1}{9} \sin 2x$. **13.** $y_p = -\frac{1}{3} \sin 2x - \frac{1}{8} \cos 3x$. **14.** $y = x^3 - 2x^2 - 12x + 2$. **15.** $y = \sin 3x - 6x \cos 3x$. **16.** $y_p = \frac{9}{32}e^{2x}(8x^3 - 6x^2 + 3x)$. **17.** $\frac{1}{4}e^{2x}(2x^3 - 3x^2 + 3x - \frac{3}{2})$.· **18.** $y_p = (-\sin x)e^x$. **19.** $y_p = \frac{9}{20}e^{2x} (\sin 2x - 2 \cos 2x)$. **20.** $y_p = \frac{1}{2} \sin x(e^{2x} - 2e^x)$. **21.** $-2e^x(\sin x + \cos x)$. **22.** $y_p = e^x(-7 \cos 3x + 3 \sin 3x)$. **23.** $y_p = \frac{1}{4}e^{3x}(2x^2 - 6x + 7)$. **24.** $y_p = x^2e^{2x}$. **25.** $y_p = 2x^3 - 12x^2 + 54x - 120$. **26.** $y_p = \cos x + 2x \sin x$. **27.** $y_p = (4x^2 - 2) \sin x + 4x \cos x$. **28.** $y_p = 12x \sin 2x + (3 - 8x^2) \cos 2x$. **30.** $y_p = \frac{1}{41}e^{2x}$.

§55, page 155

1. $y = y_c - \cos x \ln (\sec x + \tan x)$. **2.** $y = y_c + \frac{1}{2} \tan x \sin x$. **3.** $y = y_c + \frac{1}{2}a^{-2} \tan ax \sin ax$. **4.** $y = y_c + e^{-x}(x \sin x + \cos x \ln \cos x)$. **5.** $y = y_c + \sin 2x \ln (\csc 2x - \cot 2x)$. **6.** $y = y_c - \cos x \ln (\sec x + \tan x)$. **7.** $y = c_1 + c_2x^2 - 2 \ln x$. **8.** $y = y_c - e^{-2x} \ln x$. **9.** $y = y_c + x^{n+2}e^{2x}/[(n + 1)(n + 2)]$, $n \neq -1$, $n \neq -2$. **10.** $y = y_c - 18e^{2x} \ln x$. **11.** $y = y_c + 3e^{-3x} \ln x$. **12.** $y = y_c - 1 - x \ln x$. **13.** $y = y_c + 2e^{3/x}$.

§56, page 158

1. $x = 2 - (c_1 + c_2 + c_2t)e^t$, $y = (c_1 + c_2t)e^t - 2t - 4$. **2.** $x = e^{2t}(c_1 - 4t)$, $2y = (4t - c_1 + 2)e^{2t} + c_2$. **3.** $x = c_1e^t + c_2e^{-t} + c_3 \sin t + c_4 \cos t - 1$, $y = c_1e^t + c_2e^{-t} - c_3 \sin t - c_4 \cos t$. **4.** $x = (2c_1 + 2c_2t)e^t + (2c_3 + 2c_4t)e^{-t}$, $y = (c_2 - c_1 - c_2t)e^t - (c_3 + c_4 + c_4t)e^{-t}$.

5. $x = c_1e^{2t} + c_2e^{-t} + \frac{1}{2}$, $y = c_1e^{2t} - 2c_2e^{-t} + t - \frac{1}{2}$, $z = -2c_1e^t + c_2e^{-t}$. **6.** $x = (6c_2 - 2c_1 - 2c_2t)e^t - \frac{1}{3}(c_3e^{-3t/2} + 2)$, $y = (c_1 + c_2t)e^t + c_3e^{-3t/2} - t$. **7.** $y = c_1e^{\sqrt{3}t} + c_2e^{-\sqrt{3}t} + c_3 \sin (t/\sqrt{3}) + c_4 \cos (t/\sqrt{3}) + 2e^t$, $x = -16e^t + (5D - 3D^3)y$. **8.** $2x = (23 - 13t)e^t + (23 + 13t)e^{-t} - 46$, $4y = (-36 + 13t)e^t - (36 + 13t)e^{-t} + 72$. **9.** $y = c_1e^t + c_2e^{-t}$, $z = c_1e^t - c_2e^{-t}$, $x = 3c_1e^t - 3c_2e^{-t} + c_3$. **10.** $x = c_1e^t + c_2e^{-t} + \frac{1}{4}te^t$, $2y = -c_1t + (c_3 - c_2)e^{-t} - \frac{1}{4}te^t$, $z = c_1e^t - (c_2 + c_3)e^{-t} + \frac{1}{4}te^t + \frac{1}{4}e^t$.

§57, page 159

1. $y = c_1e^{3x} + c_2e^{-x} - \frac{1}{16}e^{-x}(24x^2 + 12x)$. **2.** $y = (c_1 + x^2 - \frac{1}{3}x)e^{3x} + (c_2 - x^2 - \frac{1}{3}x)e^{-3x}$. **3.** $y = c_1 \sin kx + c_2 \cos kx$. **4.** $y = c_1e^{ax} + c_2e^{-ax} + c_3 \sin ax + c_4 \cos ax$. **5.** $y = c_1e^{2x} + c_2e^{-2x} - 4$. **6.** $y = (c_1 + c_2x)e^{-x} + \frac{1}{2}x^2e^{-x}$. **7.** $y = c_1 + c_2x + c_3e^x$. **8.** $y = e^x(c_1 + c_2x + c_3x^2) + e^{-2x}(c_4 + c_5x)$. **9.** $y = e^{-x}(c_1 \sin \sqrt{2}\, x + c_2 \cos \sqrt{2}\, x)$. **10.** $y = c_1 + c_2x + c_3e^{3x} - (3x^3 + 3x^2)$. **11.** $y = c_1 + c_2e^{-2x} + c_3e^x - 2x^2 - 2x$. **12.** $y = ce^{-4x} + c_2e^{2x} - 2x + 1$. **13.** $y = 3e^{2x}$. **14.** $y = -x^2 + x + 1$. **15.** $y = x^2 + x + xe^x$. **16.** $y = e^x[(c_1 + c_2x) \sin 2x + (c_3 + c_4x) \cos 2x] + \frac{1}{9}e^x \cos x$. **17.** $y = y_c + 2x \sin 3x$. **18.** $y = y_c + 2 \sin 2x \ln (\sec 2x + \tan 2x) - 4$. **19.** $y = y_c + 2/x$. **20.** $x = c_1e^{4t} - c_2e^{8t}$, $y = 3c_1e^{4t} + c_2e^{8t}$. **21.** $y = e^{-x} - \frac{1}{2} \sin x$. **22.** $y = -5e^{2x} + 5e^{3x}$. **23.** $y = 1 + \cos x + \sin 2x$. **24.** $y = e^x(1 - 2 \sin x - \cos x)$. **25.** $y = e^{-x} + e^x(2x - 5)$. **26.** $y = e^{-x} - 6x - \sin 5x$. **27.** $y = e^x \cos 2x - e^{-x} \cos x$. **28.** $y = y_c - 2x \sin 2x$. **29.** $y = y_c + \frac{1}{42}x^7e^x$. **30.** $y = y_c + (8x^2 - 4x)e^{2x}$. **31.** $y = y_c - 3e^x \sin 3x$. **32.** $y = y_c + e^x[\sin x \ln \cos x - x \cos x + \ln (\sec x + \tan x)]$. **33.** $y = y_c + [\sin x \ln (\csc x - \cot x) - \cos x \ln (\sec x + \tan x)]e^x$. **34.** $y = y_c + 3xe^x - 4xe^{-3x/2}$. **35.** $y = 1 + (x - 1)e^x + \frac{1}{6}e^x(x^3 - 3x^2)$. **36.** $y = y_c - 2xe^x \cos x$. **37.** $y = y_c + 6xe^{-2x} \sin x$. **38.** $y = y_c + e^{2x} \ln x$. **39.** $y = y_c + \sin x \ln x$. **40.** $y = y_c + \sin x \tan x$. **41.** $y = y_c + 2e^{2x^2}$. **42.** $x = c_1e^{3t} + c_2e^{-t} + 0.4 \sin t - 0.2 \cos t$, $y = -2c_1e^{3t} + 2c_2e^{-t} + \frac{2}{5} \cos t + \frac{1}{5} \sin t$. **43.** $x = e^t(1 - c_2 - 2t) + c_3$, $y = e^t(4t + 2c_2 - 1) - c_3$, $z = e^t(1 - c_2 - 2t) + c_1$.

§59, page 162

7. $p/(p^2 - a^2)$. **8.** $a/(p^2 - a^2)$. **9.** $1/(p + a)^2$. **10.** $k/[(p - a)^2 + k^2]$.

§60, page 165

1. $p^{-1} + 6p^{-4}$. **2.** $\frac{1}{2}(p + 2)(p^2 + 4)^{-1}$. **3.** $(15 - 4p)(p^2 - 9)^{-1}$. **4.** $(13 - 3p)[(p - 3)^2 + 4]^{-1}$. **5.** $\frac{1}{144}t^6 + 7e^{at}$. **6.** $3e^{mt} + 4e^{-mt}$. **7.** $\frac{1}{10}t(5 \sin 5t - \cos 5t) + \frac{1}{50} \sin 5t$. **8.** $2 \cosh 5t + \sinh 5t$. **9.** $\cos 2t - \cos 3t - \frac{2}{3} \sin 3t + \frac{3}{2} \sin 2t$. **10.** $(10 - 3p)/[(p - 2)^2 + 16]$. **11.** $-2p^3(p^2 + 1)^{-2}$. **12.** $n!(p^{n-1} + 1)/p^{n+1}$. **13.** $n!/(p - 1)^{n+1}$. **14.** $\cos 2t - \frac{3}{4}t \sin 2t$. **15.** $\frac{1}{6}(4 \sin 2t - 4 \cos 2t - 2 \sin t + 4 \cos t)$.

§61, page 169

5. $e^{-3t} \cos t$. **6.** $(\cos 3t + \frac{2}{3} \sin 3t)$. **7.** $e^{mt} \cosh at$. **8.** $a^{-1}e^{mt} \sinh at$. **9.** $e^{-mt} \cos at$. **10.** $[ab(a^2 - b^2)]^{-1}\ e^{-mt}(a \sin bt - b \sin at)$. **11.** $5/[(p + 3)^2 + 25]$. **12.** $(p - 3)/[(p - 3)^2 + 25]$. **13.** $10(p + 3)/[(p + 3)^2 + 25]^2$. **14.** $3!/(p - 2)^4$. **15.** $1/[(p + 1)^2 - 1]$. **16.** $128/[(p + 2)^2 + 16]^2$. **17.** $be^{at} \cos kt$. **18.** t^5e^{5t}. **19.** $e^{2t}(2 \cos 6t + 3 \sin 6t)$. **20.** $e^{-3t}(\frac{1}{2} \sin 4t -$

$\cos 4t + e^t)$. **21.** $\frac{1}{10}e^{2t} - \frac{1}{30}e^{3t}(3\cos 3t - \sin 3t)$. **22.** $\frac{1}{68}e^{-4t}(e^{8t} - \cos 2t + \frac{9}{2}\sin 2t)$. **23.** $e^{-4t}[t(\sin t + \cos t) - \sin t]$.

§62, page 172

2. $y = 2e^t - 2$. **3.** $y = 2e^{-3t} - 2e^t$. **4.** $y = 3 + 3\cos 2t - \frac{3}{2}\sin 2t$. **5.** $y = \frac{5}{9} - \frac{5}{9}\cos 3t$. **6.** $y = \frac{5}{2}\sin t - \frac{5}{2}t\cos t$. **7.** $y = 2t\sin 3t + 2\cos 3t + \frac{5}{3}\sin 3t$. **8.** $y = 2 + e^{3t}(4\sin 2t - 2\cos 2t)$. **9.** $y = \frac{1}{4}(3e^t + 2e^{-t} - e^{-3t})$. **10.** $y = 3 + e^{3t}(\sin 2t - 2\cos 2t)$. **11.** $y = 12e^t - (2t^3 + 6t^2 + 12t + 12)$. **12.** $y = 2e^{-t}\cos t + te^{-t}\sin t$. **13.** $y = e^{-t}(t^3 + 7t + 2)$. **14.** $y = 10e^t - 6t^2 - 9t - 9$. **15.** $y = \frac{1}{5}e^t(7\cos t - 4\sin t + 3e^{-2t})$.

§63, page 177

1. $y = 2\cos t + \sin t$. **2.** $y = e^{3t}(\frac{36}{5}t - \frac{21}{25}) - \frac{54}{25}e^{-2t}$. **3.** $y = -\frac{2}{5}e^{-t} + \frac{2}{5}\cos 2t + \frac{4}{5}\sin 2t$. **7.** $-4(p-3)/[(p-3)^2 + 4]^2$. **8.** $[16 - 12(p-3)^2]/[(p-3)^2 + 4]^3$. **9.** $[-1 - (p-3)^2]/[(p-3)^2 - 1]^2$. **11.** $\int_0^t \tau e^{3\tau}\sin 2(t-\tau)\,d\tau$.* **12.** $\int_0^t \frac{1}{24}\cos 2(t-\tau)\tau^4 e^{2\tau}\,d\tau$. **13.** $\int_0^t a^{-1}\sin a(t-\tau)\cos b\tau\,d\tau$. **14.** $(2\,ab)^{-1}\int_0^t \sin a(t-\tau)(\tau\sin b\tau)\,d\tau$. **15.** $y = 1 - e^t + \frac{1}{3}e^{2t} - \frac{1}{3}e^{-t}$.

§64, page 180

4. $\frac{5}{12}e^{2t} - \frac{1}{3}e^t + \frac{1}{3}e^{-t} - \frac{5}{12}e^{-2t}$. **5.** $-1 + \frac{5}{2}e^{-t} - \frac{3}{2}e^{-3t}$. **6.** $e^{-2t} - \frac{1}{2}e^{-3t/2}$. **7.** $T_a^{-1}\{F(p)\} = e^{at}[\frac{1}{2}\varphi(a)t^2 + \varphi^{(1)}(a)t + \frac{1}{2}\varphi^{(2)}(a)]$. **8.** $(\frac{4}{9}t^3 - \frac{1}{9}t^2 + \frac{2}{27}t - \frac{2}{81})e^t + \frac{2}{81}e^{-t}$. **9.** $(\frac{1}{8}t^4 + \frac{1}{2}t^3 - t^2 + t)e^{-t}$. **10.** $-\frac{1}{4} + e^{2t}(\frac{2}{3}t^4 - \frac{1}{2}t^2 - \frac{1}{2}t + \frac{1}{4})$. **11.** $-2t^2 + 4t - 2 + 2e^{-2t}$. **12.** $\frac{1}{32}[(-\frac{11}{3}t^3 + \frac{7}{2}t^2 + \frac{2}{3}t + \frac{2}{9})e^{-2t} - \frac{2}{9}e^t]$. **13.** $y = 2e^{3t} + 3e^{-2t}$. **14.** $y = 2e^{-3t} + e^{-t}$. **15.** $y = 2e^t - e^{-4t}$. **16.** $y = e^{2t} + te^t$. **17.** $y = 2 - 4t + e^{2t}$. **18.** $y = 1 - e^{-t}(\frac{1}{2}t^2 + t - 1)$. **19.** $y = (t^3 + 2t)e^{-2t}$. **20.** $y = e^t(6t^2 + 4t - 6) + 2t + 6$.

§65, page 182

1. $e^{2t}(\cos 2t - 2\sin 2t)$. **2.** $b^{-1}e^{at}[bm\cos bt + (ma + n)\sin bt]$. **3.** $1 - 2e^{-t}\sin t$. **4.** $1 + e^{-t}(2\cos t - 3)$. **5.** $\frac{1}{16}e^t[1 - (4t + 1)\cos 2t + (3t + 2)\sin 2t]$. **6.** $2e^{at}[(\frac{1}{2}A_0t^2 + A_1t + A_2/2)\cos bt - (\frac{1}{2}B_0t^2 + B_1t + B_2/2)\sin bt]$. **7.** $\frac{1}{8}e^t[(6 - t^2)\sin t - 3t\cos t]$. **8.** $y = e^{2t}(\frac{7}{4}\cos 2t - 3\sin 2t + \frac{1}{4})$. **9.** $y = -3t\cos 2t + \frac{3}{2}\sin 2t$. **10.** $y = (\frac{1}{3}\sin 3t - 2\cos 3t)e^t + (3\cos 3t + \sin 3t)e^{-t}$. **11.** $\frac{1}{2}(t\cos t + \sin t)$. **12.** $\frac{1}{2}e^t(1 + t)\sin 2t$. **13.** $e^{-t}[(t - 4)\cos t - (\frac{3}{2}t + 1)\sin t + 4]$. **14.** $\frac{1}{16}[(2t^2 - 6t - 16)\cos t - (2t^2 + 10t - 6)\sin t + 16]$. **15.** $y = \frac{1}{8}e^{-3t}(5\sin 2t - 26t\cos 2t)$.

§66, page 184

2. $x = t - 2\sin t$, $y = 2t + 2\cos t$. **3.** $x = e^t$, $y = t$. **4.** $x = 3t + 2$, $y = \sin t$. **5.** $x = t^2$, $y = 2t$, $z = -t$. **6.** $x = 2t$, $y = 2t - 2$. **7.** $x = t + 2e^t$, $y = 3e^t$. **8.** $x = te^t$, $y = e^t$. **9.** $x = 1 + \sin t$, $y = \cos t$. **10.** $x = 1 + t\sin t$, $y = t\cos t$. **11.** $x = Em/(H^2e)[1 - \cos(He/m)t]$, $y = (E/H)t - Em/(H^2e)\sin(He/m)t$.

* Because of (8), two answers may be given.

§67, page 186

1. $y = cte^{-at}$. **2.** $y = \frac{1}{2}mt^2$. **3.** $y = \frac{1}{2}t + ct^3$. **4.** $y = 5t$. **5.** $y = -\frac{1}{3}t$. **6.** $y = 12t + c[t - \frac{1}{2}t^2 + \frac{1}{2}t^3/3! - \cdots + (-1)^n t^n]/[(n-1)!n!] + \cdots$. **7.** $y = t^n/(n^2 + 3n + 2)$. **8.** $y = \sin t$. **9.** $y = T^{-1}\{\varphi(p)\}$.

§68, page 190

3. $f(t) = t$, $0 < t < h$; $f(t) = h$ when $t > h$. **5.** $f(t) = t^3$, $0 \leqq t \leqq k$. **6.** $\psi(t) = 2$ when $0 < t < 2$, $\psi(t) = 0$, $2 < t < 10$, $\psi(t + 10) = \psi(t)$: $T\{f(t)\} = (1 - e^{-2p})/p$. **7.** (*a*) $\varphi(t) = 5$, $0 < t < 1$, and $\varphi(t) = 0$, $1 < t < 6$, $\varphi(t + 6) = \varphi(t)$. (*b*) $\varphi(t) = t^3$, $0 < t < 4$, and $\varphi(t) = 0$, $4 < t < 8$, $\varphi(t + 8) = \varphi(t)$. **11.** $[ap - 1 + (ap + 1)e^{-2ap}]/[p^2(1 - e^{-2ap})]$. **13.** (*a*) $k \coth(\frac{1}{2}\pi p/k)/(p^2 + k^2)$. (*b*) $[p(1 - e^{-\pi p/k}) + 2ke^{-\pi p/(2k)}]/[(p^2 + k^2)(1 - e^{-\pi p/k})]$. **14.** $[(1 - e^{-ap})p^{-2} - ap^{-1}e^{-(a+b)p}]/[1 - e^{-(a+b)p}]$. **15.** $(pe^{-9\pi p/2} - 3e^{-\pi p})/(p^2 + 9)$. **16.** (*a*) Yes, no. (*b*) Yes, $\cos(\pi + t) = -\cos t$. (*c*) Yes. $\sin(t + \pi) = -\sin t$.

§69, page 194

3. 5.8. **4.** $2e^{at}\sqrt{t/\pi}$. **5.** $\frac{27}{28}e^{at}T\{t^{7/3}\}/\Gamma(\frac{1}{3})$. **6.** $[\frac{27}{28}\int_0^t e^{a\tau}\tau^{7/3}\,d\tau]/\Gamma(\frac{1}{3})$. **10.** $a^{-1}e^{a^2t}\operatorname{erf}(a\sqrt{t})$. **11.** $[\sqrt{\pi}/(b^2 - a^2)][b^{-1}e^{b^2t}\operatorname{erf} b\sqrt{t} - a^{-1}e^{a^2t}\operatorname{erf}(a\sqrt{t})]$. **14.** $\int_0^t e^{b^2\tau}(m^2 + b^2)^{-1/2}\operatorname{erf}(\sqrt{m^2 + b^2}\,\tau)\,d\tau$. **17.** $\int_0^t e^{-a^2\tau}(m^2 - a^2)^{-1}\operatorname{erf}(\sqrt{m^2 - a^2}\sqrt{\tau})\,d\tau$. **18.** (*a*) $1/\sqrt{\pi t} + ae^{a^2t}\operatorname{erf}(a\sqrt{t})$. (*b*) $e^{a^2t}(1 - \operatorname{erf}(a\sqrt{t})$. (*c*) $\dfrac{1}{\sqrt{\pi t}} - \dfrac{2ae^{-a^2t}}{\sqrt{\pi}}\int_0^{a\sqrt{t}} e^{\lambda^2}\,d\lambda$.

§70, page 199

1. 5 ft, $\frac{1}{6}$ sec, 6 cycles/sec, $\frac{5}{72}$ sec, $\frac{11}{72}$ sec. **2.** $\frac{1}{16}\pi$ sec, $16/\pi$ cycles/sec, 13 ft. **3.** $y = 5\sin 10t + 10\cos 10t$, $5/\pi$ cycles/sec, $\frac{1}{5}\pi$/sec, $\sqrt{125}$ ft. **4.** $\frac{1}{60}$ sec, 60 cycles/sec. **6.** $k^2 < 120$. **7.** 1.99 sec, $e^{-0.05t}$, 13.9 sec. **8.** $b = \frac{1}{5}$, $c = (120\pi)^2$ nearly. **9.** $b = 0.046$, $c = 400\pi^2$ nearly.

§71, page 201

1. (*a*) 4. (*b*) 1. (*c*) 5. **2.** (*a*) Overdamped. (*b*) Critical. (*c*) Overdamped. **3.** (*a*) ± 8. (*b*) $-4\pi \times 10^6$, $4\pi \times 10^6 + \frac{1}{2}\pi$.

§74, page 207

2. $s = 0.4\sin 10t + 0.32$, 0.4 ft, $\frac{1}{5}\pi$ sec, $5/\pi$ cycles/sec. **3.** $x = e^{-0.01t}(0.0004\sin 10t + 0.4\cos 10t) + 0.32$, $\frac{1}{5}\pi$sec, $5/\pi$ cycles/sec, $e^{-0.01t}$, 69.3 sec. **4.** (*a*) $x = -0.6\sin 10t + 0.4\cos 10t + \sin 6t + 0.32$. (*b*) $\frac{1}{5}\pi$ sec, $\frac{1}{3}\pi$ cycles/sec, π. (*c*) 0.210 ft. **5.** $x = 4\sin(\sqrt{3g}\,t) - 3.46\sin(2\sqrt{g}\,t)$, 3.80 ft. **6.** $2\pi\sqrt{I/k}$. **7.** $g/(16\pi^2)$ lb-ft². **8.** 656 lb. **9.** $2\pi\sqrt{l/g}$. **10.** $x = e^{-1.08t}(0.111\sin 9.77t + \cos 9.77t)$, 0.643 sec. **11.** $x = a\cos\sqrt{g/h}\,t$. **12.** $x = \frac{1}{2}\sin(2\sqrt{g}\,t) - \sqrt{g}\,t\cos(2\sqrt{g}\,t)$, 157 ft below initial position. **13.** $\theta = e^{-0.0693t}(c_1\sin 12.6t + c_2\cos 12.6t)$,

0.5 sec. **14.** 0.886 sec, $0.886\rho^{-1/2}$ sec. **15.** 21.1 min. **16.** (*a*) 5.4×10^7 ton-ft². (*b*) 9.9 sec. **17.** $x = 48 \sin 10t + 2\sqrt{6} \sin \sqrt{600}\, t$, $y = 24 \sin 10t - 4\sqrt{6} \sin \sqrt{600}\, t$.

§75, page 213

1. $x = v_0 \cos \varphi \cdot t$, $y = v_0 \sin \varphi \cdot t - \frac{1}{2}gt^2$. **2.** $x = 130{,}000(1 - e^{-0.02t})$, $y = 155{,}500\ (1 - e^{-0.02t}) - 1{,}610t$, maximum $y = 22{,}000$ ft. **3.** $x = 49{,}800(1 - e^{-0.04t})$, $y = 174.3t - 16.1t^2$. **4.** $x = 347t$, $y = 118{,}000(1 - e^{-0.0268t}) - 1{,}200t$, 30,000 ft, 27,000 ft. **5.** $x = a \cos (\sqrt{k/m}\, t)$, $y = v_0 \sqrt{m/k} \sin (\sqrt{k/m}\, t)$. **6.** (*a*) $x = 3 \sin t$, $y = 2 \cos t - 2$. (*b*) $x = 2.12 \sin t$, $y = 2.12 \sin t + 2 \cos t - 2$. **7.** 2,120 ft. **8.** 300 ft/sec, 10,600 ft.

§77, page 220

2. $q = CE[1 - \cos (t/\sqrt{LC})]$, $i = CE/\sqrt{LC} \sin (t/\sqrt{LC})$, $2\pi\sqrt{LC}$, CE, $2\pi\sqrt{LC}$, $E\sqrt{C/L}$. **5.** $q = (\epsilon^{-5t}/1{,}600)(-\sin 200t - 40 \cos 200t) + \frac{1}{40}$, $i = 5.00\epsilon^{-5t} \sin 200t$, 0.460 sec, $\frac{1}{40}$, 0. **6.** $q = 200\epsilon^{-t/2}(-2 - t) + 400$, $i = 100t\epsilon^{-t/2}$. **7.** $q = q_0 \cos (t/\sqrt{LC})$, $i = dq/dt$. **8.** $q = -CE \cos (t/\sqrt{LC}) + CE$, $i = E\sqrt{C/L} \sin (t/\sqrt{LC})$. **9.** $q = (q_0/\omega_1)e^{-at}(\omega_1 \cos \omega_1 t + a \sin \omega_1 t)$, $i = dq/dt$.

§79, page 222

2. $i = -(E/Z^2)(X \cos \omega t - R \sin \omega t)$. (*a*) $i = -[E/(L\omega)] \cos \omega t$. (*b*) $i = (E/R) \sin \omega t$. (*c*) $q = CE \sin \omega t$. (*d*) $q = [EC/(1 + R^2C^2\omega^2)] (\sin \omega t - RC\omega \cos \omega t)$, $i = [EC\omega/(1 + RC^2\omega^2)] (\cos \omega t + RC\, \omega \sin \omega t)$. (*e*) $i = [E/(R^2 + L^2\omega^2)] (R \sin \omega t - L\omega \cos \omega t)$. (*f*) $q = (CE/(1 - LC\omega^2)] \sin \omega t$, $i = CE\omega/(1 - LC\omega^2) \cos \omega t$. **4.** $\frac{1}{2}\sqrt{13}$. **5.** $q = -0.00275(3 \sin 400t + \cos 400t)$, $i = 1.10(\sin 400t - 3 \cos 400t)$; 0.0087 coul, 3.48 amp. **6.** $10 \sin 500t\ (1 - \epsilon^{-5t})$, 0.0067 sec. **7.** 4.2×10^{-7} farad, 1.00072. **8.** 10 amp, 0.504 amp.

§80, page 226

1. $i_2 = E/R \sin \omega t$, $i_1 = [E/(L\omega)](1 - \cos \omega t)$, $i = i_1 + i_2$. **2.** $i_1 = E/R \sin \omega t$, $q = CE \sin \omega t$, $i_2 = CE\omega \cos \omega t$, $i = i_1 + i_2$. **4.** $i_2 = 2 \sin 400t$. 160, 100 $q = 40\epsilon^{-10t} + \sin 400t - 40 \cos 400t$, $i_1 = dq/dt$. **6.** (*a*) 100 per cent nearly. (*b*) 100 per cent nearly. (*c*) 89 per cent. (*d*) Less than 20 per cent. **7.** $i = \frac{1}{12}(1 - \cos 300t) + 10t \sin 300t$. **8.** $i = (C_1E_1 + C_2E_2)\omega]/[1 - L(C_1 + C_2)\omega^2][\cos \omega t - \cos t/\sqrt{L(C_1 + C_2)}]$. **10.** $q_1 = a/[\omega\omega_1(\omega_1^2 - \omega^2)]\ (\omega_1 \sin \omega t - \omega \sin \omega_1 t)$, $q_2 = Ma\omega/[L_2(\omega^2 - \omega_1^2)] (\cos \omega t - \cos \omega_1 t)$.

§81, page 229

1. 2π sec, $\frac{1}{2}\pi^{-1}$ cycle/sec, 5 ft. **2.** $x = 2e^{-t/2} \sin 3t$, 2.09 sec, 0.223. **3.** $a = 2$, $x = 2e^{-t/3} \sin 3t$, 2.09 sec. **4.** $b = 13.86$, $c = 14{,}400\pi^2$ nearly. **5.** $\frac{1}{20}w$, 93.8 ft. **6.** 0.815 sec. **7.** $q = 0.05\epsilon^{-5t} \cos 200t$, $i = e^{-5t}(-10 \sin 200t - 0.25 \cos 200t)$, $\frac{1}{100}\pi$, 0, 0. **9.** 0.000025. **10.** (*a*) $b^2 - |4ac| > 0$. (*b*) $b^2 - 4ac = 0$. (*c*) $b^2 - 4ac < 0$. (*d*) $b = 0$. **12.** Rises 2.02 sec, then falls. Speed approaches 80.5 ft/sec downward. **13.** $q = 1 - \frac{1}{2}\epsilon^{-t/2}(2 - t)$. **14.** $i = \epsilon^{-50t}(3.04 \sin 312t - \cos 312t) + 1$. **15.** $x = 2(\sin 2t - 4 \cos 2t) + \epsilon^{-t}(-2 \sin 2t + 8 \cos 2t)$. Amp $= 2\sqrt{17}$, Period $= \pi$. **16.** $i_1 = [E/(L_1\omega)](1 - \cos \omega t) + [L_2E/(R_2L_1)] \sin \omega t$, $i_2 = -[ME/(L_1R_1)] \sin \omega t$. **17.** $x = \frac{5}{2} - \frac{1}{6} \cos 8t - \frac{4}{3} \cos 4t$, $y = 2 + \frac{1}{3} \cos 8t - \frac{4}{3} \cos 4t$. **18.** $\frac{1}{10}\sqrt{100 + (0.3 - h)^2\omega^2}$.

§83, page 235

1. $y = 2x^3 + c_1x + c_2$. **2.** $y = 6 \ln x + c_1x^2 + c_2x + c_3$. **3.** $y = c_1 \ln x + c_2$. **4.** $y = c_1x + c_1^2 \ln (x - c_1) + c_2$. **5.** $y = -4x^3 + 2x - 26$. **6.** $xy = e^{-x}(c_1 - 12x) + c_2$. **7.** $y = \frac{1}{5}(3x^5 - 5x^4 + 10x - 8)$. **8.** $y = a^3 \sinh (x/a) - 2a^2x$. **9.** $x^2y = xe^x - 2e^x + c_1x + c_2$.

§84, page 237

1. $y^2 = c_1x + c_2$. **2.** $x = c_1y - \ln y + c_2, y = c_3$. **3.** $y^3 = c_1x + c_2, y = c$. **4.** $e^y(y - 1) = Rx + c$. **5.** $\sqrt{cs^2 - 1} = \pm ct + c_1$. **6.** $\ln (8e^s + \sqrt{64e^{2s} - c}) = \pm 8t + c_1$. **7.** $y = 1 + \sin \sqrt{8}\, x$. **8.** $10e^s + \sqrt{576 + 100e^{2s}} = 36e^{10t}$. **9.** $e^{-y/2} = \cos (\frac{1}{2}x)$. **11.** $y = \cosh x$.

§85, page 239

1. $y = c_1x + c_2x^{-2} + \frac{1}{4}x^2$. **2.** $y = c_1x^{-2} + \frac{1}{7}x^5$. **3.** $y = c_1 + c_2 \ln x + c_3 (\ln x)^2 + \frac{1}{27}x^3$. **4.** $y = c_1x^{-1} + c_2x^{-2} + c_3x^3$. **5.** $y = c_1x^3 + c_2x^{-3} + [1/(n^2 - 9)]x^n$. **6.** $y = c_1(x - 1) + c_2(x - 1)^2 + c_3(x - 1)^{-2} + \ln [e(x - 1)]$. **7.** $y = c_1 + c_2x + c_3 \ln x$.

§86, page 241

1. $y = e^{-x}(x^3 + c_1)$. **2.** $y = y_c + (\frac{1}{2}x^2 - \frac{1}{2}x)e^x$. **3.** $y = \{c_1 + c_2x + x^{n+2}/[(n - 1)(n + 2)]\} e^{2x}$. **4.** $y = c_1x + c_2 \sqrt{x^2 - 1}$. **5.** $xy = c_1e^{-x}(x^2 + 2x + 2) + c_2$. **6.** $y = x \sin x + c_1 \sin x + c_2 \cos x$. **7.** $y = c_1(x^2 - 1) + c_2x + 3x^2 + x^4$. **8.** $y = (x^4 + c_2)e^x + c_1(x^3 + 3x^2 + 6x + 6)$. **9.** $y = \sin x \ln [c_1 \sin x(\csc x - \cot x)^c]$.

§87, page 243

4. $xy = c_1(x^2 - 2x + 2) + c_2e^{-x}$. **5.** $y = c_1x^2 + c_2 - 12x$. **6.** $y = 12x^2 + c_1x \ln x + c_2x$. **7.** $x^2y = 6x^2 - 12x + c_1e^{-x} + c_2$. **8.** $y = e^{x^3}(2x^2 + c_1x + c_2)$. **9.** $y = (12x + c_1 \ln x + c_2) e^{-x^2}$. **10.** $x^3y = \frac{1}{3}x^3 + c_1e^{-x^3} + c_2$. **11.** $x^3y = 3x^2 + c_1e^{-x^2} + c_2$. **12.** $x^2y = x + c_1e^{-x} + c_2$. **13.** $x^2y = (x^2 + c_1)e^{x^2} + c_2$. **14.** $y = 2x^2 - 2x + 1 + c_1(2x - 1) + c_2e^{-2x}$. **15.** $y = c_1(2x - 1) + c_2e^{-2x}$. **16.** $y = c_1(x + 1)e^{-3x} + c_2e^{-2x}$. **17.** $y \sin x = \frac{1}{2}e^x(\sin x - \cos x) + c_1e^{-x}(\sin x + \cos x) + c_2$.

§88, page 244

1. $y = x^4 + c_1 \ln x + c_2$. **2.** $y = c_1xe^x + c_2$. **3.** $y = \ln |2.6 \cos (\tan^{-1} 2.4 - 10x)|$. **4.** $y = 1 - e^{-x}$. **5.** $y = c_1 \ln x + c_2x + c_3$. **6.** $y = (x^2 + c_1x + c_2)e^{x^2}$. **7.** $y = c_1\{4x^2 \ln [x/(2x + 1)] + 2x - \frac{1}{2}\} + c_2x^2$. **8.** $(v - v_1)/(v - v_0) = [r_0(r - r_1)]/[r_1(r - r_0)]$. **9.** $y = c_1 \sin (\ln x) + c_2 \cos (\ln x) + c_3x + x^2$. **10.** $4y = 5 \ln (1 + 3 \tan \frac{1}{2}x) - 5 \ln (\tan \frac{1}{2}x + 3)$. **11.** $y = c_1x + (c_2 - x^2)xe^{-x^2}$. **12.** $y = (x + c_1e^{-x} + c_2)e^{x^2/2}$. **13.** $xy = c_1(x^2 - 2x + 2) + c_2e^{-x}$. **14.** $y = c_1x^{3/2} + c_2x^{-1} - 4 \ln x + \frac{4}{3}$. **15.** $y = a^2 \sinh (x/a) - 2ax$. **16.** $y = c_1 + c_2 \sin (\sqrt{3} \ln x) + c_3 \cos (\sqrt{3} \ln x) - \frac{1}{6}(\ln x) \sin (\sqrt{3} \ln x)$. **17.** $y = c_1e^{-4x}(7x + 1) + c_2x^{3x}$. **18.** $y = c_1 + c_2 \ln x + c_3(\ln x)^2 + \frac{1}{27}x^3$.

§89, page 247

1. $(x - c_1)^2 + (y - c_2)^2 = a^2$. **2.** $(x - c_1)^2 + y^2 = c_2^2$. **3.** $y = c_1 \cosh (x/c_1 + c_2)$. **4.** $c_1x = \cosh (c_1y + c_2)$. **5.** $(c_1x + c_2)^2 = k(c_1y^2 - 1)$. **6.** $x^2 + (y - c_1)^2 = c_2^2$. **7.** $(y - c_1)^2 = 4c_2(x - c_2)$. **8.** $e^{x/a} = c_1 \sin (y/a + c_2)$.

§90, page 249

1. $2Hy = wx^2$. **2.** $H(d^2y/dx^2) = w_1 + w_2\sqrt{1 + (dy/dx)^2}$. **3.** $y = a \cosh \sqrt{w/H}\, x$. **4.*** $y = (c_1 + kt) \cosh (\sqrt{w/H}\, x) - kt$. **5.** $y = c \cosh \sqrt{w/H}\, x$. **6.** $2Hy = wlx^2$. **8.** $x = w \int_c^y [(c^2 + 2H/w - y^2)/\sqrt{4H^2 - w^2(c^2 + 2H/w - y^2)^2}]\, dy$. **9.*** Answer is like that of exercise 8 with y replaced by $y + kt$. **10.** $x_0 = a \tanh^{-1} (b/l)$, $y_0 = l \coth (c/a) = a$, where a satisfies $l^2 - b^2 = a^2 \sinh^2 (c/a)$.

§91, page 253

1. $y = w/(24EI)(2lx^3 - x^4 - l^3x)$; maximum deflection $= 5wl^4/(384EI)$. **2.** $y = P/(EI)(\frac{1}{12}x^3 - \frac{1}{16}l^2x)$; maximum deflection $= Pl^3/(48EI)$. **3.** $5wl^4/(384EI) + \frac{1}{48}Pl^3/(EI)$. **4.** (*a*) 0.889 in. (*b*) 1.067 in. (*c*) 1.956 in. **5.** (*a*) $y = P/(6EI)(-3lx^2 + x^3)$. (*b*) $y = -w/(24EI)(x^4 - 4lx^3 + 6l^2x^2)$. (*c*) y equals the sum of the y's from (*a*) and (*b*). **6.** $y = -w/(48EI)(2x^4 - 5lx^3 + 3l^2x^2)$; 0.578. **7.** $y = P/(48EI)(4x^3 - 3lx^2)$. **8.** $y_1 = Px/(18EI)(x^2 - 8a^2)$, $y_2 = P/(18EI)[x^3 - 3(x - 2a)^3 - 8a^2x]$, $(16\sqrt{6}\, Pa^3)/(81EI)$. **9.** $\frac{16}{147}Pa^3$. **10.** $kl^5/(30EI)$. **11.** $y = Pb^2x^2/(6l^3EI)[(3a + b)x - 3al]$, $x \leqq a$.

§92, page 258

2. Equation of graph $y = G/P[1 - \cos (4\pi x/l)]$ from $x = 0$ to $x = l$. **4.** Equation of graph $y = -a[1 - \cos (\frac{3}{2}\pi x/l)]$, $0 \leqq x \leqq l$.

§93, page 262

1. $[a_t, a_n] = [\frac{11}{40}\sqrt{10}, \frac{13}{40}\sqrt{10}]$. **2.** $\frac{1}{3}a^{3/2}(2 - 1/\sqrt{2})$. **3.** $a^2/\sqrt{k}$. **5.** $\rho = 2e^{2t} - e^{-2t} - \frac{1}{4}$, $\theta = 2t$; $a_\theta = 16e^{2t} + 8e^{-2t}$. **6.** 80 lb. **7.** $\rho = t + 1$, $\theta = 2t/(t + 1)$, $F_p = 4W/[g(t + 1)^3]$. **9.** $\rho = 4/(1 + \cos \theta)$.

§94, page 266

1. No, assuming that the earth is considered as a particle. **2.** Yes. Hyperbolic orbit. **3.** Because it is the least velocity a body, acted upon only by the gravitational field of the earth, must have to recede farther and farther without end from the earth. 4.74 mi/sec, 3.64 mi/sec nearly. **5.** Yes. **6.** Increases. **7.** $\rho = (9/5)/(1 + \frac{4}{5} \cos \theta)$, Period $= 30\pi = 94.2$ sec. **8.** $\rho = [a^2v_0^2/(gR^2)]/(1 - e \cos \theta)$ where $e = 1 - (av_0^2/(gR^2))$. **9.** Equation (20) if $k = gR^2$.

* c is the depth of material over the highest point of the arch and t is the thickness of the top layer.

§95, page 267

1. $e^{-r} = c_1 + c_2\theta$. **2.** $y = 5t^2, x = 5[t\sqrt{1 - t^2} + \sin^{-1} t]$. **3.** $EIy = \frac{1}{360}wx(-7l^4 + 10l^2x^2 - 3x^4)$. **5.** (a) $EIy = \frac{1}{360}wx^2(-7l^3 + 9l^2x - 2x^3)$. (b) $EIy = \frac{1}{120}wx^2(-2l^3 + 3l^2x - x^3)$. **6.** $\rho^{n+1}[\cos(n + 1)\theta + c_1] = c_2$. **8.** $x_{\max} = 2gR^2/(2gR^2/a - v_0^2), x = 2gR^2/(2gR^2/a - \frac{3}{4}v_0^2)$.

§97, page 270

1. (a) $dy/dx = z, dz/dx = -x^2z - x^3y$. (b) $dx/1 = dy/z = dz/(-Pz - Qy)$. (c) $x_1 = dx/dt, y_1 = dy/dt, y_2 = dy_1/dt, dx_1/dt = 3x - 3y - 2y_2, dy_2/dt = 3x - y_2$. **3.** $y = \frac{1}{2}x^2 \ln x - \ln x + c_1x^2 + c_2$. **4.** $y = -x + \int c_1e^{-x^2/2}\,dx + c_2$. **5.** $y_1 = dy/dx; y_i = dy_{i-1}/dx, i = 2, 3, \ldots, n - 1; a_0dy_{n-1}/dx = -a_1y_{n-1} - a_2y_{n-2} - \cdots - a_ny - f(x) = 0$. n constants.

§98, page 274

3. No. $dy_1/dx = y/x$ is not defined when $x = 0$. **4.** $x = n\pi$, all x in the intervals $(2n + 1)\pi < x < (2n + 2)\pi$, n any integer. **5.** All real roots of $f_1(x) = 0$. **6.** $y = 0$. **7.** dy/dx undefined. No, for $3y/x$ has no value at $(0,0)$. **8.** No solution guaranteed through (x_0,y_0) if $\varphi(x_0) = 0$ or if P, Q, or R is discontinuous or multiple valued at x_0. There is a unique solution $y = \psi(x)$ through (x_0,y_0) if $Q(x)/P(x)$ and $R(x)/P(x)$ are continuous and single valued in $|x - x_0| < b$, $b > 0$. **9.** Yes. $y = 2x + 5, z = -3x - 8 - 2e^x$. **10.** No. **11.** When $x = x_0$, and $d^2y/dx^2 = (2x + c_1)^{1/2} = 0$, then $c_1 = -2x_0$, and d^3y/dx^3 does not exist when $c_1 = -2x_0$. **12.** $a = 0$, $a = 1$.

§99, page 278

1. $x^2 - y^2 = c_1, x + y = c_2z$. **2.** $bx^2 - ay^2 = c_1, cy^2 - bz^2 = c_2$. **3.** $y = c_1x, 2x - 2y = z^2 + c_2$. **4.** $y^2 + z^2 = c_1, \ln c_2x = \tan^{-1}(y/z)$. **5.** $y = c_1e^x + c_2e^{-x} - x^2 - 2, z = (c_1e^x - c_2e^{-x} - 2x)/x$. **6.** $y = c_1 \sin 2x + c_2 \cos 2x + \frac{5}{2}, 5z = (2c_1 + c_2)\cos 2x + (c_1 - 2c_2)\sin 2x - 10x + \frac{5}{2}$. **7.** $x^2 - y^2 = c_1, (x + y)(z - 1) = c_2(z + 1)$. **8.** $x^2 + y^2 + z^2 = c_1y, y = c_2z$. **9.** $lx + my + nz = c_1, x^2 + y^2 + z^2 = c_2$. **10.** $x^2 - y^2 = c_1, z^2 - w^2 = c_2, x + y = c_3(z + w)$. **11.** $x - y = c_1(x - z) = c_2(y - z)$. **12.** $x + y + z = c_1, xyz = c_2$. **13.** $xy - z = c_1, x^2 - y^2 + z^2 = c$. **14.** $x - y - z = c_1, x^2 - y^2 = cz^2$. **15.** $x - y = c_1, x + y + z + w = c_2, x^2 + y^2 + z^2 + w^2 = c_3$. **16.** $y = c_1 \sin 2x + c_2 \cos 2x + \frac{5}{2}, 5z = (2c_1 + c_2)\cos 2x + (c_1 - 2c_2)\sin 2x - 10x + \frac{5}{2}$. **17.** $y = c_1 + c_2e^{2x} - 4e^x, z = -c_1 + c_2e^{2x} - 2e^x$. **18.** $y = c_1e^{2x} + c_2e^{-2x} - 3x, 3z = 3c_1e^{2x} - c_2e^{-2x} - 3 - 3x - 6x^2$. **19.** $y = c_1 \sin x + c_2 \cos x - \frac{1}{3}a \sin 2x, 2z = (c_1 - c_2)\cos x, -(c_1 + c_2)\sin x - \frac{2}{3}a(\cos 2x + \sin 2x)$.

§100, page 283

1. $y = x(z^2 + c)$. **2.** $2y - z = ce^{-x/2}$. **3.** $4xy = 2z - 1 + ce^{-2z}$. **4.** $y = xz^2(3z + c)$. **5.** $(x^2 + y^2)e^z + z = c$. **6.** $x^2y^2z + y = c$. **7.** $(x^2 - xyz)e^z = c$. **8.** $z^2(x + y) = z\cos z - \sin z + c$. **9.** $xy(1 + z)^2 = 2z^2 + \frac{4}{3}z^3 + c$. **10.** $x \sin z + y \cos z = (c + \ln \cos z)e^{-z}$. **11.** $xy + xz + yz = cx$. **12.** (a) $M_y - N_x = MN_z - NM_z$. (b) $M_y = N_x$. **13.** $ax^2 + 2(a + 2b)xy + 2by^2 - 2cx = c_1$. $ax^2 + 2(a + 2b)xy + 2by^2 - 2cx = c_1$.

§102, page 286

1. $z = mx$, $z = ny$. **2.** $xz = m$, $z^3y = n$. **3.** $x^3y^2 = m$, $x^2z = n$. **4.** $x^2 - y^2 = m$, $z^2 - y^2 = n$. **5.** $x^2 + xy + yz = c$. **6.** $z^2x - y = cx$. **7.** $xy = c(3 - z)$, and $z = 3$. **8.** $y = cx^2$. **9.** $[2x,2y,2z]$, $x = c_1y$, $x = c_2z$, $x^2 + y^2 + z^2 = c_3$. **10.** $[kx/r,ky/r,kz/r]$, $x = c_1y$, $x = c_2z$, $x^2 + y^2 + z^2 = c_3$. **11.** $[a/x,b/y,c/z]$, $bx^2 - ay^2 = c_1$, $cx^2 - az^2 = c_2$, $x^ay^bz^c = c_3$. **12.** $[yz,xz,xy]$, $x^2 - y^2 = c_1$, $x^2 - z^2 = c_2$, $xyz = c_3$. **13.** $z = c_1y$, $(x + a)r_2 - (x - a)r_1 = c_2r_1r_2$. **14.** $y = c_1x$, $z = -2y^2 - 2x^2 + c_2x^3$, $8(x^2 + y^2) + 12z = 3 + ce^{-4z}$.

§103, page 287

1. $y_1 = Dy$, $y_2 = Dy_1$, $z_1 = Dz$, $Dz_1 = -(x - 3)y_2/(x - 4)$, $Dy_2 = [x(3 - x)y_2 + (y + 2z)/(4 - x)]/[x(x - 2)(x - 4)]$; 0, 2, 4. **2.** (*a*) Yes. (*b*) No. (*c*) No. (*d*) Yes. **3.** (*a*), (*b*), (*d*). **4.** $x = c_1y$, $x = c_2z$, $k(x^2 + y^2 + z^2)^{-n} = c_3$. **5.** $x(y + z^3) = cz^2$. **6.** $y = c_1e^x + c_2e^{-x} - x - 1$, $z = c_1e^x - c_2e^{-x} - x - 1$. **7.** $x + y + z = c_1$, $x^2 + y^2 + z^2 = c_2$. **8.** $y = e^x(c_1 \sin x + c_2 \cos x) + x + 2$, $z = e^x(c_1 \cos x - c_2 \sin x) - 3x - 1$. **9.** $y = c_1z^{\sqrt{2}} + c_2z^{-\sqrt{2}}$, $x = c_1(1 + \sqrt{2})z^{\sqrt{2}} + c_2(1 - \sqrt{2})z^{-\sqrt{2}}$; $x^2 + 2xy - y^2 + z^2 = c$. **10.** $x = c_1y^3$, $y^2(4z^2 - 3x^2 - 4y^2) = c_2$; $(6x^2 + 4y^2)z^2 = z^4 + c$. **11.** $kr_1^{-1} + kr_2^{-1} = c_1$, $(x + a)r_1 + (x - a)r_2 = c_2$.

§104, page 290

1. $a_m = 5/(m + 5)!$, $m > 4$. **5.** $-3 \leqq x < 3$. **6.** $a_{2n+3} = 1/[3 \cdot 5 \cdot 7 \cdots (2n + 1)] = 2^nn!/(2n + 1)!$. **7.** $a_{2n} = 1/[2^n(n!)^2]$. **8.** $a_n = 2/[(n!(n + 1)!]$. **11.** (*a*) $-3 < x < 3$. (*b*) $|x - 3| < 2$. (*c*) All values.

§106, page 296

1. $y = c_0[1 + x + x^2/2! + \cdots + x^n/n! + \cdots] = c_0 \sum_{n=0}^{\infty} x^n/(n!)$. **2.** $y = c_0 \sum_{n=0}^{\infty} x^{2n}/(n!)$. **3.** $y = c_0[1 + x^4/(3 \cdot 4) + x^8/(3 \cdot 4 \cdot 7 \cdot 8) + \cdots] + c_1(x + x^5/(4 \cdot 5) + x^9/(4 \cdot 5 \cdot 8 \cdot 9) + \cdots]$. **4.** $y = c_0 \sum_{n=0}^{\infty} (-1)^n(2n + 1)x^{2n} + c_1 \sum_{n=0}^{\infty} (-1)^n(2n + 2)x^{2n+1}$. **5.** $y = c_1(x - x^3) + c_2 \sum_{n=0}^{\infty} 3x^{2n}/[(2n - 1)(2n - 3)]$. **6.** $y = a_0 \sum_{n=0}^{\infty} x^{4n}/[2^{2n}(2n)!] + a_2 \sum_{n=0}^{\infty} x^{4n+2}/[2^{2n}(2n + 1)!]$. **7.** $y = c_1x^4 + c_2x^{-3}$. **8.** $y = cx^9$. **9.** $y = c_1x^2 + c_2x^3$. **10.** $y = c_0 \sum_{n=0}^{\infty} (2n + 1)(x - 1)^{2n} + c_1 \sum_{n=0}^{\infty} (n + 1)(x - 1)^{2n+1}$. **11.** $y = c_0 \sum_{n=0}^{\infty} (n + 1)(2n + 1)(x + 1)^{2n} + c_1 \sum_{n=0}^{\infty} (n + 1)(2n + 3)(x + 1)^{2n+1}$. **12.** $y = c_0 + c_1x + c_0x^2/(2!) + (c_1 + c_0)x^3/3! + (3c_0 + 2c_1)x^4/4! + \cdots$. **13.** $y = c_0 + c_1x - \frac{1}{6}(c_0 + c_1)x^3 - \frac{1}{12}c_1x^4 + \frac{9}{40}(c_0 + c_1)x^5 + \cdots$ **14.** $c_0e^{-x^2/2}$, $c_2(1 - 2x^2)e^{-x^2/2}$, $c_3(x - \frac{2}{3}x^3)e^{-x^2/2}$.

§107, page 299

2. $y = a_0 \sum_{n=0}^{\infty} x^{-n}/[n!(n + 1)!]$. **3.** $y = c_0 \sum_{n=0}^{\infty} (-1)^nx^{-2n}/(2n + 1)!$. **4.** $y = c_1x$, $y = c_0 -$

$c_0 \sum_{n=1}^{\infty} (2n-2)!x^{2n}/[2^{2n-1}n!(n-1)!]$, $y = \sum_{n=1}^{\infty} (2n-2)!x^{-2n+1}/[2^{2n-1}n!(n-1)!]$. **5.** $y = c_0 \sum_{n=0}^{\infty} (-1)^n x^{-2n}/(2n)! + c_1 \sum_{n=1}^{\infty} (-1)^{n+1}x^{-2n+1}/(2n-1)!$.

§108, page 301

1. $y = a_0 \sum_{n=0}^{\infty} [(4x)^n/(2n)!] + a_1 x^{1/2} \sum_{n=0}^{\infty} (4x)^n/(2n+1)!$. **2.** $y = A(1 + 2x^2 + 3x^4 + 4x^6 + \cdots) + Bx^{-1}(1 + 3x^2 + 5x^4 + 7x^6 + \cdots)$. **3.** $y = A[(1/2) + (1 \cdot 4/5!)x^3 + (1 \cdot 4 \cdot 7/8!)x^6 + \cdots] + Bx^{-2}[1 + (2/3!)x^3 + (2 \cdot 5/6!)x^6 + \cdots]$. **4.** $y = Ax^2[1 - (2 \cdot 2/5)x + 3 \cdot 2^2/(5 \cdot 6)x^2 - \cdots] + B[(1/x^2) - 4/(3x) + \frac{2}{3}]$. **6.** $y = Ax^{-1} + B[1 + (x^2/3) + (x^4/5) + \cdots]$ **7.** $y = A(x+1) + B(x^2 + x^3 + x^4 + \cdots)$. **8.** $y = A[1 - (x^{-2}/3!) - (x^{-4}/5!) - (3x^{-6}/7!) - \cdots] + B(x - x^{-1})$. **9.** $y = Ax^2 + Bx - Bx \sum_{n=1}^{\infty} x^{2n}/(2n-1)$, $y = \sum_{n=0}^{\infty} x^{-2n-1}/(2n+3)$. **10.** $y = a_0 \sum_{n=0}^{\infty} x^{-2n-1}/[n!2^n]$, $y = a_1 \sum_{n=0}^{\infty} 2^n(n!)x^{-2n}/(2n+1)!$. **11.** $y = c_0 \sum_{n=0}^{\infty} (2n)!x^{2n+3}/[n!(n+1)!\,2^{2n}] + c_1 x$. $y = cx^2\left\{1 - \sum_{n=1}^{\infty} (2n-2)!x^{-2n}/[2^{2n-1}n!(n-1)!]\right\}$. **12.** $y = A \sum_{n=0}^{\infty} (-1)^n(a^3x^3)^n/(3n+2)! + Bx^{-1} \sum_{n=0}^{\infty} (-1)^n(a^3x^3)^n/(3n+1)! + cx^{-2} \sum_{n=0}^{\infty} (-1)^n(a^3x^3)^n/(3n)!$. **13.** $y = x^{3/2} \sum_{n=0}^{\infty} 4^{n+1}x^{2n}/\{[1 \cdot 5 \cdot 9 \ldots (4n+1)]^2(4n+5)\}$.

§109, page 304

1. $y = c_0 x - c_1 x \ln x + c_1 \left(1 + x - \sum_{n=2}^{\infty} x^n/(n-1)\right)$. **2.** $y = (c_0 + c_1 \ln x)(1 + 2x + x^2) + c_1 \left\{-3x - 3x^2 + \sum_{n=3}^{\infty} (-1)^n 2x^n/[n(n-1)(n-2)]\right\}$. **3.** $y = (c_0 + c_1 \ln x) \sum_{n=0}^{\infty} \left(\frac{x}{2}\right)^{2n} \frac{1}{(n!)^2} - c_1 \sum_{n=1}^{\infty} \left(\frac{x}{2}\right)^{2n} \frac{1}{(n!)^2} \sum_{k=1}^{n} \frac{1}{k}$. **4.** $y = (c_0 + c_1 \ln x) \sum_{n=1}^{\infty} \frac{x^n}{(n-1)!n!} + c_1 \left[1 - \sum_{n=1}^{\infty} \frac{x^n}{(n-1)!n!} \left(\frac{1}{n} + \sum_{k=1}^{n-1} \frac{2}{k}\right)\right]$. **5.** $y = (c_0 + c_1 \ln x) \sum_{n=0}^{\infty} [(-1)^n x^n/(n!)^2] - c_1 \sum_{n=1}^{\infty} [(-1)^n x^n/(n!)^2] \sum_{k=1}^{n} 2/k$. **6.** $y = (c_0 + c_1 \ln x) \sum_{n=0}^{\infty} x^{3n}/[3^{2n}(n!)^2] - 2c_1 \sum_{n=1}^{\infty} \left\{x^{3n}/[3^{2n}(n!)^2] \sum_{k=1}^{n} 1/3k\right\}$. **7.** $y = \left(c_0 - \frac{1}{2} c_1 \ln x\right) \sum_{0}^{\infty} (x/2)^{2n} (-1)^n/[n!(n+1)!] + c_1 x^{-2}\left[1 + \frac{1}{4}x^2 - \sum_{n=2}^{\infty} 2(x/2)^{2n} (-1)^n/[n!(n-1)!] \left(1/2n + \sum_{k=1}^{n-1} 1/k\right)\right]$.

8. $y = (A - B \ln x) \sum_{n=0}^{\infty} x^{-n}/[n!(n+1)!] + Bx\left[1 - x^{-1} \sum_{n=1}^{\infty} x^{-n}/[(n-1)!n!] \left(1/n + \sum_{k=1}^{n-1} 2/k\right)\right]$.

§110, page 306

1. 6. **2.** 1.10. **3.** $-2\sqrt{\pi}$. **4.** 3.33. **5.** $(2+t)(1+t)t\Gamma(t)$. **6.** $(-3+t)^{-1}(-2+t)^{-1}(-1+t)^{-1}\Gamma(t)$. **7.** $[t/(t-1)]\Gamma^2(t)$. **8.** 1.43. **9.** 1.77. **10.** 0.443. **11.** 0.310. **12.** $t > -56$. **13.** $-(2n+1) < t < -2n$, where n is zero or a positive integer. **14.** (*a*) $\frac{1}{32}\pi$. (*b*) $\frac{1}{24}$. (*c*) $1.51/\sqrt{\pi}$.

§111, page 309

1. (*a*) $y = c_1x^{1/2}[1 - (x^2/3!) + (x^4/5!) - \cdots] + c_2x^{-1/2}[1 - (x^2/2!) + (x^4/4!) - \cdots]$. (*b*) $y = c_1x^3 \sum_{r=0}^{\infty} \frac{(-1)^r(\frac{1}{2}x)^{2r}3!}{r!(r+3)!} + c_2\left(x^{-3} \ln x \sum_{r=3}^{\infty} 2\{(\frac{1}{2}x)^{2r}(-1)^r/[r!(r-3)!]\} + 2x^{-3} + \frac{1}{4}x^{-1} + \frac{1}{32}x + x^{-3} \sum_{r=3}^{\infty} \left\{2\frac{(-1)^{r+1}(\frac{1}{2}x)^{2r}}{r!(r-3)!} \sum_{n=1}^{r} \left[\frac{1}{2n} + \frac{1}{2(\bar{n}-3)}\right]\right\}\right)$, where $\bar{n}$ does not take the value 3 but takes all the other values from 1 to r. **2.** $y = c_1 \sum_{r=0}^{\infty} [(\frac{1}{2}x)^{2r}(-1)^r/(r!)^2] + c_2\left\{\ln x \sum_{r=0}^{\infty} [(\frac{1}{2}x)^{2r}(-1)^r/(r!)^2] + \sum_{r=1}^{\infty} [(\frac{1}{2}x)^{2r}(-1)^{r+1}/(r!)^2] \sum_{n=1}^{r} 1/n\right\}$. **3.** Yes.

§112, page 312

4. 1, 0, 0.0012. **5.** $2x^{-1}J_1 - J_0$; $8x^{-2}J_1 - 4x^{-1}J_0 - J_1$; $48x^{-3}J_1 - 24x^{-2}J_0 - 8x^{-1}J_1 + J_0$. Here $x \neq 0$, but the formula holds as $x \to 0$. **6.** $(3/x^2 - 1)J_{1/2} - (3/x)J_{-1/2}$; $(\frac{16}{9}x^{-2} - 1)J_{1/3} - \frac{8}{3}x^{-1}J_{-2/3}$; $(\frac{16}{9}x^{-2} - 1)J_{-1/3} + \frac{8}{3}x^{-1}J_{2/3}$. **7.** (*a*) $-J_0 + x^{-1}J_1$. (*b*) $(2x^{-2} - 1)J_1 - x^{-1}J_0$. (*c*) $(-1 + 12x^{-2} - 48x^{-4})J_1 + (-3x^{-1} + 24x^{-3})J_0$. **8.** (*a*) 0.325. (*b*) 0.210. (*c*) 0.135.

§113, page 317

1. $x^2 = \sum_{n=1}^{\infty} 2(\alpha_n^{-1} - 4\,\alpha_n^{-3})J_0(\alpha_n x)/J_1(\alpha_n)$, $J_0(\alpha_n) = 0$. **2.** $x^4 = \sum_{n=1}^{\infty} (128\alpha_n^{-5} - 32\alpha_n^{-3} + 2\alpha_n^{-1})J_0(\alpha_n x)/J_1(\alpha_n)$. **3.** $x = \sum_{n=1}^{\infty} -2\alpha_n^{-1}J_0(\alpha_n)J_1(\alpha_n x)/J_2^2(\alpha_n)$, $J_1(\alpha_n) = 0$.

§114, page 321

2. $\frac{21}{16}(11x^6 - 15x^4 + 5x^2 - \frac{5}{21})$. **3.** $\frac{1}{16}(429x^7 - 693x^5 + 315x^3 - 35x)$. **4.** k. **5.** $\frac{3}{7}P_1 + \frac{4}{9}P_3 + \frac{8}{63}P_5$. **6.** (*a*) $\frac{2}{3}$ if $n = 0$, $\frac{4}{15}$ if $n = 2$, otherwise zero. (*b*) $\frac{2}{5}$ if $n = 1$, $\frac{4}{35}$ if $n = 3$, otherwise zero. (*c*) $\frac{2}{5}$ if $n = 0$, $\frac{8}{35}$ if $n = 2$, $\frac{16}{315}$ if $n = 4$, otherwise zero. **8.** (*a*) $\frac{1}{3} + \frac{2}{3}P_2$. (*b*) $\frac{3}{5}P_1 + \frac{2}{5}P_3$. (*c*) $\frac{7}{35} + \frac{4}{7}P_2 + \frac{8}{35}P_4$. (*d*) $(\frac{1}{3}a + c)P_0 + bP_1 + \frac{2}{3}aP_2$. **9.** $P_k(x)$, $k > n$.

§116, page 325

3. $y = 1 + x + 3x^2 + \frac{1}{2}x^3 - \frac{3}{40}x^5 + \cdots$. **4.** $y = 1 + \frac{1}{2}x^2 + (1/2!)x^4/2^2 + \cdots + (1/n!)x^{2n}/2^n + \cdots$. **5.** $y = 1 + x + x^2/2 + x^3/3 + x^4/(2 \cdot 4) + x^5/(3 \cdot 5) + x^6/(2 \cdot 4 \cdot 6) + (x^7/3 \cdot 5 \cdot 7) + \cdots$.

§117, page 328

2. 70, 58.5.

§118, page 330

2. 0.677, 0.628, 0.572. **3.** $T = 50 - 10(t - 9) - \frac{10}{9}(t - 9)(t - 6)$.

§120, page 335

4. $y = 7.8167$, $y' = 10.0167$, etc. **5.** $y_0 = 1$, $y_{0.1} = 1.0151$, $y_{0.2} = 1.0403$, $y_{0.3} = 1.0759$, $y_{0.4} = 1.1219$, etc. **6.** $y_{0.5} = 1.1784$, $y_{0.6} = 1.2455$, $y_{0.7} = 1.3233$, $y_{0.8} = 1.4120$, etc. **7.** $y_{1.4} = 2.1773$, etc. **9.** $y_{0.1} = 1.1104$. **10.** $y_{0.1} = 2.0309$, $y_{0.2} = 2.0635$, $y_{0.3} = 2.0979$, $y_{0.4} = 2.1339$.

§121, page 337

2. $y'_{0.85} = 7.3587$, $y_{0.85} = 5.6587$. **3.** $y'_{1.1} = 2.3356$, $y_{1.1} = 2.0644$.

§122, page 341

4. $y_{1.5} = 1.100$, $z_{1.5} = 0.4193$, $y_{1.6} = 1.1556$, $z_{1.6} = 0.4924$. **5.** $z_3 = 2 + x - x^3/3! + 3x^5/5!$, $y_3 = 1 + 2x + \frac{1}{2}x^2 - x^4/4! + 3x^6/6!$. **6.** $y_{0.6} = 2.3748$, $z_{0.6} = 2.5659$.

§123, page 343

1. $y_{0.4} = 2.8918$. **2.** $y_{0.5} = 3.1488$, $y_{0.6} = 3.4222$. **3.** (0.1, 1.1053), (0.2, 1.2229) (0.3, 1.3552). **4.** (0.05, 1.0501, 1.0525), (0.1, 1.1004, 1.1100), (0.15, 1.1511, 1.1725), (0.25, 1.2552, 1.3128), (0.3, 1.3091, 1.3907).

§125, page 348

4. $p = q$. **5.** $x(p - q) = z$. **6.** $y(q - p) = z$. **7.** $pt = qs$. **8.** $q(r - s) + p(t - s) = 0$. **9.** $2r + s - t = 0$. **10.** $xy = z(xq - yp)$. **11.** $xy = z(xq + yp)$. **15.** 2π ft, 2 sec, $v = -\pi$ ft/sec.

§126, page 350

1. $z = x^3 + xy^2 + \varphi(y)$. **2.** $yz = x^2y + \varphi(x)$. **3.** $yz = x^2y + \varphi(x) + \psi(y)$. **4.** $z = \int\int f(x,y)\,dx^2 + x\varphi(y) + \psi(y)$. **5.** $2z = x^2 \ln y + 2axy + \varphi(x) + \psi(y)$. **6.** $z = -ye^x + e^y[y + \varphi(x)] + \psi(x)$. **7.** $2z = x^2y - 2xy + \varphi(y) + e^{-x}\psi(y)$. **8.** $4z = x^2y + \varphi(y) \ln x + \psi(y)$. **9.** $z = \varphi_1(y)e^{2x} + \varphi_2(y)e^{-2x} - 2y^2$. **10.** $z = \varphi_1(x) \cos 2y + \varphi_2(x) \sin 2y + 2x^2$. **11.** $z = \varphi_1(y)e^{3x} + \varphi_2(y)e^{-x} - xy^2$. **12.** $z = \varphi_1(x)e^{-3y} + \varphi_2(x)e^y + x^2y$. **13.** $z = \varphi_1(y)e^{5xy/2} + \varphi_2(y)e^{-xy} - \frac{1}{5}(5x + 3 - 3y^{-1})$. **14.** $z = e^{xy}[\varphi_1(x) \sin 2xy + \varphi_2(x) \cos 2xy] + x$. **15.** $z = \ln\,[e^{xy}\varphi(y) - e^{-xy}] + \psi(y)$. **16.** $2z^2 = (2x - 1)y^2 + \varphi(y)e^{-2x}$. **17.** $\pm z = \sqrt{a - \varphi(y)e^{-2x}} + \sqrt{a} \ln [\sqrt{ae^{2x} - \varphi(y)} - \sqrt{a}\, e^x| + \psi(y)$. **18.** $z = 6x^2 + e^{-y}\varphi(x) + x\psi(y) + \theta(y)$.

§127, page 354

1. $z = e^x\varphi(x - y)$. **2.** $z = x^2\varphi(y/x)$. **3.** $y + z = x\varphi[x(y - z)]$. **4.** $az = cx + \varphi(bx - ay)$. **5.** $\tan^{-1}(y/z) = \ln x + \varphi(z^2 + y^2)$. **6.** $3xz = y^2 + \varphi(xy)$. **7.** $z(y - x) = axy \ln (y/x) + (y - x)\varphi[(x - y)/(xy)]$. **8.** $x^2 - y^2 = \varphi(z^2 + 2y^2)$. **9.** $3y^2 \ln z + ax = 3y^2\varphi(xy)$. **10.** $3z = 2x^3 + \varphi(ye^{-x}) + \psi(y)$. **11.** $z = y^3\varphi(x^5y^{-3})$. **12.** $yz + 2xy^2 = \varphi(xy) + \psi(y)$. **13.** $z = y\varphi(x/y) + \psi(y)$. **14.** $xz - 2x^2y = \varphi(xy) + \psi(x)$. **15.** $x^2z - 6x^3y^2 = \varphi(xy) + \psi(x)$. **16.** $z = x\varphi(y + x^2) + \psi(x)$. **17.** $z + xy = \varphi(x^2y) + \psi(x)$. **18.** $z = y\varphi(e^x/y) + \psi(y)$. **19.** $z = \varphi(e^y/x) + \psi(y)$. **20.** (*a*) $(y + y^2)^{-2}$. (*b*) x^{-1}. **21.** (*a*) $e^{\int f(y)\,dy}$. (*b*) $e^{-\int f(x)\,dx}$. (*c*) $(xy)^k$. (*d*) x^by^a. **22.** $x = 3y^2[\varphi(xy) + a \ln |z|]$. **23.** $x = (x + y) \ln z + \varphi(x + y)$. **24.** $z = \varphi(y/x)$. **25.** $z = (1 + y)^{1/x}\varphi(x)$. **26.** $2z = x^2 + xy + \varphi(y/x) + \psi(y)$. **27.** $(y - x)^2 + 2z^2 = \varphi(x + y)$.

§128, page 356

1. $xyz = y^3 + 5x^3$. **2.** $z = 2/(xy^2 - 2y)$. **3.** $(y^2 - 2z)(y^2 - xy) = 8x^2$. **4.** $z = e^x(x - y)e^{(y-x-1)/2}$. **5.** $z = \pm e^y \sqrt{a^2 - (x - y)^2}$. **6.** $z = e^y \sin (x - y)$. **7.** $\varphi[(x - 1)/y, z/y] = 0$, $y^2 + z^2 = 25(x - 1)^2$. **8.** $z = \frac{1}{3}y^2/x + \frac{1}{6}x^2y^2 + xy + \ln (xy - 1) + c$.

§129, page 358

1. $\tau = \sum^k (a_ke^{\sqrt{k}x} + b_ke^{-\sqrt{k}x})(c_k \sin \sqrt{k}\, y + d_k \cos \sqrt{k}\, y)$; same with x and y interchanged; $z = c_1xy + c_2x + c_3y + c_4$. **2.** $z = \sum^k e^{ky}(a_ke^{\sqrt{k}x} + b_ke^{-\sqrt{k}x})$; $z = \sum^k e^{-ky}(a_k \sin \sqrt{k}\, x + b_k \cos \sqrt{k}\, x)$; $z = c_1 + c_2x$. **3.** $z = \sum^k c_ke^{k(x-y)}$. **4.** $z = \sum^k c_ke^{k(x-y)+3y}$. **5.** $z = \sum^k (a_ke^{\sqrt{k-1}x} + b_ke^{-\sqrt{k-1}x})(c_k \sin \sqrt{k}\, y + d_k \cos \sqrt{k}\, y)$; $k = 1, 0 < k < 1, k = 0, k < 0$. **6.** (*a*) $k < -\frac{1}{4}$. (*b*) $k > 1$. (*c*) $-\frac{1}{4} \leqq x \leqq 1$. **7.** $u = \sum^{k,l,m} (a_ke^{\sqrt{l}x} + b_ke^{-\sqrt{l}x})(c_ke^{\sqrt{m}y} + d_ke^{-\sqrt{m}y})$ $(g_k \sin \sqrt{l + m}\, z + f_k \cos \sqrt{l + m}\, z)$; 13 types. **9.** $z = \sum^k e^{kx-ky/2}[a_k \sin (\frac{1}{2}\sqrt{3}\, ky) + b_k \cos (\frac{1}{2}\sqrt{3}\, ky)]$.

§130, page 364

1. $z = \varphi(y + x) + \psi(y - 2x)$. **2.** $z = \varphi(y + mx) + \psi(y + nx)$. **3.** $z = x\varphi(y - mx) + \psi(y - mx)$. **4.** $z = \sum^k [c_{1k} \sin \sqrt{k}\, (x + y) + c_{2k} \cos \sqrt{k}\, (x + y)](c_{3k}e^{2\sqrt{k}x} + c_{4k}e^{-2\sqrt{k}x})$. **5.** $z = \sum^k [c_{1k} \sin \sqrt{k}\, (y + mx) + c_{2k} \cos \sqrt{k}\, (y + mx)](c_{3k}e^{2\sqrt{k}nx} + c_{4k}e^{-2\sqrt{k}nx})$. **6.** $z = y^{1/9}x^{-2/9}\varphi(xy) + \psi(yx^{-2})$. **7.** $z = x\varphi(y + x^2) + \psi(y + x^2)$. **8.** $z = \sum^k [c_{1k} \cos (\sqrt{k}\, ye^x \cos 2x) + c_{2k} \sin (\sqrt{k}\, ye^x \cos 2x)](c_{3k}e^{\sqrt{k}ye^x \sin 2x} + c_{4k}e^{-\sqrt{k}ye^x \sin 2x})$. **9.** $z = \sum^k \rho^{-1/2}(c_{1k} \sin \sqrt{k}\, \theta + c_{2k} \cos \sqrt{k}\, \theta)\ (c_{3k}\rho^{\sqrt{k+1/4}} + c_{4k}\rho^{-\sqrt{k+1/4}})$. **10.** $z = y\varphi(xe^{-y^2/2}) + \psi(xe^{-y^2/2})$. **13.** By using the substitution u arbitrary and finding v from the solution of $dy/dx - b/2a = 0$.

§131, page 369

3. $1 = (4/\pi)(\frac{1}{1}\sin x + \frac{1}{3}\sin 3x + \frac{1}{5}\sin 5x + \cdots)$. **4.** $\frac{1}{2} + (2/\pi)(\sin x + \frac{1}{3}\sin 3x + \frac{1}{5}\sin 5x + \cdots)$. **5.** $f(x) = \frac{1}{2}\pi + \pi^{-1}\sum_{m=1}^{\infty} 2(1 - \cos m\pi)m^{-2}\cos mx$. **6.** $(8/\pi)[(\sin x)/(1.3) - 2(\sin 2x)/(3.5) + 3(\sin 3x)/(5.7) - \cdots]$. **8.** $[2(1)^{-1}\pi^2 - 12(1)^{-3}]\sin x - [2(2)^{-1}\pi^2 - 12(2)^{-3}]\sin 2x + [2(3)^{-1}\pi^2 - 12(3)^{-3}]\sin 3x - \cdots, -\pi < x < \pi$.

§132, page 372

1. (*a*) $x = 2L/\pi[\sin(\pi x/L) - \frac{1}{2}\sin(2\pi x/L) + \frac{1}{3}\sin(3\pi x/L) - \cdots]$. (*b*) $x^2 = L^2/3 - 4L^2/\pi^2[1/1^2\cos(\pi x/L) - 1/2^2\cos(2\pi x/L) + 1/3^2\cos(3\pi x/L) - \cdots]$. (*c*) $F(x + k2L) = F(x)$, $F(x) = x^2$ on $-L \leqq x \leqq L$. **2.** $x^2 = 2/\pi[(\pi^2/1 - 4/1^3)\sin x - \pi^2/2\sin 2x + (\pi^2/3 - 4/3^3)\sin 3x - \pi^2/4\sin 4x + \cdots]$. **3.** $F(x) = mx(L - x)$ on $0 \leqq x \leqq L$, $F(x) = mx(L + x)$ on $-L \leqq x \leqq 0$, $F(x + k2L) = F(x)$. **4.** $mx(L^2 - x^2) = 12mL^3/\pi^3[1/1^3\sin(\pi x/L) - 1/2^3\sin(2\pi x/L) + 1/3^3\sin(3\pi x/L) - \cdots]$, $-L \leqq x \leqq L$. **5.** $|\pi + \frac{1}{2}x| = \frac{5}{4}\pi + \sum_{m=1}^{\infty}[(4\cos m\pi - 4\cos\frac{1}{2}m\pi)/(\pi m^2)\cos\frac{1}{4}mx + (4\sin\frac{1}{2}m\pi - 2\pi m\cos m\pi)/(\pi m^2)\sin\frac{1}{4}mx]$. **7.** $\cos\frac{1}{2}x = 2/\pi - 4/\pi\sum_{m=1}^{\infty}(\cos m\pi/(4m^2 - 1)\cos mx$.

§133, page 375

1. $y = 0.0232[\cos 178t\sin(\pi x/3) + 3^{-3}\cos(3\cdot 178t)\sin(3\pi x/3) + \cdots]$, about 28 vibrations per second. **2.** $y(x,0) = 8L^2m\pi^{-3}\sin(\pi x/L)$, $y_t(0,0) = 0$, $y_t(L,0) = 0$. **3.** $y(x,0) = 8L^2m\pi^{-3}k^{-3}\sin[(2k - 1)\pi xL^{-1}]$, $y_t(0,0) = 0$, $y_t(L,0) = 0$. **4.** $y_t(x,0) = 0$. **5.** $y = 12mL^3\pi^{-3}\sum_{n=1}^{\infty}(-1)^{n+1}n^{-3}\sin n\pi x/L\cos(n\pi at/L)$, $2L/a$ sec. **6.** $y(x,0) = 0$. **7.** $y = 4mL\pi^{-2}\sum_{n=1}^{\infty}(-1)^{n+1}(2n - 1)^{-2}\sin[(2n - 1)xL^{-1}]\cos[(2n - 1)\pi atL^{-1}]$, $t = 2L/a$ units of time.

§134, page 377

1. $y = t$ when $0 < t < x/a$; $y = x/a$ when $t > x/a$. **2.** $y = 0$ if $t < x$, $y = (t - x)e^{t-x}$ if $t \geqq x$. **3.** $y = 0$ if $t < 3x$, $y = (t - 3x)^3$ if $t \geqq 3x$.

§135, page 378

1. $2L/a$ units of time. **2.** $y = (0.0072/\pi^3)\sum_{n=1}^{\infty}(2n - 1)^{-3}\sin(2n - 1)\pi x/3\cos[(2n - 1)\pi 17{,}000t/3]$, 2,800 oscillations per second.

§136, page 382

1. (*a*) $\theta(x,y) = Ae^{-3y}\sin 3x$. (*b*) $2A\sum_{n=1}^{\infty}(-1)^{n+1}n^{-1}e^{-ny}\sin nx$.

2. $\theta = 4/\pi \sum_{n=1}^{\infty} (2n - 1)^{-1}e^{-\pi(2n-1)y/L} \sin [(2n - 1)\pi x/L]$.

3. $\theta = 4mL/\pi^2 \sum_{n=0}^{\infty} (2n - 1)^{-2}e^{-(2n-1)\pi y/L} \sin [(2n - 1)\pi x/L]$.

§137, page 384

1. $\theta = 1.2x + 56$ deg; $\theta = (Q - P)L^{-1}x + P$. **2.** $\theta = 56 + 1.44x - (144/\pi) \sum_{n=1}^{\infty} (2n - 1)^{-1}e^{-(2n-1)^2\pi^2a^2t/100^2} \sin [(2n - 1)\pi x/100] + 512/\pi \sum_{n=1}^{\infty} (-1)^{n+1}e^{-n^2\pi^2a^2t/100^2} \sin (n\pi x/100)$. **3.** $\theta = 60 + x - (160/\pi) \sum_{n=1}^{\infty} (2n - 1)^{-1}e^{-a^2(2n-1)^2\pi^2t/100^2} \sin [(2n - 1)\pi x/100] + 160/\pi \sum_{n=1}^{\infty} (-1)^{n-1}(n)^{-1}e^{-n^2a^2\pi^2t/100^2} \sin (n\pi x/100)$. **4.** 110°. **5.** $\theta(x,t) = 800 - 20x + \sum_{n=1}^{\infty} \{(3{,}200/(n^2\pi^2) \sin \frac{1}{2}n\pi - [1{,}600/(n\pi)]\} e^{-a^2n^2\pi^2t/40^2} \sin \frac{1}{40}n\pi x$.

§139, page 391

1. $Y_0(w)$ is infinite at $w = r = 0$. Yes. **2.** $y = 0$ at the boundary of the drum head. $2.40/a$, $5.52/a$, $8.65/a$. **3.** (*a*) It makes a complete oscillation up and down with maximum amplitude. (*b*) Same as (*a*) but with different amplitude. (*c*) Remain fixed. (*d*) Remain fixed.

§140, page 394

2. $e = -A\sqrt{L/C} \sin (\omega\sqrt{LC}\, x) \cos \omega t + B$. **4.** (*a*) $i = (A\omega/R)e^{(-\omega^2/RC)t} \sin \omega x$. (*b*) $e = (AR/\omega)e^{(-\omega^2/RC)t} \cos \omega x + B$. **5.** $e = Ax + B$, $i = -A/R$. **6.** $e = -2/\pi\, [7e^{-at} \sin (\pi x/L) - \frac{3}{2}e^{-4at} \sin (2\pi x/L) + \frac{7}{3}e^{-9at} \sin (3\pi x/L) - \frac{3}{4}e^{-16at} \sin (4\pi x/L) + \cdots]$, $i = 2/(RL)[7e^{-at} \cos (\pi x/L) - 3e^{-4at} \cos (2\pi x/L) + 7e^{-9at} \cos (3\pi x/L) - 3e^{-16at} \cos (4\pi x/L) + \cdots]$, where $a = \pi^2/(L^2RC)$, $t > 0$. **8.** $i = -A\sqrt{C/L}\, e^{-((G/C)t} \cos (\omega\sqrt{LC}\, x) \sin \omega t$.

§141, page 398

3. (*a*) $x^3 - 3xy^2 = c$, $3x^2y - y^3 = c$, $p = \rho/g[c - gy - \frac{1}{2}(x^2 + y^2)^2]$. (*b*) $x^2 + y^2 = c, y = cx$, $p = \rho/g[c - gy - \frac{1}{2}(x^2 + y^2)^{-1}]$.

Index